Fermented Foods
Part I: Biochemistry and Biotechnology

Food Biology Series

Fermented Foods
Part I: Biochemistry and Biotechnology

Editors
Didier Montet
Head, Food Safety Team, CIRAD, UMR Qualisud
Montpellier, Cedex, France

and

Ramesh C. Ray
Principal Scientist (Microbiology)
ICAR - Central Tuber Crops Research Institute
Bhubaneswar, Odisha, India

CRC Press is an imprint of the
Taylor & Francis Group, an **informa** business

A SCIENCE PUBLISHERS BOOK

Cover Credit: Clementine Kouakou

CRC Press
Taylor & Francis Group
6000 Broken Sound Parkway NW, Suite 300
Boca Raton, FL 33487-2742

First issued in paperback 2020

© 2016 by Taylor & Francis Group, LLC
CRC Press is an imprint of Taylor & Francis Group, an Informa business

No claim to original U.S. Government works

ISBN-13: 978-1-4987-4079-1 (hbk)
ISBN-13: 978-0-367-73745-0 (pbk)

Visit the Taylor & Francis Web site at
http://www.taylorandfrancis.com

and the CRC Press Web site at
http://www.crcpress.com

Preface to the Series

Food is the essential source of nutrients (such as carbohydrates, proteins, fats, vitamins, and minerals) for all living organisms to sustain life. A large part of daily human efforts is concentrated on food production, processing, packaging and marketing, product development, preservation, storage, and ensuring food safety and quality. It is obvious therefore, our food supply chain can contain microorganisms that interact with the food, thereby interfering in the ecology of food substrates. The microbe-food interaction can be mostly beneficial (as in the case of many fermented foods such as cheese, butter, sausage, etc.) or in some cases, it is detrimental (spoilage of food, mycotoxin, etc.). The *Food Biology* series aims at bringing all these aspects of microbe-food interactions in form of topical volumes, covering food microbiology, food mycology, biochemistry, microbial ecology, food biotechnology and bio-processing, new food product developments with microbial interventions, food nutrification with nutraceuticals, food authenticity, food origin traceability, and food science and technology. Special emphasis is laid on new molecular techniques relevant to food biology research or to monitoring and assessing food safety and quality, multiple hurdle food preservation techniques, as well as new interventions in biotechnological applications in food processing and development.

The series is broadly broken up into food fermentation, food safety and hygiene, food authenticity and traceability, microbial interventions in food bio-processing and food additive development, sensory science, molecular diagnostic methods in detecting food borne pathogens and food policy, etc. Leading international authorities with background in academia, research, industry and government have been drawn into the series either as authors or as editors. The series will be a useful reference resource base in food microbiology, biochemistry, biotechnology, food science and technology for researchers, teachers, students and food science and technology practitioners.

Ramesh C. Ray
Series Editor

Preface

Traditional fermented foods are not only staple foods for most of the developing countries but also significant health food for developed countries. Fermented foods are food substrates that are processed by edible microorganisms (bacteria, yeasts and molds) whose enzymes (e.g., amylases, proteases and lipases) hydrolyze the polysaccharides, proteins and lipids to non-toxic products with flavors, aromas and textures desirable to the humans. As the health functions of these foods are being rediscovered, higher throughput biotechnologies are being used to promote the fermented food industries. As a result, microorganisms, process biochemistry, manufacturing and down-stream processing, as well as bioactive metabolites released by the fermenting organisms and above all, the health functions of these foods are being extensively researched. Furthermore, the applications and progress of biochemistry and biotechnology applied to traditional fermented food systems are different from each other, as the microorganisms and the food matrices vary widely.

Part I of the book, "Fermented Foods: Biochemistry and Biotechnology" covers general aspects on microbiology, biochemistry, and biotechnological applications involving yeasts, filamentous fungi, acetic acid and lactic acid bacteria in promoting and accelerating the development of multiple functional factors in fermented foods, the release of bioactive compounds during fermentation, development of starter cultures, and metagenomics of fermented foods. The impact of yeasts on food and beverage production beyond the original and popular application in bread, beer and wine has been described. Likewise, the importance of acetic acid bacteria, which are desirable and essential for the production of vinegar and cocoa, while they are sometimes involved in foods and beverages in detrimental way, such as in wine, beer, soft drinks and fruits, has been discussed. The classification, metabolism, and applications of lactic acid bacterial group, including their antimicrobial activities (bacteriocins-based) and effects on human health have been elucidated. Two chapters are devoted specifically to fermentation of cereals and vegetables by these bacteria and the resulting health benefits. A chapter is devoted to lactic acid fermentation of seaweeds (macro-algae), which can constitute a new raw material for the production of fermented foods and feeds. Microencapsulation of probiotics provides a good promising approach to overcome their loss during detrimental environmental conditions. The various techniques used for microencapsulation of probiotics and their applications are also outlined. Despite the focus on health benefits of probiotic yeasts and lactic acid bacteria, the negative aspect is the production of biogenic amines by some strains of lactic acid bacteria that has detrimental effects on health. These amines occur in a wide variety of fermented foods such as sausage, fish products, cheese and wine. Therefore, a chapter is devoted on biogenic amine in fermented foods, as well as a chapter on antimicrobial resistance of fermentative bacteria in the context of Whole Food Chain Approach. Other aspects such as microbial fructo-oligosaccharides and their relevance to human health, and technology adapted for bio-valorization of food wastes are covered in two separate chapters. Molecular methods used to engineer fermentative microorganisms are presented in a chapter dedicated to wine biotechnology. Other applications of molecular methods, and more specifically metagenomics, are also discussed in the light of monitoring ecosystems during fermented food elaboration. The 20 chapters in this book have been authored by highly reputed international contributors having an in-depth understanding of fermented food science and technology. We believe that this book will be a useful reference book for researchers, teachers, students, nutritional and functional food experts and all those working in the field of food science and technology.

The detailed technological interventions involved in different categories of fermented foods such as fermented cereals (bread and sourdough), fermented milk products (yogurt, cheese, kefir and koumiss), fermented sausages, fermented vegetables (kimchi, sauerkraut), fermented legumes (tempeh, natto, miso), coffee and cocoa fermentations, cassava and sweet potato fermentations, are discussed in Part II of this book (Fermented Foods: Technological Interventions).

Didier Montet
Ramesh C. Ray

Contents

1

Yeasts in Fermented Foods and Beverages

Tek Chand Bhalla[1,]* *and* *Savitri*[1]

1. Introduction

For more than 6000 years, fermentation has been used as a means of improving the shelf life and adding quality to food. Earlier, the transformation of basic food materials into fermented foods was considered as a mystery. The first fermentation was probably discovered by accident when salt was added in food materials, causing the selection of some harmless microorganisms that could ferment the raw material to give nutritious and acceptable food. Since then, it has been practiced to improve both, the preservation and organoleptic characteristics of food.

Food fermentation technology utilizes the ability of growth of microorganisms on various substrates for the production of a variety of fermented foods, beverages and pickles, e.g., bread, beer, cheese, *idli, tempeh, miso*, sauerkraut, fermented sausages, etc. These fermentations involve the combined action of bacteria, yeast and fungi which may act in parallel, while others may act in a sequential manner with a changing dominant flora during the course of fermentation (Haard 1999). Among the various microorganisms, yeasts have an enormous impact on food and beverage production. They play a central role in the fermentation of foods and beverages, especially those with high carbohydrate content, as yeasts can survive and grow under stress conditions. In traditional fermentation processes, natural micro-organisms, including yeasts, are employed in the preparation and preservation of different types of foods. These processes add to the nutritive value of foods as well as enhance flavor and other desirable characteristics associated with digestibility and edibility (Kolawole et al. 2007).

[1] Department of Biotechnology, Himachal Pradesh University, Summer Hill, Shimla-171005, India.
E-mail: savvy2000in@rediffmail.com
* Corresponding author: bhallatc@rediffmail.com

Today, the impact of yeasts on food and beverage production extends beyond the original and popular notions of bread, beer and wine fermentations by *Saccharomyces cerevisiae* (Fleet 2006). A variety of important food ingredients and processing aids are now obtained from yeasts. Some yeasts are being explored for their strong antifungal action and are used as novel agents in the biocontrol of food spoilage. The probiotic activity of yeast is another remarkable aspect that is attracting the attention of academia and industry. Unfortunately, yeasts are also involved in the spoilage of many commodities, with major economic losses to many food and beverage industries, while the impact of yeasts present in foods and beverages on public health is a matter of emerging concern.

2. History

Although, both yeast and bacteria were first reported to have been observed microscopically by Antonie van Leeuwenhoek in 1680, it was not until the year 1837 that Cagniard-Latour, suggested that the fermentation was intimately associated with the yeast cells. The history of yeast association with human society is synonymous with the evolution of bread, beer and wine as global food and beverage commodities, originating about 5,000 years ago (Fleet 2006, Chapter 1 in this book). Archaeologists have found evidence for the production of a fermented beverage in China about 7000 BCE (McGovern et al. 2004), and of wine in Iran and Egypt in 6000 BCE and 3000 BCE, respectively (Cavalieri et al. 2003). It is believed that the cultivation of grapevines and the production of wine spread all over the Mediterranean Sea, moving towards Greece in 2000 BCE, Italy in 1000 BCE, Northern Europe in 100 CE and America in 1500 CE (Pretorius 2000). Yeast is probably one of the earliest domesticated organisms among various microorganisms. It was used to carry out fermentation from the early beginnings of brewing. The microorganisms, or even yeasts, were not specifically known in the Middle Ages, but it was well-known that the best beers were produced next to bakeries (Kruif 1935).

3. Yeast

The term 'yeast' was derived from the Dutch word '*gist*', which refers to the foam formed during the fermentation of beer (Foligné et al. 2010). Yeasts are heterotrophic organisms occurring naturally on the surfaces of plant tissues, including flowers and fruits, and are fairly simple in their nutritional requirements. Yeasts are unicellular, although some species may become multicellular through the formation of strings of connected budding cells known as pseudohyphae, or false hyphae. Yeasts exhibit both asexual and sexual stages in their life cycle. The asexual stage of given yeast is called the anamorph, while the sexual stage is the teleomorph. Most yeasts reproduce asexually by mitosis, and many do so by an asymmetric division called budding. The budding yeasts or true yeasts are classified in the order Saccharomycetales (Kurtzman and Fell 2005).

Vegetative growth in yeasts is either entirely unicellular or a combination of hyphal and unicellular reproduction. The most common mode of vegetative growth of yeasts is by budding, which may be blastic or thallic. Anamorphic and teleomorphic genera may grow either as a 'yeast-like' unicellular organism or as a 'mold-like' filamentous organism, a phenomenon called as dimorphism. Moreover, some species are able to form a true mycelium, while genera such as *Candida* produce a well-developed pseudomycelium, or both pseudo and true mycelium as in the case of *Candida tropicalis* (Goldman 2008). Brewing yeast does have a sex life, but reproduces primarily by budding in production conditions. A single cell may bud up to 20 times, each time leaving a scar, the counting of which indicates how senile the cell has become.

Among the yeasts belonging to the phylum *Ascomycota*, the genus *Saccharomyces* is the most studied and widely used in industrial processes. *Saccharomyces cerevisiae* (Fig. 1) is spherical or ellipsoidal in structure. Laboratory strains of *S. cerevisiae* are haploid (have one copy of each of the 16 linear chromosomes), whereas industrial strains are polyploid (i.e., have multiple copies of each chromosome) or aneuploid (varying numbers of each chromosome). Some 6000 genes have been identified in yeast and the entire genome has now been sequenced (http://www.yeastgenome.org/).

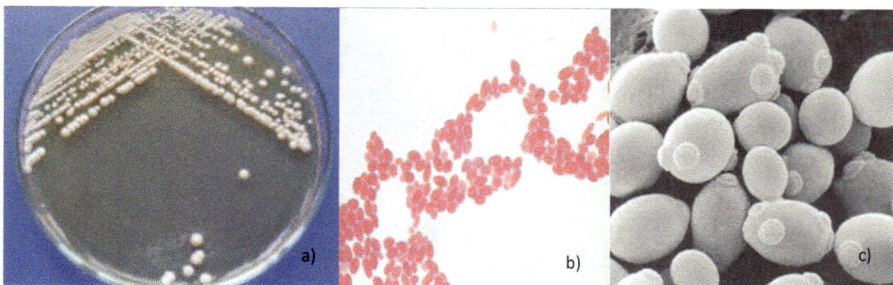

Figure 1. *Saccharomyces cerevisiae* (a) colonies on YPD agar medium, (b) microscopic view and (c) scanning electron micrograph.

3.1 Classification

Yeasts are eukaryotic microorganisms classified in the kingdom *fungi* with 1,500 species currently described, and estimated to be 1% of all fungal species (Kurtzman and Piškur 2006). They are widespread in natural environments including the normal microbial flora of humans, plants, airborne particles, water, food products, and many other ecological niches. Yeasts do not form a single taxonomic or phylogenetic group. The term 'yeast' is often taken as a synonym for *S. cerevisiae*, but the phylogenetic diversity of yeasts is shown by their placement in two separate phyla: the Ascomycota and the Basidiomycota whose vegetative growth results predominantly from budding or fission, and which do not form their sexual states within or upon a fruiting body (Kurtzman and Fell 1998). Phylogenetic analysis of the phylum Ascomycota has significantly changed yeast classification in recent years (Hibbett et al. 2007, Kurtzman et al. 2011).

3.2 Yeast Cell Structure

Yeasts consist of single cells which are smaller than animal and plant cells but larger than bacteria. There is a cell wall on the outside of the cell and the inside of the cell contains cytoplasm having a nucleus, mitochondria and ribosomes (Fig. 2). The cell wall in yeast is a two layered structure, the inner layer is made up of β-1–3-glucane linked covalently to chitin and is responsible for the stability of yeast cells. In contrast, the outer layer consists of densely packed glycosylated mannoproteins which decrease the permeability. In addition, hydrophilic attributes of the cell wall are achieved by phosphorylation of the mannoproteins (Lipke and Ovalle 1998). Due to the presence of phosphate groups attached to the mannan polysaccharides, the cell wall of yeast is negatively charged with underlying plasma membrane comprised primarily of sterols, unsaturated fatty acids, proteins and permeases.

Although, the function of the cell wall is to provide stabilization to the cell, the cell membrane, which consists of phospholipids and proteins along with sterols (mostly ergosterol and zymosterol), forms the main barrier for substrate exchange in and out of the cell. The membrane-building phospholipids vary from the outer to the inner layer (Zinser et al. 1991). Sphingolipids are mainly found on the outside and their function is to give the membrane a structure; they also have an important role in cell growth, cell regulation and have an influence on the stress reactions of the cell (Hannun and Obeid 1997). Proteins are also present in cell membrane and are spread asymmetrically over the bilayer. Intrinsic proteins are stretched through the whole

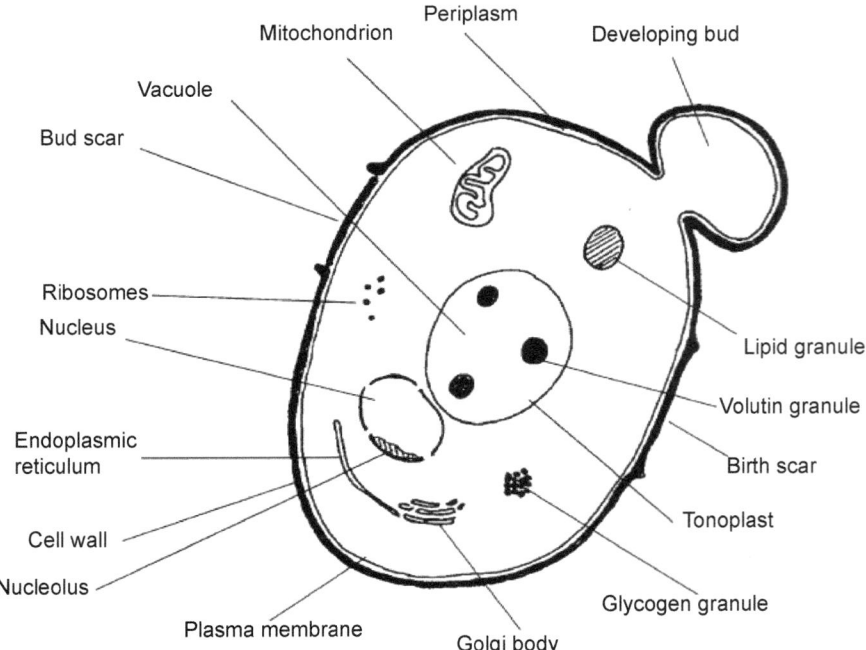

Figure 2. Diagrammatic representation of cross section of yeast cell (Source: Boulton and Quain 2001).

membrane, whereas extrinsic membrane proteins can only be found on the inner or the outer layers. Most of them are specialized to perform different tasks (Van der Rest et al. 1995). The cytoplasm of yeasts is an aqueous, slightly acidic fluid with several organelles, *viz.*, vacuoles, ribosomes, endoplasmic reticulum, golgi apparatus and the nucleus. In addition, proteins, lipid particles and glycogen can also be found. The lipid particles play a role in sterol biosynthesis and serve as lipid storage for membrane synthesis. Glycogen granules are built up at the end of fermentation; these serve as an energy reserve during storage and as starting energy for new cell cycles (Tenge 2009).

3.3 Metabolism in Yeast

Yeasts are capable of utilizing a broad spectrum of carbohydrates and sugars. They require a reduced carbon source, various minerals and a supply of nitrogen and vitamins. The key vitamin requirements are biotin, pantothenic acid and thiamine. As oxygen is required for the desaturation reactions involved in the synthesis of the lipids, relatively small quantities of oxygen must be supplied to the yeast, even when it is growing anaerobically during fermentation. Metabolism in the yeast cell (i.e., by aerobic respiration or by fermentation) is based on the concentration of sugar to which yeast is exposed. For example, when sugar concentration is high, the cell is switched into the fermentative mode, and the pyruvate is metabolized *via* acetaldehyde to ethanol, yielding 2 molecules of ATP per sugar molecule. At low sugar concentrations, the pyruvate shunts into acetyl-CoA and the respiratory chain, and this yields 32 molecules of ATP per sugar molecules. This phenomenon is known as Crabtree effect. At high sugar concentrations, the cell does not need to generate as many ATP molecules per sugar molecule, whereas if the sugar supply is limited, the yeast must utilize the sugar efficiently. The significance of this in commercial fermentation processes is clear. In brewing, where the primary requirement is a high yield of alcohol, the sugar content in the feed stock is high, whereas in the production of baker's yeast, where the requirement is a high cell yield, the sugar concentration is always kept low, but the sugar is continuously passed into the fermenter (Bamforth 2005).

3.4 Diversity of Yeasts in Nature

Various yeasts have been isolated from natural and artificial environments (Lachance et al. 2001, Lee et al. 2001). In nature, *S. cerevisiae* is present on grape surfaces at extremely low density, but becomes the predominant species during alcoholic fermentation. The most extensive work has been focused on population change and the roles of yeasts during the fermentation process of food and beverages (Strauss et al. 2001). Studies on yeast biodiversity in the natural environment have been neglected because yeasts have been regarded as playing a major role in food fermentation and a rather minor role in the biosphere as compared to other microorganisms that may act as primary producers, predators, pathogens, or important agents of nutrient cycling in natural environments (Lachance and Starmer 1998). *S. cerevisiae*, also called winemaker's bug, is so closely associated with humans that it is rarely found in environments devoid of human inhabitation. In fact, its evolutionary success can probably be explained by its relationship with humans, particularly in the production of alcoholic beverages (McGovern et al. 2004).

4. Yeasts in Fermented Foods and Beverages

Yeasts form an inevitable part of the microflora of various fermented foods and beverages and are found in a wide range of foods from plant and/or animal origin (Fig. 3), where they have a significant role in food safety and imparting organoleptic properties (Foligné et al. 2010). They play important roles in the production of many traditional fermented foods and beverages across the world (Aidoo et al. 2006), foods that signify the food culture of the regions and the community (Tamang and Fleet 2009). More than 21 major genera of functional yeasts have been reported from fermented foods and beverages; the main genera being *Brettanomyces* (*Dekkera*), *Candida*, *Cryptococcus*, *Debaryomyces*, *Galactomyces*, *Geotrichum*, *Hansenula*, *Hanseniaspora* (*Kloeckera*), *Hyphopichia*, *Kluyveromyces*, *Metschnikowia*, *Pichia*, *Rhodotorula*, *Saccharomyces*, *Saccharomycodes*, *Saccharomycopsis*, *Schizosaccharomyces*, *Torulopsis*, *Trichosporon*, *Yarrowia*, and *Zygosaccharomyces* (Kurtzman and Fell 1998, Pretorius 2000, Romano et al. 2006, Tamang and Fleet 2009). Yeasts, alone or in stable mixed populations with mycelial fungi or with bacteria (usually lactic acid), have a significant impact on food quality parameters such as taste, texture, odour and nutritive value (Table 1). Molds are relatively limited in fermented foods and beverages, and include the genera *Actinomucor*, *Mucor*, *Rhizopus*, *Amylomyces*, *Monascus*, *Neurospora*, *Aspergillus*, and *Penicillium* (Nout and Aidoo 2002).

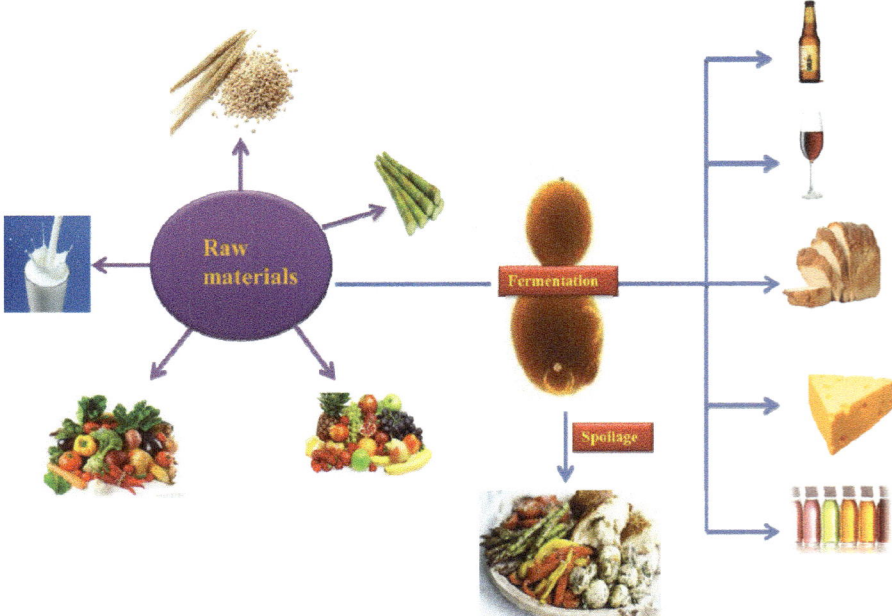

Figure 3. Role of yeast in food and beverages.

Table 1. Role of yeast in production of fermented food and beverages (modified from Romano et al. 2006).

Fermented Food/ Beverage	Yeast	Function
Beer, wine, sourdough, bread, cheese, indigenous fermented foods and beverages	*Saccharomyces* sp.	Sugar fermentation and formation of alcohol Improvement of flavor and texture Production of secondary metabolites Inhibitory effect on undesirable microorganisms
Sourdough, cocoa beans, soy sauce, indigenous fermented foods and beverages	*Candida* sp.	Production of different enzymes (protease, galactosidase and pectinase) Inhibition of undesirable organisms Secondary metabolite production
Cheese	*Clavispora lusitaniae*	Production of flavor compounds in cheese
Fresh fruits and fermented meat products	*Cryptococcus* sp.	Biocontrol agent against fungal pathogens (some species) Spoilage activity
Wine, sourdough, indigenous fermented foods and beverages	*Hansenula/Pichia* sp.	Production of volatile/aroma compounds in wine Inhibition of various moulds
Cheese	*Torulopsis* sp.	Flavour enhancement
Cheese, salami, rye sourdough	*Debaryomyces* sp.	Production of flavor compounds in cheese (nutty or malty flavor) Increase in pH Help in ripening of cheese
Cheese, salami	*Yarrowia lipolytica*	Lipolytic and proteolytic activities Reduction in fat rancidity
Fermented meat products, wine	*Rhodotorula* sp.	Lipolytic activities Production of carotenoids Spoilage of dairy foods
Cheese, cocoa beans	*Kluyveromyces* sp.	Fermentation of lactose Production of pectinase, lipase and rennet
Soy sauce	*Zygosaccharomyces* sp.	Osmotolerance

Apart from fermented foods and beverages, yeasts are also used for the production of enzymes, fine chemicals, single-cell proteins, and flavoring compounds (Gatto and Torriani 2004, Wang 2008). The non pathogenic *Saccharomyces unisporus* has shown a significant role in the ripening of cheese and production of fermented milk products such as *kefir* and *koumiss*. Two yeast species, *viz.*, *S. unisporus* and *S. florentinus* are accepted by the International Dairy Federation and the European Food and Feed Cultures Association for food and feed applications.

4.1 Significance of Yeasts in Food and Beverage Production

Out of various microorganisms used for fermentation, yeasts are the most important group being exploited for commercial purposes. They are used in food fermentation processes to modify the cheaper raw materials organoleptically, physically and nutritionally into valuable products, and have been used for millennia in bread-making and the production of alcoholic beverages. In fermented foods and beverages, yeasts play a very important role in the bio-enrichment of foods, leading to the production of proteins, vitamins, minerals, aroma, alcohols, acids and esters. These compounds play a further role in improvements in the digestibility and preservation of these foods.

The significance of yeasts in food and beverage production includes (Fleet 2006):

- Fermentation and production of fermented foods and beverages
- Production of safe and suitable ingredients (flavor compounds) and additives for food processing
- Prevention of growth of spoilage microorganisms
- As animal and human probiotic
- Recombinant yeasts as biotherapeutic agents
- Spoilage of foods and beverages
- Source of food allergens

4.2 Yeasts in Dairy Products

Although dairy products are dominated by lactic acid bacteria, there is now substantial literature describing the important role of yeasts in flavor and texture development during the maturation stage of cheese production and in the production of fermented milks such as *kefir* and *koumiss* (Fleet 1990, Frohlich-Wyder 2003). Among the dairy products, yeasts are mainly used as single starter cultures for cheese production. In most cases, yeasts are used as secondary starter cultures which facilitate the production of aroma compounds and growth of other microorganisms. *Debaryomyces hansenii*, *Yarrowia lipolytica*, *Kluyveromyces marxianus* and *S. cerevisiae* are the predominant and most important species in dairy products but *Galactomyces geotrichum*, *Candida zeylanoides* and *Pichia* species are also significant. Dairy products which derive some of their character from the activity of fermenting yeasts are *kefir* and *koumiss*. The populations and yeast flora of *kefir* have been described by Marshall et al. (1984). In some cheese varieties, especially in semi-soft cheeses with surface films, yeasts are involved in the maturation process, contributing to ripening by utilization of lactic acid, thereby increasing the pH which encourages bacterial growth and initiates the second stage of maturation. During the production of surface-ripened cheeses such as Brick, Limburger, Port Salut, Taleggio, Tilsitter, Trappist, and the Danish Danbo cheese, *D. hansenii* is the key yeast involved. The surface smear of the above varieties of cheese consists of both yeasts and bacteria. In these cheeses, yeasts initiate the ripening by hydrolysis of lactic acid, thereby increasing the pH on the cheese surface and allowing the growth of a more acid-sensitive bacterial population (Leclercq-Perlat et al. 1999, Petersen et al. 2001). Further, *D. hansenii* might produce growth factors of importance for the bacteria as well as aroma components and lipolytic and proteolytic enzymes that contribute to the ripening process (Jakobsen and Narvhus 1996). In Gorgonzola cheese, *S. cerevisiae* is the main starter culture used in the production, but it apparently also occurs spontaneously, together with other yeasts, as an integral part of the microbial population of both blue veined cheeses and some types of soft cheese (Beresford et al. 2001, Hansen and Jakobsen 2001).

Under the genus *Saccharomyces*, 10 species are accepted in dairy products, of which *S. unisporus*, also called as *Kazachstania unispora* (Lu et al. 2004), is one. *S. unisporus* is found to be the dominant yeast in traditional dairy products (Rahman et al. 2009, Yildiz 2010). A unique synchronization of *Kluyveromyces marxianus* and *S. unisporus* (*Kazachstania*) species has been found in milk-fermented products, such as *dahi, suusac, gariss, kefir, shubat*, and *koumiss* (Narvhus and Gadaga 2003, Abdelgadir et al. 2008, Rizk et al. 2008). *S. unisporus* is the slowest producer of ethanol and performs a clean fermentation in milk and whey (Montanari et al. 1996).

4.3 Yeasts in Fermentation of Meats

In fresh meat, yeasts are usually present in very low numbers, but their counts may increase during storage at low temperature and they may eventually dominate the microflora in later stages (Cook 1995). Fermentation is a crucial phase of the curing stage of meat based products (e.g., sausages) since, at this stage, the principal physical, biochemical and microbiological transformations take place. The strains of *D. hansenii, D. kloeckeri, Y. lipolytica, Citeromyces matritensis, Trichosporon ovoides* and several other species of *Candida* (*C. intermedia* or *C. curvata, C. parapilosis, C. zeylanoides*), *Pichia, Cryptococcus*, and *Rhodotorula* are the yeasts which contribute significantly to the flavor of fermented meat products and meat-flavored products (Bolumar et al. 2003, Encinas et al. 2000). Various enzymes, *viz.*, lipases and proteases, are secreted by these yeasts which contribute to the flavor of the products by offsetting the acidic pH caused by the activities of mixed bacterial starter culture through the degradation of lipids to free fatty acids and glycerol and the breakdown of nitrogenous compounds to amino acids with release of ammonia. Several recently published reports describe the isolation and characterization of several such enzymes from *D. hansenii* (Bolumar et al. 2003, Durá et al. 2002). Yeasts (*D. hansenii, Debaryomyces* species, *Y. lipolytica* and *Candida* species) play an important role in the fermentation of meat sausages and the maturation of hams where they contribute to flavor and color development in the products (Samelis and Sofos 2003). Although yeasts are not the part of the starter cultures used for salami processing, a high number of yeasts were observed during the later stages of maturation, suggesting that they play an important role in the ripening of salami (Abunyewa et al. 2000).

In addition, meaty flavor can be achieved in non-meat-derived products by the processing of aqueous yeast or yeast extracts by treatments that utilize heat in the presence of sugars (preferably xylose or lactose) with or without other amino acids such as methionine, cysteine, or cystine (Abbas 2006).

4.4 Role of Yeast in Cereal Fermentation

Cereals, in their dry state, are not subject to fermentation due to their low water content. Dried cereals contain less than 14% water which limits the growth of microorganisms and chemical changes during storage. However, on mixing grains or cereal flour with water or other water-based fluids, enzymatic changes occur that may be attributed to the enzymes inherently present in the grain and/or contributed by the microorganisms. These microorganisms can either be those present as the natural contaminating flora of the cereal, or they can be added as a starter culture (Narvhus and Sørhaug 2006).

Yeasts other than *S. cerevisiae* are found in the fermentation of various cereal products, including sourdough breads, where they affect product flavor and rheology. Prominent yeasts in sourdough fermentations include *S. exiguus*, *C. humicola/ C. milleri* and other species of *Candida* (e.g., *C. krusei/Issatchenkia orientalis*), *Torulaspora delbrueckii* and various *Pichia* species (Jenson 1998, Meroth et al. 2003, Hammes et al. 2005). *Zygosaccharomyces rouxii*, *C. versatilis*, and *C. etchellsii* are important osmotolerant species that play a key role in soy sauce fermentation (Hanya and Nakadai 2003). In sourdough, yeasts are responsible for the leavening process, while the bacteria determine the souring of the dough. The combined metabolic activities of these microorganisms lead to final products with particular sensory properties and longer shelf life. In addition, a vast range of traditional, fermented products are produced in Africa, Asia and Latin America where, along with bacteria, a diversity of yeast species make important contributions (Nout 2003). The yeasts *Geotrichum candidum*, *Torulopsis holmii*, *Torulopsis candida* and *Trichosporon pullulans* have also been identified in *idli* fermentation (Chavan and Kadam 1989).

Yeasts have well known association with bread, beer, wine and other alcoholic beverages. Many species, other than those of *Saccharomyces*, also make positive contributions to the production of fermented foods and beverages. Various species of *Hanseniaspora*, *Candida*, *Pichia*, *Metschnikowia*, *Kluyveromyces*, *Schizosaccharomyces* and *Issatchenkia* can contribute positively to the fermentation of wine from grapes and cider from apples (Fleet 1998, 2003, Pretorius 2000). *Dekkera* (*Brettanomyces*) species, in addition to *S. pastorianus* and *S. cerevisiae,* are significant in the production of some types of beer (Dufour et al. 2003), while *Schizosaccharomyces pombe* can be important in rum fermentations (Fahrasmane and Ganou-Parfait 1998).

4.5 Baker's Yeast

The first records of the use of yeast in bread making came from Ancient Egypt (Legras et al. 2007). *S. cerevisiae* is the most commonly used yeast as a leavening agent in baking, where it converts the fermentable sugars present in dough into carbon dioxide. This causes the dough to expand or rise as gas forms pockets or bubbles. When the dough is baked, the air pockets 'set', giving the baked product a soft and spongy texture. Most yeast strains used in baking are of the same species common in alcoholic fermentation. In addition, *Saccharomyces exiguus* (*S. minor*), wild yeast found on plants, fruits and grains, is occasionally used for baking. In bread making, the yeast initially respires aerobically, producing carbon dioxide and water. When the oxygen is depleted, fermentation begins, producing ethanol as a waste product; however, this evaporates during baking.

In bread making, the most important function of baker's yeasts is leavening (Paramithiotis et al. 2000), by producing CO_2 via the alcoholic fermentation of the sugars thus increasing the dough volume and giving bread its characteristic light and spongy texture. For improved performance during bread making, the yeast strains should possess different characteristics like (i) high CO_2 production, (ii) the ability to quickly start utilizing maltose when the level of glucose in the flour is depleted, (iii) the ability to store high concentrations of trehalose, which will give tolerance to freezing and to high sugar and salt concentrations, (iv) tolerance to bread preservatives and chemicals, and (v) viability and retained activity during various storage conditions. Baker's yeast also influences the development of the gluten structure in dough, brought about by expansion of the dough owing to CO_2 production. Furthermore, yeasts produce primary and secondary metabolites, such as alcohols, esters and carbonyl compounds, which contribute to the development of the characteristic bread flavor (Damiani et al. 1996, Martinez-Anaya 1996).

In dough fermentation, formation of end products of yeast metabolism varies considerably with pH. In bread, the pH is usually below 6.0; above this pH, end products such as succinate, acetic acid and glycerol, along with ethanol and CO_2, are formed. *S. cerevisiae* is also able to degrade proteins and lipids, producing several flavor compounds (Narvhus and Sørhaug 2006).

Three different types of yeasts, *viz.*, wet compressed yeast (WCY), active dry yeast (ADY) and instant yeast (ITY) are produced commercially for different purposes. Wet compressed yeast is sold as fresh yeast compressed into a square cake and this form perishes quickly, so it must be used soon after production. In contrast to this, active dry yeast has solid levels of 92–97% and has shelf life of 4–12 months. Instant yeast, also called highly active dried yeast, has high solid levels and a shelf life of up to 2 years (Joshi and Pandey 1999, Wood 1998).

4.6 Yeast in Sourdough

Sourdough is defined as an ingredient containing cereal components, liquids and active microorganisms including lactic acid bacteria and yeast (Brummer and Huber 1987). Traditional sourdough bread technology is based on a spontaneous fermentation process from lactic acid bacteria and yeast occurring naturally in flour. Dense dough promotes the growth of lactic acid bacteria, whereas soft dough promotes the development of yeast (Gobbetti et al. 1995). Numerous yeasts have been isolated from sourdoughs but only a part of them can be considered to play a substantial role in fermentation processes. Table 2 shows the different yeast species isolated from sourdough. The most often reported yeast species in sourdough, regardless of the flour type and fermentation conditions, are *S. cerevisiae* (Gobbetti et al. 1994), *S. exiguus* (Gobbetti et al. 1994, Ottogalli et al. 1996), and *C. milleri* (Meroth et al. 2003, Vernocchi et al. 2004). Less often, *C. krusei*, *Torulaspora delbrueckii*, and *Pichia saitoi* are reported (Spicher and Stephan 1993). In the course of sourdough fermentation, yeast plays several important roles; for example, it produces carbon dioxide which expands the dough, resulting

Table 2. Yeast species identified in sourdoughs (Maloney and Foy 2003).

Yeast species	Sources
Saccharomyces cerevisiae	Rye, corn and wheat sourdough
Saccharomyces chevalieri	Rye and wheat sourdough
Saccharomyces curvatus	Rye and wheat sourdough
Saccharomyces exiguus	Rye and wheat sourdough
Saccharomyces fructuum	Wheat sourdough
Saccharomyces inusitalus	San Francisco wheat sourdough
Saccharomyces panis fermentati	Rye and wheat sourdough
Candida boidinii	Rye and wheat sourdough
Candida crusei	Wheat sourdough
Candida guillermondii	Wheat sourdough
Candida milleri	Wheat sourdough
Candida norvegensis	Wheat sourdough
Hansenula anomala	Wheat sourdough
Hansenula subpelliculosa	Wheat sourdough
Pichia salvi	Rye sourdough
Torulopsis delbrueckii	Corn and rye sourdough

in the proper porosity of the crumb and the proper volume of bread. Besides, yeast produces a variety of metabolites such as aldehydes, alcohols, acids, keto acids and esters, which alone or in combination with other compounds can create specific and unique flavors of bread (Makoto et al. 1990).

4.7 Yeasts in Alcoholic Fermentation

In general, yeasts (mainly the species of *Saccharomyces*) predominate during alcoholic fermentation, where the low pH and nutritional content of the substrate/juice lends itself to yeast growth. The use of yeast for the production of alcohol is found in the earliest human records. Beer, which is produced from hopped malt extract, is usually fermented by *Saccharomyces carlsbergensis* which differs from *S. cerevisiae* in that it is able to ferment the sugar melibiose. The two main types of beer, lager and ale, are fermented with strains of *S. carlsbergensis* and *S. cerevisiae*, respectively. The yeast used in beer production is called brewer's yeast (Gibson 2010).

4.8 Brewer's Yeast

Brewing yeasts may be classified as top-fermenting and bottom-fermenting. Top fermenting yeasts tend to be less flocculent and loose clumps of cells are carried to the fermenting wort surface, adsorbed to carbon dioxide bubbles. An example of a top fermenting yeast is *S. cerevisiae*, sometimes called an 'ale yeast'. Bottom fermenting yeasts are typically used to produce lager-type beers and these flocculate and collect at the bottom towards the end of fermentation, though they can also produce ale-type beers. These yeasts ferment well at low temperatures. An example of bottom fermenting yeast is *Saccharomyces pastorianus*, formerly known as *S. carlsbergensis* (Gibson 2010).

4.9 Yeast in Wine Preparation

Yeasts are the prominent organisms used in winemaking, where they convert the sugars present in grape juice (must) into ethanol. Yeasts determine several characteristics of the wine, including the flavor, through a range of mechanisms and activities (Fleet 2003). Yeasts are usually present as normal microflora on grape skin. Studies of spontaneous fermentations have identified a genetically diverse range of yeast populations, *S. cerevisiae* being the one which changes during fermentation (Schutz and Gafner 1993).

During wine preparation, fermentation can be done with endogenous 'wild yeast', but this gives an unpredictable product, which depends upon the exact types of yeast species present. So, a pure yeast culture is usually added to the must which ensures a reliable and predictable fermentation (González et al. 2001). Till date, more than 100 strains of wine yeasts are available in the market in the form of dry powder or tablet called 'dry wine yeast'. Based on the yeast usage of 0.1–0.2 g/l wine, approximately 5000 tonnes of dry wine yeast is required to produce the total estimated quantity (3.25 x 10^{10} L) of wine produced in different countries of world. But the actual market for the dry wine yeast does not exceed 600 tonnes per annum because many wineries either use the indigenous yeasts or propagate the wine yeast in-house. Most added wine yeasts are strains of *S. cerevisiae*, but not all strains of this yeast are suitable (González et al. 2001). Different *S. cerevisiae* yeast strains have different physiological and fermentative characteristics; therefore the actual strain of yeast selected for fermentation can have a direct impact on the finished wine (Dunn et al. 2005). Some of the desirable characteristics that wine yeast should possess are listed in Table 3.

In wine, brandy and other fruit based alcoholic beverages, yeast is responsible for the production of a variety of esters, *viz.*, ethyl acetate, hexyl acetate, ethyl caproate, isoamyl acetate, ethyl caprylate and 2-phenylethyl acetate (Swiegers et al. 2005), which give the characteristic fruity odor to these beverages. Yeast cells synthesize these esters by alcohol acetyltransferases (AATases), using higher alcohols and acetyl-CoA as substrates. The *ATF1*- and *ATF2*-encoded alcohol acetyltransferases of *S. cerevisiae* are responsible for the synthesis of ethyl acetate and isoamyl acetate esters, while the *EHT1*-encoded ethanol hexanoyl transferase is responsible for synthesizing ethyl caproate (Lilly et al. 2006).

In addition, wine yeast strains have limited but varying capacities to produce aroma-enhancing thiols such as 4-mercapto-4-methylpentan-2-one (4MMP), 3-mercaptohexan-1-ol (3MH) and 3-mercaptohexyl acetate (3MHA), from their non-volatile counterparts in grape juice (Swiegers et al. 2007).

The non-*Saccharomyces* yeasts present in grape juice are *Hanseniaspora* (*Kloeckera*), *Candida*, *Pichia*, *Kluyveromyces* and *Metschnikowia*, which could proliferate to final populations of about 10^6–10^7 cfu/ml; and as the ethanol level exceeds

Table 3. Desirable characteristics of wine yeast (Bhalla et al. 2011).

A. Fermentation properties	C. Technological properties
• Rapid initiation of fermentation • Fermentation of variety of carbohydrates • High osmotolerance • High ethanol tolerance (up to 15% v/v or more) • Ability to ferment at low temperature and under pressure • Moderate biomass production	• High genetic stability • High sulfite tolerance • Low foam formation • Flocculation properties • Compact sediment • Resistance to desiccation • Resistance to copper • Zymocidal (killer) properties • Genetic marking • Low nitrogen demand • Secretion of pectinases, glucanases and proteases
B. Flavor characteristics	**D. Metabolic properties with health implications**
• Production of desirable aldehydes and ketones • Reduced sulphide/dimethyl sulphide/thiol formation • Low production of higher alcohol and volatile acidity • High glycerol production • Potential to produce b-glucosidase to enhance wine flavour • Enhanced production of desirable volatile esters	• Reduced ethyl carbamate (urea production) • Low biogenic amine formation • Least oxidation of polyphenols • Low sulfite formation

5–7%, their numbers start to decrease (Heard and Fleet 1988, Gao and Fleet 1988). Ethanol production by *S. cerevisiae* is the major factor affecting the growth of non-*Saccharomyces* yeasts. Significant research has been undertaken for the development of novel wine yeast strains that produce typical flavor profiles or increased complexity in wines (McBryde et al. 2006).

In addition, the growth of some yeasts such as *Zygosaccharomyces* and *Brettanomyces* in wine leads to its spoilage (Loureiro and Malfeito-Ferreira 2003).

5. Flavor Active Wine Yeasts

Besides the main production of ethanol and CO_2, yeast also produces a variety of flavor compounds which greatly influence the aroma and taste of wine. The aroma and flavor properties of wine can be enhanced by glycosidases, which catalyze the hydrolysis of non-volatile glycosidic precursors of the grapes. It must be underlined that yeasts involved in winemaking can be important producers of numerous enzymes. Thus, the various yeast species/strains that develop during the fermentative process metabolize grape juice components to a wide range of volatile and non-volatile end-products that contribute to the aroma and flavor characteristics of the wine. The flavor of fermented beverages such as beer, cider, sake and wine owe much to the primary fermentation yeast (*S. cerevisiae*) used in their production. The flavor compounds such as ethyl esters, acetate esters, fusel alcohols, carbonyls, and volatile fatty acids formed during yeast fermentation are also called 'yeast bouquet'. Wine strains of *Saccharomyces bayanus* produce relatively high concentrations of 2-phenylethanol and 2-phenylethyl acetate compared to other higher alcohols and acetate esters, which may enhance 'rose' and 'floral' flavors (Masneuf-Pomarede et al. 2006). Flavor active wine yeast can also be arranged based on the production of sulfur containing compounds, which are associated with 'tropical' or 'sulfidic' flavors in wine.

These are the reasons why the alcoholic beverage industry is very much interested in isolation and identification of yeast strains that give a unique flavor to the fermented product, especially to the wine. So, intensive research is going on to understand the yeast flavor characteristics and on the selection of yeast strains yielding desired flavors (Fleet 2008).

6. Yeasts in Traditional Fermentations

The preparation of many indigenous/traditional fermented foods and beverages have remained a household art in homes, villages and small-scale industries. On the contrary, the preparations of others, such as soy sauce are now carried out on a large commercial scale (Bol and de Vos 1997). Most of the traditional fermented foods are prepared by the process of solid-substrate fermentation, in which the substrate is allowed to ferment either naturally or by adding starter culture (Tamang 1998). The association of yeasts and lactic acid bacteria is often encountered or used in the production of fermented foods and beverages (Gobbetti 1998). The most well-known example of stable coexistence of yeast and lactic acid bacteria is the presence of *C. milleri* and *Lactobacillus sanfransiscensis* in San Francisco sourdough (Kline and Sugihara 1971), in Dutch wheat sourdoughs (Nout and Creemers-Molenaar 1987) and in German commercial sourdoughs (Böcker et al. 1990). Table 4 shows various yeast species isolated from traditional fermented foods.

Demuyakor and Ohta (1991) reported *S. cerevisiae* as being the predominant species in Ghanaian *pito*, alongwith *Candida* sp. and *Kluyveromyces* sp. A similar prevalence of *S. cerevisiae* (38%) in Ghanaian *pito* was observed by Sefa-Deheh et al. (1999), who also isolated *Candida tropicalis* (19%), *Torulaspora delbrueckii* (14%) and members of four other species. Sanni and Lönner (1993) examined Nigerian sorghum beer and reported *Candida* spp., *Geotrichum candidum*, *S. cerevisiae*, *Kloeckera apiculata* and *T. delbrueckii* to be the predominant yeast involved in fermentation. From *dolo* originating from Togo, *S. cerevisiae* has been found to dominate with a prevalence of 55–90% of the yeast population, whereas *Candida krusei* (70%) was dominant in one sample (Konlani et al. 1996). Similarly, *S. cerevisiae*, *Cryptococcus laurentii* and *Torulospora delbrueckii* are the yeasts isolated from *seera*, a traditional fermented food of Himachal Pradesh in India (Savitri et al. 2012). Sake is a traditional alcoholic drink in Japan, and is made from steamed rice by fermentation with *Aspergillus oryzae*, which saccharifies rice starch, and sake yeast, which converts the resultant glucose to ethanol. Although, *S. cerevisiae* var. *sake* along with other yeast strains is used in sake fermentation (Jay 1991, Azumi and Goto-Yamamoto 2001), it differs from other yeast strains as it has many characteristics which makes it suitable for sake brewing. These characteristics include a good aroma and the production of a high concentration of ethanol at low temperature (Hosaka et al. 1998).

Table 4. Some important yeast species isolated from various fermented foods.

Yeast	Fermented food	Reference
Candida boidini	*Dosa*	Sandhu and Waraich (1984)
Candida cacaoi	*Idli*	Sandhu and Waraich (1984)
Candida curvata	*Warri, bhallae*	Sandhu and Waraich (1984)
Candida etchellsii	*Miso*, soy sauce	Sandhu and Waraich (1984)
Candida famata	*Bhallae, warri*	Sandhu and Waraich (1984)
Candida fragicola	*Idli*	Sandhu and Waraich (1984)
Candida glabrata	*Idli, dosa, kodo ko jaanr*	Venkatasubbaiah et al. (1985) Thapa and Tamang (2006)
Candida guillermondii	*Injera, torani*	Padmaja and George (1999)
Candida kefyr	*Idli*	Sandhu and Waraich (1984)
Candida krusei	Punjabi *warri*, cocoa beans	Sandhu and Soni (1989)
Candida lactose	*Tape ketan*	Cronk et al. (1977)
Candida melinii	*Tape ketan*	Cronk et al. (1977)
Candida menbranafaciens	*Bhallae*	Sandhu and Waraich (1984)
Candida milleri	Sourdough	Stolz (2003)
Candida parapsilosis	*Tape ketan*, Punjabi *warri*	Cronk et al. (1977) Sandhu and Waraich (1984)
Candida pseudotropicalis	*Idli, dosa*	Sandhu and Waraich (1984)
Candida sake	*Idli*	Venkatasubbaiah et al. (1985)
Candida tropicalis	*Torani, idli, pito*	Venkatasubbaiah et al. (1985), Sefa-Deheh et al. (1999)
Candida vartiovaarai	*Warri, bhallae*	Sandhu et al. (1986)
Candida vini	*Poi*	Padmaja and George (1999)
Cryptococcus sp.	*Seera*	Savitri et al. (2012)
Debaryomyces hansenii	*Bhallae, warri*	Sandhu et al. (1986)
Debaryomyces tamari	*Idli, warri*	Sandhu and Waraich (1984)
Endomycopsis fibuliges	*Thumba, tape*	Padmaja and George (1999)
Geotrichum candidum	Nigerian sorghum beer	Sanni and Lönner (1993)
Hansenula anomala	*Bhallae, idli*, Punjabi *warri, kanji, kecap, torani*	Soni and Sandhu (1999)
Hansenula malanga or H. subpelliculosa	*Tape ketan*	Cronk et al. (1977)
Hansenula polymorpha	*Bhallae, dosa*	Sandhu and Waraich (1984)
Issatchenkia torricola (*Pichia terricola, Saccharomyces terricolus*)	*Dosa, idli*	Sandhu and Waraich (1984)
Kluyveromyces marxianus	*Bhallae, warri*	Soni and Sandhu (1990)
Pichia anomala	*Kodo ko jaanr*	Thapa and Tamang (2006)

Table 4. contd....

Table 4. contd.

Yeast	Fermented food	Reference
Pichia membranefaciens	*Warri, bhallae*	Sandhu et al. (1986)
Pichia saitoi	*Sourdough*	Stolz (2003)
Rhodotorula flava	*Doenjang*	Chang et al. (1977)
Saccharomyces bayanus	*Jalebies*	Padmaja and George (1999)
Saccharomyces bisporus	Teekwass	Hesseltine (1983)
Saccharomyces carbajali	*Pulque*	Steinkraus (1998)
Saccharomyces carlbergensis	Beer and ale	Jay (1991)
Saccharomyces cerevisiae	*Bhallae*, beer, *burukutu*, cider, *fufu, ogi, puto, dosa, idli, papdam, lao chao*, scotch whiskey, *pito, bhatooru*	Padmaja and George (1999), Batra and Milner (1974), Soni and Sandhu (1990), Demuyakor and Ohta (1991), Savitri and Bhalla (2012)
Saccharomyces ellipsoideus	Wines	Jay (1991)
Saccharomyces exiguus	San Francisco sourdough	Jay (1991)
Saccharomyces intermedium	*Tibi*	Jay (1991)
Saccharomyces rouxii	*Miso*, Punjabi *warri*, Soy sauce	Batra (1981), Wang and Hesseltine (1982), Yokotsuka (1960)
Saccharomyces sake	*Sake* (rice beer)	Jay (1991)
Saccharomyces soyae	*Miso*	Winarno et al. (1977)
Saccharomycopsis fibuligera	*Tape ketan, lao chao, kodo ko jaanr*	Hesseltine (1983), Thapa and Tamang (2006)
Saccharomycopsis malanga	*Tape ketan, lao chao*	Hesseltine (1983)
Schizosaccharomyces pombe	*Teekwass*, traditional African beverages	Hesseltine (1983)
Torula sp.	*Kefir, kumiss*	Jay (1991)
Torulopsis candida	*Idli*	Batra and Milner (1974)
Torulopsis dattila	*Doenjang*	Chang et al. (1977)
Torulopsis etchellsii	*Miso*	Padmaja and George (1999)
Torulopsis holmii	*Idli*	Venkatasubbaiah et al. (1985)
Torulospora delbrueckii	*Seera*	Savitri et al. (2012)
Zygosaccharomyces soyae	*Tauco*	Winarno et al. (1977)

7. Yeasts as Source of Ingredients and Additives for Food Processing

With the increase in consumer demand for more natural foods, there is increasing interest in using microorganisms, including yeasts, as unique sources of food ingredients and additives, such as flavors, colors, antioxidants and vitamins. Since yeasts have a positive image with consumers, they are considered as a safe source of ingredients and additives for food processing (Demain et al. 1998). Preparations of baker's and brewer's yeasts have been available for many years as dietary and nutrient supplements because of their high contents of B vitamins, proteins, peptides, amino acids and trace minerals. Yeasts are also often considered as alternative source of protein for human consumption (Harrison 1993).

Yeasts serve as a source of antioxidants, aromas, flavors, colors, vitamins, glutamic acid, and nucleotides. Flavor ingredients based on yeast extracts, yeast autolysates and dried yeast preparations represent the most commercially significant products extracted from yeasts, and are used extensively in the food industry as a source of savory, roasted, nutty, cheesy, meaty and chicken flavors. In addition, some extracts are specifically enriched in their contents of glutamic acid and nucleotides that function as strong flavor enhancers (Stam et al. 1998). Although, baker's and brewer's yeasts have been the traditional sources of these products, some other yeasts such as *C. utilis* (*Pichia jadinii*) and *K. marxianus* are also used for this purpose (Lukondeh et al. 2003).

In addition to the above mentioned compounds, yeasts are frequently mentioned as potential sources of high value aroma and flavor substances (Vandamme and Soetaert 2002) such as vanillin (*S. cerevisiae, Rhodotorula glutinis*), citronellol, linalool and geraniol (*K. marxianus*), and γ- and δ-decalactones (*Sporidiobolus sulmonicolor, Y. lipolytica*). The yeast cell wall is composed principally of b-(1→3) and b-(1→6)-glucans that have gelling, thickening and fat-sparing functional properties and offer a range of applications in food processing (Seeley 1977).

Food colorants such as astaxanthin and other carotenoid pigments (Lyons et al. 1993, Johnson and Schroeder 1995) and several vitamins (Sauer et al. 2004) can also be derived from yeasts. Carotenoids belong to the most important components in foods. They are natural colorants and, as yellow to red colors, have great influence on the acceptability of many foods. Yeasts are reliable microbial source of carotenoids, e.g., species of the genus *Rhodotorula, viz., R. glutinis, R. minuta, R. mucilaginosa, R. acheniorum* and *R. graminis* have been recognized as carotenoid producers. Other yeasts such as *Sporobolomyces roseus*, *Sporidiobolus salmonicolor* and *S. patagonicus* also produce carotenoids. Most of these yeasts are known to produce β-carotene but these also produce other carotenoids such as torulene, torularodine, and γ-carotene (Tinoi et al. 2005, Moliné et al. 2010). Moreover, some carotenoids are precursors of vitamin A; in terms of human health, they are among the bioactive compounds credited to reduce the risks of degenerative diseases such as cancer, cardiovascular diseases, macular degeneration and cataract. Yeast, *Xanthophyllomyces dendrorhous* has been reported to produce astaxanthin which is widely used a food colorant (Mata-Gómez et al. 2014).

8. Probiotic Yeasts

Probiotics are live microorganisms, mainly lactic acid bacteria, that are beneficial to the host when consumed in appropriate quantities. Although, lactic acid bacteria are widely accepted as the main probiotic species, interest in probiotic yeasts has increased (Klaenhammer 2001), especially in relation to animal feed applications. Preparations of live *S. cerevisiae* have been used as feed supplements to animal and poultry for many years, and have been reported to improve the growth and health of the animals (Lyons et al. 1993). Milk production increased by 7% in Tunisian Holstein Friesian cows whose feed was supplemented with probiotic yeast (Maamouri et al. 2014). There is increasing interest in the use of yeasts as probiotics in the aquaculture industry (Gatesoupe 1995). In humans, *S. cerevisiae* var. *boulardii*, has been successfully used over the last 20 years as an oral, biotherapeutic agent to treat patients with severe cases of diarrhea and other gastrointestinal disorders (Czerucka and Rampal 2002). Both baker's and brewer's yeasts (*S. cerevisiae*) are used as dietary supplements because of their high contents of vitamin B, proteins, peptides, amino-acids and trace minerals. Regardless of their non-human origin, such non-pathogenic yeasts fulfill the major criteria for probiotic definition (http://www.who.int/foodsafety/ fs_management_probiotic_guidelines.Pdf).

The mechanisms behind their probiotic activities are based on the secretion of proteases and other inhibitory proteins, stimulation of immunoglobulins and the elimination of toxins secreted by other microorganisms (Fooks and Gibsen 2002). Recently, yeast strains obtained from local fermented foods of Ethiopia were reported to have antimicrobial activities against *Listeria monocytogenes*, *Salmonella* spp. and *Staphylococcus aureus* (Mariam et al. 2014).

Over the past few decades, *Saccharomyces boulardii* has been extensively studied for its potential probiotic use (Buts 2009). Most studied aspects of yeasts as probiotic organisms in clinical trials are in treatment of (1) antibiotic associated diarrhea, (2) infectious diarrhea including recurrent *Clostridium difficile* (*C. difficile*) related diseases, (3) irritable bowel syndrome, and (4) inflammatory bowel diseases (Rajkowska and Kunicka-Stycznska 2010). Saurabh et al. (2012) screened around 23 indigenous yeast isolates obtained from the traditional fermented foods of the Western Himalayas for various probiotic attributes.

However, there are serious concerns about its public health safety because of the reports on the association of yeasts with cases of fungaemia (Cassone et al. 2003). Apart from health benefit and safety issues, probiotic yeasts will also require certain technological properties for use in foods, e.g. (i) to remain viable in the food, (ii) not to grow in foods, (iii) not spoil the food and (iv) not adversely affect sensory acceptability of the food (Heenan et al. 2004).

9. Yeast Starters

In fermentation, the raw materials are converted by microorganisms or their enzymes to products that have acceptable food qualities. Fermentation initiated without the use of a starter inoculum, also called spontaneous fermentation, has been applied to food preservation for millennia. However, spontaneous fermentations are neither predictable nor controllable as the natural microflora of the raw material is inefficient, uncontrollable, and unpredictable, or is destroyed altogether by the heat treatments given to the food. Modern fermentations are initiated by starter cultures consisting of different microorganisms (bacteria, yeast and molds) that are inoculated into food materials to bring about desired changes such as novel functionality, enhanced preservation, reduced food safety risks, improved nutritional or health value, enhanced sensory qualities, and increased economic value of the finished product (Kolawole et al.

2007). Since the beginning of the 1980s, the use of *S. cerevisiae* yeast starters has been extensively applied in the industrial and homemade beverage production processes. Depending upon the adaptation of microorganisms to a substrate or raw material, starter cultures are selected either as single strain or multiple strains. Currently, most of the wine production processes depend on *S. cerevisiae* strains that allow rapid and reliable fermentations, reduce the risk of sluggish or stuck fermentations and prevent microbial contaminations (Romano et al. 2003). A wine starter culture is normally able to dominate intrinsic yeasts in the grape must during fermentation (Pretorius 2000). Yeast starter cultures that are specifically selected for the winemaking process on the basis of scientifically verified characteristics typically complement and optimize the raw material quality and individual characteristics of the wine, creating a more desirable and acceptable product (Swiegers et al. 2005). It has been reported that the wines produced with selected yeast starters have a higher quality than wines produced by spontaneous fermentation (Srivastava et al. 1997).

Numerous studies have proposed the use of mycocin producing yeasts as starter cultures to prevent the growth of spoilage yeast strains and secondary fermentation of wines (Boone et al. 1990, Comitini et al. 2004). Calmette (1892) was the first to report the presence of several wild yeast species accompanied by molds *Amylomyces, Mucor, Aspergillus,* and about 30 different bacteria in starters used in India and China to produce alcohol.

Traditional fermentation starters are referred to as *chu* in Chinese, *nuruk* in Korean, *koji* in Japanese, *ragi* in Southeast Asian countries, *bakhar ranu* or *marchaar/murcha* in India (Batra and Millner 1974) and *phab* (Fig. 4) in the North Western Himalayan region of India (Thakur et al. 2004). *Saccharomyces bayanus, Candida glabrata,*

Figure 4. '*Phab*' traditional yeast based fermentation starter of the North Western Himalayas.

Pichia anomala, Saccharomycopsis fibuligera, Saccharomycopsis capsularis and *Pichia burtonii,* have been isolated from the *marcha. Hamei,* is used in the Eastern Himalayas for the preparation of alcoholic beverages, e.g., *bhaati Jaanr, aitanga* and *kodo ko jaanr* (Tamang et al. 2007). *Phabs* is an indigenous inoculum of the Trans-Himalayan Ladakh region of India and used in fermentation of two traditional barley based alcoholic beverages, *chhang* and *arrak* (Angmo and Bhalla 2014).

In addition to their numerous roles in food and beverage production, some of the yeasts play a role in the spoilage of foods and beverages, mainly those with high acidity and reduced water activity (aw). In these foods, yeasts are able to change the sensory characteristics in undesirable ways and are thus regarded as 'food spoilage yeasts'.

10. Food Spoilage Yeasts

Yeast spoilage is a constant threat in the food and beverage industries. Typically foods and beverages with high-acid, low-pH, high sugar, i.e., more than 10% w/v or high salt, i.e., more than 5% NaCl content, and products preserved with weak organic acids (e.g., sorbic, benzoic, acetic) are susceptible to spoilage by yeasts. Fruits, fruit juices, fruit drinks, fruit pulp, fruit juice concentrates, sugar and flavor syrups, confectionery products, alcoholic beverages, carbonated beverages, vegetable salads with acid dressings, salt based and acid based sauces, fermented dairy products and fermented or cured meat products are the prime candidates for yeast spoilage (Tudor and Board 1993). A list of the most common yeasts as contaminants in different food products and spoilage yeasts is given in Table 5.

The most noticeable indication of yeast spoilage is the production of excess gas, leading to the swelling of containers or cans and in extreme cases, explosions of the containers occur (Grinbaum et al. 1994). The second most apparent sign of yeast spoilage is the visible appearance of yeast colonies on the food surface, causing discoloration of the surface and formation of mucous slimes due to the production of extracellular polysaccharides. In beverages, the growth of spoilage yeast may cause the development of hazes, clouds, particulates, surface films or colonies and sediments. In some cases, spoilage by yeast is characterized by distinct off-flavors like the mouse flavor caused by *Brettanomyces intermedius* (Beech and Carr 1977), yeasty aldehyde flavor due to the formation of high levels of acetic acid, acetaldehyde (Lafon-Lafourcade 1986) and esters, including ethyl acetate (Lanciotti et al. 1998) by *Pichia* species. Off-flavor is also caused by high levels of acetoin and acetaldehyde (Romano et al. 1999) as reported in *Saccharomycodes ludwigii,* petroleum-like off-odor due to the degradation of sorbic acid to 1,3-pentadiene by *Zygosaccharomyces rouxii* and *Debaryomyces hansenii* (Casas et al. 1999), while *K. apiculata* causes off-flavors in cider comprised of high levels of esters and volatile acids (Beech and Carr 1977). In addition, the growth of lipolytic yeasts such as *Yarrowia lipolytica* on fat-rich substrates such as cheese or meat, may result in free fatty acid rancidity. Sometimes, the addition of preservatives to foods may make them more susceptible to spoilage by off-flavors.

Table 5. Different yeast species involved in food contamination and spoilage.

Most frequent contaminants	Foods	Spoilage species	Foods
Candida albidus	Fruit, vegetables and cereal grains	*Brettanomyces intermedius*	Soft drinks and fruit juices
Candida guilliermondii	Fish	*Candida dattila*	Canned fruit products
Candida parapsilosis	Fruit juices	*Candida globosa*	Fruits in sugar syrup and condensed milk
Candida tropicalis	Wine	*Candida holmii*	Soft drinks and fruit juices
Candida zeylanoides	Raw and processed poultry meat	*Candida humicola*	Refrigerated fishery products
Debaryomyces hansenii	Gherkins, fruit juices, marzipan and canned figs	*Candida krusei*	Citrus fruits, soft drinks and wine
Hanseniaspora uvarum	Wine	*Candida lactis-condensi*	Soft drinks
Issatchenkia orientalis	Meat products and processed poultry products	*Candida lipolytica*	Ready-to-drink beverages
Kluyveromyces marxianus	Fruit juices and concentrates, marzipan, salted and dry-cured meats, olives and cheeses	*Candida sake*	Unpasteurized orange juice
Pichia anomala	Grape juice concentrates	*Candida versatilis*	Wine
Pichia fermentans	Orange juice	*Cryptococcus* sp.	Cheese and frozen food products
Pichia membranifaciens	Wines and tomato sauce	*Hansenula anomala*	Italian cream-filled cakes
Rhodotorula glutinis	Strawberries	*Hansenula subpelliculosa*	Mango pickle
Rhodotorula mucilaginosa	Wines	*Kloeckera apiculata*	Figs, tomatoes, canned black cherries, strawberry toppings and yogurt
Saccharomyces cerevisiae	Grape juice concentrates, olive oil	*Pichia burtonii*	Bread and bakery products
Saccharomyces exiguus	Sugar-rich foods, maple sap, syrup, concentrated juices and condiments	*Schizosaccharomyces pombe*	Wines and high sugar syrups
Torulaspora delbrueckii	Dairy products, juices, wines and ready-to-eat (RTE) salad products	*Sporobolomyces roseus*	Proteinaceous foods and fruit products
Trichosporon pullulans	Idli, meat and different meat products	*Trichosporon cutaneum*	Fruits and vegetables, cereal grains, etc.
Zygosaccharomyces bailii	Fruit juices and concentrates, marzipan, salted, dry-cured meats, olives, cheeses, mayonnaise and salad dressings	*Zygosaccharomyces bisporus*	Mayonnaise based salad dressings
Zygosaccharomyces rouxii	Concentrated grape juice		

(Compiled from: Battey et al. 2002, Comitini et al. 2004, Fleet 2006, Lanciotti et al. 1998, Loureiro 2000, Loureiro and Malfeito-Ferreira 2003, McNamee et al. 2010, Saez et al. 2011, Saranraj and Geetha 2012, Schuller et al. 2000).

11. Health Significance of Yeasts in Food and Beverages

The presence of yeasts in foods and beverages is a subject of emerging interest in public health where in one context, the yeasts could serve as probiotic microorganisms, and in other circumstances, could lead to infections and other adverse effects to the consumer. Unlike many bacteria and viruses, yeasts are not known as aggressive infectious agents. However, some species of yeast, especially *C. albicans* and *Cryptococcus neoformans*, fall into the category of opportunistic pathogens. These cause a range of mucocutaneous, cutaneous, respiratory, central nervous, systemic and organ infections in humans (Hazen and Howell 2003). Generally, individuals with weakened health and immune function are at greater risk, including persons with AIDS and cancer, hospitalized patients and those undergoing treatments with immunosuppressive drugs, broad-spectrum bacterial antibiotics and radio chemotherapies.

12. Genetically Modified Yeasts for Food and Beverages

Initially, yeast for various purposes was obtained from breweries and distilleries, but with the growth of industrial population, there was an increase in the demand of yeast with higher yield of product and better technical performance. Considerable progress has been made in breeding new strains of yeast, and recombinant DNA techniques have been extensively used for the construction of new recombinant yeast strains, having the following desirable characteristics (Verstrepen et al. 2006):

- Improved performance and product quality
- Ability to ferment a wider range of carbohydrates
- Altered flocculation properties
- Produce products with modified flavors
- Better oligosaccharide (dextrin) utilization
- Fermentation of branched oligosaccharides and polysaccharides
- Antimicrobial properties
- Improved stress tolerance
- Improved sensory qualities of fermentation products
- Reduction in diacetyl levels in alcoholic beverages
- Improved flavor profiles of alcoholic beverages

In addition to the above properties, yeast strains are engineered with altered levels of ethanol and glycerol production, control acid levels in wine, produce decreased levels of hydrogen sulfide and ethyl carbamate and to make alcoholic beverages with increased storage/antioxidative potential. Attempts have also been made to change fermentation rates using genetically modified yeast strains engineered to produce higher levels of glycolytic enzymes (Verstrepen et al. 2006).

13. Food Yeast Collections

Research in the field of fermentation and fermented foods has led to isolation, identification and characterization of large numbers of yeast strains around the world. These yeast strains are deposited and preserved in different culture collections (Table 6). The primary aim of these culture collections is to preserve the microbial diversity, as well as to provide pure cultures for scientific research and academic purposes.

Table 6. List of some important yeast depositories and culture collections of the world.

Name of culture collection	Address and website address
American Type Culture Collection (ATCC)	12301 Parklawn Drive, Rockville, MD 20852, USA www.atcc.org.general/html
Bioresource Collection and Research Center (BCRC)	Academia Sinica, Biodiversity Research Center, 128Sec. 2, Academic Road, Nangang District Taipei, Chinese Taipei China http://www.gbif.org
Central Bureau Schimmelculture (CBS)	P.O. Box 85167, 3508 AD Utrecht, The Netherlands http://www.cbs.knaw.nl
China General Microbiological Culture Collection (CGMCC)	China General Microbiological Culture Collection, Institute of Microbiology, Chinese Academy of Sciences, No. 1, West Beichen Road, Chaoyang District Beijing, 100101, China http://www.cgmcc.net
Colecao de Culturas Tropical (CCT)	Colecao de Culturas Tropical, Fundacao Tropical de Pesquisas e Tecnologia 'André Tosello', Campinas - SP, Brazil
Spanish Type Culture Collection (CECT)	Spanish Type Culture Collection (CECT) University of Valencia, Parc Científic Universitat de València Catedrático Agustín Escardino, 9, 46980 Paterna (Valencia) Spain http://www.uv.es/uvweb/spanish-type-culture-collection/en/location-contact/contact/contact-details-1285872233467.html
Culture Collection of Yeast (CCY)	Institute of Chemistry Slovak Academy of Sciences Dúbravská cesta 9 Bratislava 845 38 Slovakia http://www.ccy.sk/
Industrial Yeast Collection (DBVPG)	Industrial Yeasts Collection DBVPG Department of Agricultural, Food and Environmental Science, University of Perugia, Borgo XX Giugno, 74 I-06121 Perugia, Italy dbvpg@unipg.it
Institute for Fermentation Osaka (IFO)	Institute for Fermentation, Osaka, Yodogawa-ku, Osaka, Japan http://trove.nla.gov.au/version/9088375

Table 6. contd....

Table 6. contd.

Name of culture collection	Address and website address
Japan Collection of Microorganisms (JCM)	Microbe Division/Japan Collection of Microorganisms RIKEN Bioresource Center, 3-1-1 Koyadai, Tsukuba, Ibaraki, 305-0074, Japan http://jcm.brc.riken.jp
Microbial Type Culture Collection (MTCC)	Institute of Microbial Technology, Sector-39 A, Chandigarh, India http://mtcc.imtech.res.in
Belgiun Coordinated Collection of Microorganisms (MUCL /BCCM)	croixduSud 2, box L7.05.06, B-1348 Louvain-La-Neuve http://bccm.belspo.be
National Bureau of Agriculturally Important Microorganisms (NBAIM)	National Bureau of Agriculturally Important Microorganisms, Kushmaur, Post Box No. 6, Mau Nath Bhanjan, Pin: 275103, Uttar Pradesh India www.nbaim.org.in
National Collection of Agricultural and Industrial Microorganisms (NCAIM)	National Collection of Agricultural and Industrial Microorganisms, Department of Microbiology and Biotechnology, University of Horticulture and Food Industry, Budapest, Hungary
National Collection of Industrial Microorganisms (NCIM)	Council of Industrial Research, National Chemical Laboratory (NCL), Dr. Homi Bhabha Road, Pune, India www.ncl-india.org/files/NCIM/
National Collection of Yeast Culture (NCYC)	National Collection of Yeast Cultures, Institute of Food Research, Norwich Research Park, Norwich, NR4 7UA, UK http://www.ncyc.co.uk
Northern Regional Research Center (NRRL)	Northern Regional Research Center, Agricultural Research Service Culture Collection, National Center for Agricultural Utilization Research, US Department of Agriculture, 1815 North University Street, Peoria, Illinois 61604, USA
Portuguese Yeast Culture Collection (PYCC)	Departamento de Ciências da Vida (DCV), Faculdade de Ciências e Tecnologia Universidade Nova de Lisboa Campus da Caparica 2829-516 Caparica Portugal http://pycc.bio-aware.com/
RCDM (Catch Rotary Culture Rev1)	MDC: Molecular Diagnostics Center, Biomolecular Technologies S.L. and Universidad Miguel Hernández, Orihuela E-03300, Alicante, Spain
All Russian Collection of Microorganisms (VKM)	Russia, 142290, Moscow Region, Pushchino, pr. Nauki, 5, IBPM Russia http://www.vkm.ru
Culture Collection of Industrial Organisms (CCIM)	Culture Collection of Industrial Microorganisms, Institute of Microbiology, 7b, A. Kadiry street, Tashkent 700128, Uzbekistan

14. Safety of Yeasts

The impact of yeasts on the quality and safety of foods and beverages is closely related to their biological activities. These activities are determined by the physical and chemical properties of the ecosystem, and by how yeasts respond according to their physiology, biochemistry and genetics. The close relationship with *S. cerevisiae* in food and beverage production over millennia supports the fact that it is safe to work with, and thus designated 'Generally Recognized as Safe' (GRAS) by the United States Food and Drug Administration (Verstrepen et al. 2006).

15. Future Prospects

A wide variety of yeasts are involved in the production of different fermented foods and beverages all over the world. The diversity of products prepared by using yeasts ranges from bread and bread like products to alcoholic beverages such as beer and wines and traditional fermented foods, flavor compounds, food ingredients, etc. Additionally, yeasts are also involved in spoilage of a variety of food products. Recently, interest in probiotic yeast for application in animal feed has also increased. Although considerable progress has been made in the isolation, identification, characterization and breeding of new strains of yeast, there is a need for further research in the areas of flavor formation in yeast, the factors affecting yeast growth, yeast flocculation and probiotic aspects. These require more precise characterization by using recent techniques of genomics, proteomics and metabolomics. Continued research in these fields will further our understanding and eventually lead to the development of yeast strains that can produce predictable products with specific metabolic profiles, thus allowing the producers to 'shape' their fermented products to suit certain consumer preferences and add value to existing fermentation. This will further enable the development of new products based on the exploitation of new and improved strains of yeasts.

Keywords: Yeast, fermentation, fermented foods, beverages, *Saccharomyces cerevisiae,* wine

References

Abbas, C.A. 2006. Production of antioxidants, aromas, colours, flavors and vitamins by yeasts. pp. 285–334. *In*: A. Querol and G.H. Fleet (eds.). The Yeast Handbook: Yeasts in Food and Beverages. Springer-Verlag, Berlin Heidelberg.

Abdelgadir, W.S., Nennis, D.S., Hamad, S.H.P. and Jakobsen, M. 2008. A traditional Sudanese fermented camel's milk product, gariss, as a habitat of *Streptococcus infantarius* subsp. *infantarius*. International Journal of Food Microbiology 127: 215–219.

Abunyewa, A.A.O., Laing, E., Hugo, A. and Viljoen, B.C. 2000. The population change of yeasts in commercial salami. Food Microbiology 17: 429–438.

Aidoo, K.E., Nout, M.J.R. and Sarkar, P.K. 2006. Occurrence and function of yeasts in Asian indigenous fermented foods. FEMS Yeast Research 6: 30–39.

Angmo, K. and Bhalla, T.C. 2014. Preparation of *phabs*—an indigenous starter culture for production of traditional alcoholic beverage, *chhang*, in Ladakh. Indian Journal of Traditional Knowledge 13: 347–351.

Azumi, M. and Goto-Yamamoto, N. 2001. AFLP analysis of type strains and laboratory and industrial strains of *Saccharomyces sensu stricto* and its application to phenetic clustering. Yeast 18: 1145–1154.

Bamforth, C.W. 2005. The science underpinning food fermentations. *In*: Food, Fermentation and Micro-organisms. Blackwell Publishing Ltd., UK.

Batra, L.R. 1981. Fermented cereals and gram legumes of India and vicinity. p. 547. *In*: M. Moo-Young and C.W. Robinson (eds.). Advances in Biotechnology. Pergamon Press, Toronto.

Batra, L.R. and Millner, P.D. 1974. Some Asian fermented foods and beverages and associated fungi. Mycologia 66: 942–950.

Battey, A.S., Duffy, S. and Schaffner, D.W. 2002. Modeling yeast spoilage in cold-filled ready-to-drink beverages with *Saccharomyces cerevisiae, Zygosaccharomyces bailii* and *Candida lipolytica*. Applied and Environmental Microbiology 1901–1906.

Beech, F.W. and Carr, J.G. 1977. Cider and perry. pp. 139–313. *In*: A.H. Rose (ed.). Economic Microbiology, vol. 1. Alcoholic beverages. Academic, London.

Beresford, T.P., Fitzsimons, N.A., Brennan, N.L. and Cogan, T.M. 2001. Recent advances in cheese microbiology. International Dairy Journal 11: 259–274.

Bhalla, T.C., Thakur, N. and Kapoor, S. 2011. Cultivation of wine yeast and bacteria. pp. 502–525. *In*: V.K. Joshi (ed.). Handbook of Enology: Principles, Practices and Recent Innovations, vol.-II. Asiatech Publishers, New Delhi.

Böcker, G., Vogel, R. and Hammes, P. 1990. *Lactobacillus sanfrancisco* als stabiles element in einem Reinzucht-Sauerteige-Präparat. Getreide Mehl und Brot 44: 269–274.

Bol, J. and de Vos, W.M. 1997. Fermented foods: An overview. pp. 45–76. *In*: J. Green (ed.). Biotechnological Innovations in Food Processing, Butterworth-Heinemann, Oxford.

Bolumar, T., Sanz, Y., Aristoy, M.C. and Toldrá, F. 2003. Purification and properties of an arginyl aminopeptidase from *Debaryomyces hansenii*. International Journal of Food Microbiology 86: 141–151.

Boone, C., Sdicu, A., Degri, R., Sanchez, C. and Bussey, H. 1990. Integration of the yeast "KI" killer toxin in gene into the genome of marked wine yeasts and its effect on vinification. Yeast 41: 37–42.

Boulton, C. and Quain, D. 2001. Brewing yeast. *In*: C. Boulton and D. Quain (eds.). Brewing Yeast and Fermentation. Blackwell Publishing Ltd.

Brummer, J.M. and Huber, H. 1987. Begriffsbestimmungen fur Vorstufen (Sauerteige, Vorteige, Quellstufen) und Weizenteigfuhrungen. Getreide Mehl und Brot 41: 110.

Buts, J.P. 2009. Twenty-five years of research on *Saccharomyces boulardii* trophic effects: Updates and perspectives. Digestive Diseases and Science 54: 15–18.

Calmette, L.C.A. 1892. Contribution à l'étude des ferments de l'amidon; la levûre chinoise. Annales de l'Institut Pasteur 6: 604–620.

Casas, E., Valderrama, M.J. and Peinado, J.M. 1999. Sorbate detoxification by spoilage yeasts isolated from marzipan products. Food Technology and Biotechnology 37: 87–91.

Cassone, M., Serra, P., Mondello, F., Giolamo, A., Scafetti, S., Pistella, E. and Venditti, M. 2003. Outbreak of *Saccharomyces cerevisiae* subtype *boulardii* fungemia in patients neighboring those treated with a probiotic preparation of the organism. Journal of Clinical Microbiology 41: 5340–5343.

Cavalieri, D., McGovern, P.E., Hartl, D.L., Mortimer, R. and Polsinelli, M. 2003. Evidence for *S. cerevisiae* fermentation in ancient wine. Journal of Molecular Evolution 57: S226–S232.

Chang, C.H., Lee, S.R., Lee, K.H., Mheen, T.I., Kwon, T.W. and Park, K.J. 1977. Fermented soybean foods. Symposium on Indigenous Fermented Foods, Bangkok, Thailand.

Chavan, J.K. and Kadam, S.S. 1989. Nutritional improvement of cereals by fermentation. CRC Critical Reviews in Food Science and Nutrition 28: 349–400.

Comitini, F., De Ingeniis, J., Pepe, L., Mannazzu, I. and Ciani, M. 2004. *Pichia anomala* and *Kluyveromyces wickerhamii* killer toxins as new tools against Dekkera/Brettanomyces spoilage yeasts. FEMS Microbiology Letters 238: 235–240.

Cook, P.E. 1995. Fungal ripened meats and meat products. p. 110. *In*: G. Campbell-Platt and P.E. Cook (eds.). Fermented Meats. Blackie, London.

Cronk, T.C., Steinkraus, K.H., Hackler, L.R. and Mattick, L.R. 1977. Indonesian *tape ketan* fermentation. Applied Environmental Microbiology 33: 1067–1073.

Czerucka, D. and Rampal, P. 2002. Experimental effects of *Saccharomyces boulardii* on diarrheal pathogens. Microbes Infect 4: 733–739.

Damiani, P., Gobbetti, M., Cossignani, L., Corsetti, A., Simonetti, M.S. and Rossi, J. 1996. The sourdough microflora. Characterization of hetero- and homofermentative lactic acid bacteria, yeasts and their interactions on the basis of the volatile compounds produced. Lebensmittel Wissenschaft und Technologie 29: 63–70.

Demain, A.L., Phaff, H.J. and Kurtzman, C.P. 1998. The industrial and agricultural significance of yeasts. pp. 13–19. *In*: C.P. Kurtzman and J.W. Fell (eds.). The yeasts—A Taxonomic Study, 4th edn. Elsevier, Amsterdam.

Demuyakor, B. and Ohta, Y. 1991. Characteristics of *pito* yeasts from Ghana. Food Microbiology 8: 183–193.

Dufour, J.P., Verstrepen, K. and Derdelinckx, G. 2003. Brewing yeasts. pp. 347–388. *In*: T. Boekhout and V. Robert (eds.). Yeasts in Food: Beneficial and Detrimental Aspects. Behr, Hamburg.

Dunn, B., Levine, R.P. and Sherlock, G. 2005. Microarray karyotyping of commercial wine yeast strains reveals shared, as well as unique, genomic signatures. BMC Genomics 6(1): 53.

Durá, M.A., Flores, M. and Toldrá, F. 2002. Purification and characterization of a glutaminase from *Debaryomyces* spp. International Journal of Food Microbiology 76: 117–126.

Encinas, J.P., López-Díaz, T.M., García-López, M.L., Otero, A. and Moreno, B. 2000. Yeast populations on Spanish sausages. Meat Science 54: 203–208.

Fahrasmane, L. and Ganou-Parfait, B. 1998. Microbial flora of rum fermentation media. Journal of Applied Microbiology 84: 921–928.

Fleet, G.H. 1990. Yeast in dairy products. Journal of Applied Bacteriology 68: 199–211.

Fleet, G.H. 1998. Microbiology of alcoholic beverages. pp. 217–262. *In*: B.J. Wood (ed.). Microbiology of Fermented Foods, Vol. 1, 2nd edn. Blackie, London.

Fleet, G.H. 2003. Yeast interactions and wine flavor. International Journal of Food Microbiology 86: 11–22.

Fleet, G.H. 2006. The commercial and community significance of yeasts in food and beverage production. pp. 1–12. *In*: A. Querol and G.H. Fleet (eds.). The Yeast Handbook: Yeasts in Food and Beverages. Springer-Verlag, Berlin Heidelberg, Germany.

Fleet, G.H. 2008. Wine yeasts for the future. FEMS Yeast Research 8: 979–995.

Foligné, B., Dewulf, J., Vandekerckove, P., Pignède, G. and Pot, B. 2010. Probiotic yeasts: Anti-inflammatory potential of various non-pathogenic strains in experimental colitis in mice. World Journal of Gastroenterology 16(17): 2134–2145.

Food and Agriculture Organization of the United Nations, World Health Organization. Guidelines for evaluation of probiotic in food. 2002. Available from: URL: http://www.who.int/foodsafety/fs_management_probiotic_guidelines.

Fooks, L.J. and Gibsen, G.R. 2002. Probiotics as modulators of gut flora. British Journal of Nutrition 88: 39–49.

Frohlich-Wyder, M.T. 2003. Yeasts in dairy products. pp. 209–237. *In*: T. Boekhout and V. Robert (eds.). Yeasts in Food: Beneficial and Detrimental Aspects. Behr, Hamburg.

Gao, C. and Fleet, G.H. 1988. The effects of temperature and pH on the ethanol tolerance of the wine yeasts, *Saccharomyces cerevisiae, Candida stellata* and *Kloeckera apiculata*. Journal of Applied Bacteriology 65: 405–410.

Gatesoupe, F.J. 1995. The use of probiotics in aquaculture. Aquaculture 180: 147–165.

Gatto, V. and Torriani, S. 2004. Microbial population changes during sourdough fermentation monitored by DGGE analysis of 16S and 26S rDNA gene fragments. Annals of Microbiology 54(1): 31–42.

Gibson, M. 2010. The Sommelier Prep Course: An Introduction to the Wines, Beers, and Spirits of the World. John Wiley and Sons, pp. 361. ISBN 978-0-470-28318-9.

Gobbetti, M. 1998. The sourdough microflora: interactions of lactic acid bacteria and yeasts. Trends in Food Science and Technology 9: 267–274.

Gobbetti, M., Corsetti, A., Rossi, J., La Rosa, F. and De Vincenzi, S. 1994. Identification and clustering of lactic acid bacteria and yeasts from wheat sourdoughs of central Italy. Italian Journal of Food Science 1: 85.

Gobbetti, M., Simonetti, M.S., Corsetti, A., Santinelli, F., Rossi, J. and Damiani, P. 1995. Volatile compound and organic acid productions by mixed wheat sourdough starters: Influence of fermentation parameters and dynamics during baking. Food Microbiology 26: 353–363.

Goldman, E. 2008. *In*: E. Goldman and L. Green (eds.). Practical Handbook of Microbiology, 2nd Edn. CRC Press.

González, T.A., Jubany, S., Carrau, F.M. and Gaggero, C. 2001. Differentiation of industrial wine yeast strains using microsatellite markers. Letters in Applied Microbiology 33(1): 71–75.

Grinbaum, A., Ashkenazi, I., Treister, G., Goldschmied-Reouven, A. and Block, C.S. 1994. Exploding bottles: Eye injury due to yeast fermentation of an uncarbonated soft drink. British Journal of Opthalmology 78: 883.

Haard, N.F. 1999. Cereals: Rationale for fermentation. pp. 15–21. *In*: Fermented Cereals a Global Perspective. FAO Agricultural Services Bulletin no. 138. Agricultural Organization of United Nations, Rome.

Hammes, W.P., Brandt, M.J., Francis, K.L., Rosenheim, J., Seitter, M.F.H. and Vogelmann, S.A. 2005. Microbial ecology of cereal fermentations. Trends in Food Science and Technology 16: 4–11.

Hannun, Y. and Obeid, L. 1997. Ceramide and the eukaryotic stress response. Biochemical Society Transactions 25: 1171–1175.

Hansen, T.K. and Jakobsen, M. 2001. Taxonomical and technological characteristics of *Saccharomyces* spp. associated with blue veined cheese. International Journal of Food Microbiology 69: 59–68.

Hanya, Y. and Nakadai, T. 2003. Yeasts and soy products. pp. 413–428. *In*: T. Boekhout and V. Robert (eds.). Yeasts in Food: Beneficial and detrimental aspects. Behr, Hamburg.

Harrison, J.S. 1993. Food and fodder yeast. pp. 399–434. *In*: A.H. Rose and J.S. Harrison (eds.). The Yeasts, vol. 3. Yeast Technology. Academic, London.

Hazen, K.C. and Howell, S.A. 2003. *Candida, Cryptococcus* and other yeasts of medical importance. pp. 1693–1711. *In*: P.R. Murray (ed.). Manual of Clinical Microbiology, 8th edn. American Society for Microbiology, Washington, DC.

Heard, G.M. and Fleet, G.H. 1988. The effects of temperature and pH on the growth of yeast species during the fermentation of grape juice. Journal of Applied Bacteriology 65: 23–28.

Heenan, CN., Adams, M.C., Hosken, R.W. and Fleet, G.H. 2004. Survival and sensory acceptability of probiotic microorganisms in a non-fermented frozen, vegetarian dessert. Lebensmittel Wissenschaft und Technologie 37: 461–466.

Hesseltine, C.W. 1983. Microbiology of oriental fermented foods. Annual Review in Microbiology 37: 575–601.

Hibbett, D.S., Binder, M., Bischoff, J.F., Blackwell, M., Cannon, P.F., Eriksson, O.E. et al. 2007. A higher-level phylogenetic classification of the Fungi. Mycological Research 111: 509–547.

Hosaka, M., Komuro, Y., Nakayama, M. et al. 1998. Fermentation of *sake moromi* prepared with *shochu*, wine, brewer's, alcohol, or sake strains of *Saccharomyces cerevisiae*. Journal of the Brewing Society of Japan 93: 833–840.

Jakobsen, M. and Narvhus, J. 1996. Yeasts and their possible beneficial and negative effects on the quality of dairy products. International Dairy Journal 6: 755–768.

Jay, J.M. 1991. Fermented foods and related products of fermentation. pp. 362–407. *In*: J.M. Jay (ed.). Modern Food Microbiology. Van Nostrand, Remheld, New York.

Jenson, I. 1998. Bread and baker's yeast. pp. 172–198. *In*: B.J. Wood (ed.). Microbiology of Fermented Foods, vol. 1, 2nd edn. Blackie, London.

Johnson, E.A. and Schroeder, W. 1995. Astaxanthin from the yeast *Phaffia rhodozyma*. Studies Mycology 38: 81–90.

Joshi, V.K. and Pandey, A. 1999. Biotechnology: Food Fermentation. pp. 1–24. *In*: V.K. Joshi and A. Pandey (eds.). Biotechnology: Food Fermentation, Microbiology, Biochemistry and Technology, Vol. 1. Educational Publishers and Distributors, New Delhi.

Klaenhammer, T. 2001. Probiotics and prebiotics. pp. 797–813. *In*: M.P. Doyle, L.R. Beuchat and T.J. Montville (eds.). Food Microbiology, Fundamentals and Frontiers, 2nd edn. American Society for Microbiology, Washington, DC.

Kline, L. and Sugihara, R. 1971. Microorganisms of the San Fransisco sourdough bread process II. Isolation and characterization of undescribed bacterial species responsible for the souring activity. Applied Microbiology 21: 459–465.

Kolawole, O.M., Kayode, R.M.O. and Akinduyo, B. 2007. Proximate and microbial analysis of *Burukutu* and *Pito* produced in Ilorin, Nigeria. African Journal of Biotechnology 6(5): 587–590.

Konlani, S., Delgenes, J.P., Moletta, R., Traore, A. and Doh, A. 1996. Isolation and physiological characterization of yeasts involved in sorghum beer production. Food Biotechnology 10: 29–40.

Kruif, P.D. 1935. *Mikrobenj ä ger*, Orell Füssli, Zurich.

Kurtzman, C.P., Fell, J.W. and Boekhout, T. 2011. Definition, classification and nomenclature of the yeasts. pp. 3–9. *In*: C.P. Kurtzman, J.W. Fell and T. Boekhout (eds.). The Yeasts a Taxonomic Study 5th Edn. Elsevier Science, Amsterdam.

Kurtzman, C.P. and Fell, J.W. 2005. Biodiversity and Ecophysiology of Yeasts. pp. 11–30. *In*: P. Gábor and C.L. de la Rosa (eds.). The Yeast Handbook. Springer, Berlin.

Kurtzman, C.P. and Fell, J.W. 1998. Definition, classification and nomenclature of the yeasts. pp. 3–5. *In*: C.P. Kurtzman and J.W. Fell (eds.). The Yeasts, A Taxonomic Study. 4th edn. Elsevier Science, Amsterdam.

Kurtzman, C.P. and Piškur, J. 2006. Taxonomy and phylogenetic diversity among the yeasts. pp. 29–46. *In*: P. Sunnerhagen and J. Piskur (eds.). Comparative Genomics: Using Fungi as Models. Springer, Berlin.

Lachance, M.A. and Starmer, W.T. 1998. Ecology and yeasts. pp. 21–30. *In*: C.P. Kurtzman and J.W. Fell (eds.). The Yeasts, a Taxonomic Study. 4th ed. Elsevier, Amsterdam.

Lachance, M.A., Starmer, W.T., Rosa, C.A., Bowles, J.M., Stuart, J., Barker, F. and Janzen, D.H. 2001. Biogeography of the yeasts of ephemeral flowers and their insects. FEMS Yeast Research 1: 1–8.

Lafon-Lafourcade, S. 1986. Applied microbiology. Experientia 42: 904–914.

Lanciotti, R., Sinigaglia, M., Gardini, F. and Guerzoni, M.E. 1998. *Hansenula anomala* as spoilage agent of cream-filled cakes. Microbiological Research 153: 145–148.

Leclercq-Perlat, M.N., Oumer, A., Bergère, J.L., Spinnler, H.E. and Corrieu, G. 1999. Growth of *Debaryomyces hansenii* on a bacterial surface-ripened soft cheese. Journal of Dairy Research 66: 271–281.

Lee, J.S., Kang, E.J., Kim, M.O., Lee, D.H., Bae, K.S. and Kim, C.K. 2001. Identification of *Yarrowia lipolytica* Y103 and its degradability of phenol and 4-chlorophenol. Journal of Microbiology and Biotechnology 11: 112–117.

Legras, J.L., Merdinoglu, D., Cornuet, J.M. and Karst, F. 2007. Bread, beer and wine: *Saccharomyces cerevisiae* diversity reflects human history. Molecular Ecology 16(10): 2091–2102.

Lilly, M., Bauer, F.F., Lambrechts, M.G., Swiegers, J.H., Cozzolino, D., Isak S. and Pretorius, I.S. 2006. The effect of increased yeast alcohol acetyltransferase and esterase activity on the flavor profiles of wine and distillates. Yeast 23: 641–659.

Lipke, N. and Ovalle, R. 1998. Cell wall architecture in yeast: new structure and new challenges. Journal of Bacteriology 180: 3735–3740.

Loureiro, V. 2000. Spoilage yeasts in foods and beverages—Characterisation and ecology for improved diagnosis and control. Food Research International 33: 247–256.

Loureiro, V. and Malfeito-Ferreira, M. 2003. Spoilage yeasts in the wine industry. International Journal of Food Microbiology 86(1–2): 23–50.

Lu, H.Z., Cai, Y., Wu, Z.W., Jia, J.H. and Bai, F.Y. 2004. *Kazachstania aerobia* spp. nov., an ascomycetous yeast species from aerobically deteriorating corn silage. International Journal of System Evolution and Microbiology 54: 2431–2435.

Lukondeh, T., Ashbolt, N.J. and Rogers, P.L. 2003. Evaluation of *Kluyveromyces marxianus* as a source of yeast autolysates. Journal of Indian Microbiology and Biotechnology 30: 52–56.

Lyons, T.P., Jacques, K.A. and Dawson, K.A. 1993. Miscellaneous products from yeasts. pp. 293–325. *In*: A.H. Rose and J.S. Harrison (eds.). The Yeasts, Vol. 5, 2nd edn. Academic, London.

Maamouri, O., Selmi, H. and M'hamdi, N. 2014. Effects of yeast (*Saccharomyces cerevisiae*) feed supplement on milk production and its composition in Tunisian Holstein Friesian cows. Scientia Agriculturae Bohemica 45(3): 170–174.

Makoto, W., Kazuro, F., Kozo, A. and Shigenori, O. 1990. Mutants of bakers' yeasts producing a large amount of isobutyl alcohol or isoamyl alcohol flavor components of bread. Applied Microbiology and Biotechnology 34: 154.

Maloney, D.H. and Foy, J.J. 2003. Yeast fermentations. p. 45. *In*: K. Kulp and K. Lorenz (eds.). Handbook of Dough Fermentations. Marcel Dekker, New York.

Mariam, S.H., Zegeye, N., Tariku, T., Andargie, E., Endalafer, N. and Aseffa, A. 2014. Potential of cell-free supernatants from cultures of selected lactic acid bacteria and yeast obtained from local fermented foods as inhibitors of *Listeria monocytogenes*, *Salmonella* spp. and *Staphylococcus aureus*. BMC Research Notes 7: 606.

Marshall, V.M., Cole, W.M. and Brooker, B.E. 1984. Observations on the structure of *kefir* grains and the distribution of the microflora. Journal of Applied Bacteriology 57: 491–497.

Martinez-Anaya, M.A. 1996. Carbohydrates and nitrogen related components in wheat sourdough processes: a review. Advances in Food Science (CTML) 18: 185–200.

Masneuf-Pomarede, I., Mansour, C., Murat, M.L., Tominaga, T. and Dubourdieu, D. 2006. Influence of fermentation temperature on volatile thiols concentrations in Sauvignon blanc wines. International Journal of Food Microbiology 108: 385–390.

Mata-Gómez, L.C., Montañez, J.C., Méndez-Zavala, A. and Aguilar, C.N. 2014. Biotechnological production of carotenoids by yeasts: an overview. Microbial Cell Factories 13: 12.

McBryde, C., Gardner, J.M., de Barros Lopes, M. and Jiranek, V. 2006. Generation of novel wine yeast strains by adaptive evolution. American Journal of Enology and Viticulture 57(4): 423–430.

McGovern, P.E., Zhang, J.Z., Tang, J.G. et al. 2004. Fermented beverages of pre- and proto-historic China. Proceedings of the National Academy of Sciences, USA 101: 17593–17598.

McNamee, C., Noci, F., Cronin, D.A., Lyng, J.G., Morgan, D.J. and Scannell, A.G. 2010. PEF based hurdle strategy to control *Pichia fermentans*, *Listeria innocua* and *Escherichia coli* k12 in orange juice. International Journal of Food Microbiology 138: 13–18.

Meroth, C., Hammes, W. and Hertel, C. 2003. Identification and population dynamics of yeasts in sourdough fermentation processes by PCR-denaturing gradient gel electrophoresis. Applied Environmental Microbiology 69: 7453–7461.

Moliné, M., Flores, M.R., Libkind, D., del Carmen, D.M., Farías, M.E. and van Broock, M. 2010. Photoprotection by carotenoid pigments in the yeast *Rhodotorula mucilaginosa*: the role of torularhodin. Photochem. Photobiol. Sci. 9: 1145–1151.

Montanari, G., Zambonelli, C., Grazia, L., Kamesheva, G.K. and Shigaeva, M.K. 1996. *Saccharomyces unisporus* as the principal alcoholic fermentation microorganism of traditional *koumiss*. Journal of Dairy Research 63: 327–331.

Narvhus, J.A. and Sørhaug, T. 2006. Bakery and cereal products. *In*: Y.H. Hui (ed.). Food Biochemistry and Food Processing. Wiley Online Library.

Narvhus, J.A. and Gadaga, T.H. 2003. The role of interaction between yeasts and lactic acid bacteria in African fermented milks: A review. International Journal of Food Microbiology 86: 51–60.

Nout, M.J.R. 2003. Traditional fermented products from Africa, Latin America and Asia. pp. 451–473. *In*: T. Boekhout and V. Robert (eds.). Yeasts in food: Beneficial and Detrimental Aspects. Behr, Hamburg.

Nout, M.J.R. and Aidoo, K.E. 2002. Asian fungal fermented food. pp. 23–47. *In*: H.D. Osiewacz (ed.). The Mycota. Springer-Verlag, New York.

Nout, M.J.R. and Creemers-Molenaar, T. 1987. Microbiological properties of some wheatmeal sourdough starters. Chemie Mikrobiologie Technologie der Lebensmittel 10: 162–167.

Ottogalli, G., Galli, A. and Foschino, R. 1996. Italian bakery products obtained with sourdough: Characterization of the typical microflora. Advances in Food Sciences 18: 131.

Padmaja, G. and George, M. 1999. Oriental fermented foods; Biotechnological approaches. pp. 143–189. *In*: S.S. Marwaha and J.K. Arora (eds.). Food Processing: Biotechnological Applications. Asiatech Publishers Inc., New Delhi.

Paramithiotis, S., Müller, M.R.A., Ehrmann, M.A., Tsakalidou, E., Seiler, H., Vogel, R. and Kalantzopoulos, G. 2000. Polyphasic identification of wild yeast strains isolated from Greek sourdoughs. Systematic and Applied Microbiology 23: 156–164.

Petersen, K.M., Møller, P.L. and Jespersen, L. 2001. DNA typing methods for differentiation of *Debaryomyces hansenii* strains and other yeasts related to surface ripened cheeses. International Journal of Food Microbiology 69: 11–24.

Pretorius, I. 2000. Tailoring wine yeast for the new millennium: Novel approaches to the ancient art of winemaking. Yeast 16: 675–729.

Rahman, N., Xiaohong, C., Meiqin, F. and Mingsheng, D. 2009. Characterization of the dominant microflora in naturally fermented camel milk shubat. World Journal of Microbiology and Biotechnology 25: 1941–1946.

Rajkowska, K. and Kunicka-Stycznska, A. 2010. Probiotic properties of yeasts isolated from chicken feces and *kefirs*. Polish Journal of Microbiology 59(4): 257–263.

Rizk, S., Maalouf, K. and Baydoun, E. 2008. The antiproliferative effect of *kefir* cell-free fraction on HuT-102 malignant T lymphocytes. Clinical Lymphoma Myeloma 9(3): 198–203.

Romano, P., Capace, A. and Jespersen, L. 2006. Taxonomic and ecological diversity of food and beverage yeasts. pp. 13–53. *In*: A. Querol and G.H. Fleet (eds.). The Yeast Handbook—Yeasts in Food and Beverages. Springer-Verlag, Germany.

Romano, P., Fiore, C., Paraggio, M., Caruso, M. and Capece, A. 2003. Function of yeast species and strains in wine flavour. International Journal of Food Microbiology 86: 169–180.

Romano, P., Marchese, R., Laurita, C., Saleano, G. and Turbanti, L. 1999. Biotechnological suitability of *Saccharomycodes ludwigii* for fermented beverages. World Journal of Microbiology and Biotechnology 15: 451–454.

Saez, J.S., Lopes, C.A., Kirs, V.E. and Sangorrin, M. 2011. Production of volatile phenols by *Pichia manshurica* and *Pichia membranifaciens* isolated from spoiled wines and cellar environment in Patagonia. Food Microbiology 28(3): 503–509.

Samelis, J. and Sofos, J.N. 2003. Yeasts in Meat and Meat Products. pp. 239–266. *In*: T. Boekhout and
    V. Robert (eds.). Yeasts in food: Beneficial and detrimental aspects. Behr, Hamburg.
Sandhu, D.K. and Soni, S.K. 1989. Microflora associated with Indian Punjabi *warri* fermentation. Journal
    of Food Science and Technology 26: 21–25.
Sandhu, D.K. and Waraich, M.K. 1984. Distribution of yeasts in indigenous fermented foods with a brief
    review of literature. Kavaka 12: 73–85.
Sandhu, D.K., Soni, S.K. and Vilkhu, K.S. 1986. Distribution and role of yeasts in Indian fermented foods.
    p. 142. *In*: R.K. Vashisht and P. Tauro (eds.). Yeast Biotechnology. HAU Press, Hissar.
Sanni, A.I. and Lönner, C. 1993. Identification of yeasts isolated from Nigerian traditional alcoholic
    beverages. Food Microbiology 10: 517–523.
Saranraj, P. and Geetha, M. 2012. Microbial spoilage of bakery products and its control by preservatives.
    International Journal of Pharmaceutical and Biological Archives 3(1): 38–48.
Sauer, M., Branduardi, P., Valli, M. and Porro, D. 2004. Production of L-ascorbic acid by metabolically
    engineered *Saccharomyces cerevisiae* and *Zygosaccharomyces bailii*. Applied Environmental
    Microbiology 70: 6086–6091.
Saurabh, A., Kanwar, S.S. and Sharma, O.P. 2012. Screening of indigenous yeast isolates obtained from
    traditional fermented foods of Western Himalayas for probiotic attributes. Journal of Yeast and Fungal
    Research 2(8): 117–126.
Savitri, and Bhalla, T.C. 2012. Characterization of *bhatooru*, a traditional fermented food of Himachal
    Pradesh: microbiological and biochemical aspects. 3 Biotech 3: 247–254.
Savitri, Thakur, N., Kumar, D. and Bhalla, T.C. 2012. Microbiological and biochemical characterization of
    Seera: A traditional fermented food of Himachal Pradesh. International Journal of Food Fermentation
    and Technology 2: 49–56.
Schuller, D., Côrte-Real, M. and Leão, C. 2000. A differential medium for the enumeration of the spoilage
    yeast *Zygosaccharomyces bailii* in wine. Journal of Food Protection 11: 1467–1609.
Schutz, M. and Gafner, J. 1993. Analysis of yeast diversity during spontaneous and induced alcoholic
    fermentations. Journal of Applied Bacteriology 75: 551–558.
Seeley, R.D. 1977. Fractionation and utilization of baker's yeast. Master Brewers Association of Americas
    Technology 14: 35–39.
Sefa-Deheh, S., Sanni, A.I., Tetteh, G. and Sakyi-Dawson, E. 1999. Yeasts in the traditional brewing of
    pito in Ghana. World Journal of Microbiology and Biotechnology 15: 593–597.
Soni, S.K. and Sandhu, D.K. 1990. Indian fermented foods: Microbiological and biochemical aspects.
    Indian Journal of Microbiology 30: 135–157.
Soni, S.K. and Sandhu, D.K. 1999. Fermented cereal products. pp. 895–949. *In*: V.K. Joshi and A. Pandey
    (eds.). Biotechnology: Food Fermentation, Microbiology, Biochemistry and Technology, vol. II.
    Educational Publishers and Distributors, New Delhi.
Spicher, G. and Stephan, H. 1993. Handbuch Sauerteig. Biologie, Biochemie, Technologie. Hamburg:
    Behr's Verlag.
Srivastava, S., Modi, D.R. and Garg, S.K. 1997. Production of ethanol from guava pulp by yeast strains.
    Bioresource Technology 60: 263–265.
Stam, H., Hoogland, M. and Laane, C. 1998. Food flavours from yeast. pp. 505–542. *In*: B.J. Wood (ed.).
    Microbiology of Fermented Foods, vol. 2, 2nd edn. Blackie, London.
Steinkraus, K.H. 1998. Bioenrichment: Production of vitamins in fermented foods. pp. 602–621. *In*: B.J.B.
    Wood (ed.). Microbiology of Fermented Foods, 2nd ed. Blackie Academic and Professional, London.
Stolz, P. 2003. Biological fundamentals of yeast and *Lactobacilli* fermentation in bread dough. pp. 23–43.
    In: K. Kulp and K. Lorenz (eds.). Handbook of Dough Fermentations. Marcel Dekker Inc., New York.
Strauss, M.L.A., Jolly, N.P., Lambrechts, M.G. and van Rensburg, P. 2001. Screening for the production of
    extracellular hydrolytic enzymes by non-*Saccharomyces* wine yeasts. Journal of Applied Microbiology
    91: 182–190.
Swiegers, J.H., Bartowsky, E.J., Henschke, P.A. and Pretorius, I.S. 2005. Yeast and bacterial modulation of
    wine aroma and flavor. Australian Journal of Grape and Wine Research 11: 139–173.
Swiegers, J.H., Capone, D.L., Pardon, K.H., Elsey, G.M., Sefton, M.A.I. and Pretorius, L.F.I.S. 2007.
    Engineering volatile thiol release in *Saccharomyces cerevisiae* for improved wine aroma. Yeast
    24: 561–574.
Tamang, J.P. 1998. Role of microorganisms in traditional fermented foods. Indian Food Industry 17: 162–167.

Tamang, J.P. and Fleet, G.H. 2009. Yeasts diversity in fermented foods and beverages. pp. 169–198. *In*: T. Satyanarayana and G. Kunze (eds.). Yeasts Biotechnology: Diversity and Applications. Springer, New York.

Tamang, J.P., Dewan, S., Tamang, B., Rai, A., Schillinger, U. and Holzapfel, W.H. 2007. Lactic acid bacteria in *Hamei* and *Marcha* of North east India, Indian Journal of Microbiology 47(2): 119–125.

Tenge, C. 2009. Yeast. *In*: H.M. Eßlinger (ed.). Handbook of Brewing: Processes, Technology, Markets. WILEY-VCH Verlag GmbH & Co. KGaA, Weinheim ISBN: 978-3-527-31674-8.

Thakur, N., Savitri and Bhalla, T.C. 2004. Characterization of traditional fermented foods and beverages of Himachal Pradesh. Indian Journal of Traditional Knowledge 3: 325–335.

Thapa, S. and Tamang, J.P. 2006. Microbiological and physio-chemical changes during fermentation of *kodo ko jaanr*, a traditional alcoholic beverage of the Darjeeling hills and Sikkim. Indian Journal of Microbiology 46: 333–341.

Tinoi, J., Rakariyatham, N. and Deming, R.L. 2005. Simplex optimization of carotenoid production by *Rhodotorula glutinis* using hydrolyzed mung bean waste flour as substrate. Process Biochemistry 40: 2551–2557.

Tudor, E.A. and Board, R.G. 1993. Food-spoilage yeasts. pp. 435–516. *In*: A.H. Rose and J.S. Harrison (eds.). The Yeasts, vol. 5. Yeast Technology, 2nd edn. Academic, London.

Van der Rest, M., Kamminga, A.H., Nakano, A. et al. 1995. The plasma membrane of *Saccharomyces cerevisiae*: structure, function and biogenesis. Microbiolgical Reviews 59: 304–322.

Vandamme, E.J. and Soetaert, W. 2002. Bioflavours and fragrances via fermentation and biocatalysis. Journal of Chemical Technology and Biotechnology 77: 1323–1332.

Venkatasubbaiah, P., Dwarkanath, C.T. and Sreenivasamurthy, V. 1985. Involvement of yeast flora in *idli* batter fermentation. Journal of Food Science and Technology 22: 88–90.

Vernocchi, P., Valmorri, S., Gatto, V., Torriani, S., Gianotti, S., Suzzi, G., Guerzoni, M.E. and Gardini, F. 2004. A survey on yeast microbiota associated with an Italian traditional sweet-leavened baked good fermentation. Food Research International 37: 469.

Verstrepen, K.J., Paul, J., Chamber, S. and Pretorius, I.S. 2006. The development of superior yeast strains for the food and beverage industries: challenges, opportunities and potential benefits. pp. 339–444. *In*: A. Querol and G.H. Fleet (eds.). The Yeast Handbook: Yeasts in Food and Beverages. Springer-Verlag Berlin Heidelberg.

Wang, H.L. and Hesseltine, C.W. 1982. Oriental fermented foods. p. 492. *In*: G. Reed (ed.). Prescott and Dunn's Industrial Microbiology. AVI Publishing Co. Inc., Westport.

Wang, Y. 2008. Mutations that uncouple the yeast *Saccharomyces cerevisiae* ATP synthase. [Ph.D. thesis]. North Chicago, Ill.: Rosalind Franklin Univ. of Medicine and Sciences. 164p. Available from: UMI Microform, ProQuest LLC, Ann Arbor, MI:48106-1346.

Winarno, F.G., Muchatadi, D., Lakshmi, B.S., Rahman, A., Swastomo, W., Zainuddin, D. and Santaso, S.N. 1977. Indonesian *tauco*. *In*: Symposium on Indigenous Fermented Foods, Bangkok, Thailand.

Wood, B.J.B. 1998. *In*: B.J.B. Wood (ed.). Microbiology of Fermented Foods, 2nd ed. Blackie Academic and Professional, London.

Yildiz, F. 2010. Development and Manufacture of yogurt and other functional dairy products. 1st ed. CRC Press.

Yokotsuka, T. 1960. Aroma and flavor of Japanese soy sauce. Advances in Food Research 10: 75–134.

Zinser, E., Sperka-Gottlieb, C.D., Fasch, E.V. et al. 1991. Phospholipid synthesis and lipid composition of subcellular membranes in the unicellular eukaryote *Saccharomyces cerevisiae*. Journal of Bacteriology 173: 2026–2034.

2

Biotechnology of Wine Yeasts

Fabienne Remize

1. Introduction

The story of biotechnology of wine yeasts started a long time ago. The selection of yeasts from wineries for commercial purposes was developed from the 1970s. The aim was to display satisfactory fermentation profiles and to ensure product quality. From the 1980s, classical breeding, interspecific breeding and mutagenesis were applied for the improvement of existing strains. This resulted in many developments: for example, better nitrogen assimilation and fermentation kinetics were obtained by random mutagenesis (Salmon and Barre 1998). These classical approaches are still used; EMS (ethyl methyl sulfonate) mutagenesis resulted in the selection of commercial wine strain variants which produce less reduced hydrogen sulfide due to mutations into genes encoding sulfide reductase subunits (Cordente et al. 2009). But genomic features of wine yeasts were rapidly identified as strong limitations for these approaches. Indeed, wine strains are mainly diploid, polyploid or aneuploid. They exhibit a chromosomal polymorphism (Bidenne et al. 1992). Chromosomal trisomies or tetrasomies result in impaired sporulation ability and in highly variable spore viability (Johnston et al. 2000). In addition, wine strains are generally homothallic, i.e., able to switch mating type when haploid, and highly heterozygous. All these features confer to wine yeast high genome plasticity and limit the stability of lineage of variants.

The 1990s have been marked by considerable efforts to solve wine making process issues and to improve wine quality by metabolic engineering of wine yeasts. This was made possible by the incredible step beyond genetic improvement by the development of recombination methods. Genetic modification tools have benefited of research in gene function understanding. Indeed, the first yeast sequenced genome was released in 1997 and revisited in 2000 (Blandin et al. 2000). Proteomic and

Université de La Réunion, UMR95, QualiSud – Démarche intégrée pour l'obtention d'aliments de qualité, ESIROI, Parc Technologique Universitaire, 2 rue Joseph Wetzell, F-97490 Sainte-Clotilde, France.
E-mail: fabienne.remize@univ-reunion.fr

transcriptomic studies resulted in a lot of useful data which were used to complete genome structure and to understand gene function. In this view, the considerable work realized on the understanding of osmotic stress response by *Saccharomyces cerevisiae* was exemplary. The genes involved in stress adaptation and the signaling pathways have been elucidated (Albertyn et al. 1994, Rep et al. 1999, Tamás et al. 2000, Pahlman et al. 2001, Rep et al. 2001). Consequently, the HOG (High Osmolarity Glycerol) pathway, a branched MAPK (Mitogen Activated Protein Kinase) pathway, has become a model to analyze systems level properties of signal transduction (Hohmann 2009).

Genetic engineering of wine yeasts was used to explore many issues. Among them, the examples which are hereby presented are representative of methods and are emblematic of research in the field.

2. Metabolic Engineering of Wine Yeasts

In the following section, the various aspects of genetic engineering of wine yeasts have been discussed.

2.1 Genetic Engineering of Malolactic Fermentation

One of the main focuses for wine making process improvement was to obtain a malolactic or a maloethanolic fermentation of wine from modified yeast. Malolactic fermentation is performed after alcoholic fermentation by a lactic acid bacterium, generally *Oenococcus oeni*. This bacterium converts the di-acid malic acid to the mono-acid lactic acid, thus leading to acidity decrease. Such a reaction is expected in most of white wines and in red wines from northern production areas of France. As a consequence of bacterial growth and metabolic activity, malolactic fermentation contributes also to wine stabilization (thanks to nutrient exhaust), and to flavor development. However, due to harsh conditions after alcoholic fermentation (ethanol level, low pH, low temperatures, sulfur dioxide and low nutrients level), bacterial implantation and growth in wine is difficult to control. As a consequence, many researches focused on the genetic engineering of yeast ability to use malic acid. A malolactic *S. cerevisiae* strain was obtained by expressing the genes encoding a malate transporter from *Schizosaccharomyces pombe* and of a malolactic enzyme from *Lactococcus lactis* (Bony et al. 1997, Volschenk et al. 1997). This strain efficiently converted malic acid to lactic acid. Similarly, an engineered *S. cerevisiae* strain, able to convert malic acid to ethanol, was obtained by the cloning of malate permease and malic enzyme genes from *Sc. Pombe* under the control of strong *S. cerevisiae* promoters (Volschenk et al. 2001). The Fig. 1 shows the two strategies, malolactic and maloethanolic, alternatively used in yeast.

Eventually, a genetically-stable malolactic yeast strain was constructed (Husnik et al. 2006). Fermentation kinetics, growth kinetics, analytical and sensorial profile of wine obtained with this strain, called ML01, were investigated. The resulting wines were compared to wines obtained by the traditional process, i.e., by successive yeast and lactic bacteria fermentations (Husnik et al. 2007). ML01 received

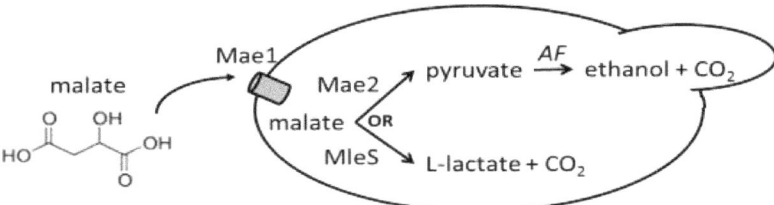

Figure 1. Malolactic and maloethanolic pathways, which were introduced in *S. cerevisiae* yeast. Mae1: *Schizosaccharomyces pombe* malic acid transporter; Mae2: *Sc. pombe* malic enzyme; MleS: *Lactococcus lactis* malolactic enzyme; AF: alcoholic fermentation.

approval from the Food and Drug Administration (USA) and the Health Canada and Environment Canada agencies. It is commercialized since 2004.

In parallel, alternative non-GM solutions were researched and co-inoculation of musts with yeast and lactic acid bacteria has attracted attention. Recently, another approach, that does not employ genetically modified microorganisms either, was investigated. It was observed that *Lactobacillus plantarum* was able to cope with wine conditions and exhibited interesting enzymatic activities to enhance wine aroma formation. *Lb. plantarum* starter cultures were selected to perform malolactic fermentation (du Toit et al. 2011, Lerm et al. 2011).

2.2 Engineering of Resveratrol Level

Health effects of a moderate consumption of wine are recognized as positive for the reduction of coronary diseases. Several wine molecules are detrimental to health, for instance ethyl carbamate, a carcinogenic molecule, biogenic amines or sulfites. Some others, however, exert beneficial effects, like resveratrol involved in the so-called 'French paradox'. Resveratrol (3, 5,4'-*trans*-trihydroxystilbene) is a phenolic compound with a demonstrated bioactivity and its main source in the diet is wine (Fernández-Mar et al. 2012). This compound is naturally present in grape, especially in the skin and seeds. As a consequence of skin maceration process, its level is higher in red wines than in white or rosé wines. Its concentration depends on many factors, among them biotic or abiotic stress exposure of vine plants, as resveratrol is a plant defense compound (phytoalexin). Its level in wine can be increased either by an increase of its release from grapes (González-Candelas et al. 2000) or by its formation by yeast from free aminoacids or derivatives. The resveratrol formation pathway is not present in yeast but was described in several plants (Fig. 2). The first studies that managed to produce resveratrol in yeast relied on over-expression of heterologous CoenzymeA ligase (C4L) gene from a hybrid poplar and resveratrol synthase (VST) gene from grapevine in a medium containing p-coumaric acid (Becker et al. 2003).

Other gene origins, other promoters, a synthetic scaffold strategy and fusion proteins were successively investigated and resulted in a further increase of resveratrol production by yeast (Shin et al. 2011, Wang and Yu 2012). The main increase in production was obtained by the adaptation of codon usage to yeast (Wang et al. 2011). The same author showed that the expression of the poorly characterized

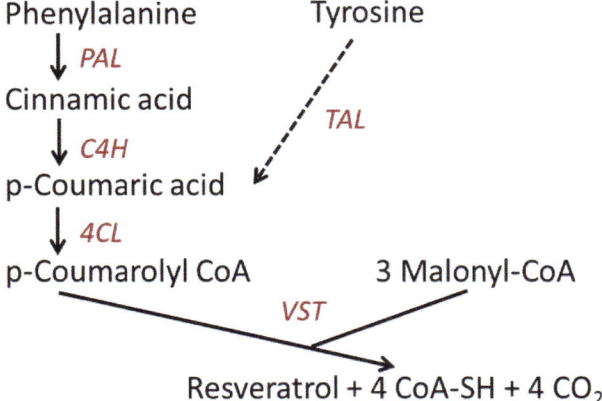

Figure 2. Resveratrol formation pathway. PAL: Phenylalanine ammonia-lyase; C4H: Cinnamate 4-hydroxylase; 4CL: CoA-ligase; VST: Resveratrol synthase; TAL: Tyrosine ammonia lyase.

arabinose transporter *araE* was able to enhance resveratrol permeability through lipid membranes in yeast, and then, to further increase resveratrol production. Resveratrol was also produced from tyrosine thanks to the additional over-expression of the gene encoding tyrosine ammonia lyase that converts the amino acid into p-coumaric acid (Shin et al. 2011, Jeandet et al. 2012). The different strategies used to produce resveratrol from yeast do not currently result in wine application. However, the progress led to issues for the industrial production of the stilbene for cosmetic or nutraceutical applications (Jeandet et al. 2012, Liu et al. 2013).

2.3 Low-Alcohol Wines

Another topic widely investigated was to produce wines with lower ethanol content. The consumer's expectation towards wines with reduced ethanol level has been reinforced by health considerations. In addition, climatic changes result in higher grape maturity at the harvest stage, and thus, in higher sugar contents that, in turn, lead to higher ethanol levels in wines. Besides, physical and chemical approaches to decrease wine ethanol content generally result in color or/and aromatic reduction. Ethanol is produced by yeast *via* the glycolytic pathway followed by alcoholic fermentation of pyruvate. The complete pathway of ethanol formation is redox balanced and produces two molecules of ATP per C6-sugar consumed. In parallel during wine fermentation, biomass is produced from a part of carbon available in sugars. Biomass formation generates reduced cofactors. Redox homeostasis under anaerobiosis is ensured by the formation of reduced by-products, thereby re-oxidizing co-factors. The main by-product is glycerol, whose content in wine generally ranges between 6 and 9 g/L (Remize et al. 2000b). Glycerol formation during wine fermentation is not only driven by redox imbalance but mainly by the osmotic stress response due to high sugar content at the beginning of the process, and further, by the increasing ethanol level (Albertyn et al. 1994, Pahlman et al. 2001, Remize et al. 2003).

Glycerol does not exhibit aromatic properties by itself, but contributes to wine sensorial characteristics by providing sweetness and fullness. Its formation at the expense of carbon flux towards ethanol synthesis has been investigated (Remize et al. 1999, Remize et al. 2001). It was established that the first enzymatic step after triose-phosphate formation was rate-limiting. The overexpression of one of the genes encoding this enzyme, glycerol-3-phosphate dehydrogenase, under the control of a strong promoter, resulted in a 1.5–2.5 fold increase in glycerol content (Remize et al. 1999). As a consequence, ethanol content was reduced by 5–10 g/L. From that first genetic engineering modification, secondary effects were deeply examined (Fig. 3). Among those, an increase in acetic acid production was demonstrated. This effect was linked to the dynamic requirement of redox equivalents during wine fermentation. The PDH by-pass was then engineered to combine the high glycerol formation with a limited increase in oxidized by-products (Remize et al. 2000a, Cambon et al. 2006). The modified yeast exhibited fermentative properties comparable to control strains but with an efficient diversion of carbons from sugar towards glycerol instead of ethanol. The ethanol content was decreased by 15–20%. The formation of another by-product, acetoin, was increased by lowering ethanol content. *BDH1*, coding for native NADH-dependent butanediol dehydrogenase, was over-expressed in a yeast strain, overexpressing glycerol-3-phosphate dehydrogenase, and inactivated

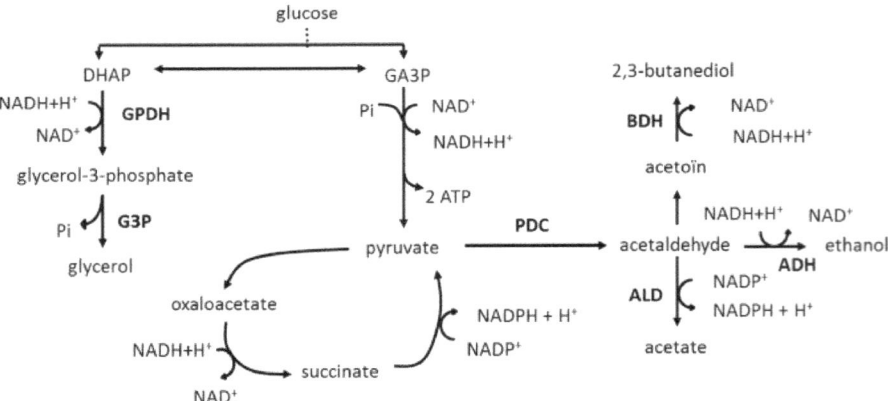

Figure 3. Glycerol formation pathway and formation of other by-products. ADH: alcohol dehydrogenase; ALD: acetaldehyde dehydrogenase; BDH: butanediol dehydrogenase; GPDH: glycerol-3-phosphate dehydrogenase; G3P: glycerol-3-phophatase; PDC: pyruvate decarboxylase; DHAP: dihydroxyacetone phosphate; GA3P: glyceraldehyde-3-phosphate.

for the *ALD6* gene encoding acetaldehyde dehydrogenase (Ehsani et al. 2009). This additional genetic modification was one step beyond for low-alcohol producing yeasts with acceptable organoleptic profiles.

2.4 Quantitative Trait Loci (QTL) Approaches

The increase of availability of genomic sequences and the development of genomic tools over the last decade increased the global understanding of wine yeast response to its changing environment. The QTL (Quantitative Trait Loci) approach has been applied to link phenotypes with genomic loci. For instance, the QTL approach has been explored in *S. cerevisiae* to increase ethanol formation yield at the expense of glycerol (Hubmann et al. 2013). A strain producing unusually low glycerol levels and a commercial strain were compared. To achieve that, haploid segregants with phenotypes similar to the parental strains were obtained and further crossed for genetic mapping of the diploid resulting strains. With that approach, a recessive allele of *SSK1*, expressing a truncated and partially mistranslated protein, was identified and used for reverse genetic engineering. While the glycerol/ethanol ratio reduction was stronger in the strain which carried the identified allele than in the deletion mutant, side effects on growth, on ethanol productivity and on osmosensitivity were low. *SSK1* encodes a protein of the osmolarity two-component signal transduction system which controls the activity of the HOG pathway.

Wine applications of QTL are still few. In a study, acetic acid production in wine was investigated by QTL mapping (Marullo et al. 2007). Acetic acid concentration is tightly related to volatile acidity and it has to be kept low. With haploïd derivatives of commercial wine strains producing very different levels of acetic acid, a major QTL was mapped. It corresponded to a variant of the *APS1* gene encoding type 1 asparaginase, which carried a mutation in the catalytic motif of the enzyme. More recently, Ambroset et al. (2011) examined the relationships between genetic polymorphism, expression QTL and phenotypic traits associated with wine fermentation by using a laboratory yeast strain and a haploid derivative of a commercial wine strain. Expression QTL aim to evidence relationships between genetic polymorphism and transcript abundance. In this study, fermentation of must by the two parental strains exhibited differences in fermentation parameters, in particular carbon dioxide production rates depending on the fermentation stage (Ambroset et al. 2011). By the original approach of combining expression QTL, genetic QTL and phenotypic traits, new associations were evidenced and an allelic variant of *ABZ1* gene was mapped from overlapping results. It was shown that this gene is involved in the control of fermentation rate through nitrogen utilization modulation.

QTL approaches have now demonstrated their interest for biotechnology of wine yeasts: they allow identifying allelic variants not directly involved in the pathway of a metabolite formation and thus define new targets for metabolic reengineering.

2.5 Evolutionary Engineering for Wine Yeast Improvement

From the 2000s, European Union positioning towards GMO was clarified. In several other countries, regulations are more permissive for GMOs. Whatever the case may be, consumers' acceptance for GMO suffers from a bad image and alternative strategies were investigated, more and more frequently. In addition, metabolic engineering failed to define genetic targets related either to phenotypes resulting from pleiotropic regulations or to phenotypes resulting from multiple cellular properties that do not share a defined molecular basis. In that perspective, evolutionary engineering was developed as alternative strategy (Çakar et al. 2005, Çakar et al. 2012). This approach is based on the generation of high clonal diversity, by classical mutagenesis, and the progressive selection of phenotype. The challenging step is phenotype selection which, contrarily to classical approaches, is performed by taking into consideration multiple characteristics.

The latest researches to engineer the ethanol levels in wines investigated adaptive evolution (Cadière et al. 2011, Kutyna et al. 2012). To achieve that, a poor carbon source, δ-gluconolactone, was used (Cadière et al. 2011). This carbon source is consumed *via* the pentose-phosphate pathway that generates reducing equivalents and thus is interconnected, *via* the cellular redox level, to glycerol formation. A long-term serial transfer procedure on δ-gluconolactone resulted in the isolation of variants with a better growth on this substrate while the growth rate on glucose was maintained. The accurate characterization of strain properties with ^{13}C flux analysis highlighted a re-routing of metabolic flux from the glycolytic pathway towards the pentose phosphate pathway. As a consequence, acetate production from the evolved strain was reduced, whereas ethanol level reduction was hardly detectable. Kutyna et al. (2012) selected a haploid variant from successive generations on a medium containing high sulfite concentration at alkaline pH. Sulfite stress was previously shown to result in increased glycerol formation. The heterozygous diploid obtained after backcrossing did not exhibit an increased glycerol production as the genetic modification was recessive. Other applications of adaptive evolution for winemaking were investigated, particularly in the enhancement of desirable aromatic properties (Cadière et al. 2012) and improvement of fermentation kinetics (Novo et al. 2014). Promising results were obtained but require further studies to result in a usable commercial starter.

3. Wine Ecosystems and Terroir

Whereas wine alcoholic fermentation is essentially performed by *S. cerevisiae* thanks to its peculiar resistance to ethanol, non-*Saccharomyces* yeasts are dominant during the first stages of sugar assimilation, and their contribution to wine complexity has been recognized for decades. It was established that yeasts in must originate from grapes and cellar equipment (Renouf et al. 2007). Yeast population on grapes averages 10^3 cfu/g. In must, its population rapidly reaches 10^8 cfu/mL and then gradually decreases to stabilize thereafter (Torija et al. 2001).

The diversity of the species present on grape berries has been described (Zott et al. 2008, Barata et al. 2012). Recent molecular approaches have allowed understanding the ecological dynamics during wine making (Esteve-Zarzoso et al. 1998, Pramateftaki et al. 2000, Schuller et al. 2004, Hierro et al. 2006). The most recent studies highlight a possible geographic effect on the yeast selection so that the most adapted to the ecological niche constitute an ecosystem fingerprint (Pramateftaki et al. 2000, Raspor et al. 2006, Dequin and Casaregola 2011, Gayevskiy and Goddard 2012, Schuller et al. 2012). Domestication of yeasts from millenniums, not only for winemaking but also for baking and brewing, has impacted genetic structures (Legras et al. 2007) and results in specific phenotypic traits shared by isolates from the same ecological niche (For a comprehensive review, see Dequin and Casaregola 2011). Starter strains, which are well-adapted to their ecosystem, could disseminate in the cellar and around, to the vines. This might then result in a diversity loss in the particular environment. A study has shown that yeast dissemination was restricted to close environments (Valero et al. 2005, Schuller et al. 2012). This evidence is strengthened by genomic characterization of yeasts that shows a relative geographic diversity within domestic wine yeasts (Martínez et al. 2007). From this point of view, the ecological adaptation of yeasts could partly explain the terroir of wines. Commercial starters which claim to respect wine terroir typicity are now developed.

Conversely, several tentative to develop non-*Saccharomyces* starters are reported (Hong and Park 2013). Starters, combining non-*Saccharomyces* and *Saccharomyces* strains, are commercialized in order to improve the aromatic complexity (See lallemandwine.com). Eventually, the use of non-*Saccharomyces* yeasts able to consume sugar by respiration in the harsh must conditions without producing off-flavors was investigated (Gonzalez et al. 2013, Quirós et al. 2014) in order to produce less ethanol in wines. Research in that direction is required to propose in the future, yeast cocktails that warrant wine quality without any loss of terroir influence.

Keywords: Wine, *Saccharomyces*, ethanol, glycerol, malolactic, antioxidant, resveratrol, genetic engineering, GMO, evolution, QTL, terroir, aromatic

References

Albertyn, J., Hohmann, S., Thevelein, J. and Prior, B. 1994. *GPD1*, which encodes glycerol-3-phosphate dehydrogenase, is essential for growth under osmotic stress in *Saccharomyces cerevisiae*, and its expression is regulated by the high-osmolarity glycerol response pathway. Molecular and Cellular Biology 14: 4135–4144.

Ambroset, C., Petit, M., Brion, C., Sanchez, I., Delobel, P., Guérin, C., Chiapello, H., Nicolas, P., Bigey, F., Dequin, S. and Blondin, B. 2011. Deciphering the molecular basis of wine yeast fermentation traits using a combined genetic and genomic approach. G3 (Bethesda) 1: 263–281.

Barata, A., Malfeito-Ferreira, M. and Loureiro, V. 2012. The microbial ecology of wine grape berries. International Journal of Food Microbiology 153: 243–259.

Becker, J., Armstrong, G., Vandermerwe, M., Lambrechts, M., Vivier, M. and Pretorius, I.S. 2003. Metabolic engineering of *Saccharomyces cerevisiae* for the synthesis of the wine-related antioxidant resveratrol. FEMS Yeast Research 4: 79–85.

Bidenne, C., Blondin, B., Dequin, S. and Vezinhet, F. 1992. Analysis of the chromosomal DNA polymorphism of wine strains of *Saccharomyces cerevisiae*. Current Genetics 22: 1–7.

Blandin, G., Durrens, P., Tekaia, F., Aigle, M., Bolotin-Fukuhara, M., Bon, E., Casarégola, S., de Montigny, J., Gaillardin, C., Lépingle, A., Llorente, B., Malpertuy, A., Neuvéglise, C., Ozier-Kalogeropoulos, O., Perrin, A., Potier, S., Souciet, J.-L., Talla, E., Toffano-Nioche, C., Wésolowski-Louvel, M., Marck, C. and Dujon, B. 2000. Genomic Exploration of the Hemiascomycetous Yeasts: 4. The genome of *Saccharomyces cerevisiae* revisited. FEBS Letters 487: 31–36.

Bony, M., Bidart, F., Camarasa, C., Ansanay, V., Dulau, L., Barre, P. and Dequin, S. 1997. Metabolic analysis of *S. cerevisiae* strains engineered for malolactic fermentation. FEBS Letters 410: 452–456.

Cadière, A., Aguera, E., Caillé, S., Ortiz-Julien, A. and Dequin, S. 2012. Pilot-scale evaluation the enological traits of a novel, aromatic wine yeast strain obtained by adaptive evolution. Food Microbiology 32: 332–337.

Cadière, A., Ortiz-Julien, A., Camarasa, C. and Dequin, S. 2011. Evolutionary engineered *Saccharomyces cerevisiae* wine yeast strains with increased *in vivo* flux through the pentose phosphate pathway. Metabolic Engineering 13: 263–71.

Çakar, Z.P., Seker, U.O.S., Tamerler, C., Sonderegger, M. and Sauer, U. 2005. Evolutionary engineering of multiple-stress resistant *Saccharomyces cerevisiae*. FEMS Yeast Research 5: 569–578.

Çakar, Z.P., Turanli-Yildiz, B., Alkim, C. and Yilmaz, U. 2012. Evolutionary engineering of *Saccharomyces cerevisiae* for improved industrially important properties. FEMS Yeast Research 12: 171–82.

Cambon, B., Monteil, V., Remize, F., Camarasa, C. and Dequin, S. 2006. Effects of *GPD1* overexpression in *Saccharomyces cerevisiae* commercial wine yeast strains lacking *ALD6* genes. Applied and Environmental Microbiology 72: 4688–4694.

Cordente, A.G., Heinrich, A., Pretorius, I.S. and Swiegers, J.H. 2009. Isolation of sulfite reductase variants of a commercial wine yeast with significantly reduced hydrogen sulfide production. FEMS Yeast Research 9: 446–459.

Dequin, S. and Casaregola, S. 2011. The genomes of fermentative *Saccharomyces*. Compte-Rendus de Biologie 334: 687–693.

Du Toit, M., Engelbrecht, L., Lerm, E. and Krieger-Weber, S. 2011. *Lactobacillus*: The next generation of malolactic fermentation starter cultures: An overview. Food Bioprocess and Technology 4: 876–906.

Ehsani, M., Fernández, M.R., Biosca, J.A., Julien, A. and Dequin, S. 2009. Engineering of 2,3-butanediol dehydrogenase to reduce acetoin formation by glycerol-overproducing, low-alcohol *Saccharomyces cerevisiae*. Applied and Environmental Microbiology 75: 3196–3205.

Esteve-Zarzoso, B., Manzanares, P., Ramon, D. and Querol, A. 1998. The role of non-*Saccharomyces* yeasts in industrial winemaking. International Microbiology 1: 143–148.

Fernández-Mar, M.I., Mateos, R., García-Parrilla, M.C., Puertas, B. and Cantos-Villar, E. 2012. Bioactive compounds in wine: resveratrol, hydroxytyrosol and melatonin: A review. Food Chemistry 130: 797–813.

Gayevskiy, V. and Goddard, M.R. 2012. Geographic delineations of yeast communities and populations associated with vines and wines in New Zealand. ISME Journal 6: 1281–90.

Gonzalez, R., Quirós, M. and Morales, P. 2013. Yeast respiration of sugars by non-*Saccharomyces* yeast species: a promising and barely explored approach to lowering alcohol content of wines. Trends in Food Sciences and Technology 29: 55–61.

González-Candelas, L., Gil, J., Lamuela-Raventós, R. and Ramón, D. 2000. The use of transgenic yeasts expressing a gene encoding a glycosyl-hydrolase as a tool to increase resveratrol content in wine. International Journal of Food Microbiology 59: 179–183.

Hierro, N., Gonzalez, A., Mas, A. and Guillamon, J.M. 2006. Diversity and evolution of non-*Saccharomyces* yeast populations during wine fermentation: Effect of grape ripeness and cold maceration. FEMS Yeast Research 6: 102–111.

Hohmann, S. 2009. Control of high osmolarity signalling in the yeast *Saccharomyces cerevisiae*. FEBS Letters 583: 4025–4029.

Hong, Y.-A. and Park, H.-D. 2013. Role of non-*Saccharomyces* yeasts in Korean wines produced from Campbell Early grapes: potential use of *Hanseniaspora uvarum* as a starter culture. Food Microbiology 34: 207–214.

Hubmann, G., Foulquié-Moreno, M.R., Nevoigt, E., Duitama, J., Meurens, N., Pais, T.M., Mathé, L., Saerens, S., Nguyen, H.T.T., Swinnen, S., Verstrepen, K.J., Concilio, L., de Troostembergh, J.-C. and Thevelein, J.M. 2013. Quantitative trait analysis of yeast biodiversity yields novel gene tools for metabolic engineering. Metabolic Engineering 17: 68–81.

Husnik, J.I., Delaquis, P.J., Cliff, M.A. and van Vuuren, H.J.J. 2007. Functional analyses of the malolactic wine yeast ML01. American Journal of Enology and Viticulture 58: 42–52.

Husnik, J.I., Volschenk, H., Bauer, J., Colavizza, D., Luo, Z. and van Vuuren, H.J.J. 2006. Metabolic engineering of malolactic wine yeast. Metabolic Engineering 8: 315–23.

Jeandet, P., Delaunois, B., Aziz, A., Donnez, D., Vasserot, Y., Cordelier, S. and Courot, E. 2012. Metabolic engineering of yeast and plants for the production of the biologically active hydroxystilbene, resveratrol. Journal of Biomedicine and Biotechnology. 2012:ID 579089. doi:10.1155/2012/579089.

Johnston, R.J., Baccari, C. and Mortimer, R.K. 2000. Genotypic characterization of strains of commercial wine yeasts by tetrad analysis. Research in Microbiology 151: 583–590.

Kutyna, D.R., Varela, C., Stanley, G.A., Borneman, A.R., Henschke, P.A. and Chambers, P.J. 2012. Adaptive evolution of *Saccharomyces cerevisiae* to generate strains with enhanced glycerol production. Applied Microbiology and Biotechnology 93: 1175–1184.

Legras, J.-L., Merdinoglu, D., Cornuet, J.-M. and Karst, F. 2007. Bread, beer and wine: *Saccharomyces cerevisiae* diversity reflects human history. Molecular Ecology 16: 2091–2102.

Lerm, E., Engelbrecht, L. and du Toit, M. 2011. Selection and characterisation of *Oenococcus oeni* and *Lactobacillus plantarum* South African wine isolates for use as malolactic fermentation starter cultures. South African Journal of Enology and Viticulture 32: 280–295.

Liu, L., Redden, H. and Alper, H.S. 2013. Frontiers of yeast metabolic engineering: Diversifying beyond ethanol and *Saccharomyces*. Current Opinion of Biotechnology 24: 1023–1030.

Martínez, C., Cosgaya, P., Vásquez, C., Gac, S. and Ganga, A. 2007. High degree of correlation between molecular polymorphism and geographic origin of wine yeast strains. Journal of Applied Microbiology 103: 2185–2195.

Marullo, P., Aigle, M., Bely, M., Masneuf-Pomarède, I., Durrens, P., Dubourdieu, D. and Yvert, G. 2007. Single QTL mapping and nucleotide-level resolution of a physiologic trait in wine *Saccharomyces cerevisiae* strains. FEMS Yeast Research 7: 941–52.

Novo, M., Gonzalez, R., Bertran, E., Martínez, M., Yuste, M. and Morales, P. 2014. Improved fermentation kinetics by wine yeast strains evolved under ethanol stress. LWT—Food Sciences and Technology. 58: 166–72.

Pahlman, A., Granath, K., Ansell, R., Hohmann, S. and Adler, L. 2001. The yeast glycerol 3-phosphatases Gpp1p and Gpp2p are required for glycerol biosynthesis and differentially involved in the cellular responses to osmotic, anaerobic, and oxidative stress. Journal of Biological Chemistry 276: 3555–3563.

Pramateftaki, P.V., Lanaridis, P. and Typas, M. 2000. Molecular identification of wine yeasts at species or strain level: A case study with strains from two vine-growing areas of Greece. Journal of Applied Microbiology 89: 236–248.

Quirós, M., Rojas, V., Gonzalez, R. and Morales, P. 2014. Selection of non-*Saccharomyces* yeast strains for reducing alcohol levels in wine by sugar respiration. International Journal of Food Microbiology 181: 85–91.

Raspor, P., Milek, D.M., Polanc, J., Mozina, S.S. and Cadez, N. 2006. Yeasts isolated from three varieties of grapes cultivated in different locations of the Dolenjska vine-growing region, Slovenia. International Journal of Food Microbiology 109: 97–102.

Remize, F., Andrieu, E. and Dequin, S. 2000a. Engineering of the pyruvate dehydrogenase bypass in *Saccharomyces cerevisiae*: role of the cytosolic Mg(2+) and mitochondrial K(+) acetaldehyde dehydrogenases Ald6p and Ald4p in acetate formation during alcoholic fermentation. Applied and Environmental Microbiology 66: 3151–9.

Remize, F., Barnavon, L. and Dequin, S. 2001. Glycerol export and glycerol-3-phosphate dehydrogenase, but not glycerol phosphatase, are rate limiting for glycerol production in *Saccharomyces cerevisiae*. Metabolic Engineering 3: 301–312.

Remize, F., Cambon, B., Barnavon, L. and Dequin, S. 2003. Glycerol formation during wine fermentation is mainly linked to Gpd1p and is only partially controlled by the HOG pathway. Yeast 20: 1243–1253.

Remize, F., Roustan, J., Sablayrolles, J., Barre, P. and Dequin, S. 1999. Glycerol overproduction by engineered *Saccharomyces cerevisiae* wine yeast strains leads to substantial changes in by-product formation and to a stimulation of fermentation rate in stationary phase. Applied and Environmental Microbiology 65: 143–149.

Remize, F., Sablayrolles, J.M. and Dequin, S. 2000b. Re-assessment of the influence of yeast strain and environmental factors on glycerol production in wine. Journal of Applied Microbiology 88: 371–378.

Renouf, V., Claisse, O. and Lonvaud-Funel, A. 2007. Inventory and monitoring of wine microbial consortia. Applied Microbiology and Biotechnology 75: 149–164.

Rep, M., Albertyn, J., Thevelein, J.M., Prior, B. and Hohmann, S. 1999. Different signalling pathways contribute to the control of *GPD1* gene expression by osmotic stress in *Saccharomyces cerevisiae*. Microbiology 145: 715–727.

Rep, M., Proft, M., Remize, F., Tamás, M., Serrano, R., Thevelein, J.M. and Hohmann, S. 2001. The *Saccharomyces cerevisiae* Sko1p transcription factor mediates HOG pathway-dependent osmotic regulation of a set of genes encoding enzymes implicated in protection from oxidative damage. Molecular Microbiology 40: 1067–1083.

Salmon, J.-M. and Barre, P. 1998. Improvement of nitrogen assimilation and fermentation kinetics under enological conditions by derepression of alternative nitrogen-assimilatory pathways in an industrial *Saccharomyces cerevisiae* strain. Applied and Environmental Microbiology 64: 3831–3837.

Schuller, D., Cardoso, F., Sousa, S., Gomes, P., Gomes, A.C., Santos, M.S. and Casal, M. 2012. Genetic diversity and population structure of *Saccharomyces cerevisiae* strains isolated from different grape varieties and winemaking regions. PLoS One 7: e32507. doi:10.1371/journal.pone.0032507.

Schuller, D., Valero, E., Dequin, S. and Casal, M. 2004. Survey of molecular methods for the typing of wine yeast strains. FEMS Microbiology Letters 231: 19–26.

Shin, S.-Y., Han, N.S., Park, Y.-C., Kim, M.-D. and Seo, J.-H. 2011. Production of resveratrol from p-coumaric acid in recombinant *Saccharomyces cerevisiae* expressing 4-coumarate:coenzyme A ligase and stilbene synthase genes. Enzyme and Microbial Technology 48: 48–53.

Tamás, M.J., Rep, M., Thevelein, J.M. and Hohmann, S. 2000. Stimulation of the yeast high osmolarity glycerol (HOG) pathway: Evidence for a signal generated by a change in turgor rather than by water stress. FEBS Letters 472: 159–165.

Torija, M.J., Rozes, N., Poblet, M., Guillamon, J.M. and Mas, A. 2001. Yeast population dynamics in spontaneous fermentations: Comparison between two different wine-producing areas over a period of three years. Antonie Van Leeuwenhoek 79: 345–352.

Valero, E., Schuller, D., Cambon, B., Casal, M. and Dequin, S. 2005. Dissemination and survival of commercial wine yeast in the vineyard: A large-scale, three-year study. FEMS Yeast Research 5: 959–69. doi:10.1016/j.femsyr.2005.04.007.

Volschenk, H., Viljoen, M., Grobler, J., Bauer, F., Lonvaud-Funel, A., Denayrolles, M., Subden, R.E. and van Vuuren, H.J.J. 1997. Malolactic fermentation by a genetically engineered strain of *Saccharomyces cerevisiae*. American Journal of Enology and Viticulture 47: 193–197.

Volschenk, H., Viljoen-Bloom, M., Subden, R.E. and van Vuuren, H.J.J. 2001. Malo-ethanolic fermentation in grape must by recombinant strains of *Saccharomyces cerevisiae*. Yeast 18: 963–970.

Wang, Y., Halls, C., Zhang, J., Matsuno, M., Zhang, Y. and Yu, O. 2011. Stepwise increase of resveratrol biosynthesis in yeast *Saccharomyces cerevisiae* by metabolic engineering. Metabolic Engineering 13: 455–463.

Wang, Y. and Yu, O. 2012. Synthetic scaffolds increased resveratrol biosynthesis in engineered yeast cells. Journal of Biotechnology 157: 258–260.3

Zott, K., Miot-Sertier, C., Claisse, O., Lonvaud-Funel, A. and Masneuf-Pomarede, I. 2008. Dynamics and diversity of non-*Saccharomyces* yeasts during the early stages in winemaking. International Journal of Food Microbiology 125: 197–203. doi:10.1016/j.ijfoodmicro.2008.04.001.

3

Filamentous Fungi in Fermented Foods

Eta Ashu,[1,#] *Adrian Forsythe,*[1,#] *Aaron A. Vogan*[1,#]
and *Jianping Xu**

1. Introduction

Fermentation has long been used by humans for preserving food and for enhancing the nutritional and gastronomical values of foods. Broadly defined, fermentation refers to any form of food processing that involves the conversion of carbohydrates to alcohols and carbon dioxide or to organic acids using fungi, bacteria, or a combination thereof, under anaerobic conditions. One of the best-known fermentations in contemporary culture is the chemical conversion of sugars into ethanol by unicellular fungi to produce alcoholic beverages such as wine, beer, and cider (Chapter 2 in this book). While filamentous fungi are known to play very important roles in medicine (e.g., for producing antibiotics), industry (e.g., for producing enzymes), agriculture (as edible mushrooms), and forestry (ectomycorrhizal fungi), their roles in food fermentation is far less appreciated. In this chapter, we provide an overview of the diversity of filamentous fungi as well as their roles in representative fermented foods from broad regions of the world. The major sections are divided according to the types (genus) of filamentous fungi. Within each genus, we describe the major fermented foods that the specific fungi are found in, including their roles in the fermentation processes. The problems and potential opportunities for future research are discussed.

[#] These three authors contributed equally to the work.
[1] Department of Biology, McMaster University, Hamilton, Ontario, L8S 4K1, Canada.
   E-mails: ashuebasi@yahoo.com; adrian.e.forsythe@gmail.com; voganaa@gmail.com
* Corresponding author: jpxu@mcmaster.ca

2. Aspergillus

In this section, we first introduce the fungal genus *Aspergillus*, including its general and diagnostic characteristics and the main species within this genus found in fermented foods. We then describe the major fermented foods (including, briefly, their protocols of preparations) where *Aspergillus* species are found.

2.1 General Introduction of the Genus Aspergillus

Aspergilli are Ascomycetes that belong to the family Trichocomaceae. They are ubiquitous filamentous fungi found in a wide range of habitats and climates. *Aspergillus* was first given its name in 1792 by Pier Antonio Micheli, a Catholic priest and botanist from Italy. He noted that fungi did not have spontaneous origins by observing that the same kind of fungi grew on slices of melon whenever spores were placed on them (Agrios 2005). In viewing the microscopic spore-bearing structure of *Aspergillus*, Micheli was reminded of an aspergillum—a device used by Roman Catholic clergies to sprinkle holy water (Bennett 2010). *Aspergillus* spore-bearing structures consist of conidiophores, phialides (spore producing cells) and vesicles. Conidiophores bud off from supporting hyphae and terminate in vesicles; vesicles in turn bear flask shaped spore producing cells called phialides (Fig. 1). Phialides can (i) directly attach themselves to the vesicle, in which case vesicles are referred to as uniseriate heads, or

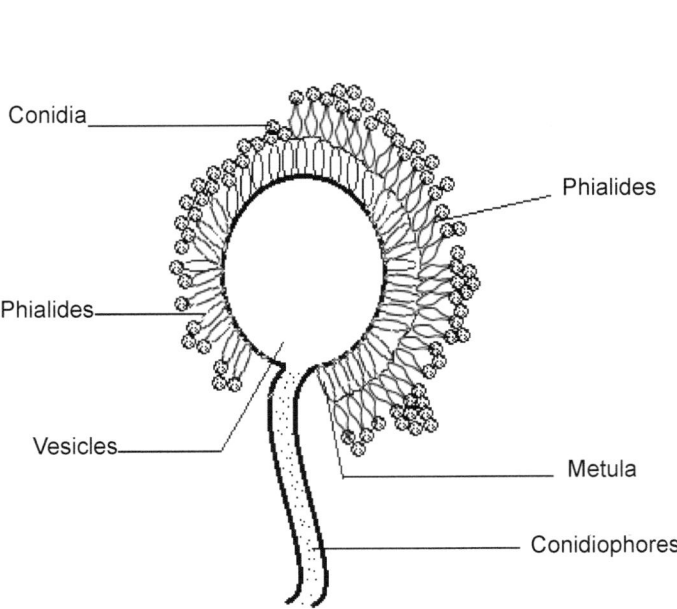

Figure 1. Spore-bearing structure of *Aspergillus* species.

(ii) attach themselves *via* a supporting cell, called metula, in which case vesicles are referred to as biseriate. Conidiophores and vesicles are very important for identification of *Aspergillus* spp. as their morphology varies from species to species.

Aspergillus spp. are common contaminants of starchy food and some of them can grow at high temperatures (e.g., at 50°C for *Aspergillus fumigatus*). They are non-fastidious and can easily grow on a diversity of laboratory media. Depending on the species and media, they can produce colonies of different colors ranging from brown to black, green or yellow (Collier et al. 1998). There are about 250 known species of *Aspergillus* (Geiser et al. 2008). However, only a few of them are being used for food fermentation. Although not often used as starter cultures, *Aspergillus* species have been found to be associated with the fermentation of many foods (Table 1). The most important *Aspergillus* species used in food fermentation have been recently sequenced and there are good prospects of genetically enhancing traditional fermentation processes. A discussion on the use of several common *Aspergillus* species in food fermentation follows.

Table 1. List of fermented foods described in this chapter along with the fungal species used in their processing.

Food	Description	Species
Cheese (Camembert and Roquefort)	Coagulated Milk Product	*Penicillium camemberti* *Penicillium roqueforti*
Gari/Fufu/Kum-kum/Bobolo/ Meyondo	Fermented Cassava Tubers	*Aspergillus flavus* *Aspergillus niger*
Katsuobushi	Shaved Smoked Tuna	*Aspergillus repens*
Sufu	Fermented Soybean Curd	*Mucor racemosus* *Mucor circinelloides* *Actinomucor taiwanensis* *Actinomucor elegans*
koji/Murcha/Loong-pang/Ragi/ Jiuqu	Starter Cultures (from rice)	*Aspergillus oryzae* *Rhizopus arrhizus*
Tempeh	Fermented Soybeans	*Rhizopus microsporus*
Meju	Starter Cultures (from soybeans)	*Aspergillus oryzae* *Aspergillus sojae*
Bread (Khamir)	Fermented Sorghum Grain	*Aspergillus niger*

2.2 Aspergillus flavus and Aspergillus niger

Both *Aspergillus flavus* and *A. niger* species are common soil inhabitants and among the most frequently isolated fungi in agricultural products. *A. niger* is recognized by its black colonies, it is a common food contaminant and causes black mold disease in certain vegetables and fruits (Samson et al. 2011). *A. flavus* can also grow on a wide range of food. However, *A. flavus* is known to produce aflatoxins, a group of highly carcinogenic substances (Hudler 2000). Aflatoxins can be present in a wide range of foods, including cereals, maize, groundnuts (peanuts), yam, cassava and dried fruit. Contamination of agricultural crops with aflatoxin causes annual losses of over US

$750 million in Africa and income losses of over US $100 million per year are reported in the United States (Leslie et al. 2008, N'dede et al. 2012). Although widely known for its ability to produce aflatoxin, *A. flavus* also has beneficial biotechnological uses; it has been substantially used in cassava fermentations for several decades.

2.2.1 Fermented cassava food products by natural strains of A. flavus and A. niger

Cassava (*Manihot esculenta* Crantz) is one of the most important major root crops in West and Central Africa where it is a significant calorie source for most households. Cassava roots are processed into numerous products by various methods which are dictated by local customs and preferences. Some well-known fungal fermented cassava products include gari, fufu, smoked cassava balls (kum-kum) and bobolo. Below we briefly introduce the processes for making these four foods.

2.2.1.1 Gari

Gari (also known as garri or garry) is routinely eaten in Nigeria, Cameroon, Ghana, Sierra Leone, Benin and Togo. There are different types of gari. Depending on its grain size, gari can be classified as coarse or fine. Gari could also be classified based on its color after processing, e.g., yellow (when it is fried with palm oil) or white (fried without palm oil). To make gari, cassava tubers are peeled, soaked and grated. The grated cassava is then tied in a permeable (cloth) bag to allow water to run out of the grated cassava, left to ferment naturally and dry for 3–7 days. The dried grated cassava is fried until very dry and brittle. Gari can be eaten as a snack by adding cold water and sugar; however, most often, gari is cooked in boiling water to make a paste. Interestingly, gari was recently used as a substrate in artificial medium for the isolation and characterization of fungi (Okorondu et al. 2013), similar to the use of potatoes (e.g., in the potato dextrose medium).

2.2.1.2 Fufu

To make fufu, cassava is peeled and soaked for about 72 hours to naturally ferment and become soft. It is then mashed and allowed to drain for a few hours in a cloth bag. The cassava mash is then sieved; fufu is prepared by putting the sieved cassava mash in boiling water and stirring thoroughly to obtain a stiff paste.

2.2.1.3 Smoked cassava balls (kum-kum)

Kum-kum is a common Cameroonian staple and is widely sold in most markets in the country. Kum-Kum is processed from cassava in the same way as fufu except that the fermented wet pulp is mashed and moulded into round balls of about 3 cm in diameter. These balls are then smoked on a platform above the fire place. Smoke from the fire darkens and dries the ball; however, the built up tar from the smoke is scraped off before consumption. Smoked cassava balls can be milled into flour and reconstituted into fufu (Numfor and Ay 1987).

2.2.1.4 Bobolo and meyondo

Bobolo and meyondo are processed in the same way as smoked cassava balls; however, the mashed fermented cassava is then wrapped in plantain or banana leaves. The leaves are firmly tied with ropes from banana or plantain suckers and boiled. Myondo has a diameter of about 1 cm and a length of 15 cm. Bobolo is much thicker, it has a diameter of about 3 cm and a length of 20 cm. Bobolo and meyondo are names peculiar to Cameroon; however, similar products are produced in the Democratic Republic of the Congo, Gabon and the Central African Republic.

2.2.2 A. flavus and A. niger as starter cultures in cassava processing

In West Africa, until recently, cassava was traditionally processed to prevent it from rotting. Given that unprocessed cassava rots about 4–7 days after it has been harvested, processing significantly increased its shelf life and reduced the cyanide content (Mahungu et al. 1987). However, with the advent of industrialization, cassava has been processed to produce larger quantities of cassava products to meet the growing food need in tropical Africa. Also, the nutritional content, acceptability, and palatability of traditionally processed cassava products have been improved upon. Fermenting cassava with *A. flavus* has shown to improve gari's color and market value (Akindahunsi and Oboh 2003), possibly due to its ability to decrease tannin in cassava (Akindahuns et al. 1999). Tannins have been associated with dull color in cassava products hence decreasing their market value (Akindahunsi and Oboh 2003). *A. flavus* also significantly increased *in vitro* multi-enzyme digestibility of cassava products (Akindahunsi and Oboh 2003). Similarly, fermenting cassava with *A. niger* was shown to improve gari's aroma and texture (Akindahunsi and Oboh 2003). *A. niger* fermented cassava products have been shown to have significantly increased ($P < 0.05$) protein [flour ($12.2 \pm 0.2\%$); gari ($7.3 \pm 0.1\%$)] and fat [flour ($5.7 \pm 0.1\%$); gari ($4.0 \pm 0.3\%$)] contents as compared to unfermented and traditionally fermented cassava products. In addition, *A. niger* fermented cassava products have very low cyanide levels (Oboh et al. 2002).

We would like to mention that even though cassava fermentation processes are well established, the use of *A. flavus* and *A. niger* to improve cassava products is still at infancy. Furthermore, the underlying genetics and physiology are still poorly understood. However, the recent availability of full genome sequences for *Aspergillus* spp. should provide valuable aid towards understanding and improving genes responsible for cassava fermentation and fermentation at large.

2.3 Aspergillus niger in Khamir (Bread) Fermentation

Khamir is a traditional Emirati breakfast bread; Khamir in the Gizan region of Saudi Arabia is prepared from fermented sorghum grain. *Sorghum* is a genus of grasses cultivated for grains and as fodder plants. *Sorghum bicolor* grows in clusters of about 2–4 meters high and is commonly cultivated for its grains. It is primarily cultivated in tropical and subtropical, arid and semiarid regions of Africa and Asia because it is drought and heat tolerant (Food and Agriculture Organization of the United Nations

1995). Khamir is prepared by mixing sorghum flour, spices (garlic, onion and lemon juice) and water; the mixture is then allowed to ferment for 24 hours and cooked on a hot tava. *A. niger* was identified as one of the fungi involved in the fermentation of sorghum grains to make khamir. Even though *Aspergillus* species haven't yet been intentionally used to improve the quality of khamir bread, earlier use of *Aspergillus* species in cassava products underlines their potential in ameliorating the digestibility, nutrient content and palatability of Khamir bread (Gassem 1999).

2.4 Aspergillus repens and Katsuobushi

Aspergillus repens is xerophilic and can tolerate conditions of high salinity by secretion of enzymes to balance salt excesses (Kelavkar and Chhatpar 1993). Because of its xerophilic property, *A. repens* is used for the moulding step of katsuobushi processing, contributing to katsuobushi digestibility and flavour improvement (Kaminishi et al. 1999). Katsuobushi is the Japanese name for dried and shaven skipjack tuna (*Katsuwonus pelamis*) and the main ingredient in many Japanese soups and sauces. To prepare katsuobushi, skipjack tuna or bonito is first beheaded from underneath the gills. All internal organs are removed and the fish is cut into fillets. Fillets are then arranged in baskets and boiled at 98°C for 60–90 minutes; fresher fish is boiled at lower temperatures. After boiling, fillet rib bones are removed and fillets are then smoked till they turn golden brown. After about 1–2 days, the fillets are smoked again for 6–8 hours, and this process is repeated 12–13 times. Fillets are then sprayed with *Aspergillus* spp. and allowed to sun-dry. The moulding and sun-drying steps require periods as long as 2 months (Toko 2013). While *A. repens* is the most commonly used *Aspergillus* species for Katsuobushi moulding, *A. ruber* is also used (Dimici and Wada 1994). Unfortunately, there are no records of start culture amounts or mold strains used in katsuobushi processing. Indeed, relatively little is known about the specific fermentation processes in katsuobushi processing.

2.5 Aspergillus oryzae and A. sojae in Koji and Related Products

Aspergillus sojae and *A. oryzae* are taxonomically classified in section Flavi (Takahashi et al. 2002) and related to *A. parasiticus* and *A. flavus*. Although molecular evidence regarding the evolutionary origin of *A. sojae* and *A. oryzae* is still debated, it is generally agreed that *A. sojae* and *A. oryzae*, which are seldom isolated in nature, are domesticated variants of *A. parasiticus* and *A. flavus* respectively (Machida et al. 2008, Takahashi et al. 2002). It is also generally accepted that neither *A. sojae* nor *A. oryzae* produce aflatoxin, hence they are widely used for the production of enzymes and fermented foods. Recognized by its olive to greyish-yellow color, *A. oryzae* has been widely used in Eastern Asian fermentation industries. *A. oryzae* was first isolated in 1876 by Ahlburg (Rokas 2009, Shurtleff and Aoyagi 2012) from food materials (rice grain, soybean and wheat) used in solid state fermentation (SSF); however, *A. oryzae* was used in koji inocula for SSF before 1876 without knowing what it was then. Koji is a Japanese name for the material (in this case, *A. oryzae*) used in SSF. Koji originated about 2000–3000 years ago in China (Rokas 2009), and was imported to Japan during the Yayoi period (B.C. 10th–A.D. 3rd). *A. oryzae* inocula have been commercially available as early as the 13–15th century (Machida et al. 2008). At present, koji is used in making soy sauce, miso and other Japanese fermented food. In addition, *A. oryzae* is also used as starter cultures in making meju, gari and fermented fish. Similarly, *A. sojae* is used to ferment soybeans for making soy sauce and with vegetables for making tsukemono (pickled vegetables), a fragrant Japanese dish (Leboffe and Pierce 2012).

2.5.1 Koji, jiuqu and their products

The word koji can refer to both the mold and the food based on which the mold is used. Different koji types are made from different food bases and molds. Both *A. oryzae* and *A. sojae* are well-known molds used in making koji from either rice, soybean or wheat bases (Fig. 2). Rice koji is prepared by soaking rice in water overnight and steaming it for about 40 minutes. The steamed rice is allowed to cool and is sprinkled with *A. oryzae* or *A. sojae* conidia; it is then carefully mixed to facilitate an even spread of conidia and kept warm by packing it in heaps and wrapping it in a cloth. It is left for several hours and then partitioned into smaller boxes upon growth of mycelium. These boxes are kept for 2 days in warm and humid conditions that favour fermentation; well-grown rice koji is white and fluffy like velvet. Rice koji can be used as starter cultures for homemade and factory-made miso (Chieko 2008). Miso is a traditional Japanese condiment produced by fermenting rice, wheat or soybeans. The resulting thick paste is used in making soups and pickling vegetables and meat.

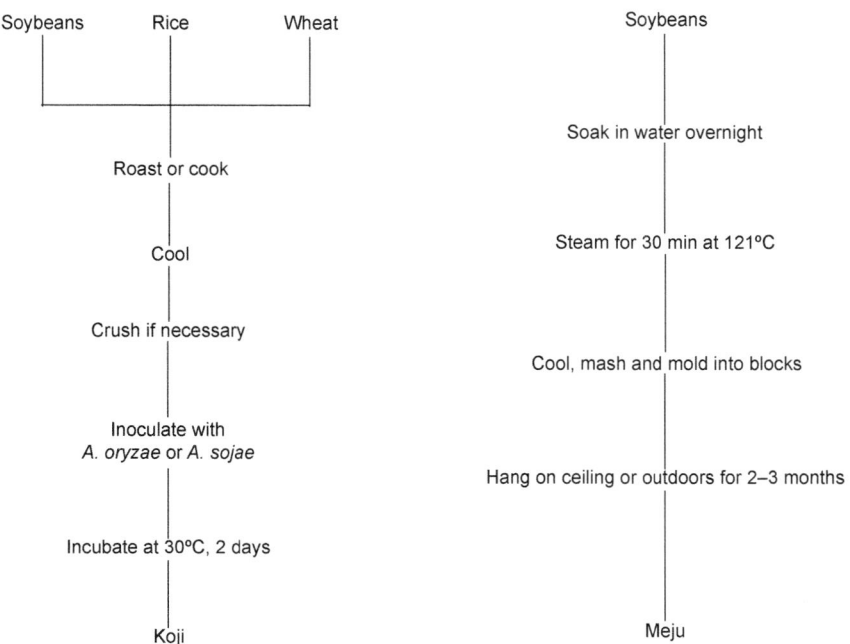

Figure 2. Flow charts for the preparation of koji (fermentation starter) and Meju.

Different from rice koji, koji for soy sauce is made by adding *A. oryzae* or *A. sojae* condia to boiled and crushed soybeans and wheat (Leboffe and Pierce 2012) and incubating it for three days. The koji is transferred to fermentation tanks, where it is mixed with water and salt to produce a mash. The wheat and soybeans mash is left to ferment during which time the mash turns into a reddish-brown semi-liquid mixture. Proteolysis by *Aspergillus* mold gives soy sauce its dark brown color. After approximately six months of fermentation, raw soy sauce is separated from mature mash by pressing the mash through layers of filtration cloth. The liquid that emerges is then pasteurized and commercialized.

Jiuqu is a Chinese equivalent of koji; it, however, predates koji and is slightly different from it (McGovern et al. 2004). Typically stored and sold in the form of dried bricks (Daqu), balls (Xiaoqu), grains (Red Yeast Rice) or powder, jiuqu is a mixture of various molds, yeasts, and bacteria cultured on a starch-rich substrate. The most common substrates used include wheat, barley, rice or peas; however, substrates used vary across China in accordance with regional preferences. Microbial communities on jiuqu also vary in accordance with regional preferences; nonetheless, A. *oryzae* is one of the most common molds found in jiuqu (Shang et al. 2012, Xie et al. 2007, Zheng et al. 2012); alongside other microorganisms, they serve two main parallel functions, fermentation and generation of enzymes. Traditionally made jiuqu was fermented spontaneously; it has, however, been discovered that using small amounts of a previously successful batch to inoculate the current one gave more consistent results (Huang 2000). Like most traditionally fermented foods and condiments in China, jiuqu production is carried out in small-scale manufacturing facilities and homes which are non-sanitized. Consequently, most jiuqu lack consistency of quality. In stark contrast, koji manufacture in Japan is highly industrialized with the use of isolated mono-cultures. A future directive for biotechnological industries will be to replicate advances made in the koji production in China.

2.5.2 Meju and other fermented food products

Meju is a Korean fermented soybean condiment used to prepare soy paste and sauces. It is prepared by moulding steamed crushed soybeans into rectangular blocks and hanging those blocks outdoors for about two months (Fig. 2). Meju is fermented spontaneously; typically it is first fermented by the bacteria *Bacillus* species, followed by *A. oryzae* (Kwon et al. 2011). In the final product, either *A. oryzae* or *Bacillus subtilis,* or both could be the dominant organism(s) (Haard et al. 1999). Even though it is debatable which species is the primary fermenter it is commonly known that changes in isoflavonoids and peptides which occur during longer fermentation periods of meju have been shown to enhance its anti-diabetic effect (Kwon et al. 2011). Both *A. oryzae* and *A. sojae* have also been used for the preparation of fermented fish meal (FFM). FFM has more vitamins and less thiobarbituric acid than non-fermented fish meals. Furthermore, the growth of broilers fed on FFM diet was superior to that of those fed the commercial broiler diet (Kato et al. 1986).

2.6 Applied Genomics of Aspergillus oryzae and A. sojae

The genome of *A. oryzae* was first sequenced in 2005 and consists of eight chromosomes with a genome size of 37.6 Mb. It has 12,000 predicted genes, 2,000–3,000 more than in *A. nidulans* and *A. fumigatus*. This increase in genome size is mainly derived from the expansion of gene families coding for secretory hydrolases, transporters and secondary metabolism genes. The expansion of genes for secretory hydrolases and transporters is likely to have contributed to *A. oryzae*'s ability to degrade various materials and excrete toxic compounds (Machida et al. 2008). Even though *A. sojae* and *A. oryzae* have highly similar genomes (*A. sojae* genome is ~ 1.9 Mb (39.5 Mb) bigger than *A. oryzae* genome); comparative genome analysis revealed that *A. sojae* possesses a single α-amyalse gene whereas *A. oryzae* has three copies. In addition, carboxypeptidase and aspartic protease genes were identified as unique to *A. sojae* while genes for cyclopiazonic acid (CPA) biosynthesis were unique to *A. oryzae* (Sato et al. 2011). Depending on the objectives, this data gives food industries the ability to choose the appropriate *Aspergillus* species and strains to use in food fermentation. The availability of full genome sequences for *A. sojae* and *A. oryzae* makes it possible to genetically delete industrially undesirable genes to further improve efficiency and safety. Recently, the disruption of *ku70* or *ku80* genes was found to enhance transformation efficiency for *A. sojae* and *A. oryzae* (Takahashi et al. 2006). This novel discovery has been applied to the deletion of a large DNA region of about 200 kb for both *A. sojae* and *A. oryzae* (Machida et al. 2008). The advent of techniques such as this will prove extremely valuable in deleting redundant genes and adding industrially useful genes to both *A. sojae* and *A. oryzae* in order to enhance productivity.

2.7 Safety Issues with Aspergillus Species in Fermented Food

Aflatoxins are a group of mutagenic and carcinogenic fungal metabolites produced by *A. flavus*, *A. parasiticus* and *A. nomius*. Aflatoxins are part of a larger group of toxic compounds called di-furanocoumarins, generally produced by plants as a defense mechanism against insect and mammal predators (Lawley 2012). Aflatoxins B1, B2, G1 and G2 are among the more frequently isolated mycotoxins in food, with B1 as the most common and most toxic. Exposure to high level aflatoxins can cause acute toxicity while moderate exposure over long periods has been implicated in liver cancer, chronic hepatitis, jaundice, cirrhosis and impaired nutrient conversion (Lawley 2013). However, the use of *Aspergillus* species in fermenting food can increase aflatoxins in food (Chikezie and Ojiako 2013).

Different standards and analytical methods have been developed to quantify aflatoxins in food. Most countries in the European Union (EU) have regulations governing aflatoxin limits. These limits vary according to the commodity range, from 2–12 μg/kg for B1 and from 4–15 μg/kg for total aflatoxins (B1, B2, G1 and G2) in nuts, fermented spices, dried fruits and cereals. In the United States, there is a limit of 20 μg/kg for total aflatoxins (B1, B2, G1 and G2) in all foods. Australia and Canada have limits of 15 μg/kg for total aflatoxins (B1, B2, G1 and G2) in nuts (Lawley 2013). Aflatoxin limit regulations in Africa need to be viewed in a local context; the establishment of mycotoxin regulations will have limited effects on population health protection since most African farmers grow agricultural produce for their own consumption. Most African countries have no laws governing aflatoxin limits (Food and Agriculture Organization of the United Nations 2004). A few studies have shown that market gari in Nigeria have safe aflatoxin levels (Jonathan et al. 2013, Ogiehor et al. 2008). However, caution must be applied when extrapolating such results to neighbouring countries such as Ghana, Cameroon, Togo, Sierra Leone, and Benin where gari production and consumption are common but different fungi and fermentation methods may be used.

The close relationship between *A. oryzae* and *A. flavus* has resulted in careful examination of *A. oryzae*'s potential to produce aflatoxins. Fermented foods produced by *A. oryzae* and *A. sojae*, a close relative to *A. oryzae*, have been shown to have no aflatoxin (Machida et al. 2008). Nonetheless, *A. oryzae* can produce other mycotoxins such as kojic, cyclopiazonic and β-nitropropionic acids, and maltoryzine (Ciegler and Vesonder 1987). Even though *A. oryzae*, and to some extent *A. sojae*, can produce kojic acid, no kojic acid has been reported in fermented foods (Manabe et al. 1984). While being highly toxic, maltoryzine is presumably produced by a single *A. oryzae* variant, IAM 2950 (Ciegler and Vesonder 1987). Given that commercial strains of *A. oryzae* and *A. sojae* apparently do not produce maltoryzine, using both species in food fermentation does not seem to pose any significant safety issues with regard to maltoryzine.

Cyclopiazonic acid appears to be toxic only in high concentrations; it is a natural contaminant of foods and is produced by *A. flavus* and *A. oryzae*. Intra-peritoneal and oral administration of cyclopiazonic acid produced degenerative changes and necrosis in the entire alimentary system and skeletal muscle of rats and dogs at levels ranging from 0.25 to > 50 mg/kg (Nuehring et al. 1985, Purchase 1971). Unfortunately, there is hardly any available literature on recommended limits of cyclopiazonic acid in food fermented by *Aspergillus* species. Like cyclopiazonic, β-nitropropionic acid has no recommended limits in fermented food; however an LD_{50} of 20–50 mg/kg has been determined in rats (Hamilton and Gould 1987). β-nitropropionic acid is produced by *A. oryzae* and acts by inhibiting succinate dehydrogenase (Penel and Kosikowski 1990). Kinosita et al. (1968) associated high stomach cancer and hepatoma in several distinct provincial regions of Japan to consumption of fermented food such as miso and katsuobushi. However, the unavailability of updated literature makes it difficult to determine if β-nitropropionic acid still has similar effects or if the food industries have found a way to eliminate or reduce it in fermented food.

The last major group of mycotoxins is the ochratoxins. Ochratoxins are secondary metabolites produced by *A. niger* strains; there are 3 types with ochratoxin A being the most relevant, while ochratoxins B and C are less prevalent. However, only 3–10% of examined *A. niger* strains produced ochratoxin A under favourable conditions (Schuster et al. 2002). It is recommended that new and unknown *A. niger* strains be tested for ochratoxin A production before they are used in fermenting food.

3. Penicillium

Penicillium is a genus of filamentous fungi in the division Ascomycota (order Eurotiales, class Eurotiomycetes, family Trichocomaceae). The genus was first described by Link (1809) in '*Observationes in Ordines Plantarum Naturales*', in which three species were described: *P. candidum*, *P. expansum*, and *P. glaucum*. There are currently 180 accepted members of the genus *Penicillium* (Samson et al. 2004), with all species exhibiting conidiophores with a characteristic brush-like structure (Link 1809). However, the number of species will likely change due to continued studies from previously under-sampled regions and analyses using polyphasic approaches (Dörge et al. 2000). Mycology literature has experienced many changes with regard to nomenclature, involving the removal of some species as well as the amalgamation of other groups. This is particularly troublesome when considering applications that require the use of specific cultures; biotechnology requires the use of proper species to generate the desired products as well as to avoid unwanted results such as mycotoxins (Miller 1994). In this section, we review and describe the major types of fermented foods that involved fungi in the genus *Penicillium*.

3.1 Milk Fermentation and Cheese

One of the earliest records of fermentation was milk fermentation, recorded in Greek literature by Homer in approximately 1184 BC (Prajapati and Nair 2008). This early account describes the formation of solids in goat milk after the addition of crushed fig branches. The solids in the milk appear due to the presence of the enzyme ficain in the sap of the fig tree (Tramper and Zhu 2011); crushing the fig branches releases the enzyme into the milk. Due to its proteolytic nature, ficain acts upon the casein milk proteins. The amino acid chains that make up these protein molecules become positively charged and cause the tertiary structure to destabilize. Amino acids are now free to interact with other hydrophobic molecules, such as the lipids present in milk. The chemical interaction of these two components creates a network of molecules resulting in a gel-like consistency, thickening the mixture and resulting in the semi-solid texture of yogurt (Slonczewski and Foster 2013). This network of molecules is held together by relatively weak bonds which are easily broken up by subtle perturbations. Movement during the transportation of milk would suffice to break these bonds.

Another accidental discovery of cheese curds was made possible through the rise of animal domestication. The herders that looked after livestock were unable to transport the liquid milk that they harvested without the convenience of modern containers. Organs harvested from a freshly slaughtered calf, in particular the 4th stomach, became the first fermentation chambers for cheese production. Once contained in the calf stomach, milk is exposed to natural digestive enzymes chymosin and pepsin (Slonczewski and Foster 2013). The chymosin specifically causes casein to cleave into both charged and water-soluble components. As with ficain, the hydrophobic nature of the molecules is key to forming cheese curds (Slonczewski and Foster 2013). Exposure, after a few hours, is sufficient to produce semi-solid curds and whey.

These early techniques for developing cheese arose without the direct application of live microorganisms. As time progressed, procedures for producing specific desired flavours or consistencies were developed. Proper maturation or ripening required the competent handling of live microbial cultures. An essential species in cheese manufacturing, *Lactobacillus*, was isolated from milk in 1878 (Prajapati and Nair 2008). The ripening and production of harder consistency cheese generally begins with

Lactobacillus and/or *Streptococcus* bacteria. Prior to this, the quality of fermentation was principally controlled by adjusting the temperature of milk. The characteristic decrease in pH during milk fermentation prevents the growth of unwanted bacteria. In turn, the oxidation of amino acids is inhibited and food quality is maintained (Slonczewski and Foster 2013).

3.1.1 Penicillium in cheese making

Compared to the early introduction of lactobacilli to cheese production, the introduction of filamentous fungi was relatively late, mostly in the production of soft cheeses. In these cases, fungi are applied after the cheese has been shaped and is thus considered a secondary culture. Mold ripened cheeses make use of rennet to produce milk solids. In some cases a starter culture, usually lactic acid bacteria (LAB) such as *Lactobacillus* spp., is used in coagulation of milk solids. Various enzymes and starter cultures, as well as the method in which secondary cultures are introduced, can result in different varieties of cheese. Mold that is superficially applied to the cheese—as is the case for Brie, Camembert, or other goat's milk cheeses—contributes to the presence and appearance of rind. Alternatively, culture injected into the body of the cheese will result in (sometimes colourful) veins showing mycelial growth throughout. Renowned cheeses such as Blue d'Auvergne, Bavarian Blue, Blue des Causses, and Roquefort, are all ripened using the injection method. Among the *Penicillium* species, *P. roqueforti* and *P. camemberti* are the dominant ones. Modern cheese producers select patented strains based on the timing of germination, colour of fungal mycelium, length of said mycelium, and the overall density of fungal colony growth (Chamba and Irlinger 2004). Below, we describe these two species and their specific involvement in cheese production.

3.1.2 Penicillium camemberti—surface mold-ripened cheeses

Camembert is a soft, mold-ripened cheese originating from the Normandy region of France, around 1790 (Robinson 1995). However, early production was limited to rural farms. The industrialization of cheese making in the 20th century led manufacturers of mold-ripened cheeses to become progressively widespread in France, with their presence in other European countries occurring later (Spinnler and Gripon 2004). France is the biggest cheese producing country in the world—it produced approximately 10, 294, 092 tonnes in 2010, over half the world total (Food and Agriculture Organization of the United Nations 2013). *P. camemberti* is the secondary culture of choice for the production of Brie and Camembert cheeses as well as several semi-hard cheese varieties.

Beginning with unpasteurized milk, the coagulation of milk solids occurs as rennet is introduced. On occasion, the cheese making begins with the addition of a starter culture rather than a proteolytic enzyme. Starter LAB's and secondary mold cultures are commonly used in conjunction; non-starter lactic acid bacteria (NSLABs) and species of yeast can also be employed—though the latter is most commonly found as a surface contaminant. The flora of surface mold-ripened cheeses can be quite complex, especially when raw milk is used (Addis et al. 2001, Lund et al. 1995). As a result, the diversity of fungal and bacterial species on mold-ripened cheeses can be high. The metabolic activity of microbial flora can generate various compounds causing changes to physical characteristics, chemical by-products, and selective anti-microbial activity. The pH of the cheese can also acts as a barrier, allowing certain microbes to grow and proliferate while limiting others (Spinnler and Gripon 2004).

After the addition of starter culture(s), the coagulum (curds) are cut and moulded, then salted (McSweeney et al. 2004). *P. camemberti* can produce abundant asexual spores, conidia (Fig. 3) (Griffin 1996). The inoculation of these spores onto the salted curds begins the ripening process. Strict control of environmental variables is essential for consistency and success; 26–28°C is the preferred temperature at the

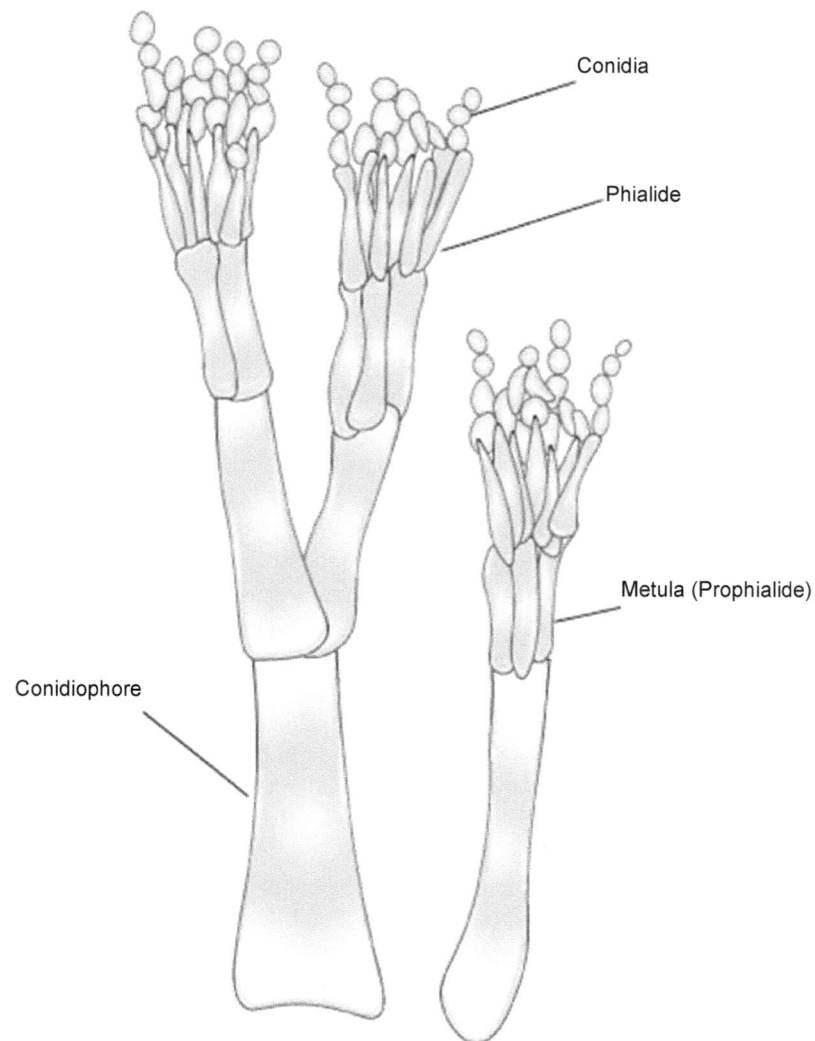

Conidia

Phialide

Metula (Prophialide)

Conidiophore

Figure 3. Representation of typical asexual structures of species in the genus *Penicillium.*

outset of the process, and then a decrease to 20°C after the curds have finished draining (Spinnler and Gripon 2004). At this time, the cheese has a low mineral content and a pH of 4.6–4.7. The cheese is then dry salted and the ripening or maturation process begins. The cheese is left for a minimum of 21 days to fully develop. Consistent temperature control (11–13°C) is required during ripening and a relative humidity of 90% is the standard (about the conditions of a typical cellar). In addition to environmental conditions, changing chemical parameters—e.g., salt, water activity (a_w, denotes the vapour pressure of water in a substance relative to the partial vapour pressure at equilibrium), pH—can greatly affect the outcome of the final product (Chamba and Irlinger 2004, Spinnler and Gripon 2004).

Camembert and Brie cheeses undergo a complex series of biochemical reactions during ripening. We briefly summarize below, the general effects. Lactate is metabolized by *P. camemberti*; this process produces $Ca_3(PO_4)_2$ as a precipitate. Soluble Ca^{2+}, PO_4^{3-}, and molecules migrate towards the cheese surface. In addition, *P. camemberti* initiates proteolysis of casein proteins. Ammonia is created as a byproduct of proteolysis and it promptly diffuses into the cheese. The interior texture softens, progressing from the surface towards core. Overall, the pH of the cheese is highest closest to the rind, and lowest near the core (McSweeney and Fox 2004, Spinnler and Gripon 2004). After sufficient maturation, the culture is scrubbed or washed off prior to marketing and consumption.

3.1.3 Penicillium roqueforti—blue-veined cheeses

Penicillium roqueforti is used as secondary culture during the ripening of blue-veined cheeses; it is also commonly found as a contaminant on other types of cheese (Lund et al. 1995). This species has had several synonyms: *P. aromaticum*, *P. gorgonzola*, *P. stilton*, *P. suaveolens*, and *P. biourgei* (Jakobsen et al. 2010). Its presence in blue cheese provides the characteristic colors, flavors, and smells that are associated with the varieties Gorgonzola, Roquefort, and Stilton. These types of cheese have existed for many years, the earliest records for Gorgonzola dating back to 879 A.D. (Robinson 1995).

The growth of filamentous fungi within the blue cheese varieties is achieved by first injecting the fungal spores into the cheese body, where it remains until the product is consumed. Strains of this species are carefully selected based on many variables, one being the colour of mycelia. Those exhibiting light blue or yellowish colours are typically chosen for making Gorgonzola and similar varieties. Much darker green strains can be found in Bleu des Causses, Stilton, Roquefort and similar cheeses (Chamba and Irlinger 2004).

P. roqueforti is able to grow under a wide range of pH, from pH 3–10. For cheese making, pH 4 and 5 are optimal (Cantor et al. 2004, Cerning et al. 1987). Certain strains of *P. roqueforti* are sensitive to salt content, affecting germination and growth in various degrees (Godinho and Fox 1981). The germination of the conidia occurs best at high water activity: $0.98–0.99a_w$ (Jakobsen et al. 2010). The interior sections of blue-veined cheeses provide ideal conditions for *P. roqueforti* to proliferate. Within these cheeses, growth can be observed 2–3 weeks post-inoculation. Around this time, the cultures become very metabolically active.

Extensive interactions occur between the fungi and constituents of the cheese. Of these interactions, the most important to ripening is proteolytic activity, which contributes to the development of the texture and water-binding capacity of the cheeses (Jakobsen et al. 2010). Proteinases such as aspartic proteinases and peptidases released from *P. roqueforti* are the main contributors of reactions occurring in the cheese (Le Bars and Gripon 1981, Cerning et al. 1987). Salt levels and pH influence proteolysis, which, in general, is less active in the heavily salted outer part (Gobbetti et al. 1997, Hewedi and Fox 1984). By the end of the ripening process, proteolytic activities should exhaust casein protein levels in the cheese (Zarmpoutis et al. 1997).

3.1.4 Fungal biotechnology in cheese production

Compared to most other fermented food discussed in this chapter, cheese making requires more structure and control. Proper controls in selecting strains as well as in harvesting fungal cultures are paramount for cheese makers. Contemporary methods involve the direct cultivation of conidia of *P. camemberti* on agar media in Roux-flasks; mainly because conidiation in liquid culture is inefficient for cheese production (Boualem et al. 2008). Indeed, most filamentous fungi have difficulty producing spores while in submerged cultures (Pascual et al. 1997). Instead, most spores are harvested from solid agar media after extensive incubation, thus limiting the large scale production of spores. In Camembert-type cheeses, conidia are spread or sprayed onto the cheese surface. Boualem et al. (2008) describe a method for generating conidia of *P. camemberti* by manipulating microcycle patterns of conidiation through environmental conditions. During a microcycle, conidiation occurs after spore germination, without the growth of intermediate hyphae, a state which is brought on by environmental stress (Boualem et al. 2008). The issue becomes more complex as Park and Yu report that gene expression, specialized cellular differentiation, and even intra/inter-cellular communications can impact conidiation in filamentous fungi (Park and Yu 2012). Ultimately, this provides cheese manufacturers with a greater set of tools to increase the efficiency and the growth of their cultures during production. However, for *P. oxalicum*, a sister species of *P. camemberti* and a useful biocontrol agent against *Fusarium oxysporum,* conidia could be produced in liquid media (Pascual et al. 1997), suggesting it is possible to produce conidia in large-scale industrial setting through liquid fermentation.

3.2 Cheese Contamination and Mycotoxins

The ripening or maturation stage of cheese is accompanied by growth of a diversity of yeasts, bacteria and filamentous fungi. Additional flora is considered to be a natural contaminant of the cheese making process. Their presence can affect proteolytic and lipolytic activities of *Penicillium* species; it is also likely to induce changes in overall product quality. For instance, yeast contamination on the surface of Camembert and blue cheeses can produce lipolytic and proteolytic enzymes and ferment residual lactose to lactic acid, leading to autolysis of yeast cells and undesirable flavours (Roostita and

Fleet 1996). With very diverse microflora, there is potential for complex ecological interactions to occur between species. It is very important that cheese manufactures are familiar with the diversity of microorganisms present on their products in order to optimize conditions to promote growth of desirable species, and to prevent the growth of undesirable ones (Addis et al. 2001, Beresford and Williams 2004).

Both the fermenting fungi and contaminants can produce mycotoxins in cheese. For example, *P. roqueforti* produces several mycotoxins, whereas *P. camemberti* strains only produce one. However, their quantities are usually very low, insufficient to cause health risk to humans (Chamba and Irlinger 2004). By comparison, mycotoxins produced by the more notorious species such as *A. flavus* cause as much as 25% of the world's agricultural yields to be inedible (Kabak et al. 2006, Sweeney and Dobson 1998). Among the *Penicillium* species capable of producing large quantities of mycotoxins, *P. crustosum* can be isolated from many crops and feeds. Both *P. crustosum* and *P. roqueforti* produce mycotoxins roquefortine C and penitrem A (Kokkonen et al. 2005) and if at high concentration, can cause a diversity of detrimental effects in humans and animals (Kabak et al. 2006). This fungus can be transmitted through the air (D'Mello and Macdonald 1997). Given these risks, Rundberget and Wilkins (2002) presented a method for the detection of fungal metabolites, indicative of mycotoxins, using liquid chromatographic–mass spectrometric (LC–MS) analysis (Kokkonen et al. 2005, Rundberget and Wilkins 2002).

4. Mucorales

The Mucorales include an evolutionary divergent group of fungi and are ecologically very diverse. They are traditionally considered basal fungi, grouped into the phylum Zygomycota. The defining characteristics of Zygomycota are a lack of complex sexual fruiting bodies (such as those observed in the Basidomycota and Ascomycota), the production of sporangiospores (asexual spores) and/or zygospores (sexual spores), as well as a lack of septa (discrete cellular divisions within hyphae). Recent molecular phylogenies have shown this group to be paraphyletic (White et al. 2006) and thus, new divisions and nomenclatures have been developed for the fungal phyla (Kwon-Chung 2012). However, many of these groups still have uncertain phylogenetic placements within the fungal kingdom, including the Mucormycotina. This group, and in particular the order Mucorales, contains some of the most well understood basal fungal species due to their prevalence as pathogens, causing infectious diseases in both plants and animals (including humans), as also their role in food spoilage. Concordantly, the majority of genome sequences for basal lineages also come from this group (Hoffmann et al. 2013). In addition to these nefarious attributes, many Mucorales are also the primary fermenting strains in a variety of foods produced throughout Asia. Among the Mucorales, the most prominent genera for fungal fermentation are *Rhizopus* and *Mucor*, which will be described below in more detail.

4.1 Mucor and Rhizopus

Formerly, both *Mucor* and *Rhizopus* were placed in the family Mucoraceae, which had no defining criteria, but rather was composed of genera that could not be placed in other well defined families. As such, the group is made up of very disparate lineages. A recent molecular phylogenetic analysis placed the genus *Rhizopus* into its own family, the Rhizopodaceae (Hoffmann et al. 2013). The genus is characterized by abundant sporangiospores, high temperature growth and the presence of small root like structures called rhizoids (Fig. 4). The genus can be delimited from other closely related genera based on the surface features of sporangiospores, the details of which were formerly used to define species as well (Seviour et al. 1983). These morphological characteristics had led to the naming of 10 species within the genus, but molecular analysis places this number at eight. The most well-defined species are *R. microsporus*, used in tempeh production; *R. stolonifer*, an important agent of food spoilage; and *R. arrhizus*, the most common cause of mucormycosis. There is currently much debate about whether *R. arrhizus* should be divided into two species, *R. oryzae* and *R. delemar*. For the purposes of this chapter, we will refer to these by their current official designation, *R. arrhizus*, but many references use the two other names (Abe et al. 2010).

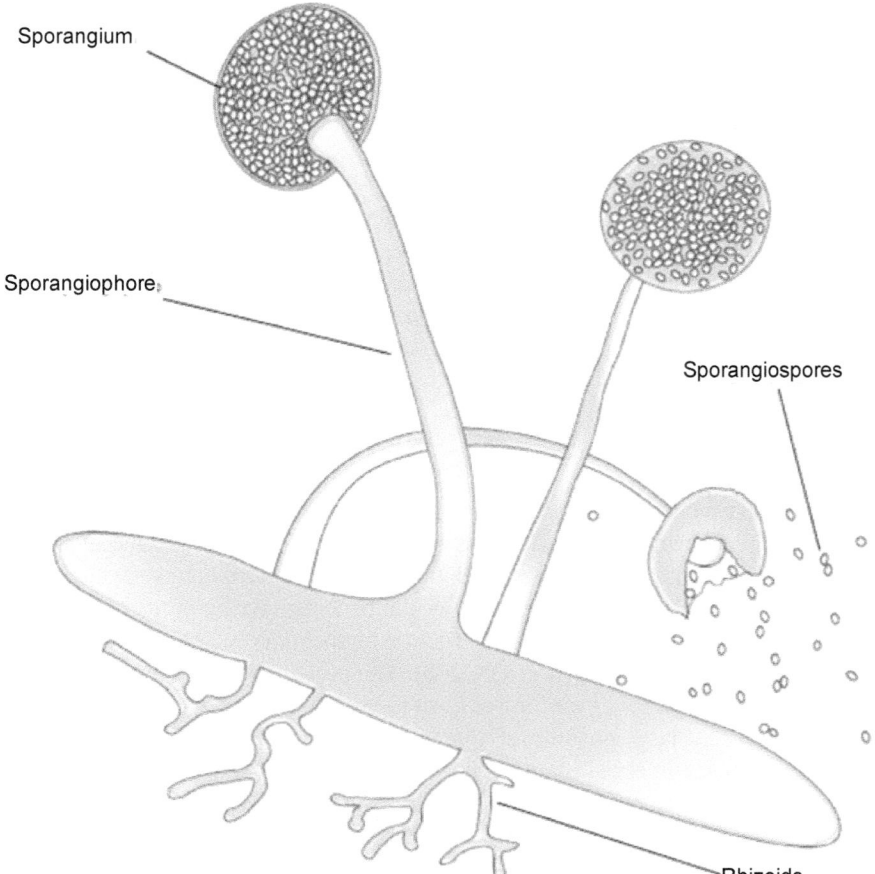

Sporangium

Sporangiophore

Sporangiospores

Rhizoids

Figure 4. Schematic diagram of *Rhizopus* sexual structures.

The genus *Mucor* shares many features with *Rhizopus,* such as being commonly associated with disease, though overall species in *Mucor* tend to grow at lower temperatures. Industrially and medically relevant *Mucor* species include: *M. racemosus* (*M. sufu*), used for the commercial production of sufu (described in detail below); *M. circinelloides*, a common human pathogen; and *M. hiemalis* and *M. indicus*, both used for the industrial production of enzymes. The current taxonomy of the group is not completely resolved, but there are some well supported clades. Of particular note is that *M. circinelloides* appears to encompass a larger species complex including several divergent lineages (Hermet et al. 2012, Li et al. 2011). The genus *Actinomucor* also falls within the Mucoraceae, differing from *Mucor* by its possession of rhizoids and the branching pattern of its sporangiophores (the structures which produce sporangiospores) (Benjamin and Hesseltine 1957). *Actinomucor* is composed of two separate species, *A. taiwanensis* and *A. elegans*, both of which have prominent roles in commercial sufu production, similar to *M. racemosus* (Han et al. 2004).

The relatively large phylogenetic distance from most model fungi (which are in Ascomycota and Basidiomycota) and the dearth of genomic information on non-dikaryotic fungi in general, limits the number of genetic tools that can be used to study and/or modify strains within the Mucorales. Despite these challenges, progress has been made on conducting genetic manipulation in this group, spurred on by the interest in these species for the industrial production of enzymes. Unfortunately, little to none of this technology has been adapted by the food industry and as a result, much of the biotechnological innovation for food production has been focused on strain selection and optimization.

Most foods fermented by Mucorales are either still transitioning from traditional production to commercial/industrial production, or have done so relatively recently. As a result, the technological advances for these foods have been limited to the development and characterization of appropriate starter cultures. Below, we will highlight some foods which have developed starter cultures using species from the Mucorales.

4.2 Tempeh

By far the most well-known product fermented by a member of the Mucorales is tempeh, an Indonesian food similar to tofu, which has recently become available throughout supermarkets around the world. It is produced by fermenting soybeans and has culinary uses akin to tofu. A variety of health benefits have been attributed to the consumption of fermented soy products, such as tempeh, likely due to the increased presence of antioxidants (Hubert et al. 2008). The most common fungi for tempeh production is *R. microsporus* var. *oligosporus* (syn. *R. oligosporus*) in part due to its ability to produce a large quantity of spores under controlled conditions, which is necessary for the commercial production of starter cultures (Thanh and Nout 2002). *R. microsporus* is able to produce relatively abundant extracellular enzymes such as amylases, lipases, and proteinases which degrade starch, fats, and proteins respectively. These enzymes have a large important commercial value and, as such, many studies have focused on optimizing their production and secretion under controlled laboratory and/or industrial conditions (Suharto et al. 2011). However, there have been little to no studies on strain identity or enzyme production to the quality of tempeh. Of the handful of studies that have strived to use biotechnological techniques to improve the value or production of tempeh, many have focused on modifying the soybeans themselves.

One of the studies, which did attempt to modify the fungal strain, strived to maximize the already high level of antioxidants in the end product. By screening 72 different isolates of *R. microsporus* they were able to isolate a strain which had a relatively higher free radical scavenging ability due to the activity of β-glucosidase (Sheih et al. 2000). Another study screened 8 strains of *R. microsporus* and *R. arrhizus* in conjunction with anaerobic fermentation and found that the amount of γ-aminobutyric acid (GABA, which has been used as an anti-anxiety drug) in tempeh could be increased over 10-fold (Aoki et al. 2003a). A follow up study found that a diet including GABA-tempeh could ameliorate hypertension in a rat model (Aoki et al. 2003b).

The beneficial factors of *Rhizopus* fermentation include making the substrate easier to digest and increasing the antioxidant levels; as such, it is possible to create fermented foods from other non soybean substrates which also show such benefits. This has become the main area of focus for biotechnological innovation of *Rhizopus* fermentation. Table 2 lists 22 other substrates which can also be used to make a suitable tempeh-like food. Of these, only two are traditionally made, the rest have been developed recently in laboratories throughout the world. The main goal behind the production of these foods is to increase the utilization of common crops grown in economically poor regions, and/or to provide a suitable protein source for areas that lack the resources to raise livestock. Additional tempehs have been made which combine soybeans with other substrates such as sunflower seeds or sesame seeds to augment their nutritional value. Despite the fact that some of these types of tempeh were developed almost 50 years ago, it does not appear as though any have seen widespread acceptance.

4.3 Sufu

Sufu is a traditional Chinese food that is also made from fermented soybeans. However, unlike tempeh, the soybeans are processed into tofu prior to the fermentation process. Various fermentation methods can be used in order to create different varieties of sufu (Han et al. 2001), but here we will focus on the methods which include mold fermentation. Traditionally, sufu was inoculated by incubating tofu with straw. It was not until the 20th century that a starter culture was isolated (Hesseltine 1965). This first isolate was identified as *Mucor* spp., but since then, *Rhizopus* and *Actinomucor* species have also been isolated from sufu (Hesseltine 1965). Multiple commercial strains have now been developed, including *A. taiwanensis*, *A. elegans*, *M. hiemalis*, *M. racemosus* (*sufu*) and *M. circinelloides* (*wutungkiao*). The taxonomy of these genera has not been well resolved, but DNA barcoding analysis suggests that *A. taiwanensis* and *A. elegans* are not separate species, nor are *M. heimalis* and *M. racemosus* (Han et al. 2004). Furthermore, a systematic study of the family Mucorales revealed that while *Rhizopus* species grouped together well as a taxonomic unit, the relationship among *Mucor* species and between *Mucor* and *Actinomucor* were not well defined (Hoffmann et al. 2013).

Table 2. List of non-soybean, tempeh-like foods fermented with *Rhizopus* spp.

Substrate	Strain	References
Barley (*Hordeum vulgare*)	*R. oligosporus* ATCC64063	Feng et al. (2005)
Pea (*Pisum sativum*)	*R. oligosporus* 2710	Nowak and Szebiotko (1992)
Chickpea (*Cicer arietinum* L.)	*R. stolonifer*	Reyes-Moreno et al. (2000)
Red Sorghum (*Sorghum bicolor*)	*R. oligosporus* NRRL2549	Hachmeister and Fung (1993)
String Bean (*Phaseolus vulgaris*)	*R. oligosporus*	Paredes-López and Harry (1989)
Fava Bean (*Vicia faba*)	*R. oligosporus* CBS338.62	Ashenafi and Busse (1991)
African Yam Beans (*Sphenostylis stenocarpa*)	*R. microsporus* var. *oligosporus* D.S.M.	Njoku et al. (1991)
Lupine Beans (*Lupinus albus* cv. *Multolupa*)	*R. oligosporus* NRRL2710	Agosin et al. (1989)
Maize (*Zea mays* L.)	*R. oligosporus*	Cuevas-Rodríguez et al. (2004)
Grass Pea (*Lathyrus sativus* L.)	*R. oligosporus* DSMZ1964	Stodolak and Starzyńska-Janiszewska (2008)
Ground Nut (*Voandzeia subterranean*)	*R. oligosporus* NRRL2549	Bhavanishankar et al. (1987)
Oat (*Avena sativa*)	*R. oligosporus* NRRL2710	Nowak (1992)
Cowpea (*Vigna unguiculata*)	*R. oligosporus*	Egounlety and Aworh (2003)
Ground Bean (*Macrotyloma geocarpa Harms*)	*R. oligosporus*	Egounlety and Aworh (2003)
Mung Bean (*Vigna radiata*)	*R. oligosporus*	Randhir and Shetty (2007)
Peanut (*Archis hypogaea*)	*R. oligosporus* NRRL2710	Matsuo (2006)
Hyancinth Bean (*Dolichos lablab*)	*R. oligosporus* MTCC556	Dhaduk et al. (2008)
Quinoa (*Chenopodium quinoa*)	*R. oligosporus* NRRL2710	Peñaloza et al. (1992)
Velvet bean (*Mucuna pruriens* var. *utilis*)	*R. oligosporus*	Egounlety and Aworh (2003)
Winged Bean (*Psophocarpus tetragonolobus*)	Traditional Fermentation	Ekpenyong and Borchers (1980)
Wheat (*Triticum* spp.)	*R. oryzae*	Wang and Hesseltine (1966)
Breadfruit Seed (*Terculia africana*)	*R. oligosporus* NRRL2710	Maduka and Njoku (2012)

A major drawback of these commercial strains is their inability to properly ferment tofu above 30°C (Yin et al. 2005) and as temperatures in South China can reach much higher than this during summer months, it limits the production of sufu. To remedy this, researchers have attempted to use *R. microsporus* var. *oligosporus* instead of *Actinomucor* species to produce sufu. As *Rhizopus* species can grow at temperatures upwards of 50°C and well defined *Rhizopus* starter cultures are readily available due to their use in tempeh, this seems like a good choice. The results found that *R. microsporus* var. *oligosporus* could ferment sufu at as high as 40°C and enzymatic profiles of the sufu were similar to that produced by *A. elegans* at 26°C. However the researchers did not assess the quality of the sufu produced, so it is still uncertain if this methodology could produce marketable sufu (Han et al. 2003).

Additional research has revealed that the commercial sufu strains are also poor at fermenting tofu when below 20°C, which is common during the winter in China (Cheng et al. 2009). Cheng et al. (2009) have attempted to remedy this issue by developing *M. flavus* as a suitable strain for sufu production. Their results show that *M. flavus* is able to grow to sufficiently high biomass between 10°C and 15°C. Furthermore, this strain can produce sufficient quantities of protease at these temperatures for the degradation of soybean proteins and β-glucosidases for the degradation of soy isoflavones, both of which are necessary for the generation of quality sufu (Cheng et al. 2009, 2011). Unfortunately, these studies suffer from the same drawback as the high temperature fermentation studies, in that no assessment of sufu quality was evaluated.

Sufu shares certain aspects with the European soft cheese Camembert (Li and Wang 2012). As Camembert is produced at very large industrial scales, it may be possible to use these same processes to produce sufu on the condition that the *Actinomucor* or *Mucor* strains used to produce sufu can be replaced with the fungal strain used to produce Camembert, *Penicillium camemberti*. With this aim in mind, Wan et al. (2011) produced sufu using both *P. camemberti* and *A. elegans*. They determined that both strains could ferment tofu under similar conditions which suggests that, indeed, a *Penicillium* species could be substituted for a traditional strain (Wan et al. 2011). However, they did not evaluate the sufu in any qualitative manner to determine if it would possess similar flavour and/or texture to traditionally produced sufu.

4.4 Amylolytic Starter Cultures

Amylolytic starter cultures (ASCs) are used throughout Asia for the production of alcoholic beverages and food. They go by many different names depending on the region of origin, such as jiuqu in China, ragi in Indonesia and murcha in India, but are all traditionally obtained from the natural fermentation of rice flour and contain a cosmopolitan mixture of bacteria, yeast, and mold (Hesseltine et al. 1988). The most notable foods made from this type of fermentation are gundruk, from Nepal and India, tapé and brem from Indonesia, and ruou nep from Vietnam. With the exception of rice wine production, most foods made from ASCs are still produced on the household scale (though in the case of Nepalese gundruk, this amounts to an impressive 2000 tons annually) (Battcock and Azam-Ali 1998). As a result, there is a lack of research into the exact fermentation process. Previous studies have largely focused on identifying the organisms present in ASCs. Most of the molds that have been identified belong to the Mucorales, including species of the genera, *Mucor, Rhizopus, Absidia* and, most prominently, *Amylomyces* (Hesseltine et al. 1988, Lee and Fujio 1999, Shrestha et al. 2002, Zheng et al. 2011).

The genus *Amylomyces* has a complex and convoluted history and taxonomy, so it is worth taking a moment to attempt to clarify the conflicting, and often times misleading, literature on this organism. *Amylomyces* was originally defined as a monotypic genus (*A. rouxii* being the only described species) by Clemette in 1892 and characterized further by Ellis et al. in 1976 (Ellis et al. 1976). It was stated that *Amylomyces* shared many characteristics with *Rhizopus*, but differed in its prevalence and the size of chlamydospores as well as in its ability to metabolize sucrose, maltose and glycerol. More recent analysis has revealed that sucrose metabolism is, in fact, quite varied across *R. arrhizus* strains, and is, quite likely, linked to enzyme stability (Watanabe and Oda 2008). With the advent of molecular techniques, it was determined that the type strain for *A. rouxii* was highly similar to multiple *R. arrhizus* strains (Abe et al. 2006) and a broader study which included additional strains found that they all clustered within *R. arrhizus* (Kito et al. 2009). These results led to the official classification of *A. rouxii* as a synonym for *R. arrhizus*. However, a variety of other studies have shown that *A. rouxii* (syn. *Mucor rouxii*) clusters with *Mucor* species, primarily *M. circinelloides* (Mamatha et al. 2010, Schwarz et al. 2006, Yang et al. 2011), rather than *Rhizopus*. A brief survey of barcode sequences of *A. rouxii* deposited in GenBank shows that indeed, there are two separate clusters of *A. rouxii*: one that groups with *R. arrhizus* and another with *M. circinelloides* (Fig. 5). Interestingly, despite the confusion surrounding the identity of *A. rouxii*, there appears to be little to no misidentification between *M. circinelloides* and *R. arrhizus*. Furthermore, the fact that the *Mucor* grouping of *A. rouxii* appears to cluster together (although with poor support), suggests that it may, in fact, represent a true taxonomic group, though any concrete conclusions will require a more thorough investigation into the genetics of *A. rouxii* and *M. circinelloides*. To add to the confusion surrounding *A. rouxii* there have also been a number of articles which have misstated the proper synonyms of various species of *Mucor*. These have included claims that *A. rouxii/ M. rouxii* are synonymous with *M. indicus* or that *M. circinelloides* is synonymous with *M. racemosus*, both of which are incorrect as exemplified in Fig. 5.

One of the hurdles to developing strains for ASC fermentation is the complex interaction between the multiple organisms present in starters. Ardhana et al. found that pure cultures of either *A. rouxii* or one of three yeast species could ferment glutinous (sticky) rice into the Indonesian dish 'ragi ketan' when they were the sole strain present. However, when any of these or even a mixture of all four was used as inoculums, the ragi ketan was still of poorer quality than that produced from a natural ragi starter. This suggests that molds such as *A. rouxii*, yeasts, and LAB are all likely necessary for proper fermentation of ragi ketan (Ardhana and Fleet 1989). However, contradictory results were obtained by Siebenhandl et al. who determined that suitable ragi ketan could be obtained from fermentation with *A. rouxii* alone (Siebenhandl 2001) and similar studies on both, ragi ketan and brem (an Indonesian cake produced in a similar manner to ragi ketan), found that a co-culture of *A. rouxii* and the yeast *Hyphopichia burtonii* could produce a product of acceptable quality (Bintari 2014, Cook et al. 1991). The Siebenhandl study used the type strain of *A. rouxii* (CSB 438.76), which clusters with *R. arrhizus*, but not all of the studies reported the strain they used, making it difficult to draw inferences between them. This highlights the importance of strain identification and selection in fermentation studies.

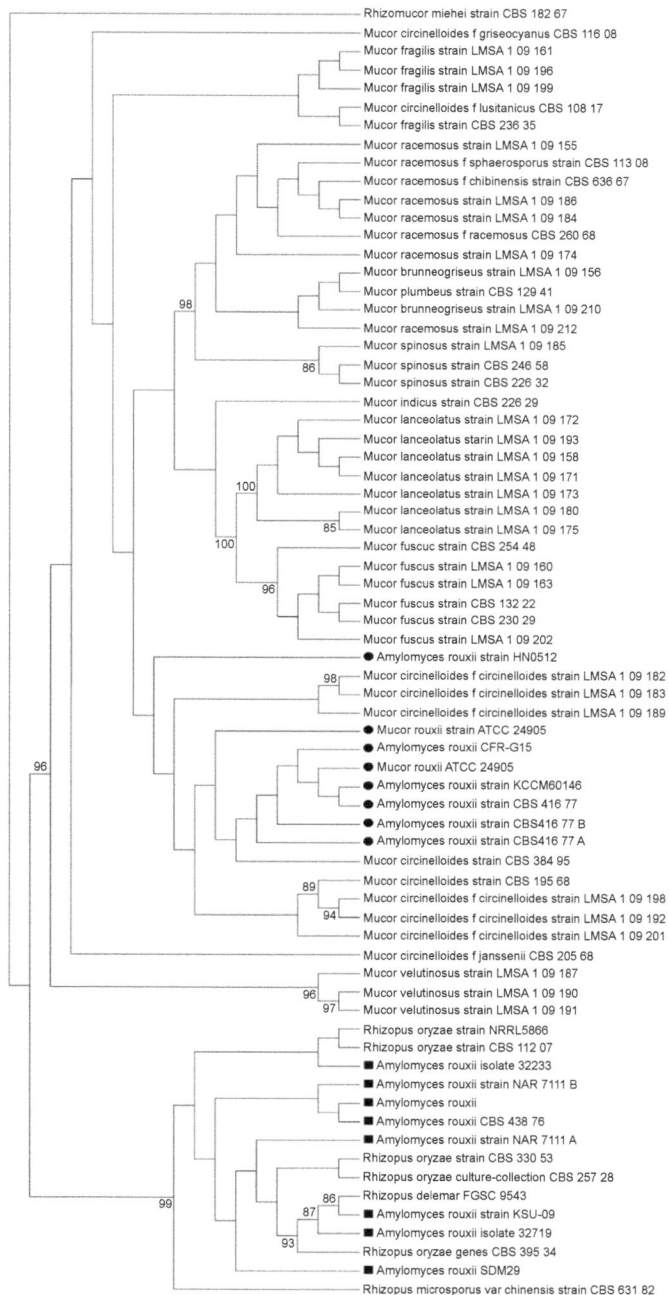

Figure 5. Maximum—Likelihood tree of barcode sequences from representative strains of *Rizopus* and *Mucor*, showing two clades of *Amylomyces rouxii*, a *Rhizopus clade* (■) and a *Mucor clade* (●). *Rhizopus oryzae* (*R. arrhizus*) sequences were reported by Abe et al. 2010, *Mucor* species reported by Hermet et al. (2012), *M. indicus* and *R. microsporus* sequences are from type specimens; *A. rouxii* sequences are from all unique sequences reported in GenBank. *Rhizomucor miehei* was set as the out group. Values at nodes show bootstrapping support, values below 80 are not shown.

There also seems to be a large variance in the quality of products produced from commercially available, but naturally produced ASCs. A study on the fermentation of the Nepalese dish poko (similar to ragi ketan) found that out of 11 different commercial starters, two produced higher quality poko than the rest. They further examined the succession of microorganisms during the fermentation process of the successful starters. They found that yeast and LAB counts increased over fermentation time, but that the concentration of mold (identified as *Rhizopus* spp.) decreased (Shrestha et al. 2002). Interestingly, a study of Nepalese gundruk (made from fermenting vegetables) found that LAB concentration stayed constant during fermentation while yeast concentration fell. They did not report any significant change in the growth of molds (Tamang and Tamang 2010). Together, these studies cast doubt on the importance of molds in the fermentation of Nepalese foods.

Currently, many commercially available starter cultures are labelled as tempeh/tapé starters. These cultures likely contain strains of *R. arrhizus*, which is commonly isolated from both tempeh and tapé. However, as described above, more efficient strains have been defined for tempeh and the tapé produced by a single pure culture is of sub-standard value. Thus, using these pure starter cultures would likely result in poorer quality product of either food. Given the wide breadth of foods fermented with ASCs, it seems likely that starter cultures that utilize pure strains will have to be developed on a substrate-by-substrate basis, i.e., different cultures will be needed to ferment sticky rice, wheat or bamboo. Additional research is required to better understand the complex interactions that occur between molds, yeasts and bacteria to move these foods from traditional, at home production, to larger scale commercial production.

While the use of Mucorales species is concentrated most intensely in Asia, they are also used to some extent in Europe. The Norwegian cheese Gamalost is ripened by the species *Mucor mucedo* in a fashion similar to how *P. camemberti* ripens Camembert (Qureshi et al. 2012). Additional efforts are being made to determine if members of the Mucorales can be used as suitable strains for the production of other European foods. Zhang and Zhao (2010) concluded that a strain of *Mucor* isolated from fermented soybeans could be used to ripen cheese and Bruna et al. (2000) found that sausages ripened with *M. racemosus* had increased sensory properties.

Overall, despite the fact that industry has intensively researched these strains for the purpose of commercial pure enzyme production, biotechnological research among the Mucorales is still in its infancy. There is still a lot of potential in the optimization of fermenting strains, without getting into genetic modification. However, many of the genera discussed in this section are amenable to genetic manipulation. The production of optimized strains could not only help the scale-up production of traditional foods to commercial production (increasing their availability in relevant regions or expanding their market globally, as has occurred with tempeh), but could also lead to the production of foods with better nutritional value or decreased fermentation time, ultimately improving the quality of life for people throughout the world.

5. Conclusions and Perspectives

In this chapter, we reviewed the major groups of filamentous fungi involved in the production of fermented foods. We described the characteristics of these fungi, including their classifications, geographic and ecological distributions, and their roles in food fermentation. The foods affected cover a broad spectrum of items, from staple crops such as rice, cassava, soybeans, to milk and fish. Though filamentous fungi are broadly distributed and have been used by people, intentionally or unintentionally, for thousands of years, our basic understanding of their genetics and physiology during food fermentation are still limited. Furthermore, there is a common theme throughout this chapter in the lack of advanced biotechnology using filamentous fungi in fermented foods. Most uses of filamentous fungi are still based on traditional approaches, e.g., foods are left outside or buried, where they eventually will be fermented. For some of the organisms such as those in Mucorales, there are many unsettled issues with regard to the taxonomy and systematics of the fermenting organisms. Furthermore, though an increasing number of fermenting fungi have been sequenced, the use of that genetic information is extremely limited so far for most species. However, the metabolic diversities revealed through genomic data mining coupled with experimental investigations of the specific candidate genes involved in fermentation should significantly enhance our ability to produce high-quality strains and starter cultures so that high-quality fermented foods could be produced more efficiently and consistently using filamentous fungi.

Acknowledgement

Research in our lab on fungi is supported by the Natural Sciences and Engineering Research Council of Canada.

Keywords: *Aspergillus*, *Aspergillus flavus*, *Aspergillus niger*, Cassava, *Manihot esculenta*, Gari, Fufu, Kum-Kum, Bobolo, Meyondo, Khamir, *Aspergillus repens*, Katsuobushi, *Aspergillus oryzae*, *A.sojae*, koji, Jiugu, Meju, *Penicillium*, *Penicillium camemberti*, Mold-Ripened Cheeses, *Penicillium roqueforti*, Blue-Veined Cheeses, Mycotoxins, *Mucor*, *Rhizopus*, Tempeh, Sufu, Amylolytic Starter Cultures

References

Abe, A., Oda, Y., Asano, K. and Sone, T. 2006. The molecular phylogeny of the genus *Rhizopus* based on rDNA sequences. Bioscience, Biotechnology, and Biochemistry 70: 2387–2393.

Abe, A., Asano, K. and Sone, T. 2010. A molecular phylogeny-based taxonomy of the genus *Rhizopus*. Bioscience, Biotechnology, and Biochemistry 74: 1325–1331.

Addis, E., Fleet, G.H., Cox, J.M., Kolak, D. and Leung, T. 2001. The growth, properties and interactions of yeasts and bacteria associated with the maturation of Camembert and blue-veined cheeses. International Journal of Food Microbiology 69: 25–36.

Agosin, E., Diaz, D., Aravena, R. and Yañez, E. 1989. Chemical and nutritional characterization of Lupine Tempeh. Journal of Food Science 54: 102–104.

Agrios, G.N. 2005. Plant Pathology. Academic Press, New York, New York.

Akindahuns, A.A., Oboh, G. and Oshodi, A.A. 1999. Effect of fermenting cassava with *Rhizopus oryzae* on the chemical composition of its flour and gari products. Rivista Italiana Delle Sostanze Grasse 76: 437–439.

Akindahunsi, A.A. and Oboh, G. 2003. Effect of fungi fermentation on organoleptic properties, energy content, and *in vitro* multienzyme digestibility of cassava products (flour & gari). Nutritional Health 17: 131–138.

Aoki, H., Uda, I., Tagami, K., Furuya, Y., Endo, Y. and Fujimoto, K. 2003a. The Production of a new tempeh-like fermented soybean containing a high level of γ-amino butyric acid by anaerobic incubation with *Rhizopus*. Bioscience, Biotechnology, and Biochemistry 67: 1018–1023.

Aoki, H., Furuya, Y., Endo, Y. and Fujimoto, K. 2003b. Effect of γ-amino butyric acid-enriched tempeh-like fermented soybean (GABA-tempeh) on the blood pressure of spontaneously hypertensive rats. Bioscience Biotechnology and Biochemistry 67: 1806–1808.

Ardhana, M.M. and Fleet, G.H. 1989. The microbial ecology of tapé ketan fermentation. International Journal of Food Microbiology 9: 157–165.

Ashenafi, M. and Busse, M. 1991. Growth of *Bacillus cereus* in fermenting tempeh made from various beans and its inhibition by *Lactobacillus plantarum*. Journal of Applied Bacteriology 70: 329–333.

Battcock, M. and Azam-Ali, S. 1998. Fermented Fruits and Vegetables: A Global Perspective (Food & Agriculture Organization).

Benjamin, C.R. and Hesseltine, C.W. 1957. The Genus *Actinomucor*. Mycologia 49: 240–249.

Bennett, J. 2010. An overview of the Genus *Aspergillus*. pp. 1–17. *In*: M. Machida and K. Gomi (eds.). *Aspergillus*: Molecular Biology and Genomics, Horizon Scientific Press, Norfolk.

Beresford, T. and Williams, A. 2004. The microbiology of cheese ripening. pp. 287–317. *In*: P.F. Fox, P.L.H. McSweeney, T.M. Cogan and T.P. Guinee (eds.). Cheese: Chemistry, Physics and Microbiology: Major Cheese Groups. Elsevier Academic Press.

Bhavanishankar, T.N., Rajashekaran, T. and Sreenivasa Murthy, V. 1987. Tempeh-like product by groundnut fermentation. Food Microbiology 4: 121–125.

Bintari, S.H. 2014. Producing solid brem using combined *Mucor rouxianus* and *Hyphopichia burtonii* inoculum. Journal of Biosciences 3: 1–6.

Boualem, K., Waché, Y., Garmyn, D., Karbowiak, T., Durand, A., Gervais, P. and Cavin, J.-F. 2008. Cloning and expression of genes involved in conidiation and surface properties of *Penicillium camemberti* grown in liquid and solid cultures. Research in Microbiology 159: 110–117.

Bruna, J.m., Fernández, M., Hierro, E.m., Ordóntez, J.a. and de la Hoz, L. 2000. Improvement of the sensory properties of dry fermented sausages by the superficial inoculation and/or the addition of intracellular extracts of *Mucor racemosus*. Journal of Food Science 65: 731–738.

Cantor, M.D., Van Den Tempel, T., Hansen, T.K. and Ardö, Y. 2004. Blue cheese. pp. 175–198. *In*: P.F. Fox, P.L.H. McSweeney, T.M. Cogan and T.P. Guinee (eds.). Cheese: Chemistry, Physics and Microbiology: Major Cheese Groups. Elsevier Academic Press.

Cerning, J., Gripon, J.C., Lamberet, G. and Lenoir, J. 1987. Les activités biochimiques des *Penicillium* utilisés en fromagerie. Le Lait 67: 3–39.

Chamba, J.F. and Irlinger, F. 2004. Secondary and adjunct cultures. pp. 191–206. *In*: P.F. Fox, P.L.H. McSweeney, T.M. Cogan and T.P. Guinee (eds.). Cheese: Chemistry, Physics and Microbiology: Major Cheese Groups. Elsevier Academic Press.

Cheng, Y.-Q., Hu, Q., Li, L.-T., Saito, M. and Yin, L.-J. 2009. Production of Sufu, a traditional Chinese fermented soybean food, by fermentation with *Mucor flavus* at low temperature. Food Science and Technology Research 15: 347–352.

Cheng, Y.-Q., Zhu, Y.-P., Hu, Q., Li, L.-T., Saito, M., Zhang, S.-X. and Yin, L.-J. 2011. Transformation of isoflavones during Sufu (a traditional Chinese fermented soybean curd) production by fermentation with *Mucor flavus* at low temperature. International Journal of Food Properties 14: 629–639.

Chieko, F. 2008. Koji, an *Aspergillus* (The Tokyo Foundation).

Chikezie, P.C. and Ojiako, O.A. 2013. Cyanide and aflatoxin loads of processed Cassava (*Manihot esculenta*) tubers (Garri) in Njaba, Imo State, Nigeria. Toxicology International 20: 261–267.

Ciegler, A. and Vesonder, F. 1987. CRC Handbook of Microbiology. Boca Raton: CRC Press.

Collier, L., Balows, A. and Sussman, M. 1998. Topley and Wilson's Microbiology and Microbial Infections. Hodder Education Publishers: London.

Cook, P.E., Owens, J.D. and Campbell-Platt, G. 1991. Fungal growth during rice tapé fermentation. Letters in Applied Microbiology 13: 123–125.

Cuevas-Rodríguez, E.O., Milán-Carrillo, J., Mora-Escobedo, R., Cárdenas-Valenzuela, O.G. and Reyes-Moreno, C. 2004. Quality protein maize (*Zea mays* L.) tempeh flour through solid state fermentation process. LWT—Food Science Technology 37: 59–67.

Dhaduk, J.J., Patel, B.G., Patel, I.N. and Rema, S. 2008. Evaluation of nutritional quality of various tempeh preparations with *Rhizopus oligosporus* in rats. Journal of Cell Tissue Research 8: 1387–1392.

Dimici, L. and Wada, S. 1994. Lipid changes in bonito meat in the Katsuobushi processing and quality assessment of the commercial product based on lipid composition. Yukagaku 43: 470–477.

D'Mello, J.P.F. and Macdonald, A.M.C. 1997. Mycotoxins. Animal Feed Science and Technology 69: 155–166.

Dörge, T., Carstensen, J.M. and Frisvad, J.C. 2000. Direct identification of pure *Penicillium* species using image analysis. Journal of Microbiological Methods 41: 121–133.

Egounlety, M. and Aworh, O.C. 2003. Effect of soaking, de-hulling, cooking and fermentation with *Rhizopus oligosporus* on the oligosaccharides, trypsin inhibitor, phytic acid and tannins of soybean (*Glycine max* Merr.), cowpea (*Vigna unguiculata* L. Walp) and groundbean (*Macrotyloma geocarpa* Harms). Journal of Food Engineering 56: 249–254.

Ekpenyong, T.E. and Borchers, R.L. 1980. The fatty acid composition of the oil of the winged bean (*Psophocarpus tetragonolobus* L.) seeds. Journal of the American Oil Chemists' Society 57(5): 147–149.

Ellis, J.J., Rhodes, L.J. and Hesseltine, C.W. 1976. The Genus *Amylomyces*. Mycologia 68: 131–143.

Feng, X.M., Eriksson, A.R.B. and Schnürer, J. 2005. Growth of lactic acid bacteria and *Rhizopus oligosporus* during barley tempeh fermentation. International Journal of Food Microbiology 104: 249–256.

Food and Agriculture Organization of the United Nations 1995. Sorghum and millets in human nutrition. FAO Food and Nutrition Series: Rome.

Food and Agriculture Organization of the United Nations 2004. Worldwide regulations for mycotoxins in food and feed in 2003. FAO Food and Nutrition Series: Rome.

Food and Agriculture Organization of the United Nations 2013. FAO Statistical Yearbook 2013. FAO Food and Nutrition Series: Rome.

Gassem, M.A. 1999. Study of the microorganisms associated with the fermented bread (khamir) produced from sorghum in Gizan region, Saudi Arabia. Journal of Applied Microbiology 86: 221–225.

Geiser, D.M., Samson, R.A., Varga, J., Rokas, A. and Witiak, S.M. 2008. A review of molecular phylogenetics in *Aspergillus*, and prospects for robust genus-wide phylogeny. p. 17. *In*: J. Varga and R.A. Samson (eds.). *Aspergillus* in the Genomic Era, Wageningen Academic Pub, Wageningen.

Gobbetti, M., Burzigotti, R., Smacchi, E., Corsetti, A. and De Angelis, M. 1997. Microbiology and biochemistry of gorgonzola cheese during ripening. International Dairy Journal 7: 519–529.

Godinho, M. and Fox, P.F. 1981. Effect of NaCl on the germination and growth of *Penicillium roqueforti*. Milchwissenschaft 4: 205–208.

Griffin, D.H. 1996. Fungal Physiology. Wiley-Liss, New York.

Haard, N.H., Odunfa, S.A., Lee, C.-H. and Quintero-Ramirez, R. 1999. Fermented cereals: A Global Perspective. Food and Agriculture Organization: Rome.

Hachmeister, K.A. and Fung, D.Y.C. 1993. Tempeh: a mold-modified indigenous fermented food made from soybeans and/or cereal grains. Critical Reviews of Microbiology 19: 137–188.

Hamilton, B.F. and Gould, D.H. 1987. Correlation of morphologic brain lesions with physiologic alterations and blood-brain barrier impairment in 3-nitropropionic acid toxicity in rats. Acta Neuropathologica 74: 67–74.

Han, B.-Z., Rombouts, F.M. and Nout, M.J.R. 2001. A Chinese fermented soybean food. International Journal of Food Microbiology 65: 1–10.

Han, B.-Z., Ma, Y., Rombouts, F.M. and Nout, M.J.R. 2003. Effects of temperature and relative humidity on growth and enzyme production by *Actinomucor elegans* and *Rhizopus oligosporus* during sufu pehtze preparation. Food Chemistry 81: 27–34.

Han, B.-Z., Kuijpers, A.F.A., Thanh, N.V. and Nout, M.J.R. 2004. Mucoraceous moulds involved in the commercial fermentation of Sufu Pehtze. Antonie Van Leeuwenhoek 85: 253–257.

Hermet, A., Méheust, D., Mounier, J., Barbier, G. and Jany, J.-L. 2012. Molecular systematics in the genus *Mucor* with special regards to species encountered in cheese. Fungal Biology 116: 692–705.

Hesseltine, C.W. 1965. A Millennium of Fungi, Food, and Fermentation. Mycologia 57: 149–197.

Hesseltine, C.W., Rogers, R. and Winarno, F.G. 1988. Microbiological studies on amylolytic oriental fermentation starters. Mycopathologia 101: 141–155.

Hewedi, M.M. and Fox, P.F. 1984. Ripening of blue cheese: Characterization of proteolysis. Milchwissenschaft 39: 198–201.

Hoffmann, K., Pawłowska, J., Walther, G., Wrzosek, M., de Hoog, G.S., Benny, G.L., Kirk, P.M. and Voigt, K. 2013. The family structure of the Mucorales: A synoptic revision based on comprehensive multigene-genealogies. Persoonia Molecular Phylogeny and Evolution of Fungi 30: 57–76.

Huang, H.T. 2000. Science and Civilisation in China: Volume 6, Biology and Biological Technology, Part 5, Fermentations and Food Science. Cambridge University Press.

Hubert, J., Berger, M., Nepveu, F., Paul, F. and Daydé, J. 2008. Effects of fermentation on the phytochemical composition and antioxidant properties of soy germ. Food Chemistry 109: 709–721.

Hudler, G.W. 2000. Magical Mushrooms, Mischievous Molds. Princeton University Press.

Jakobsen, M., Cantor, M.D. and Jespersen, L. 2010. Production of bread, cheese and meat. pp. 3–22. *In*: K. Esser, A.A. Brakhage and P.F. Zipfel (eds.). The Mycota: A Comprehensive Treatise on Fungi as Experimental Systems for Basic and Applied Research. Springer, Berlin.

Jonathan, S., Abdul-Lateef, M. and Ayansina, A. 2013. Fungal and aflatoxin detection in fresh and stored 'garri ijebu' (locally processed food). Report and Opinion 5: 13–19.

Kabak, B., Dobson, A.D.W. and Var, I. 2006. Strategies to prevent mycotoxin contamination of food and animal feed: A Review. Critical Reviews in Food Science and Nutrition 46: 593–619.

Kaminishi, Y., Egusa, J. and Kunimoto, M. 1999. Antioxidant production from several xerophilous fungi used in katsuobushi molding. Journal of the Shimonoseki University of Fisheries (Japan) 47: 113–120.

Kato, F., Nakazato, I., Murata, A., Okamoto, S. and Yone, Y. 1986. Use of waste fish for large scale production of fermented fish meal and its feed efficiency. Nippon Nōgeikagaku Kaishi 60: 287–293.

Kelavkar, U.P. and Chhatpar, D.H.S. 1993. Relations of enzymes in *Aspergillus repens* grown under sodium chloride stress. Current Microbiology 27: 157–162.

Kinosita, R., Ishiko, T., Sugiyama, S., Seto, T., Igarasi, S. and Goetz, I.E. 1968. Mycotoxins in fermented food. Cancer Research 28: 2296–2311.

Kito, H., Abe, A., Sujaya, I.N., Oda, Y., Asano, K. and Sone, T. 2009. Molecular characterization of the relationships among *Amylomyces rouxii*, *Rhizopus oryzae*, and *Rhizopus delemar*. Bioscience Biotechnology and Biochemistry 73: 861–864.

Kokkonen, M., Jestoi, M. and Rizzo, A. 2005. The effect of substrate on mycotoxin production of selected *Penicillium* strains. International Journal of Food Microbiology 99: 207–214.

Kwon, D.Y., Hong, S.M., Ahn, I.S., Kim, M.J., Yang, H.J. and Park, S. 2011. Isoflavonoids and peptides from meju, long-term fermented soybeans, increase insulin sensitivity and exert insulinotropic effects *in vitro*. Nutrition 27: 244–252.

Kwon-Chung, K.J. 2012. Taxonomy of fungi causing mucormycosis and entomophthoramycosis (zygomycosis) and nomenclature of the disease: Molecular mycologic perspectives. Clinical Infectious Diseases 54: S8–S15.

Lawley, R., Curtis, L. and Davis, J. 2012. Chemical Hazards in Food. Food Safety Info, London, UK.

Le Bars, D. and Gripon, J.C. 1981. Role of *Penicillium roqueforti* proteinases during blue cheese ripening. Journal of Dairy Research 48(03): 479–487.

Leboffe, M.J. and Pierce, B.E. 2012. Microbiology: Laboratory Theory & Application. Morton Publishing Company.

Lee, A.C. and Fujio, Y. 1999. Microflora of banhmen, a fermentation starter from Vietnam. World Journal of Microbiology and Biotechnology 15: 51–55.

Leslie, J.F., Bandyopadhyay, R. and Visconti, A. 2008. Mycotoxins: Detection Methods, Management, Public Health and Agricultural Trade (CABI).

Li, L. and Wang, J. 2012. Comparative study of chemical composition and texture profile analysis between camembert cheese and Chinese Sufu. Frontiers in Bioengineering and Biotechnology 1: 1–8.

Li, C.H., Cervantes, M., Springer, D.J., Boekhout, T., Ruiz-Vazquez, R.M., Torres-Martinez, S.R., Heitman, J. and Lee, S.C. 2011. Sporangiospore size dimorphism is linked to virulence of *Mucor circinelloides*. PLoS Pathogens 7: e1002086.

Link, H.F. 1809. Observationes in ordines plantarum naturales: Dissertatio Ima complectens anandrarum ordines epiphytas, mucedines gastromycos et fungos.

Lund, F., Filtenborg, O. and Frisvad, J.C. 1995. Associated mycoflora of cheese. Food Microbiology 12: 173–180.

Machida, M., Yamada, O. and Gomi, K. 2008. Genomics of *Aspergillus oryzae*: Learning from the history of koji mold and exploration of its future. DNA Research 15: 173–183.

Maduka, N. and Njoku, H.O. 2012. The production of tempeh-like food product using African breadfruit seeds (*Terculia africana*). Academia Arena 4: 39–43.

Mahungu, N.M., Yamaguchi, Y., Almazan, A.M. and Hahn, S.K. 1987. Reduction of cyanide during processing of cassava to some traditional African foods. Journal of Agricultural and Food Chemistry 1: 11–15.

Mamatha, S.S., Halami, P.M. and Venkateswaran, G. 2010. Identification and characterization of the n-6 fatty acid-producing *Mucor rouxii* native isolate CFR-G15. European Journal of Lipid Science and Technology 112: 380–389.

Manabe, M., Tanaka, K., Goto, T. and Matsuura, S. 1984. Producing capacity of kojic acid and aflatoxin by koji mold. Developments in Food Science 7: 4–14.

Matsuo, M. 2006. Preparation and preferences of peanut-tempeh, peanuts fermented with *Rhizopus oligosporus*. Food Science and Technology Research 12: 270–274.

McGovern, P.E., Zhang, J., Tang, J., Zhang, Z., Hall, G.R., Moreau, R.A., Nuñez, A., Butrym, E.D., Richards, M.P. and Wang, C. 2004. Fermented beverages of pre- and proto-historic China. Proceedings of the National Academy of Sciences 101: 17593–17598.

McSweeney, P.L.H., Ottogalli, G. and Fox, P.F. 2004. Diversity of cheese varieties: an overview. pp. 1–22. *In:* P.F. Fox, P.L.H. McSweeney, T.M. Cogan and T.P. Guinee (eds.). Cheese: Chemistry, Physics and Microbiology: Major Cheese Groups. Elsevier Academic Press.

McSweeney, P.L.H. and Fox, P.F. 2004. Metabolism of residual lactose and of lactate and citrate. pp. 361–371. *In:* P.F. Fox, P.L.H. McSweeney, T.M. Cogan and T.P. Guinee (eds.). Cheese: Chemistry, Physics and Microbiology: General Aspects. Elsevier Academic Press.

Miller, J.D. 1994. Mycotoxins in grain. Revista do Instituto de Medicina Tropical de São Paulo 36: 326–326.

N'dede, C.B., Jolly, C.M., Vodouhe, S.D. and Jolly, P.E. 2012. Economic risks of aflatoxin contamination in marketing of peanut in Benin. Economics Research International 2012: e230638.

Njoku, H.O., Ofuya, C.O. and Ogbulie, J.N. 1991. Production of tempeh from the African yam bean (*Sphenostylis stenocarpa* Hams). Food Microbiology 8: 209–214.

Nowak, J. 1992. Oats tempeh. Acta Biotechnologica 12: 345–348.

Nowak, J. and Szebiotko, K. 1992. Some biochemical changes during soybean and pea tempeh fermentation. Food Microbiology 9: 37–43.

Nuehring, L., Rowland, G., Harrison, L., Cole, R. and Dorner, J. 1985. Cyclopiazonic acid mycotoxicosis in the dog. American Journal of Veterinary Research 46: 1670–1676.

Numfor, F.A. and Ay, P. 1987. Postharvest Technologies of Root and Tuber Crops in Cameroon: A Survey. Ministère de l'enseignement supérieur et de la recherche scientifique, Institut de la recherche agronomique.

Oboh, G., Akindahunsi, A.A. and Oshodi, A.A. 2002. Nutrient and anti-nutrient contents of *Aspergillus niger*-fermented cassava products (flour and gari). Journal of Food Composition and Analysis 15: 617–622.

Ogiehor, I.S., Ikenebomeh, M.J. and Ekundayo, A.O. 2008. The bio-load and aflatoxin content of market gari from some selected states in southern Nigeria: Public health significance. African Health Sciences 7: 223–227.

Okorondu, S.I., Akujobi, C.O., Okorondu, J.N. and Okorondu, M.M.O. 2013. Gari agar as culture media for mycological studies. International Journal of Biological and Chemical Sciences 7: 1126–1134.

Paredes-López, O. and Harry, G.i. 1989. Changes in selected chemical and anti-nutritional components during tempeh preparation using fresh and hardened common beans. Journal of Food Science 54: 968–970.

Park, H.-S. and Yu, J.-H. 2012. Genetic control of asexual sporulation in filamentous fungi. Current Opinion in Microbiology 15: 669–677.

Pascual, S., Melgarejo, P. and Magan, N. 1997. Induction of submerged conidiation of the biocontrol agent *Penicillium oxalicum*. Applied Microbiology and Biotechnology 48: 389–392.

Peñaloza, W., Davey, C.L., Hedger, J.N. and Kell, D.B. 1992. Physiological studies on the solid-state quinoa tempeh fermentation, using on-line measurements of fungal biomass production. Journal of the Science of Food and Agriculture 59: 227–235.

Penel, A.J. and Kosikowski, F.V. 1990. Beta-nitropropionic acid production by *Aspergillus oryzae* in selected high protein and carbohydrate-rich foods. Journal of Food Protection 53: 321–323.

Prajapati, J.B. and Nair, B.M. 2008. The history of fermented foods. pp. 7–9. *In:* E.R. Farnworth, (ed.). Handbook of Fermented Functional Foods. CRC Press, Boca Raton, Florida.

Purchase, I.F.H. 1971. The acute toxicity of the mycotoxin cyclopiazonic acid to rats. Toxicology and Applied Pharmacology 18: 114–123.

Qureshi, T.M., Vegarud, G.E., Abrahamsen, R.K. and Skeie, S. 2012. Characterization of the Norwegian autochthonous cheese Gamalost and its angiotensin I-converting enzyme (ACE) inhibitory activity during ripening. Dairy Science and Technology 92: 613–625.

Randhir, R. and Shetty, K. 2007. Mung beans processed by solid-state bioconversion improves phenolic content and functionality relevant for diabetes and ulcer management. Innovative Food Science and Emerging Technologies 8: 197–204.

Reyes-Moreno, C., Romero-Urías, C., Milán-Carrillo, J., Valdéz-Torres, B. and Zárate-Márquez, E. 2000. Optimization of the solid state fermentation process to obtain tempeh from hardened chickpeas (*Cicer arietinum* L.). Plant Foods for Human Nutrition 55: 219–228.

Robinson, R.K. 1995. A Colour Guide to Cheese and Fermented Milks. Chapman & Hall, London, New York.

Rokas, A. 2009. The effect of domestication on the fungal proteome. Trends in Genetics 25: 60–63.

Roostita, R. and Fleet, G.H. 1996. The occurrence and growth of yeasts in Camembert and Blue-veined cheeses. International Journal of Food Microbiology 28: 393–404.

Rundberget, T. and Wilkins, A.L. 2002. Determination of *Penicillium* mycotoxins in foods and feeds using liquid chromatography–mass spectrometry. Journal of Chromatography A 964: 189–197.

Samson, R.A., Seifert, K.A., Kuijpers, A.F.A., Houbraken, J.A.M.P. and Frisvad, J.C. 2004. Phylogenetic analysis of *Penicillium* subgenus *Penicillium* using partial P-tubulin sequences. Studies in Mycology 49: 175–200.

Samson, R.A., Houbraken, J., Summerbell, R.C., Flannigan, B. and Miller, J.D. 2011. Common and important species of fungi and actinomycetes in indoor environments. pp. 287–292. In: B. Flannigan, R.A. Samson and J.D. Miller (eds.). Microorganisms in Home and Indoor Work Environments: Diversity, Health Impacts, Investigation and Control. CRC Press.

Sato, A., Oshima, K., Noguchi, H., Ogawa, M., Takahashi, T., Oguma, T., Koyama, Y., Itoh, T., Hattori, M. and Hanya, Y. 2011. Draft genome sequencing and comparative analysis of *Aspergillus sojae* NBRC4239. DNA Research 18: 165–176.

Schuster, E., Dunn-Coleman, N., Frisvad, J. and Dijck, P. van 2002. On the safety of *Aspergillus niger*—a review. Applied Microbiology and Biotechnology 59: 426–435.

Schwarz, P., Bretagne, S., Gantier, J.-C., Garcia-Hermoso, D., Lortholary, O., Dromer, F. and Dannaoui, E. 2006. Molecular identification of Zygomycetes from culture and experimentally infected tissues. Journal of Clinical Microbiology 44: 340–349.

Seviour, R.J., Pitt, D.E., McClure, S. and Pyle, J. 1983. A critical assessment of criteria used in the taxonomy of the genus *Rhizopus*. Canadian Journal of Botany 61: 2374–2383.

Shang, Y.-L., Chen, L.-L., Zhang, Z.-H. and Lu, J. 2012. A comparative study on the fungal communities of wheat Qu for Qingshuang-type Chinese rice wine. Journal of the Institute of Brewing 118: 243–248.

Sheih, I.C., Wu, H.Y., Lai, Y.J. and Lin, C.F. 2000. Preparation of high free radical scavenging tempeh by a newly isolated *Rhizopus* sp. R-69 from Indonesia. Food Science and Agricultural Chemistry 2: 35–40.

Shrestha, H., Nand, K. and Rati, E.R. 2002. Microbiological profile of Murcha starters and physico-chemical characteristics of Poko, a rice based traditional fermented food product of Nepal. Food Biotechnology 16: 1–15.

Shurtleff, W. and Aoyagi, A. 2012. History of Koji—Grains and/or Soybeans Enrobed with a Mold Culture (300 BCE To 2012): Extensively Annotated Bibliography and Sourcebook. Soyinfo Center.

Siebenhandl, L.N.L., D. Trimmel, E. and Berghofer, S. 2001. Studies on tapé ketan—an Indonesian fermented rice food. International Journal of Food Sciences and Nutrition 52: 347–357.

Slonczewski, J.L. and Foster, J.W. 2013. Microbiology: An Evolving Science. W.W. Norton.

Spinnler, H.E. and Gripon, J.C. 2004. Surface mould-ripened cheeses. pp. 157–174. In: P.F. Fox, P.L.H. McSweeney, T.M. Cogan and T.P. Guinee (eds.). Cheese: Chemistry, Physics and Microbiology: Major Cheese Groups. Elsevier Academic Press.

Stodolak, B. and Starzyńska-Janiszewska, A. 2008. The influence of tempeh fermentation and conventional cooking on anti-nutrient level and protein bioavailability (*in vitro* test) of grass-pea seeds. Journal of the Science of Food and Agriculture 88: 2265–2270.

Suharto, I., Prima, K.A. and Miryanti, Y.A. 2011. Technology transfer to industry of biotechnology on *Rhizopus* sp. Inoculum. Institute of Electrical and Electronics Engineers, Bali 358–362.

Sweeney, M.J. and Dobson, A.D.W. 1998. Mycotoxin production by *Aspergillus*, *Fusarium* and *Penicillium* species. International Journal of Food Microbiology 43: 141–158.

Takahashi, T., Chang, P.-K., Matsushima, K., Yu, J., Abe, K., Bhatnagar, D., Cleveland, T.E. and Koyama, Y. 2002. Nonfunctionality of *Aspergillus sojae* aflR in a strain of *Aspergillus parasiticus* with a disrupted *aflR* Gene. Applied and Environmental Microbiology 68: 3737–3743.

Takahashi, T., Masuda, T. and Koyama, Y. 2006. Enhanced gene targeting frequency in ku70 and ku80 disruption mutants of *Aspergillus sojae* and *Aspergillus oryzae*. Molecular Genetics and Genomics 275: 460–470.

Tamang, B. and Tamang, J.P. 2010. *In situ* fermentation dynamics during production of gundruk and khalpi, ethnic fermented vegetable products of the Himalayas. Indian Journal of Microbiology 50: 93–98.

Thanh, N.V. and Nout, M.J.R. 2002. *Rhizopus oligosporus* biomass, sporangiospore yield and viability as influenced by harvesting age and processing conditions. Food Microbiology 19: 91–96.

Toko, K. 2013. Biochemical Sensors: Mimicking Gustatory and Olfactory Senses. CRC Press, NewYork.

Tramper, J. and Zhu, Y. 2011. Modern Biotechnology: Panacea Or New Pandora's Box? Wageningen Academic Pub, Netherlands.

Wan, S.P., Wang, J.M., Jia, L., Zhang, Y.Y. and Lu, X.D. 2011. Utilization of *P. camemberti* in Sufu production in comparison with traditional sufu strains. Advanced Materials Research 204–210: 1143–1146.

Wang, H.L. and Hesseltine, C.W. 1966. Wheat tempeh. Cereal Chemistry 43: 563.

Watanabe, T. and Oda, Y. 2008. Comparison of sucrose-hydrolyzing enzymes produced by *Rhizopus oryzae* and *Amylomyces rouxii*. Bioscience, Biotechnology, and Biochemistry 72: 3167–3173.

White, M.M., James, T.Y., O'Donnell, K., Cafaro, M.J., Tanabe, Y. and Sugiyama, J. 2006. Phylogeny of the Zygomycota based on nuclear ribosomal sequence data. Mycologia 98: 872–884.

Xie, G., Li, W., Lu, J., Cao, Y., Fang, H., Zou, H. and Hu, Z. 2007. Isolation and identification of representative fungi from Shaoxing rice wine wheat qu using a polyphasic approach of culture-based and molecular-based methods. Journal of Institute of Brewing 113: 272–279.

Yang, S., Lee, J., Kwak, J., Kim, K., Seo, M. and Lee, Y.-W. 2011. Fungi associated with the traditional starter cultures used for rice wine in Korea. Journal of Korean Society for Applied Biological Chemistry 54: 933–943.

Yin, L., Li, L., Liu, H., Saito, M. and Tatsumi, E. 2005. Effects of fermentation temperature on the content and composition of isoflavones and β-glucosidase activity in Sufu. Bioscience, Biotechnology and Biochemistry 69: 267–272.

Zarmpoutis, I.V., McSweeney, P.L.H. and Fox, P.F. 1997. Proteolysis in blue-veined cheese: an intervarietal study. Irish Journal of Agriculture and Food Research 36: 219–229.

Zhang, N. and Zhao, X.-H. 2010. Study of *Mucor* spp. in semi-hard cheese ripening. Journal of Food Science and Technology 47: 613–619.

Zheng, X.-W., Tabrizi, M.R., Nout, M.J.R. and Han, B.-Z. 2011. Daqu—a traditional Chinese liquor fermentation starter. Journal of the Institute of Brewing 117: 82–90.

Zheng, X.-W., Yan, Z., Han, B.-Z., Zwietering, M.H., Samson, R.A., Boekhout, T. and Robert Nout, M.J. 2012. Complex microbiota of a Chinese 'Fen' liquor fermentation starter (Fen-Daqu), revealed by culture-dependent and culture-independent methods. Food Microbiology 31: 293–300.

4

Acetic Acid Bacteria in Food Fermentations

Ilkin Yucel Sengun

1. Introduction

Food fermentations that contain many kinds of microorganisms help preserve foods and provide new products. Microorganisms mainly involved in food fermentation processes are lactic acid bacteria (LAB), acetic acid bacteria (AAB), yeast and molds. Most of the food fermentations occur as traditional processes by indigenous species. However, controlled fermentation could be achieved by the use of well-defined microbial strains as starter cultures to ensure consistent product quality.

AAB occur in sugar and alcohol enriched environments, while they play a key role in the production of certain types of fermented foods. AAB are obligate aerobes that oxidize sugars, sugar alcohols, and ethanol with the production of acetic acid as the major end product. The first step of acetic acid production is the conversion of ethanol from a carbohydrate carried out by yeasts, and the second step is the oxidation of ethanol to acetic acid carried out by AAB. The most familiar and extensively used industrial application of AAB is the vinegar production. It is used as a condiment with desirable flavor and health-promoting properties. AAB also play an important role in cocoa production, which represents a significant means of income for some countries. Although growth of AAB is desirable and essential for the quality of vinegar and cocoa fermentations, they are sometimes involved in foods and beverages such as in wine, beer, soft drinks and fruits in a detrimental way. Small amounts of oxygen potentially lead to the growth of AAB in wine bottles or tanks, thereby spoiling the product. Thus, aeration strongly influences fermentation processes and influences the quality of the final product.

Ege University, Engineering Faculty, Food Engineering Department, 35100, Bornova, Izmir/Turkey.
E-mail: ilkin.sengun@ege.edu.tr

This chapter reviews the roles of AAB in food fermentation processes, highlighting AAB species found in the industrially important food products. The taxonomy of the AAB is also updated by giving general properties of the group.

2. Acetic Acid Bacteria

The general features of AAB are discussed below.

2.1 General Characteristics

Acetic acid bacteria (AAB) are Gram-negative or Gram-variable, catalase positive, oxidase negative, non-spore forming, motile (*Acetobacter* with peritrichous flagella, *Gluconobacter* with polar flagella) or non-motile, ellipsoidal to rod-shaped cells that can occur single, in pairs or chains with sizes varying between 0.4–1 µm wide and 0.8–4.5 µm long. AAB have an obligate aerobic metabolism. The growth temperature range is 5–42°C with optima between 25–30°C, although some species are recognized as thermo-tolerant. The optimum pH for the growth of AAB is 5.0–6.5, while they can grow at lower pH values between 3.0–4.0, depending on the strain (Sievers and Swings 2005).

AAB are classified in the *Acetobacteriaceae* family, in α-class of Proteobacteria with 19 genera that are currently recognized as *Acetobacter, Acidomonas, Ameyamaea, Asaia, Bombella, Commensalibacter, Endobacter, Gluconacetobacter, Gluconobacter, Granulibacter, Komagataeibacter, Kozakia, Neoasaia, Neokomagataea, Nguyenibacter, Saccharibacter, Swaminathania, Swingsia* and *Tanticharoenia* (Trček and Barja 2015).

AAB was first isolated from naturally fermented vinegar in 1837 by Friedrich Traugott Kützing. Then, Louis Pasteur determined 'mother vinegar' as the responsible mass of microorganisms from acetic acid fermentation in the year 1868. *Acetobacter* is the first genus named by Martinus Willem Beijerinck. Morphological, biochemical, and physiological criteria have been successfully used in the early years to differentiate AAB genera. Hansen (1894) proposed the first classifications, based on the occurrence of a film in the liquid media, and its reaction with iodine. Asai (1934–1935) formulated the proposal of classifying AAB into two genera: *Acetobacter* and *Gluconobacter*. Then Frateur (1950) proposed a new classification based on the catalase activity, production of gluconic acid from glucose, oxidation of acetic acid to carbon dioxide and water, oxidation of lactic acid to carbon dioxide and water, and oxidation of glycerol into di-hydroxyacetone. A major change in the classification of AAB was introduced by Yamada et al. (1997, 1998), who transferred *Acetobacter* species containing Q-10 (*A. xylinus, A. liquefaciens, A. hansenii, A. diazotrophicus* and *A. europaeus*) to the genus *Gluconacetobacter*. Then, *Gluconacetobacter* was divided into two groups: One was composed mainly of the so-called peritrichously flagellated intermediate strains, which were once named as *A. liquefaciens*, and the other was predominantly of *A. xylinus* strains. For the latter group, the name of *Komagataeibacter* was given (Yamada 2014a). At the present time, 10 species have been recognized in *Gluconacetobacter* and 15 species in *Komagataeibacter*. On the other hand, rearrangements of the group are still in progress (Yamada et al. 2012, Yamada 2014a,

2014b). For the updated taxonomy, the internet site, http://www.bacterio.net, founded by Jean P. Euzéby (Euzéby 1997), should be followed.

AAB have a respiratory metabolism characterized by incomplete oxidation in which partially oxidized organic compounds are excreted as the end product, such as aldehydes, ketones and organic acids. *Acetobacter* and *Gluconobacter* are known as main two genera and their strains are widespread in common isolation sources such as vinegar, wine, fruits and flowers. The main differences between these two genera were both, cytological and physiological. They can be differentiated by the preference for alcohol or glucose as a substrate and the ability to oxize acetic or lactic acid to carbon dioxide. *Acetobacter* species produce acetic acid from ethanol, which is catalyzed by the membrane-bound pyrroloquinoline quinone-dependent alcohol dehydrogenase and aldehyde dehydrogenase, and can oxidize acetic acid to carbon dioxide by a set of soluble $NAD(P)^+$-dependent dehydrogenases (alcohol dehydrogenase and aldehyde dehydrogenase) *via* the tri-carboxylic acid cycle. On the other hand, *Gluconobacter* species prefer sugar as the carbon source, because they can obtain energy more efficiently by the metabolization of the sugars *via* the pentose phosphate pathway. *Gluconobacter* species produce gluconic acid from glucose and sorbose from sorbitol, which are the precursors of ascorbic acid. Thus strains of *Gluconobacter oxydans*, which produce enzymes involved in amino acids synthesis such as glutamic and aspartic acids, are used industrially in the manufacture of ascorbic acid and also in the pharmaceutical industry for producing di-hydroxyacetone, miglitol's precursors and shikimate (Mamlouk and Gullo 2013). AAB can stimulate biofilm formation on different types of surfaces by producing exopolysaccharides such as dextran, levan, fructane and cellulose. All these reactions are carried out with membrane-bound dehydrogenases, mostly localized on the outer surface of the cytoplasmic membrane. The alcohol dehydrogenase activity of *Acetobacter* is more stable under acetic conditions than that of *Gluconobacter*, which explains why *Acetobacter* produces more acetic acid.

2.2 Isolation and Identification of Acetic Acid Bacteria

AAB exhibit poor growth due to large accumulations of their oxidation products in the culture medium, despite the oxidation of large amounts of the carbon source. This viable but non-cultivable (VBNC) status of AAB is the factor that limits the recovery of strains by plating techniques. Thus, cultivation of AAB is rather difficult unless optimal substrates and optimal conditions are used. Various media have been described for the isolation of AAB such as Glucose Yeast Extract $CaCO_3$ Medium (GYC), Glucose Yeast Extract Medium (GY), Yeast Extract Peptone Mannitol Medium (YPM), Acetic Acid-Ethanol Medium (AE) and Reinforced Acetic Acid-Ethanol Medium (RAE) (Sengun and Karabiyikli 2011). Among these culture media, GY and GYC are the most widely used media in the recovery of AAB strains from different sources, while AE and RAE are more suitable in promoting the growth of AAB adapted to high concentrations of acetic acid.

AAB strongly correlated at the phylogenetic level and have phenotypic characteristics that are similar to one another. Therefore, to overcome the disadvantages of culturing and identification of AAB, new techniques have been developed using molecular approaches, such as DNA-rRNA hybridization, DNA-DNA hybridization, ribosomal RNA gene sequences (5S rRNA, 16S rRNA, and 23S rRNA). Culture dependent and culture independent techniques such as Denaturing Gradient Gel Electrophoresis (DGGE-PCR), Temporal Temperature Gradient Gel Electrophoresis (TTGE-PCR), Restriction Fragment Length Polymorphism of the 16S rRNA and 16S-23S ITS regions (RFLP-PCR 16S rRNA and 16S-23S ITS), Enterobacterial Repetitive Intergenic Consensus-PCR (ERIC-PCR), Species-Specific PCR (SS-PCR), Repetitive Extragenic Palindromic-PCR (REP-PCR) and (GTG)5-PCR have proven to present high resolution capacity to determine the genetic variability of AAB (Ilabaca et al. 2008, Torija et al. 2010, Jara et al. 2013, Vegas et al. 2013, Mateo et al. 2014, Huang et al. 2014). Matrix-Assisted Laser Desorption Ionization-Time of Flight Mass Spectrometry (MALDI-TOF MS) is an alternative molecular typing method, which uses cell proteins as a target (Andrés-Barrao et al. 2013). For the quantification of AAB, Fluorescence *In Situ* Hybridization (FISH), Epi-fluorescence, Real-Time PCR (RT-PCR), Ethidium Monoazide in combination with quantitative PCR (EMA-qPCR) could be used (Shi et al. 2012, Vegas et al. 2013).

The culture-independent methods, which provide information at species level, are more accurate and represent the reality of a mixed culture much better than the classic methods. On the other hand, the classic methods are essential for future studies which will be carried out on the isolates. Thus, a polyphasic approach combining phenotypic, chemotaxonomic and genotypic data of strains is recommended to obtain phylogenetic relationships and accurate identifications of AAB.

2.3 Occurrence of Acetic Acid Bacteria

AAB are widespread in nature and are well adapted to growth in sugar rich and alcohol rich environments. AAB can be isolated from different sources and environments, such as flowers, pollen, fruits (coconut, mango, guava, sapodilla, grapes, apple, cherry, strawberry, dates, etc.) sugarcane, rice, palm, corn, and fermented products such as vinegar, cocoa, coffee, wine, beer, yogurt and curd.

A large number of strains of AAB have been isolated from tropical fruits like coconut, mango, guava, sapodilla (*Acetobacter cibinongensis, A. indonesiensis, A. lovaniensis, A. orientalis, A. orleanensis, A. syzygii, A. tropicalis, Frateuria aurantia, Gluconobacter frateurii, G. oxydans, Komagataeibacter hansenii*). Although most of the isolates belong to *Acetobacter* spp., other AAB genera, sometimes with quite unique species, could be isolated from various sources like apple (*A. malorum*), cherry (*G. cerinus*), dried fruit and palm juice (*Gluconacetobacter liquefaciens*), grapes (*A. aceti, G. oxydans, Komagataeibacter saccharivorans, K. hansenii*), strawberry (*G. frateurii*), corn roots (*G. azotocaptans*), sugarcane roots (*Gluconacetobacter diazotrophicus*), mealy bug from sugar cane (*Gluconacetobacter sacchari*), palm brown sugar (ragi) (*Kozakia baliensis*), wetland rice (*G. diazotrophicus, A. peroxydans*), Asian rice (*Nguyenibacter vanlangensis*), wild rice (*Swaminathania salitolerans*), red ginger (*Ameyamaea chiangmaiensis*) and pollen (*Saccharibacter floricola*). Besides, flowers contain diversity of AAB species such as *A. indonesiensis, A. orientalis, A. orleanensis, A. syzygii, Asaia bogorensis, As. indonesiensis, As. siamensis, G. frateurii, G. oxydans, F. aurantia* (Sengun and Karabiyikli 2011). In general,

Gluconobacter, Asaia and *Frateuria* species are found in flowers and fruits, while *Gluconacetobacter* and *Komagataeibacter* species are more common in fermented products.

In soft drinks, a low pH level and addition of chemical preservatives is not sufficient to prevent growth of spoilage microflora, including AAB. *Gluconobacter* species are the most frequently encountered cause of bacterial spoilage of soft drinks at low pH (Raspor and Goranovic 2008). AAB usually cause different organoleptic changes such as low pH and ethanol content, pack swelling, vinegary off-flavors, turbidity and ropiness by producing acetic acid in soft drinks. They can exhibit the production of extracellular substances that stimulate adhesion and flocculation. *As. lannensis* and *As. bogorensise* are the species, isolated from strawberry flavored bottled water and bottled water, respectively (Sedlackova et al. 2011, Kregiel et al. 2012).

AAB are involved in some important industrial processes. These bacteria can produce high concentrations of acetic acid from ethanol, which makes them important to the vinegar industry. AAB are also involved in the production of other foods, like palm wine, cocoa powder, nata de coco (a kind of fermented coconut), pulque (a beverage from agave) and kombucha (a kind of sweetened fermented tea). However, AAB can rarely found in yogurt, curd and also in sourdough (Mamlouk and Gullo 2013, Haghshenas et al. 2015). On the other hand, AAB are sometimes involved in foods and beverages in a detrimental way, such as in wine, coffee and beer, where the production of acetic acid is undesirable. More detailed information for the most popular fermented food products in the food industry is given below by focusing on AAB species isolated from the products.

3. AAB in Fermented Foods and Beverages

The role of AAB in food fermentations is described in the following sections.

3.1 Acetic Acid Bacteria in Vinegar

Vinegar is a kind of condiment used to afford a desired acidic flavor to foodstuffs; it is also used as a sanitizer, especially for salad vegetables. There are several types of vinegar produced worldwide. It is supplied by firms, which have large scale production and is also produced in small scales by traditional methods that change from region to region. The product has different names, taste and flavor depending on the raw material and technique used in its production (Table 1).

Even though the substrates and the end products of vinegar fermentations have some differences, the underlying mechanisms show similarities. In general, vinegar fermentation involves an activity of yeast and AAB. During production, ethanol produced by yeasts is metabolized to acetic acid by AAB. Oxidation of ethanol is also known as acetification. There are two well defined methods for vinegar production: the traditional and the submerged. In the traditional method, oxidation is started by 'seed-vinegar', also called as 'mother of vinegar'; an undefined starter culture obtained from previous vinegar and AAB grow on the media surface where oxygen concentration

Table 1. Some of the vinegar varieties produced in different countries.

Vinegar type	Raw material	Main producer
Cereal vinegars		
Beer vinegar	Barley	Germany
Black vinegar	Millet, wheat or sorghum	China, East Asia
Brewed vinegar	Rice, red rice and bran	China
Fukuyama pot vinegar	Rice and koji	Japan
Keumkangmil (Korean domestic wheat vinegar)	Wheat	Korea
Komesu (Amber rice vinegar)	Rice	Japan
Kurosu (Black rice vinegar)	Rice	Japan
Purple sweet potato *Makgeolli* vinegar	Purple sweet potato and rice	Korea
Rice vinegar	Rice or sake	Japan, China
Shanxi aged vinegar	A variety of cereal and Daqu[a]	China
Tianjin duliu mature vinegar	Rice, red sorghum and Daqu	China
Zhenjiang aromatic vinegar	Rice and starch	China
Wine vinegars		
Wine vinegar	Low alcohol wines	World wide
Sherry vinegar	Sherry wine	Spain
Palm juice vinegar	Palm wine	Africa, Asia and South America
Vegetable vinegars		
Cane vinegar	Sugar cane	Philippines
Onion vinegar	Onion	East Asia
Ophiopogon vinegar	*Radix Ophiopogon japonicus*[b]	China
Tomato vinegar	Tomato	Japan, East Asia
Yacon vinegar	*Smallanthus sonchifolius*[b]	South America
Fruit vinegars		
Balsamic vinegar	Grape must	Italy
Cider vinegar	Apple or apple wine	World wide
Coconut vinegar	Coconut	Southeast Asia
Date vinegar	Date	Middle East
Fig vinegar	Fig and grape	Turkey
Grape vinegar	Grape	Turkey, Middle East
Persimmon vinegar	Persimmon	Japan, South Korea
Sour cherry vinegar	Sour cherry	Europe, USA
Strawberry vinegar	Strawberry	Spain
Ume-su	Plum	Japan
Other vinegars		
Malt vinegar	Malt	Northern Europe
Honey vinegar	Honey	Europe, Africa, America
Spirit vinegar (White distilled vinegar)	Alcohol[c]	USA, China
Kombucha vinegar	Tea and sugar	Asia
Whey vinegar	Whey	Europe

[a] Daqu is made from barley and pea, and used as starter.
[b] Kind of plant.
[c] Alcohol that is produced from distillation of fermented sugar solutions.

is high, so this production system is called as 'surface culture method' or 'Orleans method' (for the more standardized version). The traditional method takes a long time and high quality vinegars are produced. Acidity reaches a maximum level after approximately three months. On the other hand, there are also methods that are used to reduce the acetification time by maintaining or increasing the quality of the product (Mas et al. 2014). In the submerged method, AAB are submerged in the liquid and oxygen is added constantly as air bubbles. The oxidative process occurs between the air and liquid interfaces of the air bubbles. Acetification is completed in a few days in the submerged method (Tesfaye et al. 2002, Bamforth 2008). Although short acetification time and high acidity are the advantages of submerged method, the produced vinegar is cloudier and less aromatic, as there is less opportunity for flavor development. The number of AAB, which affect the acidic value of the product, increases with the increase in fermentation time (Sengun 2013). Thus, the chemical composition and volatile components of vinegars show differences. Vinegar must contain, at least, a percentage of 4 (w/v) as acetic acid, to be legal vinegar. On the other hand, various types of vinegars have different acidic values, ranging between 6.2–14.9 in balsamic vinegar, 3.9–9.0 in cider vinegar, 4.3–5.9 in malt vinegar, 5.9–9.2 in wine vinegar, 4.1–5.3 in synthetic vinegar, as acetic acid percentage (Bamforth 2008).

AAB are the group responsible for the conversion of ethanol to acetic acid, which is important for the generation of aroma-active compounds. A certain amount of biomass is sufficient for vinegar fermentation while oxygen availability is the rate limiting factor in acetification (Qi et al. 2013). The final composition of vinegar depends on the raw material used, alcoholic fermentation and subsequent biological oxidation which occurred, and the ageing procedure employed. Aging in wood affects the color, the polyphenolic profile and the aromatic complexity of the product. Besides, the type of wood is also important, and wood sourced from cherry or oak could be preferable in vinegar aging due to their distinctive aromatic characteristics (Callejon et al. 2010). The volatile compounds of vinegar usually include acids, higher alcohols, alkanes, aldehydes, esters, ketones, phenol and terpenes. Although the compositions and concentrations of the volatile compounds in different vinegars vary by raw material and brewing process, ethyl acetate is the most prevalent ester in vinegars (Ozturk et al. 2015, Dong et al. 2016). Esters have potential influence on vinegar aroma, while iso-amyl acetate and ethyl acetate have the highest odor activity in vinegar. Acetoin is related to the cell metabolism in AAB, so it is often used as a marker for the biological origin of vinegar (Baena-Ruano et al. 2010). The population dynamics of AAB in vinegar was studied by some researchers, the studies are spread on a wide period of time with a significant gap between older papers and current AAB taxonomy (Table 2).

Different types of vinegar show individual AAB profiles. *A. pasteurianus* could be found in most of the vinegar samples; it is also possible to recover different strains within the species. *A. aceti, A. cerevisiae, A. oboediens, A. orleanensis, A. malorum, A. pomorum, A. syzygii, Ga. entanii,* and five other species *Komagataeibacter europaeus, Komagataeibacter hansenii, K. intermedius, K. saccharivorans, K. xylinus* (mentioned with their former names in the table, as given in the references) are the other species isolated in most of the vinegar fermentations. *A. pasteurianus* is commonly found in vinegars with a low concentration of acetic acid as in cereal vinegars, and *Komagataeibacter* and *Gluconacetobacter* species are more common in vinegars with

Table 2. AAB species isolated from different kinds of vinegars.

Type of vinegar	AAB species	Reference
Rice vinegar	*A. pasteurianus*	Nanda et al. 2001, Haruta et al. 2006
Fukuyama pot vinegar	*A. pasteurianus*	Okazaki et al. 2010, Furukawa et al. 2013
Shanxi aged vinegar	*A. pasteurianus, A. indonesiensis, A. malorum, A. orientalis, A. senegalensis, G. oxydans*	Wu et al. 2012
Tianjin duliu mature vinegar	*A. pasteurianus*	Nie et al. 2013
Zhenjiang aromatic vinegar	*A. pasteurianus, A. pomorum, Ga. intermedius*	Xu et al. 2011
Traditional wine vinegar	*A. pasteurianus, Ga. europaeus* *A. pasteurianus, Ga. europaeus, Ga. hansenii, Ga. intermedius, Ga. xylinus, Ga.* sp.	Ilabaca et al. 2008 Vegas et al. 2010 Vegas et al. 2013
Red wine vinegar	*Ga. europaeus, Ga. xylinus*	Fernandez-Perez et al. 2010
White wine vinegar	*A. pasteurianus, Ga. europaeus, Ga. xylinus*	Fernandez-Perez et al. 2010
Industrial vinegar	*Ga. europaeus* *Ga. europaeus, Ga. hansenii* *A. oboediens, A. pomorum* *A. intermedius* *Ga. entanii* *Ga. europaeus*	Sievers et al. 1992 Yamada et al. 1997 Sokollek et al. 1998 Boesch et al. 1998 Schüller et al. 2000 Fernandez-Perez et al. 2010
Cider vinegar	*A. pasteurianus, Ga. europaeus, Ga. hansenii, Ga. xylinus,*	Fernandez-Perez et al. 2010
Strawberry vinegar	*A. orleanensis* *A. malorum, Ga. saccharivorans, Ga. xylinus*	Mamlouk et al. 2011 Hidalgo et al. 2013a
Blueberry vinegar	*A. pasteurianus*	Hidalgo et al. 2013b
Persimmon vinegar	*A. malorun, A. pasteurianus, A. syzygii, Ga. intermedius, Ga. europaeus, Ga. saccharivorans*	Hidalgo et al. 2012

Abbreviations: *A.: Acetobacter; G.: Gluconobacter; Ga.: Gluconacetobacter.*
*Ga. europaeus, Ga. intermedius (formerly *A. intermedius*), *Ga. hansenii, Ga. saccharivorans* and *Ga. xylinus* moved recently to *Komagataeibacter* genus.

high acetic acid concentrations, such as fruit vinegars. In general, high complex strain microflora occurs in vinegars produced by biofilm on mechanical supports and on spoiled wines, whereas single dominant strains occur in submerged vinegar fermentations (Wu et al. 2010). Biofilm produced by AAB is composed mainly of cellulose, which provides a protective growth environment to survive in vinegar. *K. xylinus* is known as the main cellulose producer in vinegar. On the other hand, *Acetobacter, Acidomonas, Ameyamaea, Asaia, Gluconacetobacter, Gluconobacter, Komagataeibacter, Kozakia, Neoasaia, Tanticharoenia* was found in the biofilm of strawberry vinegar (Valera et al. 2015).

A variety of AAB species are found in raw material and their acetic acid tolerances affect population dynamics of the vinegar fermentation. The starting strains can survive at low acetic acid concentrations and high ethanol, whereas the ones that completed the acetification might be more acetic acid tolerant. At low concentrations of acetic acid, species of the genus *Acetobacter* predominate. Then *Gluconacetobacter* take over the process, with species such as *K. europaeus* or *K. intermedius* predominating, when acetic acid concentrations reach an appropriate level in wine vinegar (Mas et al. 2014). The operation length of the vinegar may also affect the diversity of AAB. Other factors affecting the acetic acid profile and the final product quality are technological processes, wood contact and aging (Hidalgo 2013a, Mas et al. 2014).

The AAB species may also affect the quality of vinegar in a negative way. Excessive amounts of aeration or shortage of alcohol causes oxidation of acetic acid to carbon dioxide and water by AAB. Another defect caused by AAB is sliminess. *A. aceti* subsp. *xylinum* is known as one of the most important species causing sliminess in vinegar. Thus, using AAB strains, which do not produce cellulose and undesirable aroma, while efficiently oxidase ethanol, produce acetic acid, and resistant to high acetic acid, low pH and bacteriophage, as a starter culture in the production of vinegar may lead to improved fermentation process and an enhanced product quality (Gullo and Giudici 2008). *A. aceti* could be used for vinegar production under controlled conditions, which gives better organoleptic properties than commercial vinegar (Zahoor et al. 2006). A mixed culture of a 'quick start' AAB, such as *A. pasteurianus* and a high acetic acid tolerance species, such as *K. europaeus*, could also be used to guarantee the best vinegar production through a rapid start and a good ending for the process (Mas et al. 2014). *A. malorum* and *K. saccharivorans* would be other appropriate choices for a good quality persimmon vinegar (Hidalgo et al. 2012). As a result, inoculation of well-defined AAB, especially using mixed inoculation systems, could be preferred for producing a high quality of vinegars.

3.2 Acetic Acid Bacteria in Cocoa

Cocoa fermentation involves an activity of different kinds of microorganisms such as AAB, LAB, spore-forming bacteria, yeasts and molds. At the beginning of cocoa fermentation, yeast and LAB are dominant under the sugar rich, acidic and anaerobic environment. AAB predominate with subsequent temperature rises; alcohol accumulates in the last stages of fermentation and all drying stages (Nielsen et al. 2014). In cocoa fermentation, the ethanol produced by yeasts is converted into acetic acid by AAB. This oxidation reaction causes an increase in the temperature and the acidity of cocoa bean mass, which start endogenous bean reactions, responsible for flavor and color formation of the product (Schwan et al. 2014). Acetic acid production during cocoa fermentation also has an anti-microbial effect on yeasts, bacteria, and filamentous fungi (Copetti et al. 2012).

Cocoa bean fermentation is generally carried out in a traditional practice following three processes in boxes, heaps and trays through a spontaneous four to six days fermentation process. Although a greater diversity of bacteria and yeasts can be observed in box fermentations, stainless steel tanks may be useful for the optimization of cocoa fermentation with starter cultures. AAB take an essential role in cocoa bean fermentation. Thus, isolation and identification of AAB during different stages of cocoa fermentation were studied by many researchers (Table 3).

Table 3. AAB species identified in different cocoa fermentations.

Country	AAB species	Fermentation method	Reference
Australia	*A. pasteurianus, As. siamensis, G. oxydans*	Box	Dircks 2009
Brazil	*A. aceti* subsp. *liquefaciens, A. pasteurianus, A. peroxydans, G. oxydans*	Box	Schwan and Wheals 2004
Brazil	*A. ghanensis/syzygii, A. pasteurianus/pomorum, A. lovaniensis/fabarum, G. entanii, Ga. europaeus/swingsii*	Box	Garcia-Armisen et al. 2010
Brazil	*A. indonesiensis/malorum/cerevisiae, A. indonesiensis, A. fabarum, A. malorum/cerevisiae, A. pasteurianus, A. peroxydans, A. orientalis, A. senegalensis, G. saccharivorans, G. oxydans*	Box	Papalexandratou et al. 2011a
Brazil	*A. pasteurianus, G. oxydans*	Box	Illeghems et al. 2012
Brazil	*A. ghanensis, A. malorum, A. orientalis, A. pomorum, A. senegalensis, A. tropicalis, Asaia* sp., *Gluconobacter* sp.	Box	Pereira et al. 2013
	A. tropicalis, A. malorum, A. pomorum	Stainless steel tank	Pereira et al. 2013
Dominican Republic	*A. lovaniensis*	Box	Galvez et al. 2007
Ecuador	*A. cibinongensis, A. fabarum, A. ghanensis, A. malorum/cerevisiae, A. orientalis, A. pasteurianus, A. senegalensis, Gluconobacter* sp., *G. oxydans*	Box	Papalexandratou et al. 2011b
	A. cibinongensis, A. fabarum, A. lovaniensis/fabarum, A. malorum/indonesiensis, A. malorum/cerevisiae, A. orientalis, A. pasteurianus, A. peroxydans, A. syzygii	Platform	Papalexandratou et al. 2011b
Ghana	*A. malorum, A. pasteurianus, A. syzigii, A. tropicalis, G. oxydans*	Heap	Nielsen et al. 2007
Ghana	*A. lovaniensis, A. pasteurianus, A. syzgii, A. tropicalis*	Not referred	De Vuyst et al. 2008
Ghana	*A. ghanensis, A. lovaniensis, A. pasteurianus, A. tropicalis, A. senegalensis, A. syzygii*	Heap	Camu et al. 2007, 2008

Table 3. contd....

Table 3. contd.

Country	AAB species	Fermentation method	Reference
Indonesia	*A. aceti, A. pasteurianus*	Box	Ardhana and Fleet 2003
Ivory Coast	*A. ghanensis, A. pasteurianus, A. senegalensis*	Vessels	Lefeber et al. 2010
Ivory Coast	*A. cerevisiae, A. lovaniensis, A. nitrogenifigens, A. pasteurianus, Acetobacter* sp., *G. xylinum*	Box	Hamdouche et al. 2015
Malaysia	*A. lovaniensis, A. rancens, A. xylinum, G. oxydans*	Box	Carr et al. 1979, Papalexandratou et al. 2013

Adapted from Schwan et al. 2014, De Vuyst et al. 2008 and Hamdouche et al. 2015.
Abbreviations: *A.: Acetobacter, As.: Asaia, G.: Gluconobacter. Ga.: Gluconacetobacter.*
* *A. xylinum* or *G. xylinum, Ga. europaeus/swingsii* are moved to *Komagataeibacter.*

The different production sites investigated show individual AAB profiles, but with *A. pasteurianus* being isolated from a majority of the samples. *A. aceti, A. tropicalis* and *G. oxydans* are the other species isolated in most of the cocoa fermentations. Other species which were less frequently isolated from cocoa fermentations are *A. cerevisiae, A. fabarum, A. ghanensis, A. lovaniensis*-like species, *A. indonesiensis, A. malorum, A. nitrogenifigens, A. lovaniensis, A. orientalis, A. peroxydans, A. pomorum, A. senegalensis, A. syzgii, Asaia* sp., *G. oxydans, G. xylinum* and *Gluconobacter* sp. and *K. xylinus* (formerly *A. xylinum* and *G. xylinum*) (Table 3).

Although cocoa bean fermentation is applied as a traditional, indigenous process all over the world, the use of starter cultures for cocoa fermentation could make the process more consistent and much faster (Schwan et al. 2014).

Different mixtures of yeasts could be used as starter cultures in cocoa fermentations, influencing the flavor of the products (Dircks 2009, Crafack et al. 2013). The use of well defined AAB species as a starter culture in cocoa fermentation could also provide some advantages. The use of the *K. xylinus* subsp. *xylinus* in cocoa fermentation reduces the pH value and increases the acetic acid level of the product, which causes less cocoa flavor production than naturally fermented cocoa beans (Samah et al. 1993). *A. pasteurianus* strains could be used in cocoa starter cultures and AAB used in combination with the mixture of yeast and LAB, in order to obtain an acceptable product (Lefeber et al. 2012). One of the most important things is the selection of appropriate strains that are capable of growing under the stressful environmental conditions of the cocoa bean fermentation. On the other hand, the knowledge about the impact of individual species of AAB on the final product quality is very limited.

3.3 Acetic Acid Bacteria in Coffee

Coffee is a kind of beverage made from coffee beans, which are obtained from the endosperm of the coffee fruits. Coffee fruits are subjected to dry, wet, or semi-dry processing to obtain green bean. Arabica coffee (*Coffea arabica*) from Brazil and Robusta coffee (*Coffea canephora*) from West Africa are generally dry processed, while Arabica coffee fruits produced in Colombia, Tanzania, Mexico, and Hawaii are generally wet processed. The coffee fermentation process is still a spontaneous, uncontrolled process, without the addition of starter cultures. During the fermentation processes, microorganisms metabolize the sugars and polysaccharides present in both, the pulp and the mucilage, and produce organic acids and other metabolites, which may affect the final sensory characteristics of the beverage (Silva 2014).

A variety of microorganisms could be found in coffee fruits, which are naturally contaminated from different sources. Microflora of coffee fruits affects the quality of the fermentation process. The microbiota involved in dry fermentation of coffee are much more complex than those in the wet fermentation process. Bacteria and yeasts represent the most frequently occurring microorganisms in coffee fermentations, while AAB are rarely isolated from the product. In the previous study, *Acetobacter* sp. was isolated from coffee plants in Brazil (Vaughn et al. 1958). Two species of *Gluconacetobacter*, associated with coffee plants, were described in Mexico as *Ga. johannae* and *Ga. azotocaptans* (Fuentes-Ramírez et al. 2001). *Ga. diazotrophicus* was also isolated from coffee plants (*Coffea arabica* L.) (Jimenez-Salgado et al. 1997, Madhaiyan et al. 2004).

Although it is possible to isolate a variety of filamentous fungi from coffee fruits, the bacteria and yeasts prevent the development of fungi during fermentation, if the fermentation process is well carried out. AAB may also be found in the dry processing of coffee, leading to excessive acetic acid production when over fermentation occurs (Silva et al. 2008).

3.4 Acetic Acid Bacteria in Wine

Wine is an alcoholic beverage made from alcoholic fermentation of grape juice or must by yeasts. It can also be produced from the fermentation of various types of fruits including apples, peaches, apricots, plums, pears, cherries, berries, etc. Yeast inoculation is a very common practice in brewing and winemaking, in order to ensure the quality and reproducibility of the final product. Grape is a raw material, which is used mainly to produce wine. AAB present in grape can survive in winemaking processes and are involved in wine spoilage. AAB can survive in the various phases of alcoholic fermentation and it is very important to control their presence and further development to obtain good quality wines. Thus, AAB diversity present in grape is important to understand with respect to the quality of wine (Table 4).

In the previous studies, only three species of AAB (*G. oxydans, A. aceti* and *A. pasteurianus*) have been reported from grapes. Nowadays, an extended number of AAB species of grapes can be identified by using molecular methods. On the other hand, grape health status is the main factor affecting the microbial ecology of grapes, increasing both microbial numbers and species diversity (Barata et al. 2012).

Different strains of AAB may be found in wine fermentations as contaminant microorganisms and affect the quality of wines in negative ways. As an exception, however, AAB are involved in the fermentation process of palm wine (Table 5). During the primary fermentation, *G. oxydans* was found as the predominant species. In the

Table 4. Acetic acid bacteria isolated from grapes of different origin.

Origin of grapes	AAB species	Reference
Australia	*As. siamensis*	Bae et al. 2006
Canary Island	*A. pasteurianus, A. tropicalis, G. japonicus, Ga. saccharivorans*	Valera et al. 2011
Chile	*A. cerevisiae, G. oxydans*	Prieto et al. 2007
France	*A. aceti, A. pasteurianus, G. oxydans*	Joyeux et al. 1984, Barbe et al. 2001
Greece	*G. cerinus*	Nisiotou et al. 2011
Portugal	*A. malorum, A. orleanensis, A. syzygii, G. oxydans, Ga. hansenii, Ga. intermedius Ga. saccharivorans*	Barata et al. 2012
South Australia	*A. malorum, Am. chiangmaiensis, As. lannaensis, As. siamensis, G. albidus, G. frateurii, G. cerinus, G. oxydans*	Mateo et al. 2014
Spain	*A. aceti, G. oxydans, Ga. hansenii*	González et al. 2005

Abbreviations: *A.: Acetobacter, Am.: Ameyamaea, As.: Asaia, G.: Gluconobacter, Ga.: Gluconacetobacter.*
*Ga. intermedius, Ga. hanseni and *Ga. saccharivorans* recently moved to *Komagataeibacter* genus.

Table 5. AAB identified in different kinds of fermented alcoholic beverages.

Source	AAB species	Reference
Austrian wine	*A. tropicalis*	Silhavy and Mandl 2006
Bottled red wine	*A. pasteurianus*	Bartowsky et al. 2003
Portugal red wine	*A. oeni*	Silva et al. 2006
Palm wine	*A. indonesiensis, A. tropicalis, G. oxydans, Ga. saccharivorans*	Ouoba et al. 2012
Red wine fermentation	*G. oxydans, Ga. hansenii, A. aceti, A. nitrogenifigens*	González 2005 Dutta and Gachhui 2006
South African red wine	*A. hansenii, A. liquefaciens, A. pasteurianus, G. oxydans*	Du Toit and Lambrechts 2002
White wine	*A. aceti, A. pasteurianus, G. oxydans*	Joyeux et al. 1984
Wine	*A. aceti, G. oxydans Ga. hansenii*	Ruiz et al. 2000
Beer	*A. estunensis, A. cerevisiae, A. orleanensis, A. pasteurianus*	Andrés-Barrao et al. 2013
Bottled dinner beer	*A. cerevisiae/A. malorum, A. fabarum, A. indonesiensis, A. persici, A. orleanensis, G. cerinus, G. cerevisiae, G. japonicus, G. oxydans, Gluconobacter* sp.	Wieme et al. 2014a
Lambic beer	*G. cerevisiae*	Wieme et al. 2014b
Pulque	*A. malorum, A. orientalis*	Escalante et al. 2008

Abbreviations: *A.: Acetobacter, G.: Gluconobacter, Ga.: Gluconacetobacter.*
*Ga. hansenii and *Ga. saccharivorans* recently moved to *Komagataeibacter* genus.

later period, anaerobic conditions favor the alcoholic fermentation by the yeasts and carbon dioxide and ethyl alcohol are produced, which cause unsuitable conditions for the growth of AAB. Although aerobic contaminants are reduced in the later period of the fermentation, any surviving AAB such as *A. aceti* species can grow during the transfer, agitation and aeration processes of wine. Tanks, barrels or bottles could be a source of contamination, causing spoilage of the wine.

Factors that are known to affect the development of AAB during the wine-making process are acidity or pH, availability of air, temperature conditions, the ethanol concentration and the amount of sulfur dioxide present. The growth of AAB cannot be stopped by any pH which is normal to wines. Besides, higher storage temperatures of wine encourage the development and the metabolism of AAB and spoilage, usually, is most rapid at 20 to 35°C. Although the levels of AAB are closely related to the dissolution of oxygen in wine, they are always present, even in the absence of oxygen, in very low levels. In the presence of air, AAB mainly *A. aceti* or *G. oxydans* may cause undesirable processes in wine such as oxidation of alcohol to acetic acid or oxidation of glucose to gluconic acid resulting in a mousy or sweet-sour taste (Machve 2009). The effects of sulfur dioxide depend on the kind of organism to be suppressed and the pH and sugar content of wine. Spoilage microorganisms in must could be inactivated by 75 to 200 ppm sulfur dioxide, while low concentrations of sulfur dioxide have minimal effect on *A. pasteurianus* strain (Du Toit et al. 2005). Besides, AAB are inhibited by over 14 to 15 percent of alcohol in wine.

Unfiltered wines are more suitable conditions for encouraging AAB growth than filtered wines. Although AAB can be prevented by filtration, there are some limitations to using this system for wine filtration such as membrane fouling, cost, and time delays (Bartowsky and Henschke 2008).

3.5 Acetic Acid Bacteria in Beer

Beer is a kind of alcoholic beverage produced from malt, hops, yeasts and water. During the fermentation process, sugar is converted to alcohol and carbon dioxide by the activity of yeasts, and the condition becomes unsuitable for microbial growth with low pH, no oxygen, high carbon dioxide, high ethanol content, and the presence of antibacterial hop compounds (Wieme et al. 2014a). However, in aerobic environments, AAB can grow in beer as spoilage microorganisms (Table 5). Some species of *Acetobacter* and *Gluconobacter* cause sourness of beer. *G. oxydans* subsp. *suboxydans* and *G. oxydans* subsp. *industrius* may produce ropiness while *A. pasteurianus* may cause turbidity and sourness in the product (Machve 2009). Other strains isolated from beer are *A. orleanensis, A. estunensis, A. cerevisiae* (Andrés-Barrao et al. 2013).

On the other hand, positive effects of AAB in lambic type beers have also been reported (Martens et al. 1997). The production of lambic beer, which is special Belgian acidic ale, consists of three stages. Yeast, LAB and, in the final stage, LAB and AAB take a role in the fermentation process. To obtain sufficient amounts of ABB is critical during the third stage of lambic beer fermentation. Thus, AAB could be immobilized on the wooden walls of the casks through which oxygen can diffuse. Pulque is a non-distilled traditional alcoholic beverage produced in the central states of Mexico and a variety of AAB species are involved in the production of Pulque (Table 5).

3.6 AAB in Dairy Products

Most dairy products are produced by the activity of LAB. The anti-microbial potential of LAB, which is obtained by the metabolites such as organic acids (lactic and acetic), hydrogen peroxide, anti-microbial enzymes, bacteriocins and reuterin is significant in the preservation and microbiological stability of fermented dairy products. On the other hand, AAB can be rarely isolated from traditional dairy products despite their nondairy primary origins. *A. indonesiensis* was isolated from yogurt, which was produced traditionally in Iran. Other AAB were isolated from Iranian curd and identified as *A. syzygii* (Haghshenas et al. 2015). There is limited knowledge about the presence and the roles of AAB in dairy products.

4. Health Effects of Acetic Acid Bacteria Involved in Fermentations

Fermented foods have attracted attention all over the world as foods that might promote longevity with their beneficial health effects. One of the well documented health effects are those of vinegar. In the classical world, Hippocrates of Kos was the first to prescribe vinegar as the main remedy against a variety of diseases, such as the common cold and cough. Bioactive components of vinegar, including acetic acid, gallic acid, catechin, ephicatechin, chlorogenic acid, caffeic acid, p-coumaric acid, and ferulic acid, show some therapeutic effects such as anti-oxidative, anti-diabetic, anti-microbial, anti-tumor, anti-obesity, anti-hypertensive, and cholesterol-lowering responses (Budak et al. 2014). Komesu and kurosu, which are traditional rice vinegars produced in Japan, are known for their health benefits such as anti-oxidative activity, anti-tumor activity and prevention of inflammation and hypertension (Murooka and Yamshita 2008). The anti-tumor activity of vinegars produced from sweet potato-shochu post-distillation slurry by *Acetobacter aceti* is also reported (Morimura et al. 2004).

Although the majority of probiotic bacteria are classified as LAB, AAB may also be classified as probiotic bacteria (Haghshenas et al. 2015). *A. aceti* and *A. syzygii* introduced as probiotics (Lefeber et al. 2011). Similarly, secretions of *A. indonesiensis* and *A. syzygii* are exhibited typical probiotic properties, such as desirable tolerance for low pH and high bile salt concentration, appropriate anti-pathogenic activity against certain pathogenic bacteria, and acceptable antibiotic susceptibility. Significant cytotoxicity of *A. syzygii* secretions against cancer cell lines is also reported (Haghshenas et al. 2015). On the other hand, more studies are needed to display the probiotic properties of AAB and related health effects of fermented foods involving this group.

5. Conclusion

Fermentation is an ancient form of bio-preservation that is common all around the world. AAB can be isolated from fermented food products in high degrees of diversity. They are involved in fermented food products such as vinegar and cocoa because of their capacity to oxidize ethanol to acetic acid and their extreme resistance to high acetic acid concentrations. Their metabolites contribute mainly to flavor and texture properties of the product, while also displaying beneficial heath effects such as anti-microbial, anti-cancerogenic, proteolytic and nutraceutical properties. On the other hand, AAB can also be found to play a detrimental role in some fermented foods such as in wine, beer and coffee. Thus, good agricultural and fermentation practices are essential to obtain high-quality fermented products.

The functional roles and the physiological adaptations of AAB species most frequently found in food fermentations have been reported by researchers; these are mainly focused on wine and vinegar associated AAB. Although yeasts are widely used as starters in the wine and beer fermentations, AAB inoculation practice in vinegar and cocoa production is still not common. It is known that isolation, cultivation and preservation difficulties restrict the usage of AAB as a starter culture in food fermentations. On the other hand, with the use of novel genomic technologies, more detailed insight has been given in some traditional fermentation. Isolation and identification of the most suitable indigenous AAB strains present in spontaneous processes and then the use of selected well-defined cultures in the fermentation processes is a good strategy for increasing process control, reducing fermentation time, and predicting and ensuring the quality and reproducibility of the final product.

Keywords: Acetic acid bacteria (AAB), food, fermentation, vinegar, cocoa, wine, coffee, therapeutic effect, Acetobacteriaceae, *Acetobacter*, *Gluconobacter*, *Gloconacetobacter, Komagataeibacter*

References

Andrés-Barraoa, C., Benaglib, C., Chappuisa, M., Péreza, R.O., Tonollab, M. and Barja, F. 2013. Rapid identification of acetic acid bacteria using MALDI-TOF mass spectrometry fingerprinting. Systematic and Applied Microbiology 36: 75–81.

Ardhana, M. and Fleet, G.H. 2003. The microbial ecology of cocoa fermentations in Indonesia. International Journal of Food Microbiology 86: 87–99.

Bae, S., Fleet, G.H. and Heard, G.M. 2006. Lactic acid bacteria associated with wine grapes from several Australian vineyards. Journal of Applied Microbiology 100: 712–727.

Baena-Ruano, S., Santos-Duenas, I.M., Mauricio, J.C. and Garcia-Garciaa, I. 2010. Relationship between changes in the total concentration of acetic acid bacteria and major volatile compounds during the acetic acid fermentation of white wine. Journal of the Science of Food and Agriculture 90: 2675–2681.

Bamforth, C.W. 2008. Food, Fermentation and Micro-organisms, John Wiley & Sons, Chichester, GBR.

Barata, A., Malfeito-Ferreira, M. and Loureiro, V. 2012. Changes in sour rotten grape berry microbiota during ripening and wine fermentation. International Journal of Food Microbiology 154: 152–161.

Barbe, J.C., De Revel, G., Joyeux, A., Bertrand, A. and Lonvaud-Funel, A. 2001. Role of botrytized grape micro-organisms in SO₂ binding phenomena. Journal of Applied Microbiology 90: 34–42.

Bartowsky, E.J., Xia, D., Gibson, R.L., Fleet, G.H. and Henschke, P.A. 2003. Spoilage of bottled red wine by acetic acid bacteria. Letters in Applied Microbiology 36: 307–314.

Bartowsky, E.J. and Henschke, P.A. 2008. Acetic acid bacteria spoilage of bottled red wine—A review. International Journal of Food Microbiology 125: 60–70.

Boesch, C., Trcek, J., Sievers, M. and Teuber, M. 1998. *Acetobacter intermedius*, sp. nov., Systematic and Applied Microbiology 21: 220–229.

Budak, N.H., Aykin, E., Seydim, A.C., Greene, A.K. and Guzel-Seydim, Z.B. 2014. Functional properties of vinegar. Journal of Food Science 79: R757–764.

Callejon, R.M., Torija, M.J., Mas, A., Morales, M.L. and Troncoso, A.M. 2010. Changes of volatile compounds in wine vinegars during their elaboration in barrels made from different woods. Food Chemistry 120: 561–571.

Camu, N., De Winter, T., Verbrugghe, K., Cleenwerck, I., Vandamme, P., Takrama, J.S., Vancanneyt, M. and De Vuyst, L. 2007. Dynamics and biodiversity of populations of lactic acid bacteria and acetic acid bacteria involved in spontaneous heap fermentation of cocoa beans in Ghana. Applied and Environmental Microbiology 73: 1809–1824.

Camu, N., González, A., De Winter, T., Van Schoor, A., De Bruyne, K., Vandamme, P., Takrama, J.S. and De Vuyst, L. 2008. Influence of turning and environmental contamination on the dynamics of lactic acid bacteria and acetic acid bacteria populations involved in spontaneous cocoa bean heap fermentation in Ghana. Applied Environmental Microbiology 74: 86–98.

Carr, J.G., Davies, P.A. and Dougan, J. 1979. Cocoa fermentation in Ghana and Malaysia. Tropical Products Institute, London.

Copetti, M.V., Iamanaka, B.T., Mororó, R.C., Pereira, J.L., Frisvad, J.C. and Taniwaki, M.H. 2012. The effect of cocoa fermentation and weak organic acids on growth and ochratoxin. A production by *Aspergillus* species. International Journal of Food Microbiology 155: 158–164.

Crafack, M., Mikkelsen, M.R., Saerens, S., Knudsen, M., Beinnow, A., Lowor, S., Takrama, J., Swiegers, J.H., Petersen, G.B., Heimdal, H. and Nielsen, D.S. 2013. Influencing cocoa flavor using *Pichia kluyveri* and *Kluyveromyces marxianus* in a defined mixed starter culture for cocoa fermentation. International Journal of Food Microbiology 167: 103–116.

De Vuyst, L., Camu, N., De Winter, T., Vandemeulebroecke, K., Perre, V.V., Vancanneyt, M., De Vos, P. and Cleenwerck, I. 2008. Validation of the (GTG)5-rep-PCR fingerprinting technique for rapid classification and identification of acetic acid bacteria, with a focus on isolates from Ghanaian fermented cocoa beans. International Journal of Food Microbiology 125: 79–90.

Dircks, H.D. 2009. Investigation into the fermentation of Australian cocoa beans and its effect on microbiology, chemistry and flavour. Ph.D. Thesis, School of Chemical Engineering, University of New South Wales, Sydney, Australia.

Dong, D., Zheng, W., Jiao, L., Lang, Y. and Zhao, X. 2016. Chinese vinegar classification via volatiles using long-optical-path infrared spectroscopy and chemometrics. Food Chemistry 194: 95–100.

Du Toit, W.J. and Lambrechts, M.G. 2002. The enumeration and identification of acetic acid bacteria from South African red wine fermentations. International Journal of Food Microbiology 74: 57–64.

Du Toit, W.J., Pretorius, I.S. and Lonvaud-Funel, A. 2005. The effect of sulphur dioxide and oxygen on the viability and culturability of a strain of *Acetobacter pasteurianus* and a strain of *Brettanomyces bruxellensis* isolated from wine. Journal of Applied Microbiology 98: 862–871.

Dutta, D. and Gachhui, R. 2006. Novel nitrogen-fixing *Acetobacter nitrogenifigens* sp. nov., isolated from Kombucha tea. International Journal of Systematic and Evolutionary Microbiology 56: 1899–1903.

Escalante, A., Giles-Gómez, M., Hernández, G., Córdova-Aguilar, M.S., López-Munguía, A., Gosset, G. and Bolívar, F. 2008. Analysis of bacterial community during the fermentation of pulque, a traditional Mexican alcoholic beverage, using a polyphasic approach. International Journal of Food Microbiology 124: 126–134.

Euzéby, J.P. 1997. List of bacterial names with standing in nomenclature: a folder available on the Internet. International Journal of Systematic Bacteriology 47: 590–592.

Fernandez-Perez, R., Torres, C., Sanz, S. and Ruiz-Larrea, F. 2010. Rapid molecular methods for enumeration and taxonomical identification of acetic acid bacteria responsible for submerged vinegar production. European Food Research and Technology 231: 813–819.

Fuentes-Ramírez, L.E., Bustillos-Crystales, R., Tapia-Hernández, A., Jimenéz-Salgado, T., Wang, E.T., Martínez-Romero, E. and Caballero-Mellado, J. 2001. Novel nitrogen-fixing acetic acid bacteria, *Gluconacetobacter johannae* sp. nov. and *Gluconacetobacter azotocaptans* sp. nov., associated with coffee plants. International Journal of Systematic and Evolutionary Microbiology 51: 1305–1314.

Furukawa, S., Watanabe, T., Toyama, H. and Morinaga, Y. 2013. Significance of microbial symbiotic coexistence in traditional fermentation. Journal of Bioscience and Bioengineering 116: 533–539.

Galvez, S.L., Loiseau, G., Paredes, J.L., Barel, M. and Guiraud, J.P. 2007. Study on the microflora and biochemistry of cocoa fermentation in the Dominican Republic. International Journal of Food Microbiology 114: 124–130.

Garcia-Armisen, T., Papalexandratou, Z., Hendryckx, H., Camu, N., Vrancken, G., De Vuyst, L. and Cornelis, P. 2010. Diversity of the total bacterial community associated with Ghanaian and Brazilian cocoa bean fermentation samples as revealed by a 16 S rRNA gene clone library. Applied Microbiology and Biotechnology 87: 2281–2292.

González, A. 2005. Application of molecular techniques for identification of acetic acid bacteria. Ph.D. Thesis, Universitat Rovira I Virgili, Tarragona, Spain.

González, A., Hierro, N., Poblet, M., Mas, A. and Guillamón, J.M. 2005. Application of molecular methods to demonstrate species and strain evolution of acetic acid bacteria population during wine production. International Journal Food Microbiology 102: 295–304.

Gullo, M. and Giudici, P. 2008. Acetic acid bacteria in traditional balsamic vinegar: phenotypic traits relevant for starter cultures selection. International Journal of Food Microbiology 125: 46–53.

Haghshenas, B., Nami, Y., Abdullah, N., Radiah, D., Rosli, R. and Khosroushahi, A.Y. 2015. Anticancer impacts of potentially probiotic acetic acid bacteria isolated from traditional dairy microbiota. LWT—Food Science and Technology 60: 690–697.

Hamdouche, Y., Guehi, T., Durand, N., Kedjebo, K.B.D., Montet, D. and Meile, J.C. 2015. Dynamics of microbial ecology during cocoa fermentation and drying: Towards the identification of molecular markers. Food Control 48: 117–122.

Haruta, S., Ueno, S., Egawa, I., Hashiguchi, K., Fujii, A., Nagano, M., Ishii, M. and Igarashi, Y. 2006. Succession of bacterial and fungal communities during a traditional pot fermentation of rice vinegar assessed by PCR-mediated denaturing gradient gel electrophoresis. International Journal of Food Microbiology 109: 79–87.

Hidalgo, C., Mateo, E., Mas, A. and Torija, M.J. 2012. Identification of yeast and acetic acid bacteria isolated from the fermentation and acetification of persimmon (Diospyros kaki). Food Microbiology 30: 98–104.

Hidalgo, C., Mateo, E., Mas, A. and Torija, M.J. 2013a. Effect of inoculation on strawberry fermentation and acetification processes using native strains of yeast and acetic acid bacteria. Food Microbiology 34: 88–94.

Hidalgo, C., Garcia, D., Romero, J., Mas, A., Torija, M.J. and Mateo, E. 2013b. *Acetobacter* strains isolated during the acetification of blueberry (*Vaccinium corymbosum* L.) wine. Letters in Applied Microbiology 57: 227–232.

Huang, C.G., Chang, M.T., Huang, L. and Chu, W.S. 2014. Utilization of elongation factor Tu gene (tuf) sequencing and species specific PCR (SS-PCR) for the molecular identification of *Acetobacter* species complex. Molecular and Cellular Probes 28: 31–33.

Ilabaca, C., Navarrete, P., Mardones, P., Romero, J. and Mas, A. 2008. Application of culture culture-independent molecular biology based methods to evaluate acetic acid bacteria diversity during vinegar processing. International Journal of Food Microbiology 126: 245–249.

Illeghems, K., De Vuyst, L., Papalexandratou, Z. and Weckx, S. 2012. Phylogenetic analysis of a spontaneous cocoa bean fermentation metagenome reveals new insights into its bacterial and fungal community diversity. Plos One 7: e38040.

Jara, C., Mateo, E., Guillamón, J.M., Mas, A. and Torija, M.J. 2013. Analysis of acetic acid bacteria by different culture-independent techniques in a controlled superficial acetification. Annals of Microbiology 63: 393–398.

Jimenez-Salgado, T., Fuentes-Ramirez, L.E., Tapia-Hernandez, A., Mascarua-Esparza, M.A., Martinez-Romero, E. and Caballero-Mellado, J. 1997. *Coffea arabica* L., a new host plant for *Acetobacter diazotrophicus*, and isolation of other nitrogen fixing acetobacteria. Applied and Environmental Microbiology 63: 3676–3683.

Joyeux, A., Lafon-Lafourcade, S. and Ribéreau-Gayon, P. 1984. Evolution of acetic acid bacteria during fermentation and storage of wine. Applied and Environmental Microbiology 48: 153–156.

Kregiel, D., Rygala, A., Libudzisz, Z., Walczak, P. and Oltuszak-Walczak, E. 2012. *Asaia lannensisethe* spoilage acetic acid bacteria isolated from strawberry flavored bottled water in Poland. Food Control 26: 147–150.

Lefeber, T., Janssens, M., Camu, N. and De Vuyst, L. 2010. Kinetic analysis of strains of lactic acid bacteria and acetic acid bacteria in cocoa pulp simulation media to compose a starter culture for cocoa bean fermentation. Applied and Environmental Microbiology 76: 7708–7716.

Lefeber, T., Gobert, W., Vrancken, G., Camu, N. and De Vuyst, L. 2011. Dynamics and species diversity of communities of lactic acid bacteria and acetic acid bacteria during spontaneous cocoa bean fermentation in vessels. Food Microbiology 28: 457–464.

Lefeber, T., Papalexandratou, Z., Gobert, W., Camu, N. and De Vuyst, L. 2012. On-farm implementation of a starter culture for improved cocoa bean fermentation and its influence on the flavor of chocolates produced thereof. Food Microbiology 30: 379–392.

Machve, K.K. 2009. Fermentation Technology, Mangalam Publishers, Delhi, India.

Madhaiyan, M., Saravananb, V.S., Jovic, D.B.S.S., Leea, H., Thenmozhid, R., Harie, K. and Sa, T. 2004. Occurrence of *Gluconacetobacter diazotrophicus* in tropical and subtropical plants of Western Ghats. India Microbiological Research 159: 233–243.

Mamlouk, D., Hidalgo, C., Torija, M.J. and Gullo, M. 2011. Evaluation and optimisation of bacterial genomic DNA extraction for no-culture techniques applied to vinegars. Food Microbiology 28: 1374–1379.

Mamlouk, D. and Gullo, M. 2013. Acetic acid bacteria: Physiology and carbon sources oxidation. Indian Journal of Microbiology 53: 377–384.

Martens, H., Iserentant, D. and Verachtert, H. 1997. Microbiological aspects of a mixed yeast-bacterial fermentation in the production of a special Belgian acidic ale. Journal of the Institute of Brewing 103: 85–91.

Mas, A., Torija, M.J., García-Parrilla, M.C. and Troncoso, A.M. 2014. Acetic acid bacteria and the production and quality of wine vinegar. The Scientific World Journal Volume 2014, Article ID 394671, 6 pages.

Mateo, E., Torija, M.J., Mas, A. and Bartowsky, E.J. 2014. Acetic acid bacteria isolated from grapes of South Australian vineyards. International Journal of Food Microbiology 178: 98–106.

Morimura, S., Kawano, K., Han, L.S., Seki, T., Shigematsu, T. and Kida, K. 2004. Production of vinegar from sweet potato-shochu post-distillation slurry and evaluation of its antitumor activity via oral administration in a mouse model (in Japanese). Seibutsu Kogaku Kaishi 82: 573–578.

Murooka, Y. and Yamshita, M. 2008. Traditional healthful fermented products of Japan. Journal of Industrial Microbiology and Biotechnology 35: 791–798.

Nanda, N., Taniguchi, M., Ujike, S., Ishihara, N., Mori, H., Ono, H. and Murooka, Y. 2001. Characterization of acetic acid bacteria in traditional acetic acid fermentation of rice vinegar (Komesu) and unpolished rice vinegar (Kurosu) produced in Japan. Applied and Environmental Microbiology 67: 986–990.

Nie, Z., Zheng, Y., Wang, M., Han, Y., Wang, Y., Luo, J. and Niu, D. 2013. Exploring microbial succession and diversity during solid-state fermentation of Tianjin duliu mature vinegar. Bioresource Technology 148: 325–333.

Nielsen, D.S., Teniola, O.D., Ban-Koffi, L., Owusu, M., Andersson, T.S. and Holzapfel, W.H. 2007. The microbiology of Ghanaian cocoa fermentations analyzed using culture-dependent and culture-independent methods. International Journal of Food Microbiology 114: 168–186.

Nielsen, D.S., Arneborg, N. and Jespersen, L. 2014. Mixed microbial fermentations and methodologies for their investigation, pp. 1–41. *In*: R.F. Schwan and G.H. Fleet (eds.). Cocoa and Coffee Fermentations (Fermented Foods and Beverages Series). CRS Press. New York.

Nisiotou, A.A., Rantsiou, K., Iliopoulos, V., Cocolin, L. and Nychas, G.J. 2011. Bacterial species associated with sound and Botrytis-infected grapes from a Greek vineyard. International Journal Food Microbiology 145: 432–436.

Okazaki, S., Furukawa, S., Ogihara, H., Kawarai, T., Kitada, C., Komenou, A. and Yamasaki, M. 2010. Microbiological and biochemical survey on the transition of fermentative processes in Fukuyama pot vinegar brewing. Journal of General and Applied Microbiology 56: 205–211.

Ouoba, L.I.I., Kando C., Parkouda, C., Sawadogo-Lingani, H., Diawara, B. and Sutherland, J.P. 2012. The microbiology of Bandji, palm wine of Borassus akeassii from Burkina Faso: identification and genotypic diversity of yeasts, lactic acid and acetic acid bacteria. Journal of Applied Microbiology 113: 1428–1441.

Ozturk, I., Caliskan, O., Tornuk, F., Ozcan, N., Yalcin, H., Baslar, M. and Sagdic, O. 2015. Antioxidant, antimicrobial, mineral, volatile, physicochemical and microbiological characteristics of traditional home-made Turkish vinegars. LWT - Food Science and Technology 63: 144–151.

Papalexandratou, Z., Camu, N., Falony, G. and De Vuyst, L. 2011a. Comparison of the bacterial species diversity of spontaneous cocoa bean fermentations carried out at selected farms in Ivory Coast and Brazil. Food Microbiology 28: 964–973.

Papalexandratou, Z., Falony, G., Romanens, E., Jimenez, J.C., Amores, F., Daniel, H.M. and De Vuyst, L. 2011b. Species diversity, community dynamics, and metabolite kinetics of the microbiota associated with traditional Ecuadorian spontaneous cocoa bean fermentations. Applied and Environmental Microbiology 77: 7698–7714.

Papalexandratou, Z., Lefeber, T., Bahrian, B., Lee, O.S., Daniel, H.M. and De Vuyst, L. 2013. *Hanseniaspora opuntiae, Saccharomyces cerevisiae, Lactobacillus fermentum*, and *Acetobacter pasteurianus* predominate during well-performed Malaysian cocoa box bean fermentation, underlining the importance of these microbial species for a successful cocoa bean fermentation process. Food Microbiology 35: 73–85.

Pereira, G.V.D., Magalhães, K.T., de Almeida, E.G., Coelho, I.D. and Schwan, R.F. 2013. Spontaneous cocoa bean fermentation carried out in a novel-design stainless steel tank: Influence on the dynamics of microbial populations and physical-chemical properties. International Journal of Food Microbiology 161: 121–133.

Prieto, C., Jara, C., Mas, A. and Romero, J. 2007. Application of molecular methods for analysing the distribution and diversity of acetic acid bacteria in Chilean vineyards. International Journal Food Microbiology 115: 348–355.

Ruiz, A., Poblet, M., Mas, A. and Guillamon, J.M. 2000. Identification of acetic acid bacteria by RFLP of PCR-amplified 16S rDNA and 16S–23S rDNA intergenic spacer. International Journal of Systematic and Evolutionary Microbiology 50: 1981–1987.

Qi, Z., Yang, H., Xia, X., Xin, Y., Zhang, L., Wang, W. and Yu, X. 2013. A protocol for optimization vinegar fermentation according to the ratio of oxygen consumption versus acid yield. Journal of Food Engineering 116: 304–309.

Raspor, P. and Goranovic, D. 2008. Biotechnological applications of acetic acid bacteria. Critical Reviews in Biotechnology 28: 101–124.

Samah, O.A., Puteh, M.F., Selemat, J. and Alimon, H. 1993. Fermentation products in cocoa beans inoculated with *Acetobacter xylinum*. ASEAN Food Journal 8: 22–25.

Schüller, G., Hertel, C. and Hammes, W.P. 2000. *Gluconacetobacter entanii* sp. nov., isolated from submerged high-acid industrial vinegar fermentations. International Journal of Systematic and Evolutionary Microbiology 50: 2013–2020.

Schwan, R.F. and Wheals, A.E. 2004. The microbiology of cocoa fermentation and its role in chocolate quality. Critical Reviews in Food Science and Nutrition 44: 205–22.

Schwan, R.F., Pereira, G.V.D.M. and Fleet, G.H. 2014. Microbial activities during cocoa fermentation. pp. 129–192. *In*: R.F. Schwan and G.H. Fleet (eds.). Cocoa and Coffee Fermentations (Fermented Foods and Beverages Series). CRS Press. New York.

Sedlackova, P., Cerovsky, M., Horskava, I. and Voldrich, M. 2011. Cell surface characteristic of *Asaia bogorensise* spoilage microorganism of bottled water. Czech Journal of Food Sciences 29: 457–461.

Sengun, I.Y. and Karabiyikli, S. 2011. Importance of acetic acid bacteria in food industry. Food Control 22: 647–656.

Sengun, I.Y. 2013. Microbiological and chemical properties of fig vinegar produced in Turkey. African Journal of Microbiology Research 7: 2332–2338.

Shi, H., Xu, W., Trinh, Q., Luo, Y., Liang, Z., Li, Y. and Huang, K. 2012. Establishment of a viable cell detection system for microorganisms in wine based on ethidium monoazide and quantitative PCR. Food Control 27: 81–86.

Sievers, M., Sellmer, S. and Teuber, M. 1992. *Acetobacter europaeus* sp. nov., a main component of industrial vinegar fermenters in central Europe. Systematic and Applied Microbiology 15: 386–392.

Sievers, M. and Swings, J. 2005. Family Acetobacteraceae. pp. 41–95. *In*: G.M. Garrity (ed.). Bergey's Manual of Systematic Bacteriology, 2nd edition. Vol. 2. Springer, New York.

Silhavy, K. and Mandl, K. 2006. *Acetobacter tropicalis* in spontaneously fermented wines with vinegar fermentation in Austria. Mitteilungen Klosterneuburg 56: 102–107.

Silva, L.R., Cleenwerck, I., Rivas, R., Swings, J., Trujillo, M.E., Willems, A. and Velázquez, E. 2006. *Acetobacter oeni* sp. nov., isolated from spoiled red wine. International Journal of Systematic and Evolutionary Microbiology 56: 21–24.

Silva, C.F., Batista, L.R., Abreu, L.M., Dias, E.S. and Schwan, R.F. 2008. Succession of bacterial and fungal communities during natural coffee (*Coffea arabica*) fermentation. Food Microbiology 25: 951–957.

Silva, C.F. 2014. Microbial activity during coffee fermentation. pp. 398–430. *In*: R.F. Schwan and G.H. Fleet (eds.). Cocoa and Coffee Fermentations (Fermented Foods and Beverages Series). CRS Press. New York.

Sokollek, S.J., Hertel, C. and Hammes, W.P. 1998. Description of *Acetobacter oboediens* sp. nov. and *Acetobacter pomorum* sp. nov., two new species isolated from industrial vinegar fermentations. International Journal of Systematic Bacteriology 48: 935–940.

Trček, J. and Barja, F. 2015. Updates on quick identification of acetic acid bacteria with a focus on the 16S–23S rRNA gene internal transcribed spacer and the analysis of cell proteins by MALDI-TOF mass spectrometry 196: 137–144.

Tesfaye, W., Morales, M.L., García-Parrilla, M.C. and Troncoso, A.M. 2002. Wine vinegar: technology, authenticity and quality evaluation. Trends in Food Science and Technology 13: 12–21.

Torija, C., Mateo, E., Guillamon, J.M. and Mas, A. 2010. Identification and quantification of acetic acid bacteria in wine and vinegar by TaqMan-MGB probes. Food Microbiology 27: 257–265.

Valera, M.J., Federico Laich, F., Sara, S., González, S.S., Torija, M.J., Mateo, E. and Mas, A. 2011. Diversity of acetic acid bacteria present in healthy grapes from the Canary Islands. International Journal of Food Microbiology 151: 105–112.

Valera, M.J., Torija, M.J., Mas, A. and Mateo, E. 2015. Acetic acid bacteria from biofilm of strawberry vinegar visualized by microscopy and detected by complementing culture-dependent and culture-independent techniques. Food Microbiology 46: 452–462.

Vaughn, R.H., Camargo, R.D.E., Fallange, H., Mello Ayres, G. and Sergedello, A. 1958. Observations on the microbiology of the coffee fermentation in Brazil. Food Technology 12: 57. Supplement.

Vegas, C., Mateo, E., González, A., Jara, C., Guillamon, J.M., Poblet, M., Torija, M.J. and Mas, A. 2010. Population dynamics of acetic acid bacteria during traditional wine vinegar production. International Journal of Food Microbiology 138: 130–136.

Vegas, C., González, A., Mateo, E., Mas, A., Poblet, M. and Torija, M.J. 2013. Evaluation of representativity of the acetic acid bacteria species identified by culture-dependent method during a traditional wine vinegar production. Food Research International 51: 404–411.

Wieme, A.D., Spitaels, F., Aerts, M., Bruyne, K.D., Landschoot, A.V. and Vandamme, P. 2014a. Identification of beer-spoilage bacteria using matrix-assisted laser desorption/ionization time-of-flight mass spectrometry. International Journal of Food Microbiology 185: 41–50.

Wieme, A.D., Spitaels, F., Balzarini, T., Cleenwerck, I., Van Landschoot, A., De Vuyst, L. and Vandamme, P. 2014b. *Gluconobacter cerevisiae* sp. nov., isolated from the brewery environment. International Journal of Systematic and Evolutionary Microbiology 64: 1134–1141.

Wu, J.J., Gullo, M., Chen, F.S. and Giudici, P. 2010. Diversity of *Acetobacter pasteurianus* strains isolated from solid-state fermentation of cereal vinegars. Current Microbiology 60: 280–286.

Wu, J.J., Mac, Y.K., Zhang, F.F. and Chen, F.S. 2012. Biodiversity of yeasts, lactic acid bacteria and acetic acid bacteria in the fermentation of 'Shanxi aged vinegar', a traditional Chinese vinegar Food Microbiology 30: 289–297.

Xu, L., Huang, Z., Xiaojun, Z., Li, Q.Z., Lu, Z., Shi, J., Xu, Z.Z. and Ma, Y. 2011. Monitoring the microbial community during solid-state acetic acid fermentation of Zhenjiang aromatic vinegar. Food Microbiology 28: 1175–1181.

Yamada, Y., Hoshino, K. and Ishikawa, T. 1997. The phylogeny of acetic acid bacteria based on the partial sequences of 16S ribosomal RNA: the elevation of the subgenus *Gluconoacetobacter* to the generic level. Bioscience Biotechnology Biochemistry 61: 1244–1251.

Yamada, Y., Hoshino, K. and Ishikawa, T. 1998. *Gluconoacetobacter* nom. corrig. (*Gluconoacetobacter* [sic]). Validation of publication of new names and new combinations previously effectively published outside the IJSB. List no. 64. International Journal of Systematic Bacteriology 48: 327–328.

Yamada, Y., Yukphan, P., Lan Vu, H.T., Muramatsu, Y., Ochaikul, D., Tanasupawat, S. and Nakagawa, Y. 2012. Description of *Komagataeibacter* gen. nov., with proposals of new combinations (*Acetobacteraceae*). The Journal of General and Applied Microbiology 58: 397–404.

Yamada, Y. 2014a. Genera and species in acetic acid bacteria: The past, present and future. 3rd International Conference on Clinical Microbiology & Microbial Genomics. September 24–26, 2014, Valencia Convention Centre, Spain.

Yamada, Y. 2014b. Transfer of *Gluconacetobacter kakiaceti*, *Gluconacetobacter medellinensis* and *Gluconacetobacter maltaceti* to the genus *Komagataeibacter* as *Komagataeibacter kakiaceti* comb. nov., *Komagataeibacter medellinensis* comb. nov. and *Komagataeibacter maltaceti* comb. nov., International Journal of Systematic and Evolutionary Microbiology 64: 1670–1672.

Zahoor, T., Siddique, F. and Farooq, U. 2006. Isolation and characterization of vinegar culture (*Acetobacter aceti*) from indigenous sources. British Food Journal 108: 429–439.

Acetic Acid Bacteria
Prospective Applications in Food Biotechnology

Corinne Teyssier[1],* and *Yasmine Hamdouche*[1]

1. Introduction

Acetic Acid Bacteria (AAB) play important roles in the food and beverage industry everywhere in the world. They present large interests in food and beverage production and, in particular, in vinegar making, mainly due to their ability to oxidize ethanol to acetic acid. Conversely, they are also responsible for spoilage effects in some products such as wine. This chapter describes, briefly, the biotechnological applications of AAB in the food and beverage industry.

2. AAB Taxonomy

AAB belong to the family *Acetobacteraceae* in the *Alphaproteobacteria* phylum. In the past, AAB comprised two mains genera (*Acetobacter* described in 1898 and *Gluconobacter* introduced in 1935). Then, the genus *Gluconacetobacter* was created in 1997 in order to group *Acetobacter* species containing Q-10 ubiquinone as the major quinone component (Yamada et al. 1997). In recent years, the number of genus expanded very quickly due to the development of molecular phylogenetic analyses and many changes occurred in AAB taxonomy. Up to date, AAB are classified in 17 genera: *Acetobacter, Gluconobacter, Acidomonas, Gluconacetobacter, Asaia, Endobacter, Kosakia, Swaminathania, Saccharibacter, Neoasaia, Granulibacter,*

[1] Corinne Teyssier, UMR Qualisud, University of Montpellier, Yasmine Hamdouche, UMR Qualisud, Cirad, TA B-95/16, 73, rue Jean-François Breton, 34398 Montpellier cedex 5, France.
* Corresponding author: corinne.teyssier@cirad.fr

Tanticharoenia, Ameyamaea, Neokomagataea, Komagataeibacter, Nguyenibacter, and *Swingsia* (Malimas et al. 2013) and recently, two more genera have been added: *Bombella* and *Commensalibacter* (Trček and Barja 2015).

A brief overview of food-associated literature shows that only three genera are mostly involved in the food industry: *Acetobacter, Gluconobacter* (*G.*), and *Gluconacetobacter* (*Ga.*).

3. Morphology and Physiology

AAB are strictly aerobic rods or ellipsoidal shaped cells with Gram negative or Gram variable staining. The non-spore-forming cells have peritrichous or polar flagella when motile. They are catalase positive and oxidase negative. AAB are mesophilic bacteria with an optimum growth temperature range of 25°C to 30°C. Although the optimum pH growth range is 5.0 to 6.5, AAB can grow at lower pH values between 3.0 and 4.0 (Holt et al. 1994, Sievers and Swings 2005). According to the species or strains, AAB are able to transform most of the sugars and alcohols into organic acids such as acetic, citric, fumaric, lactic, malic, pyruvic, tartaric and succinic acids. Two types of membrane-bound enzymes are involved in the conversion of ethanol into acetic acid: alcohol dehydrogenase (ADH) and aldehyde dehydrogenase (ALDH) (Matsushita et al. 1994). The genus *Acetobacter* oxidizes alcohol preferentially over glucose whereas *Gluconobacter* preferentially oxidizes glucose more readily than ethanol (Gullo and Guidici 2008).

4. Culture, Identification and Molecular Typing of AAB Strains

AAB are considered as fastidious bacteria because they are difficult to isolate, cultivate and maintain in pure culture. The main media used for their culture are as follows: acetic acid bacteria (AAB) medium, *Acetobacter* medium, Glucose Yeast Extract $CaCO_3$ (GYC) medium, Yeast Extract Peptone Mannitol (YPM) medium and Acetic Acid Ethanol (AE) medium (Hommel et al. 2014). Many media derive from the previous media with changes only in the proportion of the different components. AAB culture on media containing $CaCO_3$ results in $CaCO_3$ degradation, forming a characteristic clear halo around the colonies due to the acid production (Cleenwerck and deVos 2008). The incubation time at 30°C on each medium varied from 24 h to 72 h according to the species and strains.

Classic biochemical identification systems commonly used for Gram-negative bacilli identification such as API 20E and API20NE systems (BioMérieux) or automated systems like VITEK 2 with ID-GNB card (BioMérieux) give unsatisfactory results for most AAB because the characteristics of these bacteria are absent from the database of these identification systems (Alauzet et al. 2010).

Molecular approaches are highly recommended for the identification of AAB. Amplification and sequencing of the 16S rRNA gene give excellent results for the identification at the genus level but are not sufficient to discriminate species in the genus (Alauzet et al. 2010). Many previous studies showed that the 16S rRNA gene is not a useful tool for species identification because the 16S rRNA gene present in some cases allows polymorphism between species (Alauzet et al. 2010) or macro (Teyssier et al. 2003) or micro (Marchandin et al. 2003) heterogeneities which are detrimental to identification. Therefore, multilocus sequence-based analyses of housekeeping genes were performed to obtain a successful affiliation at the species level. Cleenwerck et al. (2010) defined three housekeeping genes (*dnaK, groEL* and *rpoB*) useful for the species differentiation in the AAB group such as *Acetobacter, Gluconobacter*, and *Gluconacetobacter*. This multilocus approach allowed for the reclassification of species which had been mis-identified (i.e., *Acetobacter xylinus* subsp. *sucrofermentans* as *Gluconacetobacter sucrofermentans*) (Cleenwerck et al. 2010), and the description of novel species: *Gluconacetobacter medellinensis* (Castro et al. 2013), *Gluconacetobacter maltaceti* (Slapsak et al. 2013), and *Acetobacter sicerae* (Li et al. 2014).

Strains belonging to the same AAB species could be compared with conventional molecular techniques such as Pulsed-Field Gel Electrophoresis (PFGE), or repetitive sequence-based PCR (rep-PCR) using Enterobacterial Repetitive Intergenic Consensus (ERIC) sequences, Repetitive Extragenic Palindromic (REP) elements (Versalovic et al. 1991) or $(GTG)_5$ sequences (Versalovic et al. 1994) with conclusive and robust results. In recent years, PFGE was considered as the gold standard for AAB typing (Alauzet et al. 2010). Fernandez-Perez et al. (2010) showed that PFGE of *SpeI*-digested DNA and ERIC-PCR analysis were suitable tools for typing 103 AAB isolates from vinegars. ERIC-PCR analysis was also used to detect intra-specific variability from AAB strains isolated from Shanxi aged vinegar, a traditional Chinese cereal-based vinegar (Wu et al. 2012), present in healthy grapes from the Canary Islands (Valera et al. 2011) recovered from healthy, mold-infected and rot-affected grapes from South Australian vineyards (Mateo et al. 2014). $(GTG)_5$-rep-PCR permitted the study of AAB population during traditional wine vinegar production (Vegas et al. 2010) and during fermentation of cocoa beans from Ghana (De Vuyst et al. 2008).

4.1 Methods to Detect AAB and Analyze Inter-specific Diversity in Foods and Beverages

Culture-based studies comprise both phenotypic (cultural, physiological and biochemical analyses) and genotypic (16S rRNA or multilocus sequencing) approaches. Phenotypic and genotypic methods are still used to investigate cultivable microbiota of fermented foods. For example, the presence of *Acetobacter fabarum* was detected at high concentration (5.0 x 10^6 CFU/mL) in some Tibetan kefir grains samples (Gao et al. 2012a). Ouoba et al. (2012) detected AAB in the range of 1.2 x 10^5 and 1.0 x 10^6 CFU/g in Bandgi, a palm wine of *Borassus akeassi*, from Burkina Faso. Among these AAB strains, nine *Acetobacter* species with *Acetobacter indonesiensis* as predominant species, *Gluconobacter oxydans* strains and *Gluconacetobacter saccharivorans* were isolated.

It is generally accepted that it is very difficult to obtain correct culture from AAB strains originating from wines and vinegars. In recent years, culture-independent methods have been used primarily to assess the microbial diversity of fermented foods and beverages. These metagenomic methods circumvent the problems of Viable But Non Cultivable (VBNC) AAB. Methods based on the amplification by PCR of a short hyper variable region of the 16S rRNA gene followed by a Denaturing Gradient Gel Electrophoresis (PCR-DGGE) or by a Temporal Temperature Gel Electrophoresis (PCR-TTGE) are demonstrated to be efficient and reliable for analysis of bacterial diversity in foods: raw milk (Quigley et al. 2011), vinegar (Ilabaca et al. 2008), fruits (Le Nguyen et al. 2008a, El Sheika et al. 2012), fishes (Le Nguyen et al. 2008b), etc. Thus, PCR-DGGE analysis was applied to study the importance and diversity of AAB in vinegar production (De Vero et al. 2006), in various Philippine fermented food products (Dalmacio et al. 2011), in cocoa beans from different geographical regions (Papalexandratou et al. 2011a, Lefeber et al. 2011, Hamdouche et al. 2015), in fermented onion products (Cheng et al. 2014) and during malolactic fermentation of Tempranillo wines (Ruiz et al. 2010). Real-time quantitative methods associated with TaqMan Minor Groove Binder (MGB) probes have also been successfully applied not only to study, with high accuracy, AAB diversity but also to quantify them in food products such as wine and vinegars (Torija et al. 2010, Valera et al. 2013).

The next-generation sequencing (NGS) methods such as pyrosequencing analysis (Margulies et al. 2005) are currently the most promising, accurate and reliable approaches to analyze the microbial communities because they produce thousands of genetic sequences at the same time, including sequences from uncultivable micro-organisms and those from micro-organisms present in very small numbers. Usually, the generated sequences correspond to short hyper variable regions of the bacterial 16S rRNA gene. This high-throughput sequencing method was applied to various fermented food ecosystems such as fermented dairy products (Masoud et al. 2012, Aldrete-Tapia et al. 2014), plant- (Bessmeltseva et al. 2014, Jeong et al. 2013), meat-, and fish-derived (Jung et al. 2013, Kim et al. 2014) fermented foods (Mayo et al. 2014). NGS were successfully used to detect AAB in various fermented foods such as tarag (a fermented milk of cows from Mongolia and northwest China) (Sun et al. 2014), Brazilian kefir grains (Leite et al. 2012), and water kefir from multiple sources (Marsh et al. 2013). NGS methods confirmed the presence of *Acetobacter pasteurianus* during cocoa bean fermentation process (Illeghems et al. 2012).

Besides, Matrix-Assisted Laser Desorption/Ionization Time-Of-Flight Mass Spectrometry (MALDI-TOF MS) analysis appears to be an appropriate technique for routine microbial quality control, in particular for AAB responsible for spoilage in the beer industry (Wieme et al. 2014).

5. Natural Habitats of AAB

AAB are present everywhere in the environment, mainly in fruits and flowers. For example, the recently described *Gluconobacter kanchanaburiensis*, *Gluconobacter wancherniae*, *Gluconobacter nephelii*, were isolated respectively from a spoiled fruit of *Artocarpus heterophyllus* in Thailand (Malimas et al. 2009), from an unknown yellow fruit in Thailand (Yukphan et al. 2010), and from a tropical fruit of *Nephelium lappaceum* (Kommanee et al. 2011). Similarly, *Acetobacter okinawensis*, *Acetobacter papayae* and *Acetobacter persicus* were recovered respectively from stems of sugarcane, various fruits and a flower in Japan (Lino et al. 2012a). The very recently described genus *Neokomagataea* comprises two species *N. thailandica* and *N. tanensis*, respectively isolated from the flowers of lantana and candel bush collected in Thailand (Yukphan et al. 2011). AAB may also propagate in food materials which contain sugar or in fermented products which contain alcohol. Sixty-four AAB strains belonging to *Acetobacter* (45 isolates), *Gluconobacter* (11 isolates) and *Gluconacetobacter* (8 isolates) genera were collected from Indonesian fruits, flowers and fermented foods (Yamada et al. 1999). An earlier study underscored that *Acetobacter* and *Gluconacetobacter* stains were recovered primarily from fermented foods, while *Gluconobacter* strains were collected mainly from fruits and flowers (Lisdiyanti et al. 2003). AAB are commonly recovered from various vinegars such as traditional kaki vinegar in Japan (Lino et al. 2012b), coconut water vinegar in Sri Lanka (Perumpuli et al. 2014), Shanxi aged vinegar, a traditional cereal vinegar in China (Wu et al. 2012), and of course, wine vinegar (Mas et al. 2014), among others. AAB were also recovered from fermented beverages such as water-kefir from United Kingdom, Canada, and the United States as a minor component (Marsh et al. 2013), but in higher amounts from other water-kefir analyzed from various sources in Germany (Gulitz et al. 2011). AAB were also present in other fermented beverages such as kombucha tea samples (Marsh et al. 2014). Two very recently described species of *Acetobacter*, *A. sicerae* and *A. lambici*, were isolated from cider and kefir respectively (Li et al. 2014) and from lambic beer (Spitaels et al. 2014). *Acetobacter* sp. strains and, in particular, *Acetobacter farinalis* was recovered from different fermented rice flours (khao-khab) and related products (Tanasupawat et al. 2011a,b).

AAB are present at all stages of wine-making, from the mature grape, through vinification to storage (Bartowsky and Henschke 2008). Indeed, AAB were isolated from grapes of South Australian vineyards (Mateo et al. 2014) and from healthy grapes from the Canary Islands (Valera et al. 2011). AAB were responsible of spoilage in grape wine (Bartowsky and Henschke 2008) and also in other fermented beverages such as beer, or soft drinks (Raspor and Goranovic 2008). The presence of AAB during cocoa fermentation was implied in the development of flavor in cocoa products (Lefeber et al. 2012).

6. Importance and Roles of AAB in Food and Beverage Production

AAB play not only beneficial but also undesirable roles in food and beverage production. This section presents the importance and role of AAB in foods and beverages highlighting, in particular, their worldwide economic impact (Table 1). The presence of AAB in three food products has been outlined: vinegar, wine and cocoa.

6.1 AAB and Vinegar Production

The earliest industrial application of AAB in food production corresponds to vinegar making. Vinegar is defined as the result of acetous fermentation of natural alcoholic substrates by AAB. In fact, AAB are involved secondarily after the conversion of ethanol from a carbohydrate source performed by yeasts. Various substrates could be used to produce vinegars: fruits (apple wine, grape wine, fermented juices such

Table 1. Main species of AAB recovered from foods and beverages of worldwide interest.

Food	Species	References
Vinegar (from different sources)	*A. aceti, A. cerevisiae, A. estunensis, A. malorum, A. oeni, A. pasteurianus, A. pomorum, G. oxydans, Ga. entanii, Ga. europaeus, Ga. hansenii, Ga. intermedius, Ga. liquefaciens, Ga. oboediens, Ga. xylinus*	Sokollek et al. 1998, Nanda et al. 2001, De Vero et al. 2006, Haruta et al. 2006, Wu et al. 2010
Wine	*A. aceti, A. pasteurianus, A. tropicalis, G. oxydans, Ga. liquefaciens*	Yamada et al. 1997, Du Toit and Lambrechts 2002, Bartowsky et al. 2003, Gonzalez et al. 2004, Silva et al. 2006, Silhavy and Mandl 2006
Fermented cocoa	*A. aceti, A. ghanensis, A. lovaniensis, A. pasteurianus, A. senegalensis, A. syzygii, Ga. xylinus, Ga. xylinus* subsp. *sucrofermentans*	De Vuyst et al. 2008, Camu et al. 2007, 2008, Lefeber et al. 2011, 2012, Papalexandratou et al. 2011a,b,c, 2013, Hamdouche et al. 2015
Coffee	*Ga. Azotocaptans, Ga. diazotrophicus, Ga. johannae*	Jimenez-Salgado et al. 1997, Fuentes-Ramírez et al. 2001, Madhaiyana et al. 2004

* *A.: Acetobacter, G.: Gluconobacter, Ga.: Gluconacetobacter.*

as peaches and blueberries), starch (malt, rice), and spirit (Raspor and Goranovic 2008). A wide variety of vinegars are produced around the world but wine vinegar is predominant in continental Europe. Two well-defined methods could be used to produce vinegar: the traditional or 'surface method' and 'the submerged method' (Budak et al. 2014). In the 'submerged method', commonly used for commercial vinegars, AAB are submerged in the liquid substrate and constant aeration is provided to accelerate the fermentation; vinegars are ready after 24–48 hours of fermentation. The traditional method corresponds to a 'surface method', in which the AAB culture occurs on the surface of wood barrels in direct contact with atmospheric oxygen. To initiate the process, a starter culture, called 'seed vinegar' or 'mother of vinegar', obtained from a previous vinegar batch should be added. This method is time-consuming (taking several weeks) but permits the development of metabolic by-products, increasing the sensorial complexity of vinegars, thus leading to high quality vinegars. The AAB isolated during vinegar fermentations belong mainly to the genera *Acetobacter*, *Gluconobacter* and *Gluconacetobacter* (Table 1). Among them, *Acetobacter aceti*, *Acetobacter malorum*, *Acetobacter pasteurianus*, *Acetobacter pomorum*, *Gluconobacter oxydans*, *Gluconacetobacter europaeus*, *Gluconacetobacter hansenii*, *Gluconacetobacter intermedius*, *Gluconacetobacter oboediens*, and *Gluconacetobacter xylinus* were the more frequently recovered species (De Vero and Giudici 2008, Budak et al. 2014). *A. pasteurianus* was usually predominant in the production of vinegars with low acetic acid content (4–7% w/v) whereas *Gluconacetobacter* species (*Ga. xylinus*, *Ga. europaeus*, *Ga. intermedius*, and *Ga. oboediens*) were used to produce high acidity industrial vinegars (up to 15–20%) (Vegas et al. 2010). Trček et al. (2007) showed that *Ga. europaeus* adapts to high acetic acid concentrations by modifying the phospholipid contents of their cytoplasmic membrane. For mid-acidity vinegar such as traditional balsamic vinegar, *A. pasteurianus*, *Ga. europaeus*, and *Ga. xylinus* were predominant species. Thus, different AAB were associated during vinegar production in order to optimize the properties of each species; *Acetobacter* spp. are better acid producers but in contrast, they can fully oxidize acetic acid to CO_2 and H_2O whereas *Gluconobacter* is unable to. The greater tolerance of *Gluconacetobacter* to acetic acid could explain its contribution in the end of processes when acetic acid content is higher. Depending on the strains, secondary fermentations often associated with acetic acid fermentation lead to typical aroma and flavor (Raspor and Goranovic 2008). Some *Ga. europaeus* strains used for rice vinegar production are known to produce an unfavorable flavor compound, acetoin. A genetic approach was successful applied to reduce acetoin production by using the *Ga. europaeus* strain KGMA4004 (Akasaka et al. 2013). With reference to wine vinegar-associated AAB, *A. pasteurianus* appears to be predominant throughout the process (Ilabaca et al. 2008, Vegas et al. 2010). The results of AAB monitoring during traditional wine vinegar production highlighted the predominance of *A. pasteurianus* throughout the process. *Ga. europaeus,* however, increased when the acetic acid concentration was above 6% (w/v). Different AAB strains followed each other according to the acetic acid concentration (Vegas et al. 2010). AAB diversity studies performed throughout the fermentation process lets us envisage the use of starter culture to improve the vinegar process control and sensory qualities.

6.2 AAB and Grape Wine Production

Wine is most often obtained by the alcoholic fermentation of grapes. Microorganisms responsible for wine fermentations are mainly yeasts present on grapes. Many microorganisms (yeasts, acetic acid bacteria and lactic acid bacteria) present on the grapes can also survive and grow in wine, constituting the wine microbial consortium (Barata et al. 2012a). The proportion of these microorganisms depends on the grape ripening stage and on the availability of nutrients.

Microorganisms such as AAB are responsible of wine spoilage which could affect wine quality. However good manufacturing practices could avoid proliferation of AAB and the production of acetic acid could be limited. For example, anaerobic conditions of the alcoholic fermentation do not permit the growth of AAB whereas higher temperature of wine storage and higher wine pH promote the growth and metabolism of AAB (Sengun and Karabiyikli 2011). Addition of sulfites in wine represents an efficient method to decrease the number of detrimental microorganisms. In a recent approach, promoting the reduction of chemicals, grape and wine phenolic extracts were successfully tested for their antimicrobial activities against frequently occurring wine spoilage AAB (Pastorkova et al. 2013). Among AAB, *Gluconobacter*, *Acetobacter* and *Gluconacetobacter* are the mains genera involved in wine spoilage (Table 1). *Acetobacter* spp. are the dominant AAB in the final stage of sour rot while the genera *Gluconacetobacter* and *Gluconobacter* are mostly recovered from sound berries (Barata et al. 2012b). It was commonly admitted that *G. oxydans* was preferentially detected in grape must whereas *A. aceti* and *A. pasteurianus* were detected in wine. Progress performed in the intra-specific identification of AAB could partially explain that other species are currently showed to be associated with grapes or wines. These 'novel' species often depended on the geographical origin of wine (Silva et al. 2006, Valera et al. 2011). Recent results obtained from AAB isolated from grapes of South Australian vineyards demonstrated the presence of *Gluconobacter* genus with predominance of *G. cerinus* but also strains belonging to the species *Asaia siamensis* were isolated (Mateo et al. 2014). Wine-associated AAB were also analyzed for their potential role in malolactic fermentation resulting in main spoilage of wine caused by the well-known lactic acid bacterium *Oenococcus oeni* (Ruiz et al. 2010). Thus AAB community associated to grape wine varies according to the state of grape but also to the vineyards contributing to the characteristics of wine.

6.3 AAB and Cocoa Fermentation

Cocoa beans are the major raw material for chocolate production. Spontaneous fermentation of mucilaginous pulp that surrounds beans is crucial for the development of precursors of chocolate flavor (Thompson et al. 2013). Although industry of cocoa represents tremendous income for many countries, little is known about this traditional process of fermentation. Three groups of microorganisms are successively involved in 2–8 days fermentation. First, yeasts produce ethanol from carbohydrates of the pulp in absence of oxygen. The increased pH supports the growth of lactic acid bacteria, which produce lactic acid from sugars. Then, AAB whose presence is facilitated by aeration during mixing oxidize alcohol to acetic acid resulting in the death of seed embryo and liberation of endogenous enzymes implied in flavor precursors. Finally, sun drying of beans stops fermentation and permits to eliminate the detrimental acetic acid. The activity of AAB is crucial to obtain high-quality cocoa. *Acetobacter pasteurianus* but also others *Acetobacter* species (*A. syzygii*, *A. ghanensis*, *A. senegalensis, A. tropicalis*, etc.) as *Ga. medellinensis* (formerly *Ga. xylinus*) and *G. oxydans* are the main representatives of the AAB communities (Table 1) (De Vuyst et al. 2008, Papalexandratou et al. 2011a,b,c and 2013, Hamdouche et al. 2015). Monitoring of AAB population during fermentation and sun drying of cocoa from Ivory Coast showed a succession of different AAB species: *A. pasteurianus* and *A. syzygii* were dominant in the first half of fermentation whereas *A. tropicalis* was dominant during the later stages. Presence of *Ga. xylinus* was only detected during sun drying step (Hamdouche et al. 2015). Cocoa beans origin and fermentation practices in the farm impacted microbial community in particular AAB present during fermentation (De Vuyst et al. 2008, Papalexandratou et al. 2011a,b,c and 2013, Hamdouche et al. 2015). Currently, many studies examined the microbial communities associated to cocoa fermentation in different countries in order to choice the well-adapted strains and to produce functional starter-culture (Pereira et al. 2012, Lefeber et al. 2012, Illeghems et al. 2013, Moens et al. 2014). AAB together with yeasts and lactic acid bacteria will be certainly part of this starter-culture. The objectives are to control and improve the cocoa fermentation and to obtain the best quality cocoa.

7. Other Roles of AAB in Biotechnology

AAB play also a role in the production of commercially important molecules used in the food and pharmaceutical industries. Among the AAB, *G. oxydans* is known to be resourceful in particular for its enzymes, which find applications in biotechnology. For example, metabolic bioengineering research on *G. oxydans* will soon lead to replacing two time-consuming chemical processes used to produce xylitol and vitamin C (L-ascorbic acid). Xylitol represents a potential alternative to the utilization of sucrose because xylitol: (i) has the same degree of sweetness, (ii) is appropriate for diabetics, (iii) is not involved in dental caries. Xylitol can be obtained chemically but an approach involving *G. oxydans* was proposed to limit chemicals and to reduce costs. *G. oxydans* presents the ability to produce xylitol from D-arabitol due to the presence of a xylitol dehydrogenase. A two-step process was described to produce xylitol directly from D-glucose: a recombinant strain of *G. oxydans* called PXPG (the strain expresses *xdh* together with *gdh*) was used to efficiently transform xylitol from D-arabinose, generated previously by osmotic yeasts (Zhang et al. 2013). For several decades, a two-step fermentation route involving three microorganisms, including *G. oxydans*, was used at an industrial scale for the production of 2-Keto-L-gulonic acid, the direct precursor of L-ascorbic acid, from D-sorbitol. In order to reduce costs and simplify the process but preserve the excellent yield, a recombinant strain of *G. oxydans* named WSH-003, possessing genes encoding many different enzymes, was constructed (Gao et al. 2012b, Gao et al. 2014) and tested for its utilization in a one-step process.

Other *G. oxydans* strains were chosen because of the tremendous potential in terms of enzymes and their applications in industries. Mao et al. (2012) isolated from *G. oxydans* DSM-2003, a novel intracellular dextrinase, which catalyzed the synthesis of dextran from maltodextrin. Navarro-Gonzalez et al. (2012) reported the presence of a novel enzyme, arylesterase, isolated from the wine-associated *G. oxydans* strain 621H. This enzyme could be a promising biocatalyst in the variation of wine-aromas. However, the hopes riding on *G. oxydans* for the production of D-xylonate directly from D-xylose (and originating from plant biomass hydrolysates) remain unsuccessful because of reduced yields as compared to those from others microorganism (Toivari et al. 2012).

Gluconacetobacter hansenii was also studied for its ability to produce cellulose, which could be used for a variety of applications (Mohite and Patil 2014).

8. AAB as Human Opportunistic Pathogen: Consequences on Human Health?

In recent years, members of the genera *Acetobacter*, *Asaia*, *Gluconobacter* and *Granulibacter* have been recognized as emerging opportunistic human pathogens (see review of Alauzet et al. 2010). AAB such as *Acetobacter indonesiensis* (Bittar et al. 2008), *Acetobacter cibinongensis*, *Acetobacter lannaensis*, *Asaia bogorensis* and *Gluconobacter* sp. were largely recovered from bloodstream infections or from respiratory tracts of cystic fibrosis patients (Bittar et al. 2008, Gouby et al. 2007, Abdel-Haq et al. 2009, Alauzet et al. 2010). Considering that AAB were able to grow in an acidic environment, it could explain the fact that they may specifically colonize and potentially infect the acidic liquid of respiratory tract of cystic fibrosis patients. Little is known about the mode of transmission of these bacteria to man, but food and beverages represent a potential source of contamination. A case of bacteremia reported in an intravenous drug abuser suggests that the origin of contamination results from the use of acidic substances like vinegar or lemon juice to dilute heroin (Alauzet et al. 2010). Vigilance should be observed mainly for the immune-compromised person, given that many AAB clinical strains were multi-resistant to antibiotics (Alauzet et al. 2010).

9. Conclusion

AAB are used worldwide to produce foods and beverages. They are also involved in food and pharmaceutical industries. The tremendous progress made in the molecular methods of identification have resulted in significant advances in knowledge about AAB; this will permit better monitoring and control.

Keywords: Acetic Acid Bacteria, food, fermentation

References

Abdel-Haq, N., Savacsan, S., Davis, M., Asmar, B.I., Painter, T. and Salimnia, H. 2009. *Asaia lannaensis* bloodstream infection in a child with cancer and bone marrow transplantation. Journal of Medical Microbiology 58: 974–976.

Akasaka, N., Sakoda, H., Hidese, R., Ishii, Y. and Fujiwara, S. 2013. An efficient method using *Gluconacetobacter europaeus* to reduce an unfavorable flavor compound, acetoin, in rice vinegar production. Applied and Environmental Microbiology 79: 7334–7342.

Alauzet, C., Teyssier, C., Jumas-Bilak, E., Gouby, A., Chiron, R., Rabaud, C., Counil, F., Lozniewski, A. and Marchandin, H. 2010. *Gluconobacter* as well as *Asaia* species newly emerging opportunistic human pathogens among acetic acid bacteria. Journal of Clinical Microbiology 48: 3935–3942.

Aldrete-Tapia, A., Escobar-Ramírez, M.C., Tamplin, M.L. and Hernández-Iturriaga, M. 2014. High-throughput sequencing of microbial communities in Poro cheese, an artisanal Mexican cheese. Food Microbiology 44: 136–141.

Barata, A., Malfeito-Ferreira, M. and Loureiro, V. 2012a. The microbial ecology of wine grape berries. International Journal of Food Microbiology 153: 243–259.

Barata, A., Malfeito-Ferreira, M. and Loureiro, V. 2012b. Changes in sour rotten grape berry microbiota during ripening and wine fermentation. International Journal of Food Microbiology 154: 152–161.

Bartowsky, E.J., Xia, D., Gibson, R.L., Fleet, R.L. and Henschke, P.A. 2003. Spoilage of bottled red wine by acetic acid bacteria. Letters in Applied Microbiology 36: 307–314.

Bartowsky, E.J. and Henschke, P.A. 2008. Acetic acid bacteria spoilage of bottled red wine: A review. International Journal of Food Microbiology 125: 60–70.

Bessmeltseva, M., Viiard, E., Simm, J., Paalme, T. and Sarand, I. 2014. Evolution of bacterial consortia in spontaneously started rye sourdoughs during two months of daily propagation. Plos One 9(4): e95449.

Bittar, F., Reynaud-Gaubert, M., Thomas, P., Boniface, S., Raoult, D. and Rolain, J.M. 2008. *Acetobacter indonesiensis* pneumonia after lung transplant. Emerging Infectious Diseases 14: 997–998.

Budak, N.H., Aykin, E., Seydim, A.C., Greene, A.K. and Guzel-Seydim, Z.B. 2014. Functional properties of vinegar. Journal of Food Science 79: 757–764.

Camu, N., De Winter, T., Verbrugghe, K., Cleenwerck, I., Vandamme, P., Takrama, J.S., Vancanneyt, M. and De Vuyst, L. 2007. Dynamics and biodiversity of populations of lactic acid bacteria and acetic acid bacteria involved in spontaneous heap fermentation of cocoa beans in Ghana. Applied and Environmental Microbiology 73: 1809–1824.

Camu, N., González, Á., De Winter, T., Van Schoor, A., De Bruyne, K., Vandamme, P., Takrama, J.S., Addo, S.K. and De Vuyst, L. 2008. Influence of turning and environmental contamination on the dynamics of populations of lactic acid and acetic acid bacteria involved in spontaneous cocoa bean heap fermentation in Ghana. Applied and Environmental Microbiology 74: 86–98.

Castro, C., Cleenwerk, I., Trček, J., Zuluaga, R., De Vos, P., Caro, G., Aguirre, R., Putaux, J.L. and Gañán, P. 2013. *Gluconacetobacter medellinensis* sp. nov., cellulose- and non-cellulose producing acetic acid bacteria isolated from vinegar. International Journal of Systematic and Evolutionary Microbiology 63: 1119–1125.

Cheng, L., Luo, J., Li, P., Yu, H., Huang, J. and Luo, L. 2014. Microbial diversity and flavor formation in onion fermentation. Food and Function 5: 2338–2347.

Cleenwerck, I. and De Vos, P. 2008. Polyphasic taxonomy of acetic acid bacteria: An overview of the currently applied methodology. International Journal of Food Microbiology 125: 2–14.

Cleenwerck, I., De Vos, P. and De Vuyst, L. 2010. Phylogeny and differentiation of species of the genus *Gluconacetobacter* and related taxa based on multilocus sequence analyses of housekeeping genes and reclassification of *Acetobacter xylinus* subsp. *sucrofermentans* as *Gluconacetobacter sucrofermentans* (Toyosaki et al. 1996) comb. nov., International Journal of Systematic and Applied Microbiology 60: 2277–2283.

Dalmacio, L.M., Angeles, A.K., Larcia, L.L., Balolong, M.P. and Estacio, R.C. 2011. Assessment of bacterial diversity in selected Philippine fermented food products through PCR-DGGE. Beneficial Microbes 2: 273–281.

De Vero, L., Gala, E., Gullo, M., Solieri, L., Land, S. and Giudici, P. 2006. Application of denaturing gradient gel electrophoresis (DGGE) analysis to evaluate acetic acid bacteria in traditional balsamic vinegar. Food Microbiology 23: 809–813.

De Vero, L. and Guidici, P. 2008. Genus-specific profile of acetic acid bacteria by 16S rDNA PCR-DGGE. International Journal of Food Microbiology 125: 96–101.

De Vuyst, L., Camu, N., De Winter, T., Vandemeulebroecke, K., Van de Perre, V., Vancanneyt, M., De Vos, P. and Cleenwerck, I. 2008. Validation of the (GTG)$_5$-rep-PCR fingerprinting technique for rapid classification of acetic acid bacteria, with a focus on isolates from Ghanaian fermented cocoa beans. International Journal of Food Microbiology 125: 79–90.

Du Toit, W.J. and Lambrechts, M.G. 2002. The enumeration and identification of acetic acid bacteria from South African red wine fermentations. International Journal of Food Microbiology 74: 57–64.

El Sheikha, A.F., Durand, N., Sarter, S., Okullo, J.B.L. and Montet, D. 2012. Study of the microbial discrimination of fruits by PCR-DGGE: Application to the determination of the geographical origin of *Physalis* fruits from Colombia, Egypt, Uganda and Madagascar. Food Control 24: 57–63.

Fernández-Pérez, R., Torres, C., Sanz, S. and Ruiz-Larrea, F. 2010. Strain typing of acetic acid bacteria responsible for vinegar production by the submerged elaboration method. Food Microbiology 27: 973–978.

Fuentes-Ramirez, L.E., Bustillos-Cristales, R., Tapia-Hernández, A., Jiménez-Salgado, T., Wang, E.T., Martinez-Romero, E. and Caballero-Mellado, J. 2001. Novel nitrogen-fixing acetic acid bacteria *Gluconacetobacter johannae* sp. nov. and *Gluconacetobacter azotocaptans* sp. nov., associated with coffee plants. International Journal of Systematic and Evolutionary Microbiology 51: 1305–1314.

Gao, J., Gu, F., Abdella, N.H., Ruan, H. and He, G. 2012a. Investigation on culturable microflora in Tibetan kefir grains from different areas from China. Journal of Food Science 77: 425–433.

Gao, L., Zhou, J., Liu, J., Du, G. and Chen, J. 2012b. Draft genome sequence of *Gluconobacter oxydans* WSH-003, a strain that is extremely tolerant of saccharides and alditols. Journal of Bacteriology 194: 4455–4456.

Gao, L., Hu, Y., Liu, J., Du, G., Zhou, J. and Chen, J. 2014. Stepwise metabolic engineering of *Gluconobacter oxydans* WSH-003 for the direct production of 2-keto-L-gulonic acid from D-sorbitol. Metabolic Engineering 24: 30–37.

Gonzalez, A., Hierro, N., Poblet, M., Rozes, N., Mas, A. and Guillamon, J.M. 2004. Application of molecular methods for the differentiation of acetic acid bacteria in a red wine fermentation. Journal of Applied Microbiology 96: 853–860.

Gouby, A., Teyssier, C., Vecina, F., Marchandin, H., Granolleras, C., Zorgniotti, I. and Jumas-Bilak, E. 2007. *Acetobacter cibinongensis* bacteremia in human. Emerging Infectious Diseases 13: 784–785.

Gulitz, A., Stadie, J., Wenning, M., Ehrmann, M.A. and Vogel, R.F. 2011. The microbial diversity of water kefir. International Journal of Food Microbiology 151: 284–288.

Gullo, M. and Giudici, P. 2008. Acetic acid bacteria in traditional balsamic vinegar: phenotypic traits relevant for starter cultures selection. International Journal of Food Microbiology 125: 46–53.

Hamdouche, Y., Guehi, T., Durand, N., Kedjebo, K.B.D., Montet, D. and Meile, J.C. 2015. Dynamics of microbial ecology during cocoa fermentation and drying: Towards the identification of molecular markers. Food Control 48: 117–122.

Haruta, S., Ueno, S., Egawa, I., Hashiguchi, K., Fujii, A., Nagano, M., Ishii, M. and Igarashi, Y. 2006. Succession of bacterial and fungal communities during a traditional pot fermentation of rice vinegar assessed by PCR-mediated denaturing gradient gel electrophoresis. International Journal of Food Microbiology 109: 79–87.

Holt, J.G., Krieg, N.R., Sneath, P.H.A., Staley, J.T. and Williams, S.T. 1994. Genus *Acetobacter* and *Gluconobacter*. pp. 71–84. *In*: Bergey's Manual of Determinative Bacteriology (ed.). Maryland, U.S.A.

Hommel, R.K. 2014. *Acetobacter*. pp. 3–10. *In*: Encyclopedia of Food Microbiology—Second edition. (ed.) Academic Press, Elsevier.

Ilabaca, P., Navarrete, P., Mardones, J. and Romero, A. 2008. Mas application of culture culture-independent molecular biology based methods to evaluate acetic acid bacteria diversity during vinegar processing. International Journal of Food Microbiology 126: 245–249.

Illeghems, K., De Vuyst, L., Papalexandratou, Z. and Weckx, S. 2012. Phylogenetic analysis of a spontaneous cocoa bean fermentation metagenome reveals new insights into its bacterial and fungal community diversity. Plos One 7(5): e38040.

Illeghems, K., De Vuyst, L. and Weckx, S. 2013. Complete genome sequence and comparative analysis of *Acetobacter pasteurianus* 386B, a strain well-adapted to the cocoa bean fermentation ecosystem. BMC Genomics 14: 526–540.

Jeong, S.H., Lee, S.H., Jung, J.Y., Choi, I.J. and Jeon, C.O. 2013. Microbial succession and metabolite changes during long-term storage of kimchi. Journal of Food Science 78(5): M763–9.

Jimenez-Salgado, T., Fuentes-Ramirez, L.E., Tapia-Hernandez, A., Mascarua-Esparza, M.A., Martinez-Romero, E. and Caballero-Mellado, J. 1997. *Coffea arabica* L., a new host plant for *Acetobacter diazotrophicus*, and isolation of other nitrogen fixing acetobacteria. Applied and Environmental Microbiology 63: 3676–3683.

Jung, J.Y., Lee, S.H., Lee, H.J. and Jeon, C.O. 2013. Microbial succession and metabolite changes during fermentation of saeu-jeot: Traditional Korean salted seafood. Food Microbiology 34: 360–368.

Kim, H.J., Kim, M.J., Turner, T.L., Kim, B.S., Song, K.M., Yi, S.H. and Lee, M.K. 2014. Pyrosequencing analysis of microbiota reveals that lactic acid bacteria are dominant in Korean flat fish fermented food, gajami-sikhae. Bioscience Biotechnology and Biochemistry 78: 1611–1618.

Kommanee, J., Tanasupawat, S., Yukphan, P., Malimas, T., Muramatsu, Y., Nakagawa, Y. and Yamada, Y. 2011. *Gluconobacter nephelii* sp. nov., an acetic acid bacterium in the class *Alphaproteobacteria*. International Journal of Systematic and Evolutionary Microbiology 61: 2117–2122.

Lefeber, T., Gobert, W., Vrancken, G., Camu, N. and De Vuyst, L. 2011. Dynamics and species diversity of communities of lactic acid bacteria and acetic acid bacteria during spontaneous cocoa bean fermentation in vessels. Food Microbiology 28: 457–464.

Lefeber, T., Papalexandratou, Z., Gobert, W., Camu, N. and De Vuyst, L. 2012. On-farm implementation of a starter culture for improved cocoa bean fermentation and its influence on the flavor of chocolates produced thereof. Food Microbiology 30: 379–392.

Leite, A.M.O., Mayo, B., Rachid, C.T.C.C., Peixoto, R.S., Silva, J.T., Paschoalin, V.M.F. and Delgado, S. 2012. Assessment of the microbial diversity of Brazilian kefir grains by PCR-DGGE and pyrosequencing analysis. Food Microbiology 31: 215–221.

Le Nguyen, D.D., Gemrot, E., Loiseau, G. and Montet, D. 2008a. Determination of citrus fruit origin by using 16S rDNA fingerprinting of bacterial communities by PCR-DGGE: An application on clementine from Morocco and Spain. Fruits 63: 3–9.

Le Nguyen, D.D., Ngoc, H.H., Dijoux, D., Loiseau, G. and Montet, D. 2008b. Determination of fish origin by using 16S rDNA fingerprinting of bacterial communities by PCR-DGGE: An application on Pangasius fish from Viet Nam. Food control 19: 454–460.

Li, L., Wieme, A., Spitaels, F., Balzarini, T., Nunes, O.C., Manaia, C.M., Van Landschoot, A., De Vuyst, L., Cleewerck, I. and Vandamme, P. 2014. *Acetobacter sicerae* sp. nov., isolated from cider and kefir, and identification of species of the genus *Acetobacter* by *dnaK*, *groEL* and *rpoB* sequence analysis. International Journal of Systematic and Evolutionary Microbiology 64: 2407–2415.

Lino, T., Suzuki, R., Kosako, Y., Ohkuma, M., Komagata, K. and Uchimura, T. 2012a. *Acetobacter okinawensis* sp. nov., *Acetobacter papayae* sp. nov. and *Acetobacter persicus* sp. nov.; novel acetic acid bacteria isolated from stems of sugarcane, fruits, and a flower in Japan. Journal of General and Applied Microbiology 58: 235–243.

Lino, T., Suzuki, R., Tanaka, N., Kosako, Y., Ohkuma, M., Komagata, K. and Uchimura, T. 2012b. *Gluconacetobacter kakiaceti* sp. nov., an acetic acid bacterium isolated from a traditional Japanese fruit vinegar. International Journal of Systematic and Evolutionary Microbiology 62: 1465–1469.

Lisdiyanti, P., Katsura, K., Potacharoen, W., Navarro, R.R., Yamada, Y., Uchimura, T. and Komagata, K. 2003. Diversity of acetic acid bacteria in Indonesia, Thailand, and the Philippines. Microbiology and Culture Collections 19: 91–98.

Madhaiyana, M., Saravananb, V.S., Jovic, D.B.S.S., Leea, H., Thenmozhid, R., Harie, K. et al. 2004. Occurrence of *Gluconacetobacter diazotrophicus* in tropical and subtropical plants of Western Ghats, India. Microbiological Research 159: 233–243.

Malimas, T., Yukphan, P., Lundaa, T., Muramatsu, Y., Takahashi, M., Kaneyasu, M., Potacharoen, W., Tanasupawat, S., Nakagawa, Y., Suzuki, K.-I., Tanticharoen, M. and Yamada, Y. 2009. *Gluconobacter kanchanaburiensis* sp. nov., a brown pigment-producing acetic acid bacterium for Thai isolates in the *Alphaproteobacteria*. Journal of General and Applied Microbiology 55: 247–254.

Malimas, T., Chaipitakchonlatarn, W., Thi Lan Vu, H., Yukphan, P., Muramatsu, Y., Tanasupawat, S., Potacharoen, W., Nakagawa, Y., Tanticharoen, M. and Yamada, Y. 2013. *Swingsia samuiensis* gen. nov., sp. nov., an osmotolerant acetic acid bacterium in the *Proteobacteria*. Journal of General and Applied Microbiology 59: 375–384.

Mao, X., Wang, S., Kan, F., Wei, D. and Li, F. 2012. A novel dextran dextrinase from *Gluconobacter oxydans* DSM-2003: purification and properties. Applied Biochemistry and Biotechnology 168: 1256–1264.

Marchandin, H., Teyssier, C., Siméon de Buochberg, M., Jean-Pierre, H., Carrière, C. and Jumas-Bilak, E. 2003. Intra-chromosomal heterogeneity between the four 16S rRNA gene copies in the genus *Veillonella*: Implications for phylogeny and taxonomy. Microbiology UK 149: 1493–1501.

Margulies, M. et al. 2005. Genome sequencing in microfabricated high-density picolitre reactors. Nature 437: 376–380.

Marsh, A.J., O'Sullivan, O., Hill, C., Ross, R.P. and Cotter, P.D. 2013. Sequence-based analysis of the microbial composition of water kefir from multiple sources. FEMS Microbiology Letter 348: 79–85.

Marsh, A.J., O'Sullivan, O., Hill, C., Ross, R.P. and Cotter, P.D. 2014. Sequence-based analysis of the bacterial and fungal compositions of multiple kombucha (tea fungus) samples. Food Microbiology 38: 171–178.

Mas, A., Torija, M.J., García-Parrilla, M.C. and Troncoso, A.M. 2014. Acetic acid bacteria and the production and quality of wine vinegar. The Scientific World Journal 2014: 394671.

Masoud, W., Vogensen, F., Lillevang, S., Al-Soud, W., Søresen, S. and Jakobsen, M. 2012. The fate of indigenous microbiota, starter culture, *Escherichia coli*, *Listeria innocua* and *Staphylococcus aureus* in Danish raw milk and cheese determined by pyrosequencing and quantitative real time (qRT)-PCR. International Journal of Food Microbiology 153: 192–202.

Mateo, E., Torija, M.J., Mas, A. and Bartowsky, E.J. 2014. Acetic acid bacteria isolated from grapes of South Australian vineyards. International Journal of Food Microbiology 178: 98–106.

Matsushita, K., Toyama, H. and Adachi, O. 1994. Respiratory chains and bioenergetics of acetic acid bacteria. Advances in microbial physiology 36: 247–301.

Mayo, B., Rachid, C.T., Alegria, A., Leite, A.M., Peixoto, R.S. and Delgado, S. 2014. Impact of next-generation techniques in food microbiology. Current genomics 15: 293–309.

Moens, F., Lefeber, T. and De Vuyst, L. 2014. Oxidation of metabolites highlights the microbial interactions and role of *Acetobacter pasteurianus* during cocoa bean fermentation. Applied Environmental Microbiology 80: 1848–1857.

Mohite, B.V. and Patil, S.V. 2014. A novel biomaterial: bacterial cellulose and its new era applications. Biotechnology and Applied Biochemistry 61: 101–110.

Nanda, K., Taniguchi, M., Ujike, S., Ishihara, N., Mori, H., Ono, H. and Murooka, Y. 2001. Characterization of acetic acid bacteria in traditional acetic acid fermentations of rice vinegar (Komesu) and unpolished rice vinegar (Kurosu) produced in Japan. Applied and Environmental Microbiology 67: 986–990.

Navarro-Gonzales, I., Sanchez-Ferrer, A. and Garcia-Carmona, F. 2012. Molecular characterization of a novel arylesterase from the wine-associated acetic acid bacterium *Gluconobacter oxydans* 621H. Journal of Agricultural and Food Chemistry 60: 10789–10795.

Ouoba, L.I.I., Kando, C., Parkouda, C., Sawadogo-Lingani, H., Diawara, B. and Sutherland, J.P. 2012. The microbiology of Bandji, palm wine of *Borassus akeassii* from Burkina Faso: identification and genotypic diversity of yeasts, lactic acid and acetic acid bacteria. Journal of Applied Microbiology 113: 1428–1441.

Papalexandratou, Z., Camu, N., Falony, G. and De Vuyst, L. 2011a. Comparison of the bacterial species diversity of spontaneous cocoa bean fermentations carried out at selected farms in Ivory Coast and Brazil. Food Microbiology 28: 964–973.

Papalexandratou, Z., Falony, G., Romanens, E., Jimenez, J.C., Amores, F., Daniel, H.-M. and De Vuyst, L. 2011b. Species diversity, community dynamics, and metabolite kinetics of the microbiota associated with traditional Ecuadorian spontaneous cocoa beans fermentations. Applied and Environmental Microbiology 77: 7698–7714.

Papalexandratou, Z., Vrancken, G., De Bruyne, K., Vandamme, P. and De Vuyst, L. 2011c. Spontaneous organic cocoa bean box fermentations in Brazil are characterized by a restricted species diversity of lactic acid bacteria and acetic acid bacteria. Food Microbiology 28: 1326–1338.

Papalexandratou, Z., Lefeber, T., Bahrim, B., Lee, O.S., Daniel, H.-M. and De Vuyst, L. 2013. *Hanseniospora opuntiae, Saccharomyces cerevisiae, Lactobacillus fermentum,* and *Acetobacter pasteurianus* predominate during well-performed Malaysian cocoa bean box fermentations, underlining the importance of these microbial species for a successful cocoa bean fermentation process. Food Microbiology 35: 73–85.

Pastorkova, E., Zakova, T., Landa, P., Nokakova, J., Vadlejch, J. and Kokoska, L. 2013. Growth inhibitory effect of grape phenolics against wine spoilage yeasts and acetic acid bacteria. International Journal of Food Microbiology 161: 209–213.

Pereira, G.V., Da Cruz Pedrozo Miguel, M.G., Ramos, C.L. and Schwan, R.F. 2012. Microbiological and physiological characterization of small-scale cocoa fermentations and screening of yeast and bacterial strains to develop a defined starter culture. Applied and Environmental Microbiology 78: 5395–5405.

Perumpuli, P.A., Watanabe, T. and Toyama, H. 2014. Identification and characterization of thermotolerant acetic acid bacteria strains isolated from coconut water vinegar in Sri Lanka. Bioscience Biotechnology and Biochemistry 78: 533–541.

Quigley, L., O'Sullivan, O., Beresford, T.P., Ross, R.P., Fitzgerald, G.F. and Cotter, P.D. 2011. Molecular approaches to analysing the microbial composition of raw milk and raw milk cheese. International Journal of Food Microbiology 150: 81–94.

Raspor, P. and Goranovic, D. 2008. Biotechnological applications of acetic acid bacteria. Critical Reviews in Biotechnology 28: 101–124.

Ruiz, P., Seseña, S., Izquierdo, P.M. and Palop, M.L. 2010. Bacterial biodiversity and dynamics during malolactic fermentation of Tempranillo wines as determined by a culture-independent method (PCR-DGGE). Applied Microbiology and Biotechnology 86: 1555–1562.

Sengun, I.Y. and Karabiyikli, S. 2011. Importance of acetic acid bacteria in food industry. Food Control 22: 647–656.

Sievers, M. and Swings, J. 2005. Family II. *Acetobacteraceae.* pp. 41–95. *In*: G. Garrity, D.L. Brenner, N.R. Krieg and J.T. Staley (eds.). Bergey's Manual of Systematic Bacteriology. vol. 2. Springer New York.

Silhavy, K. and Mandl, K. 2006. *Acetobacter tropicalis* in spontaneously fermented wines with vinegar fermentation in Austria. Mitteilungen Klosterneuburg 56: 102–107.

Silva, L.R., Cleenwerck, I., Rivas, R., Swings, J., Trujillo, M.E., Willems, A. and Velázquez, E. 2006. *Acetobacter oeni* sp. nov., isolated from spoiled red wine. International Journal of Systematic and Evolutionary Microbiology 56: 21–24.

Slapšak, N., Cleenwerck, I., De Vos, P. and Trček, J. 2013. *Gluconacetobacter maltaceti* sp. nov., a novel vinegar producing acetic acid bacterium. Systematic and Applied Microbiology 36: 17–21.

Sokollek, S.J., Hertel, C. and Hammes, W.P. 1998. Description of *Acetobacter oboediens* sp. nov. and *Acetobacter pomorum* sp. nov., two new species isolated from industrial vinegar fermentations. International Journal of Systematic Bacteriology 48: 935–940.

Spitaels, F., Wieme, A.D., Janssens, M., Aerts, M., Daniel, H.M., Van Landschoot, A., De Vuyst, L. and Vandamme, P. 2014. The microbial diversity of traditional spontaneously fermented lambic beer. Plos One 9(4): e95384.

Sun, Z., Liu, W., Bao, Q., Zhang, J., Hou, Q., Kwok, L., Sun, T. and Zhang, H. 2014. Investigation of bacterial and fungal diversity in Tarag using high-throughput sequencing. Journal of Dairy Science 97: 6085–96.

Tanasupawat, S., Kommanee, J., Yukphan, P., Muramatsu, Y., Nakagawa, Y. and Yamada, Y. 2011a. *Acetobacter farinalis* sp. nov., an acetic acid bacterium in the *Alphaproteobacteria*. Journal of General and Applied Microbiology 57: 159–167.

Tanasupawat, S., Kommanee, J., Yukphan, P., Nakagawa, Y. and Yamada, Y. 2011b. Identification of *Acetobacter* strains from Thai fermented rice products based on the 16S rRNA gene sequence and 16S–23S rRNA gene internal transcribed spacer restriction analyses. Journal of the Science of Food and Agriculture 91: 2652–2659.

Teyssier, C., Marchandin, H., Siméon de Buochberg, M., Ramuz, M. and Jumas-Bilak, E. 2003. Atypical 16S rRNA gene copies in *Ochrobactrum intermedium* strains reveal a large genomic rearrangement by recombination between *rrn* copies. Journal of Bacteriology 185: 2901–2909.

Thompson, S.S., Miller, K.B., Lopez, A. and Camu, N. 2013. Cocoa and coffee. pp. 881–889. *In*: M.P. Doyle and R.I. Buchanan (eds.). Food Microbiology: Fundamentals and Frontiers. ASM Press, Washington, DC.

Toivari, M.H., Nygard, Y., Penttilä, M., Ruohonen, L. and Wiebe, M.G. 2012. Microbial D-xylonate production. Applied Microbiology and Biotechnology 96: 1–8.

Torija, M.J., Mateo, E., Guillamon, J.M. and Mas, A. 2010. Identification and quantification of acetic acid bacteria in wine and vinegar by TaqMan-MGB probes. Food Microbiology 27: 257–265.

Trček, J. and F. Barja. 2015. Identification of acetic acid bacteria with a focus on the 16S–23S rRNA gene internal transcribed spacer and the analysis of cell proteins by MALDI-TOF mass spectrometry - Int. J. Food Microbiol. 196: 137–144.

Trček, J., Jernejc, K. and Matsushita, K. 2007. The highly tolerant acetic acid bacterium *Gluconacetobacter europaeus* adapts to the presence of acetic acid by changes in lipid composition, morphological properties and PQQ-dependent ADH expression. Extremophiles 11: 627–635.

Valera, M.J., Laich, F., Gonzalez, S.S., Torija, M.J., Mateo, E. and Mas, A. 2011. Diversity of acetic acid bacteria present in healthy grapes from the Canary Islands. International Journal of Food Microbiology 151: 105–112.

Valera, M.J., Torija, M.J., Mas, A. and Mateo, E. 2013. *Acetobacter malorum* and *Acetobacter cerevisiae* identification and quantification by real-time PCR with TaqMan-MGB probes. Food Microbiology 36: 30–39.

Vegas, C., Mateo, E., Gonzalez, A., Jara, C., Guillamon, J.M., Poblet, M., Torija, M.J. and Mas, A. 2010. Population dynamics of acetic acid bacteria during traditional wine vinegar production. International Journal of Food Microbiology 138: 130–136.

Versalovic, J., Koeuth, T. and Lupski, J.R. 1991. Distribution of repetitive DNA sequences in eubacteria and application to fingerprinting of bacterial genomes. Nucleic Acids Research 19: 6823–6831.

Versalovic, J., Schneider, M., De Bruijn, FJ. and Lupski, J.R. 1994. Genomic fingerprinting of bacteria using repetitive sequence-based polymerase chain reaction. Methods in Molecular and Cellular Biology 5: 25–40.

Wieme, A.D., Spitaels, F., Aerts, M. and De Bruyne, K. 2014. Identification of beer-spoilage bacteria using matrix-assisted laser desorption/ionization time-of-flight mass spectrometry. International Journal of Food Microbiology 185: 41–50.

Wu, J.J., Gullo, M., Chen, F.S. and Giudici, P. 2010. Diversity of *Acetobacter pasteurianus* strains isolated from solid-state fermentation of cereal vinegars. Current in Microbiology 60: 280–286.

Wu, J.J., Ma, Y.K., Zhang, F.F. and Chen, F.S. 2012. Biodiversity of yeasts, lactic acid bacteria and acetic acid bacteria in the fermentation of 'Shanxi aged vinegar' a traditional Chinese vinegar. Food Microbiology 30: 289–297.

Yamada, Y., Hoshino, K. and Ishikawa, T. 1997. The phylogeny of acetic acid bacteria based on the partial sequence of 16S ribosomal RNA: The elevation of subgenus *Gluconoacetobacter* to the generic level. Bioscience Biotechnology and Biochemistry 61: 1244–1251.

Yamada, Y., Hosono, R., Lisdyanti, P., Widyastuti, Y., Saono, S., Uchimura, T. and Komagata, K. 1999. Identification of acetic acid bacteria isolated from Indonesian sources, especially of isolates classified in the genus *Gluconobacter*. Journal of General and Applied Microbiology 45: 23–28.

Yamada, Y. 2000. Transfer of *Acetobacter oboediens* Sokollek et al. 1998 and *Acetobacter intermedius* Boesch et al. 1998 to the genus *Gluconacetobacter* as *Gluconacetobacter oboediens* comb. nov. and *Gluconacetobacter intermedius* comb. nov., International Journal of Systematic and Evolutionary Microbiology 50: 2225–2227.

Yukphan, P., Malimas, T., Lundaa, T., Muramatsu, Y., Takahashi, M., Kaneyasu, M., Tanasupawat, S., Nakagawa, Y., Suzuki, K.-I., Tanticharoen, M. and Yamada, Y. 2010. *Gluconobacter wancherniae* sp. nov., an acetic acid bacterium from isolates in the *Alphaproteobacteria*. Journal of General and Applied Microbiology 56: 67–73.

Yukphan, P., Malimas, T., Muramatsu, Y., Potacharoen, W., Tanasupawat, S., Nakagawa, Y., Tanticharoen, M. and Yamada, Y. 2011. *Neokomagataea* gen. nov., with descriptions of *Neokomagataea thailandica* sp. nov. and *Neokomagataea tanensis* sp. nov., osmotolerant acetic acid bacteria of the α-*Proteobacteria*. Bioscience Biotechnology and Biochemistry 75: 419–426.

Zhang, J., Li, S., Xu, H., Zhou, P., Zhang, L. and Ouyang, P. 2013. Purification of xylitol dehydrogenase and improved production of xylitol by increasing XDH activity and NADH supply in *Gluconobacter oxydans*. Journal of Agricultural and Food Chemistry 61: 2861–2867.

6

Lactic Acid Bacteria
General Characteristics, Food Preservation and Health Benefits

Pratima Khandelwal,[1,]* Frédéric Bustos Gaspar,[2,3]
Maria Teresa Barreto Crespo[2,3] and R.S. Upendra[1]

1. Introduction

It is a clearly acknowledged fact that lactic acid bacteria (LAB) is the one of the most widely used group of microorganisms in the production of varieties of fermented foods, worldwide, of both dairy and non-dairy origin. This group continues to be one of the most studied groups of bacteria till today (Konings et al. 2000). LAB habitats have been associated with materials of plant origin, human and animal cavities (mouth, genital, and intestinal and respiratory tracts), water, fruit/vegetable juices, fermented foods (dairy products, meat, fish, vegetables, fruits, silage, and beverages), as well as spoiled food, sewage, and decomposing plant materials (König and Fröhlich 2009). LAB, in the form of starter cultures, are essential for many industrial processes in the food industry, mainly for the processing of dairy and meat fermented products, the fermentation of cereals and vegetables, and brewing and wine making. The use of LAB in the cheese production industry represents a market of 55 billion Euros (mainly considering *Lactococcus* and *Lactobacillus*), in the yogurt and fresh dairy products 25 billion Euros (mainly considering *Streptococcus* and *Lactobacillus*), and in the production of probiotics, it represents a market of 20 billion Euros (mainly considering *Lactobacillus* and *Bifidobacterium*) (Konings et al. 2000, König and Fröhlich 2009, de Vos 2011, Douillard and de Vos 2014).

[1] Department of Biotechnology, New Horizon College of Engineering, Bangalore, India, 560103.
[2] iBET – Instituto de Biologia Experimental e Tecnológica, Apartado 12, 2781-901 Oeiras, Portugal.
[3] Instituto de Tecnologia Química e Biológica, Universidade Nova de Lisboa, Av. da República, 2780-157 Oeiras, Portugal.
* Corresponding author: pratima2k1@gmail.com

In this chapter, LAB relevant to food fermentation processes have been evaluated in terms of classification, metabolism, physiology, and applications. Their antimicrobial activities and effect on human health, as well as the products in which they are a fundamental part of the fermentation process, are discussed.

2. Characterizing LAB

In the following sections, the classifications and physiology of LAB are discussed in brief.

2.1 The LAB Group

LAB comprise an ecologically diverse group of microorganisms united by the formation of lactic acid as the primary metabolite of sugar metabolism (Davis et al. 1986, 1988, Lonvaud-Funel 1999, Liu and Pilone 2000, Liu 2002). They were accepted as a group of bacteria that could be studied together in the first quarter of the 20th century, after the classical publication, '*The Butter Aroma Bacteria*' by Orla-Jensen in 1919 (Orla-Jensen et al. 1926). But, LAB are not a homogenous group. They share a low G + C content (Amann et al. 1995), are Gram-positive, non-spore forming, facultative anaerobic, rod shaped (bacillus), or spherical (coccus) microorganisms (Garvie 1984), with genome sizes that can range from 1.8 to 3.2 Mb (Salvetti et al. 2013, Douillard and de Vos 2014). Acid tolerance ability provides LAB the ability to outcompete other bacteria in a natural fermentation, as they can withstand the increased acidity from organic acid production, a property that makes them important in the final phases of many food fermentation processes, when other microorganisms are inhibited by the low pH. These bacteria are involved in the fermentation of a wide range of substrates, i.e., milk, fruit, vegetables, cereals, meat, and fish. They appear in the gut of many mammals and in the external environment. The fact that they can survive in so many different substrates is a clear indication of the diversity of their metabolism. Presently, the classification of LAB inserts them into six families, *Aerococcaceae, Carnobacteriaceae, Enterococcaceae, Leuconostocaceae, Lactobacillacea*, and *Streptococcaceae*. Among them, the four families *Enterococcaceae, Leuconostocaceae, Lactobacillacea*, and *Streptococcaceae* gather the majority of the genera and species that are essential to food fermentations and are used as starter cultures. The other two families have LAB that can be related more to food spoilage, as well as related to environments but not related to food production. Among the four families, seven genera standout: *Enterococcus*, *Oenococcus* and *Leuconostoc, Lactobacillus, Pediococcus, Lactococcus* and *Streptococcus*, from each of the four families, respectively. Considering these genera, there are nearly 415 species that can be mentioned as belonging to the group LAB (Euzéby 1997, Parte 2014). However, some of the genera of those families, like *Enterococcus, Streptococcus, Lactococcus* or *Carnobacterium*, have species that have also been associated as human and animal pathogens and, as such, should be looked at according to their history of safe use before being considered for food fermentation. If LAB are considered *sensu lato*, another genus has to be and is often included in the group, the genus *Bifidobacterium* that comprises 50 species. Bifidobacteria are Gram-positive, hetero-fermentative, non-motile, non-spore forming bacteria that have lactic acid as the main end-product of fermentation. Besides lactic acid production, the genus *Bifidobacterium* is included because some species have been used as probiotics for a long time now. Nevertheless, they are phylogenetically distinct from the above mentioned LAB, as they have a higher G + C content (42–67%) (Pokusaeva et al. 2011). The genus *Bifidobacterium* belongs to the family *Bifidobacteriaceae*, order Bifidobacteriales, sub-class Actinobacteridae, class Actinobacteria, phylum Actinobacteria. In the last years, numerous studies have been published on the phylogenetical relationships among low G + C (34–51%) LAB species. Although based on a different set of genes, these studies show that LAB are phylogenetically close and differ mainly due to the gain of novel genes or the loss/decay of ancestral genes. Also, the presence of plasmids and megaplasmids are very important in terms of food production as they carry additional genes involved in metabolic pathways (Zhang et al. 2011, Douillard and de Vos 2014). Comparative and functional genomics analysis has been performed to deepen knowledge of the metabolism and physiology of LAB when used in food production, as well as of its relationship with humans (Siezen et al. 2004, Makarova et al. 2006, Canchaya et al. 2006, Makarova and Koonin 2007, Claesson et al. 2008, Schroeter and Klaenhammer 2009, Lukjancenko et al. 2012, Bottacini et al. 2014). The comparison of their metabolic pathways has also been done, as this is another very important feature for the application of these bacteria in food products (Siezen et al. 2004, Wels et al. 2011, Salvetti et al. 2013). LAB rapidly consume the fermentable carbohydrates of diverse food matrices and convert them to organic acids, namely lactic acid. This rapid acidification enables LAB to outcompete other microorganisms in the food matrix. This low pH and the production of antimicrobial compounds have a double result: (1) extension of the shelf life of the product, due to the inhibition of the growth of spoilage microorganisms, and (2) improvement of the safety of the product, due to the inhibition of pathogenic microorganisms.

LAB metabolic activities range from the breakdown of carbohydrates, to that of proteins, peptides, and fats. The regulatory networks that establish relations between the activities of transcription factors and the genes they control have been studied using transcriptome data. Biological functions such as sugar, energy and nitrogen metabolism, and stress response, were revealed for *Lactobacillus plantarum*, a highly interconnected regulatory network (Wels et al. 2011). Later, a comparative genomic approach for reconstructing the transcriptional regulatory networks (TRNs) in LAB belonging to the Lactobacillales order of the Firmicutes phylum was performed by Ravcheev and the research group, revealing interesting trends in the evolution of TRNs and individual transcriptional factors regulons (Ravcheev et al. 2013). Some of the main biological activities that are relevant to the fermentation of food products are summarized below.

2.2 Carbohydrate Metabolism

The breakdown of sugars through different catabolic pathways provides energy to the bacterial cells in the form of ATP and other reducing equivalents. The best-characterized pathways for sugar catabolism in bacteria are the Embden-Meyerhof-Parnas (EMP), the pentose phosphate (PP), and the Entner-Doudoroff (ED) pathways (Ray and Joshi 2014).

2.2.1 Homo-fermentative microorganisms

Lactic acid is the major product of glycolysis, or EMP pathway, in carbohydrate fermentation. This pathway is characterized by the formation of fructose1, 6-bisphosphate that is split into di-hydroxy acetone phosphate and glyceraldehydes 3-phosphate. Before entering glycolysis, hexoses or pentoses are generally phosphorylated when being transported inside the cell by the Phosphoenol pyruvate (PEP)-dependent carbohydrate phosphotransferase system (PTS). Homo-lactic organisms ferment one mole of glucose to two moles of pyruvate. These two moles of pyruvate are further reduced to two moles of lactic acid by the oxidation of NADH, yielding two net moles of ATP per mole of glucose consumed (Wisselink et al. 2002, Gaspar et al. 2013). *Enterococcus faecalis* has also the ED pathway, which was considered to be restricted to Gram-negative bacteria; however, current studies indicate that it is widely distributed from Archaea to Eukarya (Conway 1992, Ramsey et al. 2014). The overall scheme of the ED and EMP pathways is quite similar: 6-carbon sugars are primed by phosphorylation and subsequently cleaved by the aldolase enzyme into two 3-carbon intermediates. The two distinctive key enzymes to the ED pathway are (i) 6-phosphogluconate dehydratase which catalyses dehydration of 6-phosphogluconate to form 2-keto-3-deoxy-6-phosphogluconate (KDPG), and (ii) KDPG aldolase, which cleaves KDPG to pyruvate and glyceraldehyde-3-phosphate, the latter being further catabolized through the EMP pathway and the tri-carboxylic acid (TCA) cycle (Conway 1992, Peekhaus and Conway 1998). Homo-fermentative LAB use fructose 6-phosphate as substrate for mannitol biosynthesis. Moreover, in the fermentation of mannitol or glucose under nutrient-limited conditions, for instance by *Enterococcus faecalis*, glycolysis may lead to a mixed acid fermentation, leading to acetate, ethanol, diacetyl, acetoin, 2,3-butanediol, and in some cases mannitol (Ramsey et al. 2014).

2.2.2 Hetero-fermentative microorganisms

Hetero-fermentative LAB use the pentose-phosphate (6-phosphogluconate/phosphoketolase) pathway for carbohydrate fermentation (Ray and Joshi 2014). They lack the enzyme fructose1, 6-bisphosphate aldolase; therefore, glucose 6-phosphate is oxidised to 6-phosphogluconate. Under anaerobic conditions, 1 mole of glucose gives 1 mole of lactic acid, 1 mole of ethanol, and 1 mole of CO_2. One mole of ATP is generated per mole of glucose. With the conversion of acetyl phosphate to acetate, instead of ethanol, an extra ATP can be produced (Wisselink et al. 2002, Gaspar et al. 2013). In contrast to homo-fermentative LAB, hetero-fermentative LAB do not use fructose 6-phosphate for mannitol biosynthesis; they use fructose.

Bifidobacteria degrade hexoses through a particular metabolic pathway, the 'bifid shunt', where fructose 6-phosphate phosphoketolase is the key player. The 'bifid shunt' is quite different from the other glycolytic pathways since it allows bifidobacteria to produce more energy in the form of ATP than the fermentative pathways mentioned above. It yields 2.5 ATP molecules from 1 mole of glucose, as well as 1.5 mole of acetate and 1 mole of lactate (Fushinobu 2010, Pokusaeva et al. 2011).

Amylolytic LAB have the ability to breakdown starch molecules into hexose sugars by virtue of possessing α-amylase. It is discussed extensively in the next chapter by Panda and Ray.

2.3 Proteolytic Metabolism

LAB that are relevant to food fermentations are able to produce proteinases (or) peptidases. LAB are fastidious microorganisms that require an exogenous source of amino acids or peptides. In dairy products, the proteolytic system plays an important role because it enables bacteria to grow in milk. The utilization of casein, or other proteinaceous substrates, is initiated by cell-envelope proteinases that lead to the production of oligopeptides which are later taken up by the cells via peptide transport systems. Inside the cells, oligopeptides are degraded to shorter peptides and amino acids by intracellular peptidases. Amino acids can be further converted into flavor compounds such as aldehydes, alcohols, and ester. The proteolytic systems not only allow bacteria to grow, but also change the sensorial properties of the metabolized matrices, giving the final food products unique organoleptic properties (Kunji et al. 1996, Savijoki et al. 2006, Liu et al. 2010).

2.4 Lypolytic Metabolism

The breakdown of lipids into fatty acids and glycerol is performed by LAB through its intracellular and extracellular lipases. LAB can perform unique fatty acid transformation reactions, such as isomerization, hydration, dehydration, and saturation. These unique functions will lead to new uses of these bacteria in food industry and for probiotic purposes (Ogawa et al. 2005). Lipases can be involved in the development of flavor of fermented food products and can be associated with different health benefits for the host (Hayek and Ibrahim 2013).

2.5 Production of Aroma Compounds

The metabolic activities of LAB also lead to the formation of flavor due to the accumulation of volatile and non-volatile aroma compounds and the formation of compounds related to taste (bitterness, umami, sourness, and saltiness). These aroma and taste related compounds are alcohols, aldehydes, ketones, fatty acids, esters, and sulphur compounds (Engels et al. 1997, Ray and Joshi 2014), whose formation has been recently revised (Smid and Kleerebezem 2014). All the three main food constituents, proteins, carbohydrates, and lipids, give precursors for flavor formation. The subsequent conversion reactions are pathway dependent, rather than single-enzyme step dependent. Recent omics studies and the development of genome metabolic models (Smid and Hugenholtz 2010, O'Flaherty and Klaenhammer 2011) allowed a deeper understanding of aroma formation. Also they allowed the confirmation that metabolic capacities, in particular flavour development, are very diverse within a species or even a strain, and as such the search for specific interesting metabolic characteristic for a food product has to be done strain by strain.

3. Bacteriocins Isolated From LAB

Food-grade LAB, as well as other LAB, Gram-positive and also Gram-negative bacteria, have the ability to produce bacteriocins (Table 1). Bacteriocins are small peptides with anti-bacterial properties, generally produced by both Gram-positive and Gram-negative bacteria. These comprise a huge family of extracellularly released heat-stable ribosomally synthesized proteinaceous molecules that have antibacterial activity towards closely related strains and to which the producer cell expresses a degree of specific immunity, although there are an increasing number of bacteriocins reported to have broad range of antimicrobial activity. These antimicrobial peptides have huge potential as both food preservatives, and as next-generation antibiotics targeting the multiple-drug resistant pathogens. Discovery of novel bacteriocins system employs molecular mass analysis of supernatant from the candidate strain, coupled with a statistical analysis of their antimicrobial spectra that can successfully discriminate novel variants of known bacteriocins (Perez et al. 2014).

3.1 Bacteriocin Classification

The bacteriocin classification scheme is still under debate with new groups being proposed, based on the large number of new bacteriocins that have been identified and characterized and the evolving definition for these antimicrobial peptides. These classification strategies offer appealing and concise groupings for bacteriocin research, but the diversity of bacteriocins and lack of sufficient structural information for many bacteriocins render these systems imperfect (Snyder and Worobo 2013). The different classes in which the very diverse and heterogeneous bacteriocins may be classified include Class I bacteriocins, or lantibiotics (lanthionine-containing antibiotics), which are small peptides (< 5 kDa) that possess unusual post-translationally modified residues such as lanthionine or 3-methyllanthionine; and Class II bacteriocins, or the non-lantibiotics, that are small (< 10 kDa), heat-stable non-lantibiotics, which do not undergo extensive post-translational modifications (Cotter et al. 2005, Heng et al. 2007, Perez et al. 2014). This second group can be further subdivided into four subclasses:

- Class IIa—'pediocin-like' bacteriocins with a distinct conserved sequence (YGNGVXC) in the N-terminal region that is responsible for their high potency against the food pathogen *Listeria monocytogenes*;
- Class IIb—two-component bacteriocins that require both peptides to work synergistically to be fully active;
- Class IIc—circular bacteriocins with N- and C-termini covalently linked giving the peptide an extremely stable structure, which has also been described as meriting its own distinct group (Heng and Tagg 2006); and

Table 1. Bacteriocin applications in food preservation and biomedicine.

Sl. No.	Bacteriocins class	Bacteriocins	Organism/strain	Applications in Foods	References
1.	Class I	Nisin A	*Streptococcus lactis*	Active against methicillin-resistant *Staphylococcus aureus* biofilm used in processed cheese, meats, beverages, etc.	Okuda et al. 2013 http://bactibase.pfba-lab-tun.org/
2.		Nisin Z	*Lactococcus lactis* subsp. (*Streptococcus lactis*)	Active against Gram-positive bacteria: *Enterococcus, Listeria*. Used in processed cheese, meats, beverages, etc.	http://bactibase.pfba-lab-tun.org/
3.		Nisin Q	*Lactococcus lactis* 61-14, an LAB isolated from a Japanese river	Active against wide range of Gram-positive bacteria including *Bacillus* sp., *Listeria monocytogene, Micrococcus* sp. Less susceptible to oxidation	Yoneyama et al. 2008, Zendo et al. 2003
4.		Nisin F	*Lactococcus lactis* F10	Active against *Staphylococcus aureus, Staphylococcus carnosus*	http://bactibase.pfba-lab-tun.org/BAC146
5.		Nisin U	*Lactococcus uberis* ATCC 27958	Active against *Streptococcus pyogenes* (10/11), *Streptococcus uberis* (20/26), *Streptococcus agalactiae* (2/4), *Streptococcus dysgalactiae* (3/4), *Staphylococcus simulans* (1/1), *Staphylococcus cohnii* (1/1)	http://bactibase.pfba-lab-tun.org/BAC147
6.	Class IIa	Pediocin PA-1/AcH	*Pediococcus pentosaceus* BCC 3772	Used as starter culture for *Nham*, a traditional Thai fermented pork sausage, effectively controlled the growth of *L. monocytogenes* without compromising the quality of *Nham*. Incorporated into a biocomposite packaging film, the initial load of *L. monocytogenes* on the meat surface was significantly reduced	Iwatani et al. 2013 Woraprayote et al. 2013
7.		Pediocin 34	*Pediococcus pentosaceous* 34, a bacteriocinogenic strain was an isolate from cheddar cheese	A greater antibacterial effect was observed when the bacteriocins (Pediocin and Enterocin) were combined in pairs, indicating that the use of more than one LAB bacteriocin in combination has a higher antibacterial action than when used individually against variants of *L. monocytogenes* ATCC 53135	Kaur et al. 2013 Gupta et al. 2010
8.		Enterocin FH99	*Enterococcus faecium* FH99		

#	Class	Bacteriocin	Source organism	Description	Reference
9.	Class IIb	Sakacins A and P	*Lactobacillus sakei*	They are small, heat-resistant peptides that are not post-translationally modified. Active against other LAB but particularly effective against *L. monocytogenes* which is commonly reported to contaminate ready-to-eat refrigerated food products	Ennahar et al. 2000, Chen and Hoover 2003
10.		Leucocin A	*Leuconostoc gelidum*		
11.		Enterocins A and P	*Enterococcus faecium*		
12.		Carnobacteriocin	*Carnobacterium* sp.		
13.		Lactococcin Q	*Lactobacillus lactis* QU 4, an LAB isolated from corn	Lactococcins Q have very narrow and specific antimicrobial spectra, with antimicrobial activity	Nissen-Meyer et al. 1992, Zendo et al. 2006
14.		Enterocin X	*Enterococcus faecium* KU-B5	When equimolar concentrations of these peptides were tested against a panel of indicator strains, the combined antimicrobial activity was not uniformly enhanced	Hu et al. 2010
15.		Enterocin NKR-5-3AZ	*Enterococcus faecium* NKR-5-3, an LAB isolated from the Thai fermented fish Pla-ra	Showed very strong microbial activity (in nanomolar range) against *Listeria* spp. and other Gram-positive species	Himeno et al. 2012, Ishibashi et al. 2012
16.	Class IIc	Lactocyclicin Q	*Lactococcus* sp. QU 12	Exhibit high stability against thermal, pH, and proteolytic enzyme stresses, making them very potent food preservatives and stable antimicrobial agents	Sawa et al. 2009
17.		Leucocyclicin Q	*Leuconostoc mesenteroides* TK41401	Exhibit high stability against thermal, pH, and proteolytic enzyme stresses, making these very potent food preservatives and stable antimicrobial agents	Masuda et al. 2011
18.	Class IId	Lacticin Q	*Lb. lactis* QU 5, an LAB isolated from corn	Show very strong antimicrobial activity (at nanomolar concentrations) as well as high stability against various stresses. Exert strong bactericidal activities against methicillin-resistant *Staphylococcus aureus* (MRSA) strain both in its planktonic and biofilm cells stage	Fujita et al. 2007, Okuda et al. 2013
19.		Lacticin Z	*Lb. lactis* QU 14, was isolated from horse intestine	Show very strong antimicrobial activity (at nanomolar concentrations) as well as high stability against various stresses	Iwatani et al. 2007

Table 1. contd....

Table 1. contd.

Sl. No.	Bacteriocins class	Bacteriocins	Organism/strain	Applications in Foods	References
20.		Weissellicins Y and M	*Weissella hellenica* QU 13, an LAB isolated from Japanese pickles, Takana-zuke	Both bacteriocins showed broad antimicrobial spectra with especially high antimicrobial activity against species, which contaminate pickles such as *Bacillus coagulans*. The stability of weissellicin M against pH and heat was distinctively higher than that of weissellicin Y	Masuda et al. 2012
21.	Class III	Helveticin J	*Lb. helveticus* 481	The antimicrobial compound was active at neutral pH under aerobic or anaerobic conditions, was sensitive to proteolytic enzymes and heat (30 min at 100°C)	Joerger et al. 1986
22.		Lactacin B	*Lb. acidophilus*	Useful in the fields of food preservation, safety, health care, and pharmaceutical applications. The inhibition activity of these substances has been reported to be strain-dependent	Barefoot and Klaenhammer 1983, Muriana and Klaenhammer 1991
23.		Lactacin F	*Lb. acidophilus* 11088 (NCK88)	More heat resistant and exhibit a broader spectrum of activity	Muriana and Klaenhammer 1991
24.	Class IV	Leuconocin S	*Leuconostoc paramesenteroides* strain isolated from retail meat	Active inhibitors of *L. monocytogenes, Aeromonas hydrophila, Staphylococcus aureus, Yersinia enterocolitica* and *Clostridium botulinum* strains. Resistance to lipase, amylase, trypsin and more heat sensitive	Robert et al. 1996
25.		Leuconocin J	*Leuconostoc* sp. J2 strain was originally isolated from Kimchi, which is a traditional Korean fermented Chinese cabbage	Inhibitory to growth of Gram-positive bacteria, including *Staph. aureus* and *L. monocytogenes*	Choi et al. 1999
26.		Lactocin 27	*Lb. helveticus* Strain LP27	A glycopeptides show bacteriostatic effect against closely related bacteria and also inhibit sensitive strains by the inhibition of protein synthesis	Upreti 1994

- Class IId—unmodified, linear, non-pediocin-like bacteriocins that include sec dependent and leaderless bacteriocins (Cotter et al. 2005, Heng et al. 2007, Perez et al. 2014).
- The Class III include large (> 30 kDa) heat-labile non-lantibiotics while complex bacteriocins containing glycol and/or lipid moieties, are put under Class IV. However, the (former) Class III group has been proposed to be reclassified as bacteriolysins, since they are lytic enzymes rather than peptides (Heng et al. 2007).

3.2 Bacteriocin Mode of Action

Bacteriocins antagonize sensitive cells through different and distinctive mechanisms. Although structure–function relationships have only been determined for particular bacteriocins and to varying degrees, examples of bacteriocins targeting the cell wall, cell membrane, nucleic acids, or enzymes have been established (Snyder and Worobo 2013). Most bacteriocins act by disrupting the integrity of the membrane of target sensitive cells, mainly by pore formation but sometimes, also by interfering with the synthesis of the membrane. Small pore formation causes leakage of low molecular weight compounds, leading to dissipation of the proton motive force that is deleterious to cells, while larger pores may lead to the immediate and also deleterious loss of metabolites (McAuliffe et al. 2001). The interaction between a bacteriocin and a target cell may involve attractive but non-specific forces, with a positively charged bacteriocin and negatively charged bacterial membrane surface; however, it is believed that most bacteriocins bind specific receptors on sensitive cell surfaces (Hassan et al. 2012). Some of the identified bacteriocin target receptors on sensitive cells include proteins of the sugar transporter mannose-phosphotransferase system (in the case of Class IIa bacteriocins), while for some lantibiotics (Class I bacteriocins) the cell wall precursor molecule, lipid II, has been identified as the docking molecule on target cells (Kjos et al. 2011b). Nisin is synthesized as pro-nisin inside the bacterial cell and then released in the outer layer as a peptide containing 34 amino acids. The mechanism of action for nisin, the most extensively studied bacteriocin and the only one approved for food applications (de Arauz et al. 2009), has been characterized in great detail. It involves a specific receptor, lipid II, to which nisin binds *via* the lantibiotic ring structures in the N-terminal part of the peptide. This leads to the formation of lethal pores as well as the inhibition cell wall formation of target cells by interfering with the biosynthesis of peptidoglycan layer, a mechanism independent of the pore-forming activity that involves the relocation of lipid II molecules into patches outside their functional location (Hassan et al. 2012). According to Thomas et al. (2000), nisin inactivate endospores by means of preventing post-germination swelling and subsequent spore outgrowth.

Bacteriocins can have narrow or broad activity spectra. As the role of previously mentioned bacteriocin docking molecule lipid II is common in all bacteria, lantibiotics (Class I bacteriocins) like nisin have a relatively broad inhibitory spectrum, including a number of different genera of Gram-positive bacteria (Hassan et al. 2012). However, the activity can also be very specific, having a very narrow target spectrum, since they employ specific receptors on the target cell surfaces (Kjos et al. 2011b), being active against only a few genera/species closely related to the bacteriocin producer (Kjos et al. 2011a). Bacteriocin target specificity and circular bacteriocins; they have also been described as concentration dependent, with a nonspecific activity at higher bacteriocin concentrations and a specific activity at lower bacteriocin concentrations (Gabrielsen et al. 2014). This specific activity at lower bacteriocin concentrations make bacteriocins often very potent, acting at pico- to nano-molar concentrations, whereas micro-molar concentrations are required for the activity of eukaryotic anti microbial peptides (Hassan et al. 2012). One of the best studied and recognized bacteriocin, nisin, is produced by *Lactococcus lactis* and considered to be fully safe for humans. Nisin is widely used in the food industry, as a preserving agent (Schillinger et al. 1996). Next to nisin, Pediocin PA-1/AcH obtained from *Pediococcus* spp. and enterocin AS-48 (class IIc) from *Enterococcus faecalis* are the bacteriocin most likely to be used for bio-preservatives. Other bacteriocins including lactacin B, plantaricin, helveticin, acidophilin, bulgaricin and sakacin, synthesized by different LAB have yet to be utilized commercially. Apart from this, another non-protein bacteriocin, reuterin (chemically 3-hydroxy propionaldehyde), is synthesized by *Lactobacillus reuteri*. It is a broad spectrum bacteriocin, active against both Gram-positive and Gram-negative bacteria, as well as yeasts and molds. It is very stable over a wide range of pH.

3.3 Bacteriocin Applications

The ability to extracellularly release bacteriocins may allow the producing bacteria to compete with other bacteria for common resources, hence allowing the producers to dominate or establish growth in certain ecological niches (Cotter et al. 2005, Kjos et al. 2011a, 2011b). This feature confers to bacteriocin, especially from LAB, a potential to be utilized as both, natural food preservatives and as therapeutic antibiotics (Perez et al. 2014). Regarding their food-preservative properties, bacteriocin-producing LAB strains can be used as a part of, or adjuncts, to starter cultures for fermented foods in order to improve safety and quality (Rattanachaikunsopon and Phumkhachorn 2010), allowing bio-preservation, shelf life extension, and control of fermentation microflora (Table 1) (Balciunas et al. 2013) by preventing the growth of food-borne pathogens or food-spoiling bacteria including *Listeria, Bacillus, Clostridium* and *Staphylococcus* (Kjos et al. 2011a, Gabrielsen et al. 2014). This follows the consumer trends for healthier, 'natural' and 'traditional' foods, processed without any addition of chemical preservatives, which are now becoming more attractive (Balciunas et al. 2013). LAB bacteriocins, besides being by-products of Generally Regarded as Safe (GRAS) LAB, are especially attractive for various food applications because they are colourless, odourless, tasteless, inherently tolerant to high thermal stress and are known for their activity over a wide pH range (Perez et al. 2014), all positive attributes for their possible use as an ingredient in food production.

For clinical applications, bacteriocins have been presented as a viable alternative to traditional antibiotics due to the afore mentioned characteristics of some bacteriocins, such as potency, stability and specificity, against other bacteria and clinical pathogens, including multi-antibiotic resistant (MDR) strains (Cotter et al. 2012, Perez et al. 2014). The relatively narrow spectrum of some bacteriocins is sometimes a highly appreciated feature as it provides an excellent means to direct activity toward certain pathogens without disturbing the commensal bacterial flora. Although the application of specific bacteriocins might inevitably lead to the development of some sort of resistance (Kjos et al. 2011a, Cotter et al. 2012), an advantage bacteriocins have over classical antibiotics is that, due to their proteinaceous nature, they are easily degraded by proteolytic enzymes. Together with a fast acting mechanism that is active even at extremely low concentrations, bacteriocin fragments do not persist in their environment, minimizing the opportunity for target strains to interact with the degraded fragments, which is the common starting point in the development of antibiotic resistance Perez et al. (2014). Perhaps the most significant advantage of bacteriocins over conventional antibiotics, which are secondary metabolites made by multienzyme complexes, is their primary metabolite nature since they are ribosomally synthesized peptides and have relatively simple biosynthetic mechanisms (Kjos et al. 2011a, 2011b, Perez et al. 2014). This makes them easily amenable, through bioengineering, to increase either their bioactivity, stability, or specificity towards target microorganisms (Perez et al. 2014, Gabrielsen et al. 2014). Besides bioengineering, a more complete characterization of the bacteriocin repertoire with techniques like genome context analysis (Makarova et al. 2006) or molecular mass analysis of candidate strains' supernatants coupled with a statistical analysis of their antimicrobial spectra (Perez et al. 2014), can help identify novel or variants of known bacteriocins. The increasing number of reports of new Gram-positive bacteriocins with unique properties indicates that there is still a lot to learn about this family of peptide antibiotics; in addition, they show an extraordinary potential as the next era antibiotics (Kjos et al. 2011a, Perez et al. 2014).

In the food industry, bacteriocins have huge potential in the bio-preservation of various foods, either alone, or in combination with other methods of preservation, which is known as hurdle technology. The antimicrobial activity of many bacteriocins, especially the class IIa bacteriocins, against the highly pathogenic and food-borne *L. monocytogenes*, offers an ideal solution to the problem caused by this sturdy pathogen, which is commonly reported to contaminate ready-to-eat refrigerated food products (Chen and Hoover 2003). The pediocin PA-1/AcH producing *Pediococcus pentosaceus* BCC 3772, when used as starter culture for *Nham*, a traditional Thai fermented pork sausage, effectively controlled the growth of *L. monocytogenes* without compromising the quality of *Nham* (Kingcha et al. 2012).

Bacteriocins can be incorporated into a food through numerous ways; the direct use of bacteriocin-producing culture in fermented products to produce the bacteriocin *in situ* is the most suitable method. Other methods include using a purified or semi-purified bacteriocin preparation as a food ingredient or incorporating bacteriocin in form of a thin coating onto a layer of packaging material, i.e., bioactive packaging, a process that can protect the food from external microbial contaminants (Woraprayote et al. 2013).

3.4 Fast Tracking of Novel Bacteriocins

Efforts have been made to improve the existing bacteriocin screening and identification systems to make them more efficient in discovering novel bacteriocins. Perez et al.

have reported a rapid detection system which employs molecular mass analyses using electrospray ionization liquid chromatography/mass spectrometry (ESI-LC/MS) in combination with principal component analyses (PCA) of the antimicrobial activity spectrum of each bacteriocin producing LAB strain. This will able to rapidly identify potential novel bacteriocin-producing LAB strains and discard those that produce known bacteriocins, thus accelerating the discovery of novel bacteriocins including variants of those already reported (Perez et al. 2014). Recently, bacteriocins are receiving increasing attention due to their many applications, ranging from their initial application in strategies for food preservation to more recent proposed uses in biomedical strategies aimed at fighting certain bacterial infections (Zineb et al. 2013).

3.5 Emerging Aspects of Bacteriocins

An interesting natural antimicrobial product has been recently reported by Antoniewski and the group in 2011. The said product consists of lantibiotic nisin (2–40 ppm) component (intended to be used in self-stable water-based emulsions, such as salad dressings) and an un-dissociated organic acid/salt (sodium acetate/calcium acetate, etc.). It was found to inhibit a broad range of spoilage organisms including LAB, and fungi, such as generic yeast and acid-tolerant yeast (such as *Zygosaccharomyces bailii*) in food products. They reported that an undissociated organic acid/salt of about 3% of total weight Ca-Acetate with an undissociated organic acid/salt ion of at least 0.5% and the nisin component can be derived from a 6% total weight liquid cultured milk acidified with glacial acetic acid (Antoniewski et al. 2011). A lantibiotic polypeptide called bisin, isolated from a probiotic culture of *Bifidobacterium longum* has been shown to be an effective shelf life extender for dairy products by inhibiting the growth of both gram-positive bacteria, such as *Lactobacillus* spp., *Lactococcus* spp., *Streptococcus* spp., *Staphylococcus* spp., or *Bacillus* spp., and gram-negative bacteria such as *E. coli, Serratia* spp., *Proteus* spp. or *Salmonella* spp. (O'Sullivan and Lee 2011). Visser and Haan (2011) have reported newly formulated liquid nisin compositions at low amounts, found to be effective against gram positive bacteria which can be used as preservative agents in food products. Good physical, chemical and microbiological stability of these compositions makes them suitable for extended storage strategies, thereby providing long shelf-lives. In addition, the low turbidity of these combinations makes them suitable for use in food applications, where the low turbidity of additives is of paramount importance (Visser and Haan 2011). It is well-known that bacteriocins produced by LAB can improve the shelf life of foods. A method has been proposed that increases the efficiency of bacteriocins with a whey-degrading medium, making it as an effective food preservative. In addition, the application of the cell suspension can improve the shelf life of foods while suppressing any change in the original flavour of the foods caused by starters, such as *Lactobacillus bulgaricus* (Isawa and Kamijo 2011).

4. LAB in Health

LAB impart several health benefits when they are consumed in prescribed doses. Some of them are considered as probiotics.

4.1 Probiotics

Probiotics consumed as part of food were defined in 2001 (FAO/WHO 2001) as 'living bacteria that, when administered in adequate amounts, confer a health benefit on the host'. Probiotics may also include bio-therapeutic agents and beneficial microorganisms not used in food (Pineiro and Stanton 2007).

For the use of probiotics, which are mainly members of the genera Lactobacillus and Bifidobacterium, in food, important criteria have been documented, in particular that they should not only be capable of surviving passage through the digestive tract (by exhibiting acid and bile tolerance), but also have the capability to proliferate in the gut (FAO/WHO 2006, Pineiro and Stanton 2007). However, it is the specificity of action, not the source of the microorganism, which is most important (de Vrese et al. 2008). Hence, for their evaluation in food, it is the ability to remain viable at the target site and be effective that should be verified for each strain (FAO/WHO 2002).

4.2 LAB as Probiotics

Using LAB as probiotics is a simple way to get many health benefits. Some of the benefits of LAB and their fermentation products are improved digestibility, stimulation of the peristaltic movement of the intestines, improvement of blood circulation, normalization of acidity of gastric (stomach) juices, and help in the maintenance of correct body pH (acid-alkaline balance).

Raw vegetables cultured by LAB are especially good for people taking antibiotics or birth control bills, pregnant women, and diabetics. These are all high-risk groups for Candida and other yeast disorders (Ouwehand et al. 2002). In research papers, LAB have been demonstrated to confer anti-mutagenic effects thought to be due to their ability to bind and detoxify heterocyclic amines (carcinogenic substances formed in cooked meat). Animal studies have demonstrated that LAB can protect against colon cancer in rodents (Brady et al. 2000). Most human trials have found that LAB may exert anti-carcinogenic effects by decreasing the activity of an enzyme called ß-glucuronidase, which can regenerate carcinogens in the digestive system. Lower rates of colon cancer among higher consumers of fermented dairy products have been observed in some population studies (Wollowski et al. 2001).

Lactobacilli play a role in decreasing the absorption of cholesterol in the intestine by de-conjugating the intestinal bile salts to form bile acids, thereby inhibiting the micelle formation that would lead to absorption of cholesterol. The cholesterol that enters the intestine through the enterohepatic circulation is treated similarly. Some, but not all, human trials have shown that dairy foods fermented with LAB can produce modest reductions in total and LDL cholesterol levels in those with normal levels to begin with; however, trials in hyperlipidemic subjects are needed (Sanders 2000). Several small clinical trials have shown that the consumption of milk fermented with various strains of LAB can result in modest reductions in blood pressure. It is thought that this is due to the ACE-inhibitory peptides produced during fermentation (Wollowski et al. 2001). Nevertheless these studies suggested that probiotics can be a suitable approach to begin therapy (Aggarwal et al. 2013).

The role of gut microbiome-host metabolic signals to trigger health and disease are one of the results of the studies on human gut microbiota. This symbiosis can have local metabolic effects or systemic metabolic effects. Firmicutes and Bacteroidetes are the most represented bacterial divisions found in the human gut. Shifts in gut bacterial composition can cause disease or disease-protecting effects (Holmes et al. 2011). The effects of gut microbiota, or the ingestion of the right probiotics, in supporting fat metabolism can have an important role in obesity. Obesity is one of the modern causes of morbidity and mortality. When examining the faecal gut microbiota of obese volunteers that participated in weight loss program for a year, Turnbaugh et al. found that obese subjects had a higher proportion of Firmicutes and lower proportions of Bacteroidetes than the lean controls. They observed that subjects who lost weight had a reduced representation of the Firmicutes and an increased number of Bacteroidetes and that these changes were irrespective of the diet and in proportion to the amount of weight which was lost. Gut microbiota influences the energy harvest as well as the genes that regulate energy expenditure and storage. Being the obese microbiome, it is more capable of harvesting energy from the diet, thereby the gut microbiota contributes to the pathophysiology of obesity (Turnbaugh et al. 2006, Aggarwal et al. 2013). The intake of LAB can reduce abdominal visceral and subcutaneous fat areas, body weight and body mass (Kadooka et al. 2010).

Moreover, it was observed that the administration of dahi (an Indian fermented product similar to yogurt), which contained *Lactobacillus acidophilus*, *Lb. casei*, and *Lb. lactis* to fructose induced diabetic rats decreased the accumulation of glycogen in the liver. Knowing that individuals who have essential hypertension are at a high risk of developing diabetes, a new strategy that can be employed in the prevention or delay of diabetes and the subsequent reduction in the incidence of hypertension could be the consumption of probiotics or probiotics-based fermented foods (Aggarwal et al. 2013). It was also demonstrated in a trial, that the consumption of *Bifidobacterium lactis* twice daily, in 25 volunteers with a minimum age of 60 years, for six weeks, improved the immune response without the release of inflammatory cytokines, thus reducing the onset of systematic inflammatory induced diabetes (Arunachalam et al. 2000).

LAB are thought to have several presumably beneficial effects on the immune function. There is evidence to suggest that they may improve immune function by increasing the number of IgA-producing plasma cells, increasing or improving phagocytosis, as well as increasing the proportion of T cells and natural killer cells (Ouwehand et al. 2002, Reid et al. 2003). Clinical trials have also demonstrated that LAB as probiotics may decrease the incidence of respiratory tract infections in children (Hatakka et al. 2001). Probiotics were better than the placebo in reducing the number of participants experiencing episodes of acute upper respiratory tract infections, the rate ratio of episodes of acute upper respiratory tract infections, and therefore better in reducing antibiotic use. This indicates that probiotics may be more beneficial than a placebo for preventing acute upper respiratory tract infections. However, the results have some limitations and there was no data for older people (Hao et al. 2011).

LAB may also play a role in reducing dental cavities (caries) in children (Näse et al. 2001). Clinical trials have demonstrated that probiotics may aid in the treatment of *Helicobacter pylori* infections, responsible for peptic ulcers in adults, when used in combination with standard medical treatments (Hamilton-Miller 2003).

LAB-containing foods and supplements have been shown to be effective in the treatment and prevention of acute diarrhoea, decreasing the severity and duration of rotavirus infections in children, as well as antibiotic-associated and traveller's diarrhoea in adults (Cremonini et al. 2002). LAB-containing foods and supplements have also been found to modulate inflammatory and hypersensitivity responses by due regulation of cytokine function. Clinical studies suggest that they can prevent reoccurrences of inflammatory bowel disease in adults (Reid et al. 2003), as well as improve milk allergies (Kirjavainen et al. 2003), and decrease the risk of atopic eczema in children (Kalliomäki et al. 2003).

The safety of *Lactobacillus, Bifidobacterium, Saccharomyces, Streptococcus, Enterococcus,* and/or *Bacillus* strains used as probiotic agents to reduce the risk of, prevent, or treat disease was assessed by Hempel et al. They concluded that the available evidence does not indicate an increased risk; however, rare adverse events are difficult to assess, and despite the substantial number of publications, the current literature is not equipped well enough to answer questions on the safety of probiotic interventions with confidence (Hempel et al. 2011).

5. Conclusion and Future Perspectives

Finally, are 'Lactic acid bacteria the bugs of the millennium'? They are certainly one of the big generous bugs of the millennium due to the diversity of metabolism that makes them the most important group of bacteria in the production of traditional and modern fermented foods. Their potential use in health and preventing diseases when used as probiotics, as well as their effect in protection of food products against microbial contaminants, has been amply demonstrated.

Acknowledgement

The authors are thankful to Dr. Ami Patel, Assistant Professor, Dairy Microbiology Dept., Mansinhbhai Institute of Dairy and Food Technology, Mehsana 384002, Gujarat, India for critically reading the manuscript and making valuable suggestions.

Keywords: Lactic acid bacteria (LAB), probiotics, bacteriocins, fermented food products

References

Aggarwal, J., Swami, G. and Kumar, M. 2013. Probiotics and their effects on metabolic diseases: An update. Journal of Clinical and Diagnostic Research 7: 173–177.

Amann, R.I., Ludwig, W. and Schleifer, K.H. 1995. Phylogenetic identification and *in situ* detection of individual microbial cells without cultivation. Microbiological Reviews 59: 143–169.

Antoniewski, M.N., Kennett, C.A., McIlroy, M.A., Zheng, Z., Moca, J.G. and Kelly-Harris, S.E. 2011. Natural Antimicrobial Composition. US20110053832A1, 2011.

Arunachalam, K., Gill, H.S. and Chandra, R.K. 2000. Enhancement of natural immune function by dietary consumption of *Bifidobacterium lactis* (HN019). European Journal of Clinical Nutrition [Internet] 54: 263–267.

Balciunas, E.M., Castillo Martinez, F.A., Todorov, S.D., Franco, B.D.G. de M., Converti, A. and Oliveira, R.P. de S. 2013. Novel biotechnological applications of bacteriocins: A review. Food Control 32: 134–142.

Barefoot, S.F. and Klaenhammer, T.R. 1983. Detection and activity of lactacin B, a bacteriocin produced by *Lactobacillus acidophilus*. Applied and Environmental Microbiology 45: 1808–1815.

Bottacini, F., O'Connell Motherway, M., Kuczynski, J., O'Connell, K.J., Serafini, F., Duranti, S., Milani, C., Turroni, F., Lugli, G.A. and Zomer, A. et al. 2014. Comparative genomics of the *Bifidobacterium breve* taxon. BMC Genomics 15: 170.

Brady, L.J., Gallaher, D.D. and Busta, F.F. 2000. The role of probiotic cultures in the prevention of colon cancer. Journal of Nutrition 130: 410S–414S.

Canchaya, C., Claesson, M.J., Fitzgerald, G.F., van Sinderen, D. and O'Toole, P.W. 2006. Diversity of the genus *Lactobacillus* revealed by comparative genomics of five species. Microbiology 152: 3185–3196.

Chen, H. and Hoover, D.G. 2003. Bacteriocins and their food applications. Comparative Review in Food Science and Food Safety 2: 82–100.

Choi, H.-J., Lee, H.-S., Her², S., Oh, D.-H. and Yoon, S.-S. 1999. Partial characterization and cloning of leuconocin J, a bacteriocin produced by *Leuconostoc* sp. J2 isolated from the Korean fermented vegetable Kimchi. Journal of Applied Microbiology 86 (2): 175–18.

Claesson, M.J., van Sinderen, D. and O'Toole, P.W. 2008. *Lactobacillus* phylogenomics—Towards a reclassification of the genus. International Journal of Systematic and Evolutionary Microbiology 58: 2945–2954.

Conway, T. 1992. The Entner-Doudoroff pathway: History, physiology and molecular biology. FEMS Microbiology Letters 103: 1–27.

Cotter, P.D., Hill, C. and Ross, R.P. 2005. Food microbiology: Bacteriocins: Developing innate immunity for food. Nature Reviews Microbiology 3: 777–788.

Cotter, P.D., Ross, R.P. and Hill, C. 2012. Bacteriocins—A viable alternative to antibiotics? Nature Reviews Microbiology 11: 95–105.

Cremonini, F., Di Caro, S., Nista, E.C., Bartolozzi, F., Capelli, G., Gasbarrini, G. and Gasbarrini, A. 2002. Meta-analysis: The effect of probiotic administration on antibiotic-associated diarrhoea. Alimentary Pharmacology & Therapeutics 16: 1461–1467.

Davis, C.R., Lee, T.H., Wibowo, D. and Fleet, G.H. 1986. Growth and metabolism of lactic acid bacteria during fermentation and conservation of some Australian wines. Food Technology in Australia 38: 35–40.

Davis, C.R., Wibowo, D., Fleet, G.H. and Lee, T.H. 1988. Properties of wine lactic acid bacteria: Their potential enological significance. American Journal of Enology and Viticulture 39: 137–142.

de Arauz, L.J., Jozala, A.F., Mazzola, P.G. and Vessoni Penna, T.C. 2009. Nisin biotechnological production and application: a review. Trends in Food Science & Technology 20: 146–154.

de Vos, W.M. 2011. Systems solutions by lactic acid bacteria: From paradigms to practice. Microbial Cell Factories 10 (Suppl.) 1: S2.

de Vrese, M. and Schrezenmeir, J. 2008. Probiotics, prebiotics, and synbiotics. Advances in Biochemical Engineering/Biotechnology 111: 1–66.

Douillard, F.P. and de Vos, W.M. 2014. Functional genomics of lactic acid bacteria: From food to health. Microbial Cell Factories 13 (Suppl.) 1: S8.

Engels, W.J.M., Dekker, R., De Jong, C., Neeter, R. and Visser, S. 1997. A comparative study of volatile compounds in the water-soluble fraction of various types of ripened cheese. International Dairy Journal 7: 255–263.

Ennahar, S., Deschamps, N. and Richard, J. 2000. Natural variation in susceptibility of *Listeria* strains to class IIa bacteriocins. Current Microbiology 41: 1–4.

Euzéby, J.P. 1997. List of bacterial names with standing in nomenclature: a folder available on the Internet. International Journal of Systematic Bacteriology 47: 590–592.

FAO/WHO. 2001. Joint FAO/WHO expert consultation on evaluation of health and nutritional properties of probiotics in food including powder milk with live lactic acid bacteria. Córdoba, Argentina.

FAO/WHO. 2002. Joint FAO/WHO working group report on drafting guidelines for the evaluation of probiotics in food 1–11.

FAO/WHO. 2006. Probiotics in food. Rome: FAO/WHO.

Fujita, K., Ichimasa, S., Zendo, T., Koga, S., Yoneyama, F., Nakayama, J. and Sonomoto, K. 2007. Structural analysis and characterization of lacticin Q, a novel bacteriocin belonging to a new family of unmodified bacteriocins of Gram-positive bacteria. Applied and Environmental Microbiology 73: 2871–2877.

Fushinobu, S. 2010. Unique sugar metabolic pathways of bifidobacteria. Bioscience, Biotechnology and Biochemistry 74: 2374–2384.

Gabrielsen, C., Brede, D.A., Nes, I.F. and Diep, D.B. 2014. Circular bacteriocins: Biosynthesis and mode of action. Applied and Environmental Microbiology 80: 6854–6862.

Garvie, E.I. 1984. Separation of species of the genus *Leuconostoc* and differentiation of the *Leuconostocs* from other lactic acid bacteria. pp. 147–178. *In*: T. Bergan (ed.). Methods in Microbiology, Vol. 16 Elsevier, Amsterdam.

Gaspar, P., Carvalho, A.L., Vinga, S., Santos, H. and Neves, A.R. 2013. From physiology to systems metabolic engineering for the production of biochemicals by lactic acid bacteria. Biotechnology Advances 31: 764–788.

Gupta, H., Malik, R.K., De, S. and Kaushik, J.K. 2010. Purification and characterization of enterocin FH 99 produced by a faecal isolate *Enterococcus faecium* FH 99. Indian Journal of Microbiology 50: 145–155.

Hamilton-Miller, J.M.T. 2003. The role of probiotics in the treatment and prevention of *Helicobacter pylori* infection. International Journal of Antimicrobial Agents 22: 360–366.

Hao, Q., Lu, Z., Dong, B.R., Huang, C.Q. and Wu, T. 2011. Probiotics for preventing acute upper respiratory tract infections. The Cochrane Database of Systematic Reviews. CD006895.

Hassan, M., Kjos, M., Nes, I.F., Diep, D.B. and Lotfipour, F. 2012. Natural antimicrobial peptides from bacteria: Characteristics and potential applications to fight against antibiotic resistance. Journal of Applied Microbiology 113: 723–736.

Hatakka, K., Savilahti, E., Pönkä, A., Meurman, J.H., Poussa, T., Näse, L., Saxelin, M. and Korpela, R. 2001. Effect of long term consumption of probiotic milk on infections in children attending day care centres: Double blind, randomized trial. British Medical Journal 322: 1327.

Hayek, S.A. and Ibrahim, S.A. 2013. Current limitations and challenges with lactic acid bacteria: A review. Food and Nutrition Sciences 4: 73–87.

Hempel, S., Newberry, S., Ruelaz, A., Wang, Z., Miles, J., Suttorp, M.J., Johnsen, B., Shanman, R., Slusser, W. and Fu, N. et al. 2011. Safety of probiotics to reduce risk and prevent or treat disease. 11 ed. Southern California Evidence-based practice center under Contract No. 290-2007-10062-I, editor. Rockville, MD: Evidence Report/Technology Assessment.

Heng, N.C.K. and Tagg, J.R. 2006. What's in a name? Class distinction for bacteriocins. Nature Publishing Group 4.

Heng, N.C.K., Wescombe, P.A., Burton, J.P., Jack, R.W. and Tagg, J.R. 2007. The diversity of bacteriocins in gram-positive bacteria. pp. 45–92. *In*: M.A. Riley and M.A. Chavan (eds.). Bacteriocins Springer, Berlin Heidelberg.

Himeno, K., Fujita, K., Zendo, T., Wilaipun, P., Ishibashi, N., Masuda, Y., Yoneyama, F., Leelawatcharamas, V., Nakayama, J. and Sonomoto, K. 2012. Identification of enterocin NKR-5-3C, a novel class IIa bacteriocin produced by a multiple bacteriocin producer, *Enterococcus faecium* NKR-5-3. Bioscience Biotechnology Biochemistry 76: 1245–1247.

Holmes, E., Li, J.V., Athanasiou, T., Ashrafian, H. and Nicholson, J.K. 2011. Understanding the role of gut microbiome-host metabolic signal disruption in health and disease. Microbes and Metabolism 19: 349.

Hu, C.B., Malaphan, W., Zendo, T., Nakayama, J. and Sonomoto, K. 2010. Enterocin X, a novel two-peptide bacteriocin from *Enterococcus faecium* KU-B5, has an antibacterial spectrum entirely different from those of its component peptides. Applied and Environmental Microbiology 76: 4542–4545.

Isawa, K. and Kamijo, M. 2011. Method for culturing lactic acid bacterium and method for producing fermented milk. Trend in Microbiology 19: 349–359.

Ishibashi, N., Himeno, K., Fujita, K., Masuda, Y., Perez, R.H., Zendo, T., Wilaipun P., Leelawatcharamas, V., Nakayama, J. and Sonomoto, K. 2012. Purification and characterization of multiple bacteriocins and an inducing peptide produced by *Enterococcus faecium* NKR-5-3 from Thai fermented fish. Bioscience Biotechnology Biochemistry 76: 947–953.

Iwatani, S., Horikiri, Y., Zendo, T., Nakayama, J. and Sonomoto, K. 2013. Bi-functional gene cluster lnq BCDEF mediates bacteriocin production and immunity with differential genetic requirements. Applied and Environmental Microbiology 79: 2446–2449.

Iwatani, S., Zendo, T., Yoneyama, F., Nakayama, J. and Sonomoto, K. 2007. Characterization and structure analysis of a novel bacteriocin, lacticin Z, produced by *Lactococcus lactis* QU 14. Bioscience Biotechnology and Biochemistry 71: 1984–1992.

Joerger, M.C. and Klaenhammer, T.R. 1986. Characterization and purification of helveticin J and evidence for a chromosomally determined bacteriocin produced by *Lactobacillus helveticus* 481. Journal of Bacteriology 167(2): 439–446.

Kadooka, Y., Sato, M., Imaizumi, K., Ogawa, A., Ikuyama, K., Akai, Y., Okano, M., Kagoshima, M. and Tsuchida, T. 2010. Regulation of abdominal adiposity by probiotics (*Lactobacillus gasseri* SBT2055) in adults with obese tendencies in a randomized controlled trial. European Journal of Clinical Nutrition 64: 636–643.

Kalliomäki, M., Salminen, S., Poussa, T., Arvilommi, H. and Isolauri, E. 2003. Probiotics and prevention of atopic disease: 4-year follow-up of a randomised placebo-controlled trial. Lancet 361: 1869–1871.

Kaur, G., Singh, T.P. and Malik, R.K. 2013. Antibacterial efficacy of Nisin, Pediocin 34 and Enterocin FH99 against *Listeria monocytogenes* and cross resistance of its bacteriocin resistant variants to common food preservatives. Brazilian Journal of Microbiology 44, 1: 63–71.

Kingcha, Y., Tosukhowong, A., Zendo, T., Roytrakul, S., Luxananil, P., Chareonpornsook, K., Valyaseyi, R., Sonomoto, K. and Visessanguan, W. 2012. Anti-listeria activity of *Pediococcus pentosaceus* BCC 3772 and application as starter culture for *Nham*, a traditional fermented pork sausage. Food Control 25: 190–196.

Kirjavainen, P.V., Salminen, S.J. and Isolauri, E. 2003. Probiotic bacteria in the management of atopic disease: Underscoring the importance of viability. Journal of Pediatric Gastroenterology and Nutrition 36: 223–227.

Kjos, M., Borrero, J., Opsata, M., Birri, D.J., Holo, H., Cintas, L.M., Snipen, L., Hernandez, P.E., Nes, I.F. and Diep, D.B. 2011a. Target recognition, resistance, immunity and genome mining of class II bacteriocins from Gram-positive bacteria. Microbiology 157: 3256–3267.

Kjos, M., Nes, I.F. and Diep, D.B. 2011b. Mechanisms of resistance to bacteriocins targeting the mannose phosphotransferase system. Applied and Environmental Microbiology 77: 3335–3342.

Konings, W.N., Kok, J., Kuipers, O.P. and Poolman, B. 2000. Lactic acid bacteria: The bugs of the new millennium. Current Opinion in Microbiology 3: 276–282.

König, H. and Fröhlich, J. 2009. Lactic acid bacteria. pp. 3–29. *In*: H. König, G. Unden and J. Fröhlich (eds.). Biology of Microorganisms on Grapes, in Must and in Wine, Springer, Berlin Heidelberg.

Kunji, E.R., Mierau, I., Hagting, A., Poolman, B. and Konings, W.N. 1996. The proteolytic systems of lactic acid bacteria. Antonie van Leeuwenhoek 70: 187–221.

Liu, M., Bayjanov, J.R., Renckens, B., Nauta, A. and Siezen, R.J. 2010. The proteolytic system of lactic acid bacteria revisited: A genomic comparison. BMC Genomics 11: 36.

Liu, S.-Q. and Pilone, G.J. 2000. An overview of formation and roles of acetaldehyde in winemaking with emphasis on microbiological implications. International Journal of Food Science & Technology 35: 49–61.

Liu, S.Q. 2002. Malolactic fermentation in wine—Beyond deacidification. Journal of Applied Microbiology 92: 589–601.

Lonvaud-Funel, A. 1999. Lactic acid bacteria in the quality improvement and depreciation of wine. Antonie van Leeuwenhoek 76: 317–331.

Lukjancenko, O., Ussery, D.W. and Wassenaar, T.M. 2012. Comparative genomics of Bifidobacterium, Lactobacillus and related probiotic genera. Microbial Ecology 63: 651–673.

Makarova, K., Slesarev, A., Wolf, Y., Sorokin, A., Mirkin, B., Koonin, E., Pavlov, A., Pavlova, N., Karamychev, V. and Polouchine, N. 2006. Comparative genomics of the lactic acid bacteria. Proceedings of the National Academy of Sciences, USA 103: 15611–15616.

Makarova, K.S. and Koonin, E.V. 2007. Evolutionary genomics of lactic acid bacteria. Journal of Bacteriology 189: 1199–1208.

Masuda, Y., Ono, H., Kitagawa, H., Ito, H., Mu, F., Sawa, N., Zendo, T. and Sonomoto, K. 2011. Identification and characterization of leucocyclicin Q, a novel cyclic bacteriocin produced by *Leuconostoc mesenteroides* TK41401. Applied and Environmental Microbiology 77: 8164–8170.

Masuda, Y., Zendo, T., Sawa, N., Perez, R.H., Nakayama, J. and Sonomoto, K. 2012. Characterization and identification of weissellicin Y and weissellicin M, novel bacteriocins produced by *Weissella hellenica* QU 13. Journal of Applied Microbiology 112: 99–108.

McAuliffe, O., Ross, R.P. and Hill, C. 2001. Lantibiotics: Structure, biosynthesis and mode of action. FEMS Microbiology Reviews 25: 285–308.

Muriana, P.M. and Klaenhammer, T.R. 1991. Purification and partial characterization of lactacin F, a bacteriocin produced by *Lactobacillus acidophilus* 11088. Applied and Environmental Microbiology 57: 114–121.

Näse, L., Hatakka, K., Savilahti, E., Saxelin, M., Pönkä, A., Poussa, T., Korpela, R. and Meurman, J.H. 2001. Effect of long-term consumption of a probiotic bacterium, *Lactobacillus rhamnosus* GG, in milk on dental caries and caries risk in children. Caries Research 35: 412–420.

Nissen-Meyer, J., Holo, H., Havarstein, L.S., Sletten, K. and Nes, I.F. 1992. A novel lactococcal bacteriocin whose activity depends on the complementary action of two peptides. Journal of Bacteriology 174: 5686–5692.

O'Flaherty, S. and Klaenhammer, T.R. 2011. The impact of omic technologies on the study of food microbes. Annual Review of Food Science and Technology 2: 353–371.

Ogawa, J., Kishino, S., Ando, A., Sugimoto, S., Mihara, K. and Shimizu, S. 2005. Production of conjugated fatty acids by lactic acid bacteria. Journal of Bioscience and Bioengineering 100: 355–364.

Okuda, K., Zendo, T., Sugimoto, S., Iwase, T., Tajima, A., Yamada, S., Sonomoto, K. and Mizunoe, Y. 2013. Effects of bacteriocins on methicillin-resistant *Staphylococcus aureus* biofilm. Antimicrobial Agents Chemotherapy 57: 5572–5579.

Orla-Jensen, S., Orla-Jensen, A.D. and Spur, B. 1926. The Butter aroma bacteria. Journal of Bacteriology 12: 333–342.

O'Sullivan, D.J. and Lee, J.H. 2011. Lantibiotics and uses thereof. US7960505, 2011.

Ouwehand, A.C., Salminen, S. and Isolauri, E. 2002. Probiotics: An overview of beneficial effects. Antonie van Leeuwenhoek 82: 279–289.

Parte, A.C. 2014. LPSN—List of prokaryotic names with standing in nomenclature. Nucleic Acids Research 42: D613–6.

Peekhaus, N. and Conway, T. 1998. What's for dinner?: Entner-Doudoroff metabolism in *Escherichia coli*. Journal of Bacteriology 180: 3495–3502.

Perez, R.H., Zendo, T. and Sonomoto, K. 2014. Novel bacteriocins from lactic acid bacteria: various structures and applications. Microbial Cell Factories 13: S3.

Pineiro, M. and Stanton, C. 2007. Probiotic bacteria: legislative framework—Requirements to evidence basis. Journal of Nutritional Research 137: 850S–853S.

Pokusaeva, K., Fitzgerald, G.F. and van Sinderen, D. 2011. Carbohydrate metabolism in Bifidobacteria. Genes & Nutrition 6: 285–306.

Ramsey, M., Hartke, A. and Huycke, M. 2014. The physiology and metabolism of enterococci. p. 106. *In*: M.S. Gilmore, D.B. Clewell, Y. ike and N. Shankar (eds.). Enterococci: From Commensals to Leading Causes of Drug Resistant Infection. Massachusetts Eye and Ear Infirmary, Boston.

Rattanachaikunsopon, P. and Phumkhachorn, P. 2010. Lactic acid bacteria: Their antimicrobial compounds and their uses in food production. Annals of Biological Research 1: 218–228.

Ravcheev, D.A., Best, A.A., Sernova, N.V., Kazanov, M.D., Novichkov, P.S. and Rodionov, D.A. 2013. Genomic reconstruction of transcriptional regulatory networks in lactic acid bacteria. BMC Genomics 14: 94.

Ray, R.C. and Joshi, V.K. 2014. Fermented foods: Past, present and future scenario. pp. 1–36. *In*: R.C. Ray and D. Montet (eds.). Microorganisms and Fermentation of Traditional Foods. CRC Press, Boca raton, Florida.

Reid, G., Jass, J., Sebulsky, M.T. and McCormick, J.K. 2003. Potential uses of probiotics in clinical practice. Clinical Microbiology Reviews 16: 658–672.

Baker, Robert C., Winkowski, Karen, Montville, Thomas J. 1996. pH-Controlled fermentors to increase production of leuconocin S by *Leuconostoc paramesenteroides*. Process Biochemistry 31(3), 1996: 225–228.

Salvetti, E., Fondi, M., Fani, R., Torriani, S. and Felis, G.E. 2013. Evolution of lactic acid bacteria in the order Lactobacillales as depicted by analysis of glycolysis and pentose phosphate pathways. Systematic and Applied Microbiology 36: 291–305.

Sanders, M.E. 2000. Considerations for use of probiotic bacteria to modulate human health. Journal of Nutrition 130: 384S–390S.

Savijoki, K., Ingmer, H. and Varmanen, P. 2006. Proteolytic systems of lactic acid bacteria. Applied Microbiology and Biotechnology 71: 394–406.

Sawa, N., Zendo, T., Kiyofuji, J., Fujita, K., Himeno, K., Nakayama, J. and Sonomoto, K. 2009. Identification and characterization of lactocyclicin Q, a novel cyclic bacteriocin produced by *Lactococcus* sp. strain QU 12. Applied and Environmental Microbiology 75: 1552–1558.

Schillinger, U., Geigen, R. and Holzapfel, W.H. 1996. Potential of antagonistic microorganisms and bacteriocins for the biological preservation of foods. Trends in Food Science & Technology 7: 158–164.

Schroeter, J. and Klaenhammer, T. 2009. Genomics of lactic acid bacteria. FEMS Microbiology Letters 292: 1–6.

Siezen, R.J., van Enckevort, F.H.J., Kleerebezem, M. and Teusink, B. 2004. Genome data mining of lactic acid bacteria: The impact of bioinformatics. Current Opinion in Biotechnology 15: 105–115.

Smid, E.J. and Hugenholtz, J. 2010. Functional genomics for food fermentation processes. Annual Review of Food Science and Technology 1: 497–519.

Smid, E.J. and Kleerebezem, M. 2014. Production of aroma compounds in lactic fermentations. Annual Review of Food Science and Technology 5: 313–326.

Snyder, A.B. and Worobo, R.W. 2013. Chemical and genetic characterization of bacteriocins: Antimicrobial peptides for food safety. Journal of the Science of Food and Agriculture 94: 28–44.

Thomas, L.V., Clarkson, M.R. and Delves-Broughton, J. 2000. Nisin. pp. 463–524. *In*: Naidu, A.S. (ed.). Natural Food Antimicrobial Systems. CRC Press, Boca Raton, FL.

Turnbaugh, P.J., Ley, R.E., Mahowald, M.A., Magrini, V., Mardis, E.R. and Gordon, J.I. 2006. An obesity-associated gut microbiome with increased capacity for energy harvest. Nature 444: 1027–1031.

Upreti, G.C. 1994. Lactocin 27, a bacteriocin produced by homofermentative *Lactobacillus helveticus* Strain LP27. pp. 331–352. *In*: Vuyst, Luc De and Vandamme, Erick, J. (eds.). Bacteriocins of Lactic Acid Bacteria: Microbiology, Genetics and Applications, Springer, New York.

Visser, J.M.J. and Haan, B.R. 2011. Liquid nisin compositions. Dsm Ip Assets. EP2309881-A1, April 20.

Wels, M., Overmars, L., Francke, C., Kleerebezem, M. and Siezen, R.J. 2011. Reconstruction of the regulatory network of *Lactobacillus plantarum* WCFS1 on basis of correlated gene expression and conserved regulatory motifs. Microbial Biotechnology 4: 333–344.

Wisselink, H.W., Weusthuis, R.A., Eggink, G., Hugenholtz, J. and Grobben, G.J. 2002. Mannitol production by lactic acid bacteria: A review. International Dairy Journal 12: 151–161.

Wollowski, I., Rechkemmer, G. and Pool-Zobel, B.L. 2001. Protective role of probiotics and prebiotics in colon cancer. American Journal of Clinical Nutrition 73: 451S–455S.

Woraprayote, W., Kingcha, Y., Amonphanpokin, P., Kruenate, J., Zendo, T., Sonomoto, K., Benjakul, S. and Visessanguan, W. 2013. Anti-listeria activity of poly (lactic acid)/sawdust particle biocomposite film impregnated with pediocin PA-1/AcH and its use in raw sliced pork. International Journal of Food Microbiology 167: 229–235.

Yoneyama, F., Fukao, M., Zendo, T., Nakayama, J. and Sonomoto, K. 2008. Biosynthetic characterization and biochemical features of the third natural nisin variant, nisin Q, produced by *Lactococcus lactis* 61-14. Journal of Applied Microbiology 105: 1982–1990.

Zendo, T., Fukao, M., Ueda, K., Higuchi, T., Nakayama, J. and Sonomoto, K. 2003. Identification of the lantibiotic nisin Q, a new natural nisin variant produced by *Lactococcus lactis* 61-14 isolated from a river in Japan. Bioscience Biotechnology Biochemistry 67: 1616–1619.

Zendo, T., Koga, S., Shigeri, Y., Nakayama, J. and Sonomoto, K. 2006. Lactococcin Q, a novel two-peptide bacteriocin produced by *Lactococcus lactis* QU 4. Applied and Environmental Microbiology 72: 3383–3389.

Zhang, Z.-G., Ye, Z-Q., Yu, L. and Shi, P. 2011. Phylogenomic reconstruction of lactic acid bacteria: an update. BMC Evolutionary Biology 11: 1.

Zineb, B., Inmaculada, Fernandez-No, Mebrouk, K., Karola, B., Pilar, C.-M. and Jorge, B.-V. 2013. Recent patents on bacteriocins: food and biomedical applications. Recent Patents on DNA & Gene Sequences 7: 66–73.

Amylolytic Lactic Acid Bacteria
Microbiology and Technological
Interventions in Food Fermentations

Smita H. Panda[1],* *and Ramesh C. Ray*[2]

1. Introduction

Amylolytic lactic acid bacteria (ALAB) have been employed in food processing industries for the production of food additives such as organic acids (i.e., lactic acid) and enzymes (i.e., α-amylase). With the advances in biotechnology, the amylase applications of ALAB have been expanded to other sectors such as bio-pharmaceuticals, brewing and distillery (Bhanwar and Ganguli 2014, Hattingh et al. 2015). ALAB are also employed in preparing high energy density (ED) cereal-based foods for improving dietary starch utilization in infants and small children (Panda et al. 2008, Songré-Ouattara et al. 2009). ALAB are generally considered safe for use in food, and several technologies based on the reduction of swelling of starch granules obtained by partial starch hydrolysis by ALAB or dextrinization have been developed to increase ED of cereal gruels with appropriate semi-liquid to liquid consistency. Nguyen et al. (2007a, 2007b) showed that ALAB could be used as starter cultures for starch hydrolysis to produce high ED gruels by fermentation of pre-heated rice-soybean slurries in combination with high pressure homogenization or spray drying. Similarly, fermentation and starch hydrolysis of a pre-cooked African pearl millet (*Pennisetum glaucum*)-groundnut slurry inoculated with an ALAB, *Lactobacillus plantarum*, isolated from *Ben-saalga*, a traditional pearl millet-based gruel from Burkina Faso, was developed for young children (Songre-Ouattara et al. 2009).

[1] Department of Zoology, North Orissa University, Baripada 757003, Odisha, India.
[2] ICAR—Central Tuber Crops Research Institute, Bhubaneswar 751019, Odisha, India.
   E-mail: rc_rayctcri@rdiffmail.com
* Corresponding author: panda.smita@gmail.com

In this chapter, we briefly review the microbiology, technological aspects of ALAB and their role/applications in food fermentation, with emphasis on traditional fermented foods.

2. Lactic Acid Bacteria

Lactic acid bacteria (LAB) constitute a diverse group of Gram positive microorganisms which are generally non-sporulating, non-motile, catalase negative and devoid of cytochrome. They comprise both cocci and bacilli belonging to *Enterococcus, Lactobacillus, Streptococcus, Pediococcus, Leuconostoc,* etc. (Hofvendahl and Hahn-Hagerdal 2000, Mazzoli et al. 2014, Chapter 7 in this book). LAB are the most promising microorganisms for bio-refineries converting starchy waste biomass into industrially important products (i.e., lactic acid, LA) (Berlec and Strukelj 2009). LAB produce lactic acid as an anaerobic product of glycolysis with high yield and productivity. LAB are grouped as either homo-fermentative or hetero-fermentative based on the fermentation end product. Homo-fermentative LAB possess aldolase enzymes and produce lactic acid as the major end product. However, hetero-fermentative LAB produce acetic acid, water, ethanol, along with lactic acid; therefore, the maximum yield of lactic acid reaches only 0.5 g/g or 1.0 mol/mol glucose moiety (Abdel-Rahman et al. 2011, 2013). Recently, metabolic engineered *Lactobacillus plantarum* was reported to metabolize pentose to lactic acid in a homo-fermentative pathway (Okano et al. 2009b, Hattingh et al. 2015). Most LAB are considered to be safe for industrial lactic acid production as they have a long history of use in food and beverage without any adverse health effects on consumers. Commercially, important LAB strains such as *Lb. plantarum, Lb. rhamnosus, Lb. acidophilus,* have been useful due to their high acid tolerance and their ability to be engineered for the production of D- or L-lactic acid (Kyla-Nikkila et al. 2000, Abdel-Rahman et al. 2013).

3. Amylolytic Lactic Acid Bacteria

Amylolytic LAB (ALAB) mostly belong to the genera *Lactobacillus, Lactococcus, Streptoccocus, Pediococcus, Carnobacterium* and *Weissella* (Bhanwar and Ganguli 2014). ALAB produce starch-modifying enzymes (amylases), which hold a broad spectrum of applications as compared to chemical hydrolysis of starch in food processing industries (Petrova et al. 2013, Smerilli et al. 2015). ALAB are mainly distributed in amylaceous fermented foods, i.e., fermented meals based on manioc (cassava), sorghum, rice, millet, maize, taro, etc. (Yousif et al. 2010, Turpin et al. 2011). Other sources of ALAB are fermented cereals, beverages such as maize-based *Pozol* (Daiz-Ruiz et al. 2003), beer malt (Bohak et al. 1998) and the probiotic drink, *Boza* (Petrova et al. 2013). ALAB are also found in the digestive tract of animals, as well as in plant matrices and plant wastes (Giraud et al. 1991, Oguntoyinbo 2007). Amylolytic strains of *Lactobacillus fermentum* were isolated for the first time from Benin maize sourdough, ogi and mawe (Agati et al. 1998). Similarly, a most recently isolated ALAB strain, *Lactobacillus paracasei* B41, was the first amylolytic representative of *Lb. casei* group (Petrova et al. 2012, Bhanwar and Ganguli 2014). ALAB are mostly found in amylaceous foods where, due to the latter's high starch content, they can easily convert them to mono- and di-saccharides and make them available for lactic acid fermentation (Fig. 1) (Guyot et al. 2000). ALAB can change the microstructure of starch and induce their amylography and viscosity characteristics (Demiate et al. 2000, Putri et al. 2011). Recently, Sanni et al. (2002) described amylolytic strains of *Lb. plantarum* and *Lb. fermentum* strains in various Nigerian traditional amylaceous fermented foods. Further, *Lb. amylophilus* was the first ALAB reported in 1979 by Nakamura and Crowell (1979).

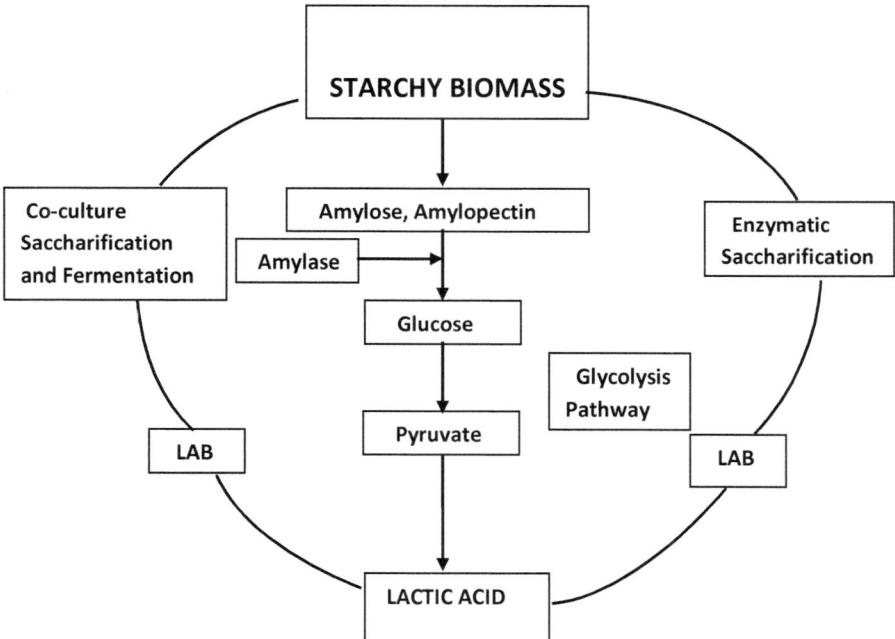

Figure 1. Schematic representation of amylolytic bacterial lactic acid fermentation.

4. Amylolytic Enzymes in LAB

The amylolytic activity of LAB is a major characteristic for fermentation of starch to lactic acid. The starch modifying enzymes in ALAB belong to α-amylase, maltogenic amylases (MAases), amylopullanases, pullulanases, neopullulanases, glycogen phosphorylases and 1,6-glucosidases (Petrova et al. 2013). Basically, there are four groups of starch converting enzymes: (1) endo-amylases, (2) exo-amylases, (3) debranching enzymes, and (4) transferases. A number of these starch converting enzymes belong to a single family: the α-amylase family or GH13 family of glycosyl hydrolases. Alpha (α)-amylases are extracellular enzymes that catalyze the hydrolysis of α-1, 4-glycosidic linkages in starch liberating linear and branched oligosaccharides of various chain lengths, as well as glucose (Sharma and Satyanarayana 2013). *AmyA* genes are mainly responsible for amylolytic activity in ALAB. The *AmyA* genes of *Lb. amylovorous, Lb. planatrum*A6 and *Lb. manihotivorans* share 98% identity and are completely different from *AmyA* genes found in other lactobacilli (Morlon-Guyot et al. 2001, Petrova et al. 2013). The enzymes are organized in two functional domains: catalytic domain (amino acids 1–474) and starch binding domain (SBD) (amino acids, 475–953) (Rodriguez-Sanoja et al. 2005). The catalytic domain contains all conserved regions for GH13 family (Janecek 2002). It consists of tandem repeat units of 91 amino acids each; four repeats in *Lb. manihotivorans* and five repeats in *Lb. amylovorus*. The role of SBD is adsorption onto raw starch granules; it increases the concentration at the active site of the enzyme, which allows microorganisms to degrade non-soluble starch (Rodriguez-Sanoja et al. 2005). Therefore, *Lactobacillus* which lacks SBD would be unable to degrade the raw starch. An important property of extracellular amylases is the existence of a signal peptide that ensures the transport of the polypeptide out of the cell before the enzyme's maturation. Many genes responsible for the growth of LAB have been found on mobile elements such as plasmids or transposons. The analysis of the phylogenetic relationships strongly supports the hypothesis of a horizontal gene transfer between the remotely related bacteria (Wasko et al. 2011). The analysis of the orthologs involved in starch hydrolysis by the genus *Lactobacillus* reveals that the amylase genes are actually present in the genome, but a majority of them are not expressed due to mutation damages in the promoter region, i.e., in the amylase catalytic domain. With long-term propagation in dairy habitats, which does not involve amylolytic enzymes, the responsible genes were converted to pseudogenes since they were not subjected to pressure of environment (Petrova et al. 2013). The above phenomenon was observed in *Streptococcus thermophilus*, the strain LMD-9, which harbours a large number of pseudogenes, indicating it has undergone major reductive evolution with loss of carbohydrate metabolic genes found in their streptococcal counterparts (Goh et al. 2011).

5. Lactic Acid and Its Importance

Lactic acid (2-hydroxypropionic acid) is the most widely occurring carboxylic acid (John et al. 2009). It is a valuable chemical with a wide variety of applications in the food, pharmaceutical, textile and cosmetic industries (Petrova et al. 2013). Lactic acid can be produced commercially both by chemical synthesis and microbial fermentation. In the case of chemical synthesis, lactic acid is obtained in a racemic DL mixture, whereas microbial fermentation produces the stereo-isomer (Datta et al. 1993). There are two optically active isomers of lactic acid, i.e., L(+) and D(–) forms. L(+) forms are easily metabolized by humans and preferred for the production of lactic acid, whereas D(–) forms are harmful to humans when present in high levels (Reddy et al. 2008). The most common ALAB employed for the production of lactic acid are from the genus *Lactobacillus*, i.e., *Lb. amylophilus, Lb. pentosus, Lb. paracasei* and *Lb. planatrum* (Petrova and Petrov 2012, Petrova et al. 2013). Lactic acid is also used as an emulsifying and moisturizing agent in cosmetic and tanning industries (Papagianni 2012). Another interesting application of lactic acid is the synthesis of ethyl lactate, which is used as biodegradable solvent (Nampoothiri et al. 2010). Recently, worldwide demand for lactic acid has increased because of its use as a building block for the synthesis of plastic polymers, i.e., polylactates (John et al. 2007, Mazzoli et al. 2014).

Many researchers have attempted direct lactic acid production from starchy materials, and several wild amylolytic LAB have been isolated in different environments (Narita et al. 2004, Mazzoli et al. 2014). *Lactobacillus lactis, Lb. delbrueckii* and *Enterococcus faecalis* have been used for the production of lactic acid from molasses (Wee et al. 2004, Joshi et al. 2010). *Lactobacillus helveticus* is generally the preferred microorganism for lactic acid production at pH 5.8 and temperature 42°C. Currently, various studies have been focused on engineered microorganisms in order to meet the commercial requirements, including improved optical purity of the product, reduction of nutritional supply, improved yield and productivity (Wang et al. 2014). Recently, *Saccharomyces cerevisiae* and *Lactococcus lactis* have been genetically engineered to obtain the pure form of lactic acid isomer (Mazzoli et al. 2014, Shinkawa et al. 2011).

6. Molecular Methods for Characterization of ALAB

ALAB are mostly characterized and identified by a polyphasic approach based on phenotypic and genotypic methods such as carbohydrate fermentation patterns using API 50CH (Lee et al. 2001, Giraud et al. 1991, Petrova et al. 2013), ribotyping (Diaz-Ruiz et al. 2003), intergenic transcribed spacers-PCR/restriction fragment length polymorphism (ITS-PCR/RFLP) (Sawadogo-Lingari et al. 2007), pulse-field gel electrophoresis (PFGE), ARDRA, rep-PCR, RAPD-PCR (Petrova et al. 2008), multipex-PCR and 16S rRNA gene sequencing (Turpin et al. 2011, Petrova and Petrov 2012). Most of the amylase properties are studied by Zymograms or enzyme's N-terminal sequencing (Calderon-Santoyo et al. 2003, Petrova et al. 2013). The molecular analyses of the responsible genes include PCR amplification, sequencing and sequence comparison. However, the investigations of ALAB are, so far, focused mainly on the enhancement of starch fermentation process and only a few strains were studied for their amylolytic enzyme (Reddy et al. 2008).

7. Solid-state and Submerged Fermentation for LA Production Using ALAB

In the field of fermentation technology, fermentation conditions such as pH, temperature and inoculum volume are considered important factors for cell growth, lactic acid concentration, lactic acid productivity and lactic acid yield. In most cases, glucose is the preferred carbon source for lactic acid fermentation by ALAB. However, cheap and widely existing renewable materials such as starch (wheat starch, corn starch, sago starch and potato starch) and lignocellulose (woody materials, crop residues) are recognized to meet the requirements for the biotechnological production of lactic acid, economically and efficiently. Several fermentation modes for lactic acid production by various ALAB producing strains, such as batch fermentation, fed-batch fermentation and solid state fermentation, have been investigated. Solid state fermentation (SSF) process is defined as 'the growth of microorganisms on moist solid materials in absence of free-flowing water' (Moo-Young et al. 1983, Reddy et al. 2008). In most cases, SSF is used for the production of industrial enzymes. The use of agricultural starchy wastes involves a two-step process of liquefaction and saccharification followed by *Lactobacillus* fermentation, which is often expensive (Linko and Javanainen 1996, Panda and Ray 2008). Amylase producing *Lactobacillus* spp. (*Lb. plantarum, Lb. amylophilus* and *Lb. amylovorus*) are often used for direct fermentation of starchy materials into lactic acid; this can lead to significant reduction in operating cost (Wee et al. 2006, Vishnu et al. 2002). A study was made to develop a novel technology for L(+) lactic acid production by SSF using *Lb. plantarum* MTCC 1407 for which cassava bagasse (a major waste release to the environment during the extraction of starch) was selected in lieu of glucose as the carbon source. In this study, the optimum incubation period, temperature and pH were 120 hr, 35°C and 6.5, respectively. Maximum starch conversion by *Lb. plantarum* MTCC 1407 to lactic acid was 63.3%. The organism produced 29.86 g of lactic acid from 60 g of starch in 100 g of cassava bagasse (Ray et al. 2009). The lactic acid yield was 49.76%. Utilization of starchy agro-wastes like cassava bagasse for lactic acid production by ALAB has dual advantages, i.e., cheaper production cost as compared with refined substrate and an appropriate technology for value addition of agro-wastes and environmental waste management. In another study by Panda and Ray (2008), lactic acid production was carried out by *Lb. plantarum* MTCC 1407 in semi-solid fermentation using sweet potato (*Ipomoea batatus* L.) flour. Sweet potato flour is a cheap source of carbohydrate and other nutrients and easily available material. *Lb. plantarum* MTCC 1407 strain is amylolytic and it could convert raw starch present in sweet potato flour to lactic acid in a single step fermentation process. The organism produced 23.86 g of lactic acid (43.4%) from 55 g of starch present in 100 g of sweet potato flour, showing 56% conversion after 120 hr of incubation (Panda and Ray 2008). A yield of 41 g/LA in mineral liquid medium was reported by Fu and Mathews (1999), using a strain of *Lb. plantarum* ATCC 21028. Similar results of lactic acid yield on starchy substrates were reported for *Lb. casei* (John et al. 2007), *Lb. amylophilus* GV6 (Reddy et al. 2008, Naveena et al. 2004) and *Enterococcus faecalis* (Oh et al. 2005). Naveena et al. (2005) reported that *Lb. amylophilus* GV6 was found to produce 36 g of lactic acid from 54.4 g starch present in 100 g of wheat bran with a lactic acid yield of 77.6%. Yoshida et al. (2011) reported co-fermentation with glucose, xylose and arabinose for homo-D lactic acid production by *Lb. plantarum* NCIM B 8826. Further, Abdel-Rahman et al. (2011) also isolated a wild type LAB, *Enerococcus mundtii* QU25, which was able to produce optically pure L-lactic acid from glucose, cellobiose and xylose with a yield of 0.9 g.

Soluble starchy substrates available in the form of agricultural wastes, soluble pure and crude starches are utilized in submerged fermentation. In a study by Narita et al. (2004), it was found that *Streptococcus bovis* 148 could directly produce lactic acid from starch and maximum lactic acid of 14.2 g/L was obtained. A novel starch degrading strain of *Lb. casei* was constructed by genetically displaying α-amylase form *S. bovis* 148 strain with a FLAG peptide tag (AmyAF) (Narita et al. 2006, Reddy et al. 2008). In the process of fermentation using AmyAF displaying *Lb. casei* cells, 50 g/L of soluble starch was reduced to 13.7 g/L, and 21.8 g/L of lactic acid was produced within 24 hr.

In all the above studies, different renewable starch sources were used for the high production of lactic acid. However, Panda et al. (2008) reported *Lb. plantarum* MTCC 1407 as a source of α-amylase in submerged fermentation using sweet potato flour. Since sweet potato is a starchy crop, it was assumed that the above strain, by virtue of possessing α-amylase activity, could convert sweet potato starch into glucose and finally to lactic acid. Statistical optimization, i.e., Response Surface Methodology (RSM), was used to optimize different fermentation parameters like incubation period, temperature and pH that affected α-amylase production. RSM is a tool which is used to overcome a number of process optimization problems. RSM has already been successfully applied for optimization of media and culture conditions in cultivation process for the production of primary and secondary metabolites, i.e., amino acid, ethanol and enzymes (He et al. 2004, Panda et al. 2008). The experimental results showed that the optimum pH, temperature and incubation period were 7.6, 37°C and 36 hr, respectively (Fig. 2A,B,C). The purified enzyme (by ammonium sulphate precipitation) had a molecular mass of 75,450 Da in SDS-PAGE (Fig. 3). The α-amylase production was 4022 U/min/ml. The application of α-amylase produced by *Lb. plantarum* MTCC 1407 was also studied. The crude enzyme of 2 ml (= 8044 units) could convert 56% of the starch present in sweet potato flour to

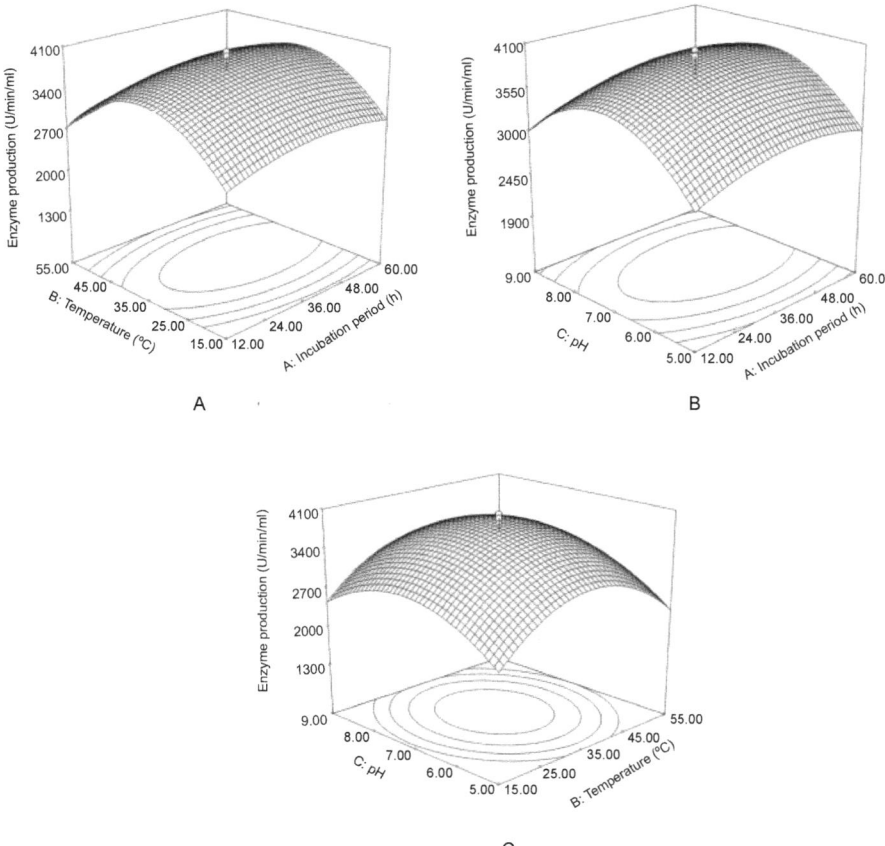

Figure 2. Statistical optimization of enzyme production using RSM (A) temperature and incubation period; (B) pH and incubation period and (C) pH and temperature (Panda et al. 2008).

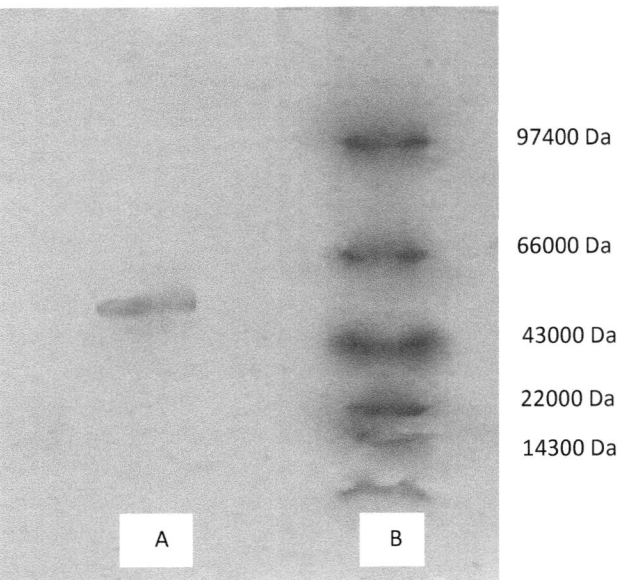

Figure 3. Determination of molecular mass by SDS-PAGE. (A) Standard protein markers (PMW-B); (B) α-amylase by *Lb. plantarum* MTCC 1407 (Panda et al. 2008).

lactic acid after 120 hr of incubation. To conclude, the above study revealed that *Lb. plantarum* MTCC 1407 is a potential strain for extracellular α-amylase production in submerged fermentation. It can be used in the production of starch-based fermented foods, feed and pharmaceuticals.

8. ALAB in Traditional Fermented Foods

Amylolytic lactic acid bacteria (ALAB) have repeatedly been isolated from traditional cereal- or cassava-based fermented foods (Nwankwo et al. 1989, Johansson et al. 1995, Morlon-Guyot et al. 1998, Olympia et al. 1995, Sanni et al. 2002). Due to the ability of their α-amylases to partially hydrolyze raw starch (Rodriguez-Sanoja et al. 2000), ALAB could ferment different types of amylaceous raw material, such as corn (Nakamura 1981), potato (Chatterjee et al. 1997), or cassava (Giraud et al. 1994). In a study conducted on cassava fermentation to produce gari in Benin, *Lb. plantarum* was the most abundantly isolated species, followed by *Leuconostoc fallax* and *Lb. fermentum* (Kostinek et al. 2005). The involvement of ALAB in cassava fermentation into foods and beverages such as agbelima, gari, fufu, lafun, sour starch, etc. have been elaborately discussed by Ray and Ravi (2005) and Ray and Ward (2006). The list of different ALAB involved in fermented foods is given in Table 1.

Nine strains of ALAB that hydrolyze starch were isolated from *burong isda*, an indigenous fermented food made from fish and rice in the Philippines. Conventional taxonomic and DNA-DNA reassociation studies indicated that all these isolates belong to *Lb. plantarum*. Each of these isolates harbored more than 10 plasmid species with molecular sizes of 2 to 60 kb. The amylolytic activity of the strain L137, one of the

Table 1. Compilation of different ALAB isolated or employed in fermented food and beverages.

Product	Substrate	Microflora	Country	Reference
Amasi	milk (cow)	*Lactobacillus* spp.	Africa	Chelule et al. 2010
Garris	milk (camel)	*Lb.fermentum, Lb. plantarum*	Africa	Narvhus and Gadaga 2003
Rob	milk (camel)	*Lb.fermentum, Lb. acidophilus*	Middle and Western Sudan	Narvhus and Gadaga 2003
Boza	barley, oats, rye, millet, maize, wheat, rice	*Lb.fermentum, Lb. plantarum, Leuconostoc mesenteriodes*	Turkey, Romania, Bulgaria	Akpinar-Bayizit et al. 2010
Ben-saalga	pearl millet	*Lb. plantarum, Lb. pentosus, Lb.fermentum*	Burkina Faso, Africa	Pulido et al. 2012
Bushera	sorghum	*Lb. brevis, Lactococcus*	Uganda	Muianja et al. 2003
Kisra	sorghum, maize	*Lb. plantarum, Lb.fermentum*	Sudan	Franz et al. 2014
Togwa	maize, sorghum, ginger millet	*Lactobacillus* spp.	Tanzania, Zimbabwe	Kitabatake et al. 2003
Kefir	milk	*Lb. brevis, Lb. plantarum*	Svaneti, Georgia	Lifeway 2014
Kombucha	tea leaves	*Lactobacillus* spp.	Northeast China	BevNet 2011
Hardaliya	red grape	*Lactobacillus pontis, Lb. pseudoplantarum*	Turkey	Arici and Coskun 2001
Capers	capers berry	*Lb. plantarum, Lb. pentosus, Lb.fermentum*	Greece, Italy, Turkey, Morocco, and Spain	Pulido et al. 2012
Salgam	carrots, turnip	*Lb. plantarum, Lb.fermentum, Lb. brevis*	Adana, Mersin (Turkey)	
Tursu	cucumbers, cabbage, tomatoes	*Lb. plantarum, Leuconostoc mesenteriodes*	Turkey	Kabak and Dobson 2011
Guedj	fish	*Lactococcus lactis*	Ghana, Mali	Franz et al. 2014
Bonome	fish	*Lactobacillus* spp.	Ghana	Franz et al. 2014
Kocho	banana	*Lactobacillus* spp.	Ethiopia	Franz et al. 2014
Lafun	cassava	*Lactobacillus fermentum, Lb. plantarum*	Nigeria, Benin	Nout 2009
Plaa-som, Burong isda	fish	*Lb.fermentum, Lb. plantarum, Lactobacillus* spp.	Thailand	Phonyiam and Yunchalard 2008

ALAB = Amylolytic Lactic Acid Bacteria; *Lb.* = *Lactobacillus*.

isolates, was lost by treatment with novobiocin at 43% frequency, concomitant with the curing of a 33-kb plasmid, pLTKl3; this suggested that pLTK13 carries a gene necessary for the synthesis of amylolytic enzyme. An acidophilic starch-hydrolyzing enzyme, secreted from L137 cells, was purified 46-fold with specific activity of 44 units/mg protein. The enzyme was shown to have a molecular mass of about 230 kDa and the optimum temperature and pH for the enzyme reaction with soluble starch were 35°C and 3.8–4.0, respectively. The enzyme hydrolyzed soluble starch, amylopectin, glycogen, pullulan and, to a small extent, amylose, while it exerted no activity on dextran and cyclodextrins (Olympia et al. 1995).

Plaa-som is a Thai fermented fish product consisting of whole fish or fish fillets mainly from a freshwater fish, *Barbodes gonionotus*, raw or cooked rice, crushed fresh garlic, and salt. LAB was found to be the most dominant group of microorganisms in *plaa-som* (Phonyiam and Yunchalard 2014).

9. Application of ALAB in Starch-Based Food Processing

ALAB are mainly used in starch processing industries for the production of lactic acid. ALAB's ability of direct conversion of starch to lactic acid, unifying the steps of saccharification and fermentation in a single process, make it more economically attractive. ALAB are employed in lactic acid fermentation which helps in maintaining the nutritional value of foods, the durability, consistency, aroma and taste of end products. It is also used in the manufacturing of various fermented food and beverages (Blandino et al. 2003). Depending on the bio-safety of species and the strain composition of ALAB, they can be applied for the development of probiotic fermented meals and beverages (Turpin et al. 2011, Petrova et al. 2010, Petrova and Petrov 2011, 2012) and hyperallergenic children's foods (Brown and Valiere 2004, Nyugen et al. 2007a,b). Extracellular amylases produced by ALAB are also applied as sourdough starters, especially to retain the original taste and aroma in bakery products (Reddy et al. 2008). The above process improves characteristics of bread; it also extends the period of bread's life by the suppression of undesired microflora and by the shortening of the amylopectin chain length due to formation of oligosaccharides. The bacterial maltogenic and acidic α-amylase with intermediate thermostability have been reported to act as anti-staling agents, thereby reducing the crumb firmness during storage (Sharma and Satyanarayana 2013). Acidic α-amylases are also used in the removal of starch from beer, fruit juices and smoothies. The acid stable and Ca^{+2} independent α-amylases are preferred over the Ca^{+2} dependent amylases in starch processing because the later are active at 95°C and at pH 6.8. Thus, the above process cannot be performed at low pH (3.2–4.5), the pH of native starch, particularly during bread baking (Shivaramakrishnan et al. 2006). Many amylase preparations are available from various enzyme manufactures for specific applications such as starch saccharification, bioethanol production, textile designing, and laundry, dish-washing detergents, etc. (Pandey et al. 2000, Rao and Satyanarayana 2007). Currently α-amylases from ALAB are gaining importance in biopharmaceutical applications and their demand is expected to rise in the coming future.

9.1 ALAB in Cereal-Based Gruel

In many developing countries, cereal-based gruels are used as complementary foods for young children. To obtain gruels with a semi-liquid consistency after cooking, suitable for feeding young children, the starchy flour is traditionally diluted with a large quantity of water. At a low flour content (5–10 g dry matter/100 ml), the gruel has a free-flowing consistency and is easy to swallow but its energy density (20–40 kcal/100 g) is lower than the minimum value of 84 kcal/100 g of gruel recommended for children aged 9–11 months fed at a rate of 2 meals/day in addition to average breast milk intake (Dewey and Brown 2003). At higher dry matter content, the gruel is stiff, due to starch gelatinization and is thus unsuitable for feeding young children. Now, technologies have been developed employing ALAB for preparing high energy density semi-liquid complementary foods such as gruel from cereals or cereal-legume mixtures (Fig. 4). The use of ALAB as a functional starter culture could provide an alternative to the use of traditional or industrial malt by combining amylase production and acidification using the same strain, which would improve control of fermentation, and lead to products of more standard quality.

The ability of ALAB (*Lb. plantarum* A6) to ferment rice-soybean mixtures prepared at high dry matter content in order to obtain new functional foods was studied (Nguyen et al. 2007a). Combining this with spray-drying, as a post-treatment

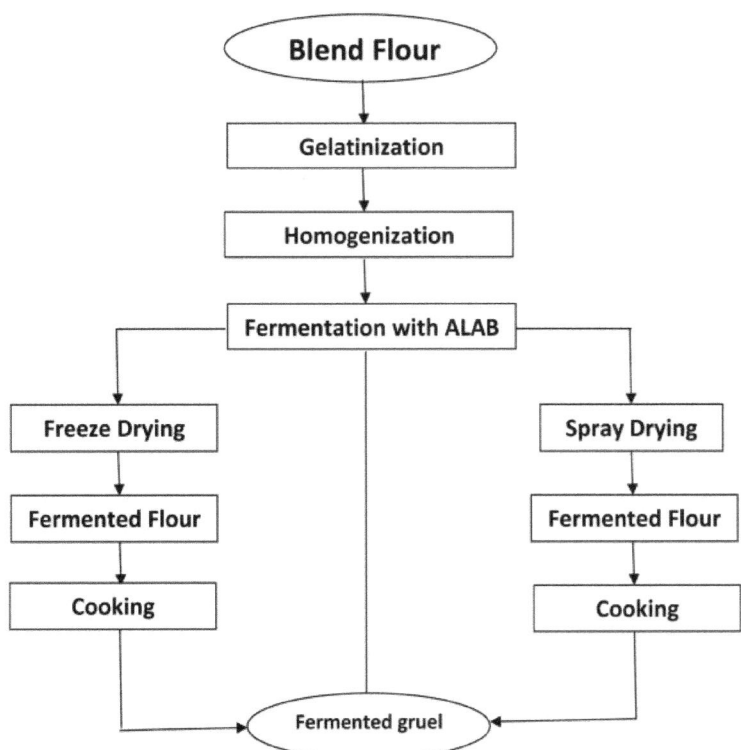

Figure 4. Flow-chart for preparation of fermented gruel using ALAB (amylolytic lactic acid bacteria) (Nguyen et al. 2007a).

after fermentation by strain A6, had the additional advantage of enabling the use of spray-dried fermented flour to prepare gruels with higher dry matter content (30–33%), corresponding to 128–140 kcal/100 g of gruel.

Fermentation and starch hydrolysis of pre-cooked pearl millet–groundnut slurry inoculated with ALAB or by back slopping was investigated as a substitute for the addition of malt in the preparation of infant gruels (Trèche and Mouquet-Rivier 2008). In Burkina Faso, the ALAB collection strain *Lb. plantarum* A6, and the endogenous microflora provided by back slopping were more efficient in acidifying and partially hydrolyzing starch in the millet-groundnut slurry than *Lb. plantarum* 6.1, isolated from the traditional process. Large amounts of maltotriose and maltotetraose accumulated in slurry fermented by strain A6. No accumulation of maltose was observed, which could be an advantage in preventing the growth of microbial contaminants such as yeasts. Starch hydrolysis in the millet-groundnut slurry inoculated with strain A6, or by back slopping, enabled the preparation of high-energy density gruels (84.7 ± 4.4 and 80.4 ± 23.8 kcal/100 g of gruel, respectively) of liquid consistency (Songré-Ouattara et al. 2009). In a subsequent study, gruels tailored to school-age children and made of soy milk and rice flour, with or without total dietary fiber from passion fruit by-products, were fermented by ALAB strains (*Lb. fermentum* Ogi E1 and *Lb. plantarum* A6), by commercial probiotic bacteria strains (*Lb. acidophilus* L10, *Lb. casei* L26 and *Bifidobacterium animalis* subsp. *lactis* B94) and by co-cultures made of one amylolytic and one probiotic strain. The combination of amylolytic and probiotic bacteria strains reduced the fermentation time of the gruels as well as increased the α-amylase activity. The addition of passion fruit fiber exerted less influence on the apparent viscosity of the fermented products than the composition of the bacterial cultures (Espirito-Santo et al. 2014).

9.2 ALAB in Lacto-pickling

Lactic acid fermentation of vegetable products applied as preservation methods for the production of finished and half finished products is considered as important technology for growing quantities of raw materials processed in the food industry. ALAB are employed in preparation of vegetable lacto-pickles (Fig. 5), lacto-juices and smoothies. Sweet potato, rich in anthocyanin or β-carotene pigments, was pickled by lactic acid fermentation by brining the cut and blanched roots in 2–10% NaCl solution and subsequently inoculated with ALAB strain *Lb. plantarum* MTCC 1407 for 28 days (Panda et al. 2007). The final product, with 8 and 10% brine, had a pH of 2.9–3.0, titratable acidity of 2.9–3.7 g/kg, LA of 2.6–3.2 g/kg and starch of 58–68 g/kg on fresh weight basis. Sensory evaluation rated the sweet potato lacto-pickle acceptable based on texture, taste, aroma, flavor and aftertaste (Sivakumar et al. 2010).

10. Perspectives and Conclusion

Malt or germinated cereals are very efficient in partially hydrolyzing starch; they are used in traditional brewing processes to prepare local beers from different cereals (e.g., sorghum, maize), and are recommended in the preparation of infant gruels.

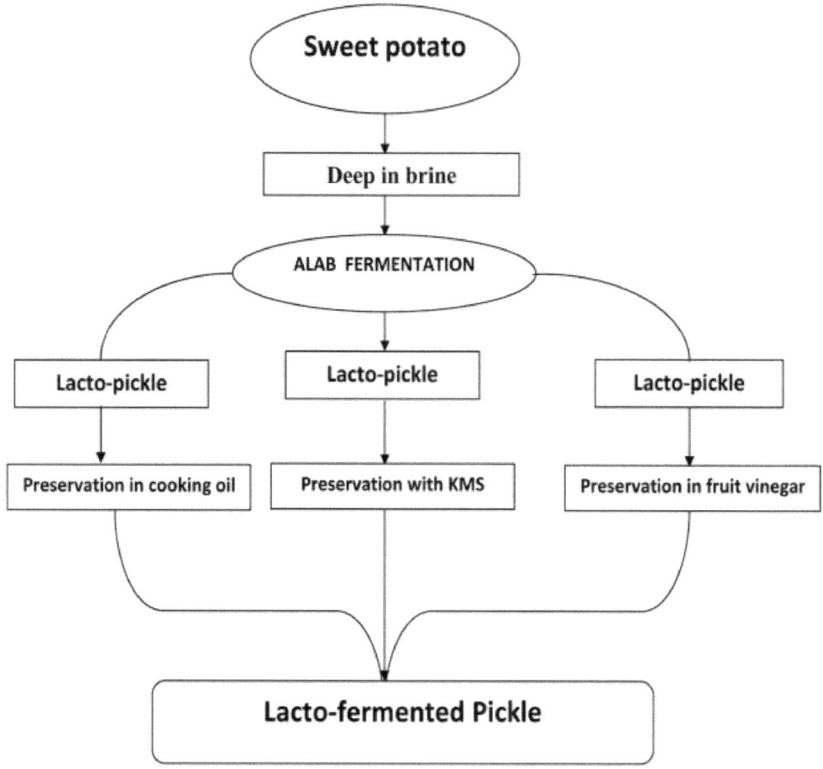

Figure 5. Flow-chart for preparation sweet potato lacto-pickle (Panda et al. 2007). ALAB = Amylolytic lactic acid bacteria; KMS = Potassium meta-bisulphite.

Nevertheless, the use of efficient ALAB such as *Lb. plantarum* A6 as starter cultures could replace the addition of malt and be used for a variety of cereal-based gruels and other functional foods. However, ALAB starters still need to be tested on different types of starchy foods. Likewise, ALAB are equally efficient in fermenting starchy vegetables and roots such as carrot, beetroot and sweet potato into lacto-pickles. ALAB possess amylase activity which exhibits a significant role in intestinal tracts of mammals and human beings. They degrade the starch present in food to lactic acid and fermentable mono-saccharides that can easily be assimilated in the body, thus improving utilization of dietary starch, and enhancing digestion. Some of the ALAB show a high potential for bio-refinery applications. Significant advances have been made in recent years in the development of molecular tools for engineering the metabolic pathways of ALAB, optimization of fermentation processes, understanding of enzymes and other biochemical systems, and of metabolic pathways relevant for fermented beverage and food applications.

Keywords: Amylolytic lactic acid bacteria, cereal-based gruel, homo-lactic fermentation, *Lactobacillus plantarum, Lactobacillus acidophilus, Lb. amylovorus, Lactobacillus rhamnosus, Lb. manihotivorans,* lacto-pickling, starch

References

Abdel-Rahman, M.A., Yukihiro, T. and Kenji, S. 2011. Lactic acid production from lignocellulose-derived sugars using lactic acid bacteria: Overview and limits. Journal of Biotechnology 156: 286–301.

Abdel-Rahman, M.A., Tashiro, Y. and Sonomoto, K. 2013. Recent advances in lactic acid production by microbial fermentation processes. Biotechnology Advances 31: 877–902.

Agati, V.J.P., Guyot, J., Morlon-Guyot, P. and Talamond, D.J. 1998. Isolation and characterization of new amylolytic strains of *Lactobacillus fermentum* from fermented maize doughs (*mawe* and *ogi*) from Benin. Journal of Applied Microbiology 85: 512–20.

Akpinar-Bayizit, A., Yilmaz-Ersan, L. and Ozcan, T. 2010. Determination of boza's organic acid composition as it is affected by raw material and fermentation. International Journal of Food Properties 13(3): 648–656.

Arici, M. and Coskun, F. 2001. *Hardaliye*: Fermented grape juice as a traditional Turkish beverage. Food Microbiology 18: 417–421.

Berlec, A. and Strukelj, B. 2009. Novel applications of recombinant lactic acid bacteria in therapy and in metabolic engineering. Recent Patent in Biotechnology 3: 77–87.

BevNet. 2011. The *kombucha* crisis: One year later. BevNet.

Bhanwar, S. and Ganguli, A. 2014. Amylase and galactosidase production on potato starch waste by *Lactococcus lactis* subsp. lactic isolated from pickled yam. Journal of Scientific and Industrial Research 73: 324–330.

Blandino, A., Al-Aseeri, M., Pandiella, S., Cantero, D. and Webb, C. 2003. Cereal-based fermented foods and beverages. Food Research International 36: 527–543.

Bohak, I., Back, W., Richter, L. and Ehrmann, M. 1998. *Lactobacillus amylolyticus* sp. nov., isolated from beer malt and beer wort. Systematic and Applied Microbiology 21: 360–364.

Brown, A.C. and Valiere, A. 2004. The medicinal uses of poi. Nutrition Clinical Care 7: 69–74.

Calderon-Santoyo, M., Loiseau, G., Rodriguez Sanoja, R. and Guyot, J.P. 2003. Study of starch fermentation at low pH by *Lactobacillus fermentum* OgiE1 reveals some coupling between growth and alpha-amylase production at pH 4.0. International Journal of Food Microbiology 80: 77–87.

Chatterjee, M., Chakrabarty, S.L., Chattopadhyay, B.D. and Mandal, R.K. 1997. Production of lactic acid by direct fermentation of starchy wastes by an amylase-producing *Lactobacillus*. Biotechnology Letter 19: 873–874.

Chelule, P.K., Mokoena, M.P. and Gqaleni, N. 2010. Advantages of traditional lactic acid bacteria fermentation of food in Africa. pp. 1160–1167. *In*: A. Mendez-Vilas (ed.). Current Research Technology and Education Topics in Applied Microbiology and Microbial Biotechnology, Formatex, Spain.

Datta, R.S., Sai, P.T., Patric, B., Moon, S.H. and Frank, J.R. 1993. Technological and economical potential of polylactic acid and lactic acid derivatives. International Congress on Chemicals from Biotechnology, Hannover, Germany, pp. 1–8.

Demiate, L.M., Dupuy, G.N., Huvenne, J.P. and Cereda, M.P. 2000. Relationship between baking behavior of modified cassava starches and starch chemical structure determined by FTIR spectroscopy. Carbohydrate Polymer 42: 149–158.

Dewey, K.G. and Brown, K.H. 2003. Update on technical issues concerning complementary feeding of young children in developing countries and implications for intervention programs. Food and Nutrition Bulletin 24(1): 5–28.

Dı́az-Ruiz, G., Guyot, J.P., Ruiz-Teran, F., Morlon-Guyot, J. and Wacher, C. 2003. Microbial and physiological characterization of weakly amylolytic but fast-growing lactic acid bacteria: A functional role in supporting microbial diversity in *pozol*, a Mexican fermented maize beverage. Applied and Environmental Microbiology 69: 4367–4373.

Espirito-Santo, A.P., Mouquet-Rivier, C., Humblot, C., Cazevieille, C., Icard-Vernière, C., Soccol, C.R. and Guyot, J.P. 2014. Influence of co-fermentation by amylolytic *Lactobacillus* strains and probiotic bacteria on the fermentation process, viscosity and microstructure of gruels made of rice, soy milk and passion fruit fiber. Food Research International 57: 104–113.

Franz, C.M.A.P., Hucha, M., Mathara, M.J., Abriouel, H., Benomar, N., Reid, G., Galvez, A. and Holzapfel, W.H. 2014. African fermented foods and probiotics. International Journal of Food Microbiology 190: 84–96.

Fu, W. and Mathews, A.P. 1999. Lactic acid production from lactose by *Lactobacillus plantarum*: Kinetic model and effects of pH, substrate and oxygen. Journal of Biochemistry and Engineering 3: 163–170.

Giraud, E., Brauman, A., Kekele, S., Lelong, B. and Raimbault, M. 1991. Isolation and physiological study of an amylolytic strain of *Lactobacillus plantarum*. Applied Microbiology and Biotechnology 36: 379–383.

Giraud, E., Champailler, A. and Raimbault, M. 1994. Degradation of raw starch by a wild amylolytic strain of *Lactobacillus plantarum*. Applied and Environmental Microbiology 60: 4319–4323.

Goh, Y.J., Goin, C., O'Flaherty, S., Altermann, E. and Hutkins, R. 2011. Specialized adaptation of a lactic acid bacterium to the milk environment: The comparative genomics of *Streptococcus thermophilus* LMD-9. Microbe Cell Factory 10: S22.

Guyot, J.P., Calderon, M. and Morlon-Guyot, J. 2000. Effect of pH control on lactic acid fermentation of starch by *Lactobacillus manihotivorans* LMG 18010T. Journal of Applied Microbiology 88: 176–82.

He, G.Q., Kong, Q. and Ding, L.X. 2004. Response surface methodology for optimizing the fermentation medium of *Clostridium butyricum*. Letter in Applied Microbiology 39: 363–368.

Hattingh, M., Alexander, A., Meijering, I., van Reenen, C.A. and Dicks, L.M.T. 2015. Amylolytic strains of *Lactobacillus plantarum* isolated from barley. African Journal of Biotechnology 14(14): 310–318.

Hofvendahl, K. and Hahn-Hägerdal, B. 2000. Factors affecting the fermentative lactic acid production from renewable resources. Enzyme Microbial Technology 26: 87–107.

Johansson, M.L., Sanni, A., Lonner, C. and Molin, G. 1995. Phenotypically-based taxonomy using API 50 CH of lactobacilli from Nigerian *Ogi*, and the occurrence of starch fermenting strains. International Journal of Food Microbiology 25: 159–168.

John, R.P., Anisha, G.S., Nampoothiri, K.M. and Pandey, A. 2009. Direct lactic acid fermentation: Focus on simultaneous saccharification and lactic acid production. Biotechnology Advances 27: 145–152.

John, R.P., Nampoothiri, M.K. and Pandey, A. 2007. Polyurethane foam as an inert carrier for the production of L(+) lactic acid by *Lactobacillus casei* under solid state fermentation. Letter in Applied Microbiology 44: 582–587.

Joshi, D.S., Singhvi, M.S., Khire, J.M. and Gokhale, D.V. 2010. Strain improvement of *Lactobacillus lactis* for D-lactic acid production. Biotechnology Letter 32: 517–20.

Kabak, B. and Dobson, A.D.W. 2011. An introduction to the traditional fermented foods and beverages of Turkey. Critical Reviews in Food Science and Nutrition 51: 248–260.

Kitabatake, N., Gimbi, D.M. and Oi, Y. 2003. Traditional nonalcoholic beverage, *Togwa*, in East Africa, produced from maize flour and germinated finger millet. International Journal of Food Sciences and Nutrition 54: 447–455.

Kostinek, M., Specht, I., Edward, V.A., Schillinger, U., Hertel, C., Holzapfel, W.H. and Franz, C.M.A.P. 2005. Diversity and technological properties of predominant lactic acid bacteria from fermented cassava used for the preparation of gari, a traditional African food. Systematic and Applied Microbiology 28: 527–540.

Kyla-Nikkila, K., Hujanen, M., Leisola, M. and Palva, A. 2000. Metabolic engineering of *Lactobacillus helveticus* CNRZ32 for production of pure L(+) lactic acid. Applied Environmental Microbiology 66: 3835–3841.

Lee, H., Se, G. and Carter, S. 2001. Amylolytic cultures of *Lactobacillus acidophilus*, potential probiotics to improve dietary starch utilization. Journal of Food Science 66: 338–344.

Lifeway. 2014. Lifeway Foods Inc., Financial results. Lifeway Foods Inc.

Linko, Y. and Javanainen, P. 1996. Simultaneous liquefaction saccharification and lactic acid fermentation on barley starch. Enzyme Microbial Technology 19: 118–23.

Mazzoli, R., Bosco, F., Mizrahi, I. and Bayer, E.A. 2014. Towards lactic acid bacteria-based biorefineries. Biotechnology Advances 32: 1216–1236.

Moo-Young, M., Moreira, A.R. and Tengerdy, R.P. 1983. Principles of solid state fermentation. pp. 117–44. *In*: J.E. Smith, D.R. Berry and B. Kristiansen (eds.). The Filamentous Fungi, London.

Morlon-Guyot, J., Guyot, J.P., Pot, B., Jacobe de Hault, I. and Raimbault, M. 1998. *Lactobacillus manihotivorans* sp. nov., a new starch-hydrolyzing lactic acid bacterium isolated from cassava sour starch fermentation. International Journal Systematic Bacteriology 48: 1101–1109.

Morlon-Guyot, J., Mucciolo-Roux, F., Rodriguez-Sanoja, R. and Guyot, J.P. 2001. Characterization of the *Lactobacillus manihotivorans* alpha-amylase gene. DNA Sequence 12: 27–37.

Muianja, C.M.B.K., Narvhus, J.A., Treimo, J. and Langsrud, T. 2003. Isolation, characterization and identification of lactic acid bacteria from *bushera*: A Ugandan traditional fermented beverage. International Journal of Food Microbiology 80: 201–210.

Nakamura, L.K. 1981. *Lactobacillus amylovorus*, a new starch-hydrolyzing species from cattle waste corn fermentation. International Journal of Systematic and Bacteriology 31: 56–63.

Nakamura, L.K. and Crowell, C.D. 1979. *Lactobacillus amylolyticus*. A new starch hydrolyzing species from swine waste corn fermentation. Developments in Industrial Microbiology 20: 531–540.

Narita, J., Nakahara, S., Fukuda, H. and Kondo, A. 2004. Efficient production of L-(+)-lactic acid from raw starch by *Streptococcus bovis* 148. Journal of Bioscience Bioengineering 97: 423–425.

Narita, J., Okano, K., Kitao, T. and Ishida, S. 2006. Display of alpha-amylase on the surface of *Lactobacillus casei* cells by use of the PgsA anchor protein, and production of lactic acid from starch. Applied and Environmental Microbiology 72: 269–275.

Narvhus, J.A. and Gadaga, T.H. 2003. The role of interaction between yeasts and lactic acid bacteria in African fermented milks: A review. International Journal of Food Microbiology 86: 51–60.

Naveena, B.J., Altaf, Md., Bhadrayya, K., Madhavendra, S.S. and Reddy, G. 2005. Direct fermentation of starch to L (+) lactic acid in SSF by *Lactobacillus amylophilus* GV6 using wheat bran as support and substrate medium optimization using RSM. Process Biochemistry 40: 681–690.

Naveena, B.J., Altaf, Md., Bhadrayya, K., Madhavendra, S.S. and Reddy, G. 2004. Production of L(+) lactic acid by *Lactobacillus amylophilus* GV6 in semi-solid state fermentation using wheat bran. Food Technology and Biotechnology 42: 147–152.

Nguyen, T.T., Loiseau, G., Icard-Verniere, C. and Rochette, I. 2007. Effect of fermentation by amylolytic lactic acid bacteria, in process combinations, on characteristics of rice/soybean slurries: A new method for preparing high energy density complementary foods for young children. Food Chemistry 100: 623–631.

Nguyen, T.T., Loiseau, G., Icard-Verniere, C., Rochette, I., Trèche, S. and Guyot, J.P. 2007a. Effect of fermentation by amylolytic lactic acid bacteria, in process combinations, on characteristics of rice/ soybean slurries: a new method for preparing high energy density complementary foods for young children. Food Chemistry 100: 623–631.

Nguyen, T.T., Guyot, J.P., Icard-Verniere, C., Rochette, I. and Loiseau, G. 2007b. Effect of high pressure homogenization on the capacity of *Lactobacillus plantarum* A6 to ferment rice/soybean slurries to prepare high energy density complementary food. Food Chemistry 102: 1288–1295.

Nout, M.J.R. 2009. Rich nutrition from the poorest—Cereal fermentations in Africa and Asia. Food Microbiology 26: 685–692.

Nwankwo, D., Anadu, E. and Usoro, R. 1989. Cassava-fermenting organisms. MIRCEN Journal 5: 169–179.

Oguntoyinbo, F.A. 2007. Identification and functional properties of dominant lactic acid bacteria isolated at different stages of solid state fermentation of cassava during traditional gari production. World Journal of Microbiology and Biotechnology 23: 1425–1432.

Oh, H., Wee, Y.J., Yun, J.S., Han, S.H., Jung, S. and Ryu, H.W. 2005. Lactic acid production from agricultural resources as cheap raw materials. Bioresource Technology 96: 1492–1498.

Okano, K., Yoshida, S., Yamda, R., Tanaka, T., Ogino, C., Fukuda, H. and Kondo, A. 2009b. Improved production of homo-d-lactic acid via xylose fermentation by introduction of xylose assimilation genes and redirection of the phosphoketolase pathway to pentose phosphate pathway in L-lactate dehydrogenase gene-deficient *Lactobacillus plantarum*. Applied Environmental Microbiology 75: 7858–7861.

Olympia, M., Fukuda, H., Ono, H., Kaneko, Y. and Takano, M. 1995. Characterization of starch-hydrolyzing lactic acid bacteria isolated from a fermented fish and rice food, '*Burong Isda*' and its amylolytic enzyme. Journal of Fermentation and Bioengineering 80: 124–130.

Panda, S.H., Swain, M.R., Kar, S. and Ray, R.C. 2008. Statistical optimization of α-amylase production by *Lactobacillus plantarum* using response surface methodology. Polish Journal of Microbiology 57: 149–155.

Panda, S.H., Parmanick, M. and Ray, R.C. 2007. Lactic acid fermentation of sweet potato (*Ipomoea batatas* L.) into pickles. Journal of Food processing and Preservation 31(1): 83–101.

Panda, S.H. and Ray, R.C. 2008. Direct conversion of raw starch to lactic acid by *Lactobacillus plantarum* MTCC 1407 in semi-solid fermentation using sweet potato (*Ipomoea batatus* L.) flour. Journal Scientific and Industrial Research 67: 531–537.

Pandey, A., Nigam, P., Soccol, C.R., Soccol, V.T., Singh, D. and Mohan, R. 2000. Advances in microbial amylases. Biotechnology and Applied Biochemistry 31: 135–152.

Papagianni, M. 2012. Metabolic engineering of lactic acid bacteria for the production of industrially important compounds. Computational and Structural Biotechnology Journal 3: [e201210003].

Petrov, K., Urshev, Z. and Petrova, P. 2008. L(+)-Lactic acid production from starch by a novel amylolytic *Lactococcus lactis* subsp. *lactis* B84. Food Microbiology 25: 550–557.

Petrova, P., Petrov, K. and Stoyancheva, G. 2013. Starch-modifying enzymes of lactic acid bacteria—structures, properties, and applications. Starch/Starke 65: 34–47.

Petrova, P. and Petrov, K. 2012. Direct starch conversion into L(+)-Lactic acid by a novel amylolytic strain of *Lactobacillus para-casei* B41. Starch/Starke 64: 10–17.

Petrova, P., Emanuilova, M. and Petrov, K. 2010. Amylolytic *Lactobacillus* strains from Bulgarian fermented beverage, *Boza*. Journal of Bioscience 65: 218–224.

Petrova, P. and Petrov, K. 2011. Antimicrobial activity of starch-degrading *Lactobacillus* strains isolated from *boza*. Biotechnology and Biotechnological Equipment 25: 114–116.

Phonyiam, N. and Yunchalard, S. 2014. Amylolytic lactic acid bacteria recovered and screened from *plaa-som* and its amylolytic enzyme. Journal of Biotechnology 136: S745.

Pulido, R.P., Benomar, N., Cañamero, M.M., Abriouel, H. and Gálvez, A. 2012. Fermentation of caper products. pp. 201–208. *In*: Y.H. Hui (ed.). Handbook of Plant-based Fermented Food and Beverage Technology, CRC Press, Boca Raton, Florida.

Putri, W.D.R., Haryadi, D.W., Marseno and Cahyanto, M.N. 2011. Effect of biodegradation by lactic acid bacteria on physical properties of cassava starch. International Food Research Journal 18(3): 1149–1154.

Rao, J.L.U.M. and Satyanarayana, T. 2007. Improving production of hyper-thermostable and high maltose-forming α-amylase by an extreme thermophile *Geobacillus thermoleovorans* using response surface methodology and its applications. Bioresource Technology 98: 345–52.

Ray, R.C. and Ravi, V. 2005. Post harvest spoilage of sweet potato in tropics and control measures. Critical Review in Food Science Nutrition 45: 623–644.

Ray, R.C. and Ward, O.P. 2006. Post harvest microbial biotechnology of tropical root and tuber crops. pp. 345–395. *In*: R.C. Ray and O.P. Ward (eds.). Microbial Biotechnology in Horticulture, Volume 1, Science Publishers Inc., Enfield, NH.

Ray, R.C., Sharma, P. and Panda, S.H. 2009. Lactic acid production from cassava fibrous residue using *Lactobacillus plantarum* MTCC1407. Journal of Environmental Biology 30(5): 847–852.

Reddy, G., Md. Altaf, Naveena, B.J., Venkateshwar, M. and Vijay, K.E. 2008. Amylolytic bacterial lactic acid fermentation: A review. Biotechnology Advances 26: 22–34.

Rodriguez-Sanoja, R., Morlon-Guyot, J., Jore, J., Pintado, J., Juge, J. and Guyot, J.P. 2000. Comparative characterization of complete and truncated forms of *Lactobacillus amylovorus* α-amylase and the role of the C-terminal direct repeats in raw starch binding. Applied and Environmental Microbiology 66: 3350–3356.

Rodrıguez-Sanoja, R., Ruiz, B., Guyot, J.P. and Sanchez, S. 2005. Starch-binding domain affects catalysis in two *Lactobacillus* α-amylases. Applied and Environmental Microbiology 71: 297–302.

Sanni, A., Morlon-Guyot, J. and Guyot, J.P. 2002. New efficient amylase-producing strains of *Lactobacillus plantarum* and *L. fermentum* isolated from different Nigerian traditional fermented foods. International Journal of Food Microbiology 72: 53–62.

Sawadogo-Lingani, H., Lei, V., Diawara, B. and Nielsen, D.S. 2007. The biodiversity of recombinant lactic acid bacteria in *dolo* and *pitowort* for the production of sorghum beer. Journal of Applied Microbiology 103: 765–777.

Sharma, A. and Satyanarayana, T. 2013. Microbial acid-stable α-amylases: Characteristics, genetic engineering and applications. Process Biochemistry 48: 201–211.

Shinkawa, S., Okano, K., Yoshida, S., Tanaka, T., Ogino, C., Fukuda, H. and Kondo, A. 2011. Improved homo L-lactic acid fermentation from xylose by abolishment of the phosphoketolase pathway and enhancement of the pentose phosphate pathway in genetically modified xylose-assimilating *Lactococcus lactis*. Applied Microbiology and Biotechnology 91: 1537–1544.

Shivaramakrishnan, S., Gangadharan, D., Nampoothiri, M.K., Soccol, C.R. and Pandey, A. 2006. Amylases from microbial sources—an overview on recent developments. Food Technology Biotechnology 44(2): 173–184.

Sivakumar, P.S., Panda, S.H., Ray, R.C., Naskar, S.K. and Bharathi, L.K. 2010. Consumer acceptance of lactic acid fermented sweet potato pickle. Journal of Sensory Studies 25: 706–719.

Songré-Ouattara, L.T., Mouquet-Rivier, C., Icard-Vernière, C., Rochette, I., Diawara, B. and Guyot, J.P. 2009. Potential of amylolytic lactic acid bacteria to replace the use of malt for partial starch hydrolysis to produce African fermented pearl millet gruel fortified with groundnut. International Journal of Food Microbiology 130: 258–264.

Smerilli, M., Neureiter, M., Wruz, S., Hass, C., Fruhauf, S. and Fuchs, W. 2015. Direct fermentation of potato starch and potato residues to lactic acid by *Geobacillus stearothermophilus* under non-sterile conditions. Journal of Chemical Technology and Biotechnology Doi: 10.1002/jctb.4627.

Trèche, S. and Mouquet-Rivier, C. 2008. Use of amylases in infant food. pp. 213–245. *In*: R. Porta, P. Di Pierro and L. Mariniello (eds.). Recent Research Developments in Food Biotechnology. Enzymes as Additives or Processing Aids. Research Signpost, Trivandrum.

Turpin, W., Humblot, C. and Guyot, J.P. 2011. Genetic screening of functional properties of lactic acid bacteria in a fermented pearl millet slurry and in the metagenome of fermented starchy foods. Applied and Environmental Microbiology 77: 8722–8734.

Vishnu, C., Seenayya, G. and Reddy, G. 2002. Direct fermentation of various pure and crude starchy substrates to L(+) lactic acid using *Lactobacillus amylophilus* GV6. World Journal of Microbiology and Biotechnology 18: 429–433.

Wang, Y., Tashiro, Y. and Sonomoto, K. 2014. Fermentative production of lactic acid from renewable materials: Recent achievements, prospects, and limits. Journal of Bioscience and Bioengineering 119: 10–18.

Wasko, A., Polak-Berecka, M. and Targonski, Z. 2010. A new protein of alpha amylase activity from *Lactococcus lactis*. Journal of Microbiology and Biotechnology 20: 1307–1313.

Wee, Y.J., Kim, J.N., Yun, J.S. and Ryu, H.W. 2004. Utilization of sugar molasses for economical L(+) lactic acid production by batch fermentation of *Enterococcus faecalis*. Enzyme Microbial Technology 35: 568–573.

Wee, Y.J., Kim, J.N. and Ryu, H.W. 2006. Biotechnological production of lactic acid and its recent applications. Food Technology and Biotechnology 44: 163–172.

Yoshida, S., Okano, K., Tanaka, T., Ogino, C. and Kondo, A. 2011. Homo-D-lactic acid production from mixed sugars using xylose-assimilating operon-integrated *Lactobacillus plantarum*. Applied Microbiology and Biotechnology 92(1): 67–76.

Yousif, N.M.K., Huch, M., Schuster, T. and Cho, G.S. 2010. Diversity of lactic acid bacteria from *Hussuwa*, a traditional African fermented sorghum food. Food Microbiology 27: 757–768.

8

Lactic Acid Bacteria as Functional Starter in Food Fermentations

Frédéric Bustos Gaspar[1,2] and Maria Teresa Barreto Crespo[1,2,*]

1. Introduction

About 9000 years ago, without being aware of it, our ancestors consumed food products that were produced by natural microbial fermentation, such as honey, fruits, and even beverages made from cereals. Since then, mankind has been able to experience the preservative, organoleptic, analgesic and mentally stimulating qualities of fermented foods and beverages (Steinkraus 2002).

The study of fermented foods began in the seventeenth century with the microscopical observations of microorganisms (then called 'animalcules') from pepper liquor by Robert Hooke and Antony van Leeuwenhoek, the two 'fathers' of microbiology (Gest 2004). Later, Louis Pasteur cleverly demonstrated the anaerobic lifestyle and defined fermentation as life without air, true to the metabolic point of view that fermentation is the process of splitting complex organic compounds into their simpler constituents, accompanied by the production of energy, in the absence of oxygen (Bourdichon et al. 2012). Fermentation may also be defined, from a biotechnological perspective, as the bulk growth of a microorganism or a group of microorganisms on a culture medium. Food can also be a culture medium for the growth of microorganisms, and therefore be fermented. In food fermentation,

[1] iBET – Instituto de Biologia Experimental e Tecnológica, Apartado 12, 2781-901 Oeiras, Portugal.
[2] Instituto de Tecnologia Química e Biológica, Universidade Nova de Lisboa, Av. da República, 2780-157 Oeiras, Portugal.
  E-mail: fgaspar@itqb.unl.pt
* Corresponding author: tcrespo@itqb.unl.pt

food commodities are produced from different raw materials, transformed by microorganisms to give new food products, which can have an increased nutritional value and an improved bioavailability of nutrients (Granier et al. 2013).

Along the centuries, and throughout the world, a large number of different raw materials, such as cereals, fruits, vegetables, dairy, meat, fish, shellfish, and seafood, have been used to produce a huge variety of fermented products, to either eat or drink (Rodríguez et al. 2009, Hurtado et al. 2012, Marshall and Mejia 2012, Di Cagno et al. 2013, Smid and Kleerebezem 2014, Seo et al. 2014).

These fermented foods and drinks are produced through spontaneous (without the use of a starter culture) or controlled fermentations (using a starter culture). In many controlled fermentations, food matrices are first treated to eliminate or control the natural microbiota, and then the starter culture is added to carry out the fermentation. Historically, this was accomplished by inoculating the fresh material with a portion of fermented product, a technique known as back-slopping (Pfeiler and Klaenhammer 2007). Starter cultures, in an industrial sense, are composed of a defined microorganism or group of microorganisms that carry out a fermentation in a given substrate, by accelerating and steering this fermentation. The starter cultures, or the spontaneous inocula, ferment sugars, decrease pH (through organic acid production which, combined with the production of antimicrobial inhibitory components, can reduce or prevent the growth of undesirable microbiota) and also improve the nutritional attributes of the food product (Waters et al. 2015).

Lactic Acid Bacteria (LAB) have always been the major group of bacteria used in the production of fermented foods and beverages, maintaining a long history of safe use. They constitute a large group of non-sporulating Gram-positive, catalase negative rods and cocci that produce mainly lactic acid as the major metabolite of the carbohydrate fermentation. The main role of LAB in food fermentation is the rapid acidification of the food matrix and the outgrowth over other microorganisms. This process is mainly caused by the fast conversion of fermentable carbohydrates into lactic acid, as also other organic acids, such as acetic acid, formic acid, and ethanol (Martinussen et al. 2013). LAB are aerotolerant anaerobic bacteria that, in some cases, have complex nutritional requirements, especially for amino acids and vitamins. The main genera comprising LAB are *Aerococcus, Alloiococcus, Carnobacterium, Dolosigranulum, Enterococcus, Globicatella, Lactobacillus, Lactococcus, Lactosphaera, Leuconostoc, Mellissococcus, Oenococcus, Pediococcus, Streptococcus, Tetragenococcus, Vagococcus, Weissella,* and bifidobacterium genus (Leroi 2010, Ruas-Madiedo et al. 2012). Knowledge about LAB starter cultures has been the target of many publications and reviews (Hammes 1990, Buckenhüskes 1993, Heller 2001, Hansen 2002, Tamime 2002).

This chapter will focus on the concept of LAB functional starter cultures, on the market and consumer demands directing the search for new applications, and on the scientific and technological trends shaping their selection and development.

2. LAB Functional Starter Cultures

A starter culture is a culture of microorganisms, either pure or mixed, used to bring a desirable change in a finished product. LAB have been used as starters for a long time. In this part of the chapter the functionalities of LAB starter cultures will be highlighted.

2.1 Starter Cultures

Traditional fermented food products rely on spontaneous fermentation associated to the microflora present on raw materials and on materials that they are exposed to during production. They are still being produced all over the world due to the low technological input required, their shelf life, and their uniqueness in terms of taste, aroma, and texture. The main drawback associated with this type of fermentation is the risk of having losses in production due to anomalous fermentations. Back-slopping, a technique used in some cases to replace the spontaneous fermentation, lowers the risk of fermentation failure, while unintentionally leading to a selection of microorganisms. This technique is still used in the production of sourdough, sauerkraut, and some dairy products. The industrial production of fermented foods required a very high degree of control over fermentations and, as such, industry started the use of defined starter cultures, with either a single or a mixture of microorganisms. This tremendous progress in terms of processing allowed the production of a product with similar quality, organoleptic, and nutritional characteristics over the years.

2.2 LAB Functional Starter Cultures

Functional starter cultures can be defined as starters that confer at least one inherent functional property to the food product they are fermenting (Leroy and De Vuyst 2014).

Therefore, LAB functional starter cultures are the ones that guarantee that the anticipated functional property or properties are expressed in the final product, whether composed of a single microorganism or of a mixture of microorganisms or in co-culture with the natural microbiota already present in the fermentation process. The acquired food functionality, or functionalities, can be related to its safe intake, to general or specific effects on the consumer's health, to the improvement of nutritional composition, digestibility and anti-nutritive compounds, or even to the reduction of wastes. The reduction of wastes is seen as functionality related to the opportunity of creating new products from unused parts of food materials.

2.2.1 Functionality to improve food intake safety

The drop in pH, due to the production of organic acids like lactic, acetic and formic acids, is the most obvious aspect of LAB's ability to reduce or prevent the growth of undesirable microflora from the early stages of fermentation, and therefore to preserve food. Food preservation can also be achieved by LAB triggered nutrient depletion, as well as by strain specific production of inhibitors like ethanol, carbon dioxide, acetoin, diacetyl, acetaldehyde, or hydrogen peroxide (in aerobic growth). The balance between their presence and the potential negative changes in sensory properties of the products has, nevertheless, to be carefully considered.

LAB have also the ability to produce different types of specific antimicrobial compounds to inhibit the activity of potential spoilage and pathogenic organisms. The selection of LAB capable of bacteriocin production, antimicrobial peptides and proteins ribosomally synthesised, has been the target of much research. Some of these bacteriocins, namely nisin, are already approved for use as food additive as they are very attractive alternatives to conventional antimicrobials (Rattanachaikunsopon and Phumkhachorn 2010, Zendo 2013).

In what constitutes antifungal activity, it has been observed that LAB also produce antifungal metabolites such as propionate, phenyllactate, hydroxyphenyllactate, several cyclic dipeptides, 3-hydroxy fatty acids, and inhibitory peptides smaller than 10 kDa (Lavermicocca et al. 2000, Gänzle 2009, Prema et al. 2010, Gerez et al. 2013). LAB also have the ability to produce other types of antimicrobial compounds. One of these is reuterin, produced by *Lactobacillus reuteri*, when glucose and glycerol are present in the culture medium (Talarico et al. 1988, Ström et al. 2002, Rattanachaikunsopon and Phumkhachorn 2010). Reuterin is an equilibrium mixture of monomeric, hydrated monomeric, and cyclic dimeric forms of β-hydroxypropionaldehyde that has a large spectrum of activity, ranging from Gram-positive and Gram-negative bacteria to fungi and protozoa.

The first antibiotic described to be produced by LAB is reutericyclin (Gänzle et al. 2000). Reutericyclin is a negatively charged highly hydrophobic antagonist agent against Gram-positive bacteria (Rattanachaikunsopon and Phumkhachorn 2010). In 2013, Crowley et al. published a review on the bioactive metabolites of LAB, their applications in food systems as well as their interactions with target fungi (Crowley et al. 2013). The term 'green preservatives' was used in 2012 to reinforce the use of LAB as natural elements in the combat against fungi in the food and feed industry (Pawlowska et al. 2012, Bevilacqua et al. 2015). Recently, the importance of LAB starter cultures as "green preservatives" was further emphasised, as they were the target of research aimed at studying the bioprotection of crops (Oliveira et al. 2014).

2.2.2 Functionality to improve the host's health

Besides the quality improvements due to the prevention of spoilage and safety enhancement by the inhibition of pathogenic microorganisms, LAB fermented foods have numerous physiological effects on the host. The use of LAB as probiotics, living microorganisms that confer a beneficial effect on the host when administered in proper amounts, is the main health related use of this group of bacteria. Probiotics are not within the scope of this chapter.

LAB beneficial effects on hosts are not just limited to their use as probiotics. In fact, LAB isolated from kimchi were reported to have anti-adipogenic effects through the inhibition of adipocyte differentiation, along with reduction in intracellular triglyceride accumulation. They also reduced GPDH activity and downregulated the adipogenic transcription factor, leading to suppression of adipogenic-specific gene mRNA molecules in 3T3-L1 adipocytes. Furthermore, LAB were also found to lower the inflammatory response. Taken together, the results indicate that LAB from kimchi may play an important role in inhibition of adipogenesis and might have further implications for *in vivo* anti-obesity and anti-inflammatory effects (Park et al. 2014).

The inhibition of angiotensin-converting enzyme (ACE) is an effective target to manage essential hypertension, a major global health issue that elevates the risk for stroke, as well as cardiovascular and kidney diseases. An evaluation of a probiotic LAB strain or of fermenting co-cultures of selected microorganisms, LAB and yeast, for their proteolytic activity and capability to produce fermented food products enriched with ACE-inhibitory peptides without impacting the product flavor has also been done successfully (Chen et al. 2014, Chaves-López et al. 2014).

Fermented foods and beverages have the potential to influence brain health, directly, due to the functional compounds produced by microbial metabolism, and indirectly by the ways in which the fermented foods or beverages influence our own microbiota. Our own intestinal microbiota profile may be influenced by lactoferrin, bioactive peptides, phytochemicals, and unique flavonoids, resulting from LAB food fermentation. At least to some degree, microbiota may control the extent to which dietary item may mitigate inflammation and oxidative stress (Selhub et al. 2014). Several experimentally examined mechanisms were also proposed, whereby beneficial microbes could influence mood or fatigue (Logan et al. 2003, Logan and Katzman 2005). Moreover, controlled fermentations may also increase the content of specific nutrient and phytochemicals in foods, having the ability to ultimately affect mental health (Selhub et al. 2014). Therefore, the consumption of fermented foods may be particularly relevant to the emerging research that links traditional dietary practices and positive mental health effects.

Anabolic LAB products such as conjugated linoleic acids (CLA) can also affect human metabolism. Some CLA isomers can reduce cancer cell viability, atherosclerosis, body fat, and ameliorate insulin resistance (Ogawa et al. 2005, Pessione 2012). The ability for CLA production catalysed by multicomponent linoleate isomerase in lactobacilli exhibited variations, but was successfully confirmed for some strains. Food-derived lactobacilli with CLA production ability offer novel opportunities for functional foods development (Yang et al. 2014).

Additionally, knowledge regarding the human microbiome is increasing, particularly for the GI tract (Turnbaugh et al. 2007, Kau et al. 2011, González et al. 2014). Little is known about the relationship between the ways in which LAB interact with each other and on the food product during production, and the ways in which they interact as a group with our own microbiota after ingestion; this opens a whole new area of potential LAB functional starter cultures applications.

2.2.3 Functionality to improve food's nutritional quality

Functional starter cultures have also been used to improve the nutritional quality of food by increasing the nutritional value and digestibility, and by reducing the effect of anti-nutritional compounds.

LAB fermentation directly affects nutrient availability, in a broad sense, by hydrolysing carbohydrates and non-digestible oligosaccharides into functional compounds. In fact, microorganisms contain certain enzymes, such as cellulases, which are not synthesised by humans. Microbial cellulases hydrolyse cellulose into sugars, which are then readily digestible by humans. Similarly, pectinases soften the texture of foods and liberate sugars for digestion. These LAB hydrolases allow for fermented foods to have an improved nutritional value; they are also often more easily digestible than unfermented foods, reducing any associated discomfort (Paredes López 1992, Kovač and Raspor 1997, Waters et al. 2015).

LAB can also be an important source of vitamins in some fermented food products. *Pulque*, a fermented plant sap from Mexico, can have increased vitamin contents, from 5 to 29 for thiamine, 54 to 515 for niacin and 18 to 33 for riboflavin in milligrams of vitamins per 100 g of *pulque*, during fermentation (Steinkraus 1992). Likewise, *idli*, a product consumed in India, is also high in thiamine and riboflavin due to LAB fermentation (Marshall and Mejia 2012). Nutrition enrichment can even be more pinpointed, such as being related to the enrichment in a micronutrient. An example of this type of nutritional enrichment is the selenium enrichment by LAB and bifidobacteria performed by transformation of inorganic selenium to organic and elemental forms. Selenium compounds like selenocysteine, selenomethionine and methylated selenium are safe and have the necessary bioavailability for human and animal consumption (Pophaly et al. 2014).

The nutritional status of food products is often counteracted by the presence of antinutrient components such as phytates, tannins, protease inhibitors, and polyphenols (Reddy and Pierson 1994, Onyango et al. 2005). Antinutritive factors from cereals, including maize, sorghum or millet, and legumes reduce the bioavailability of minerals such as calcium, iron, potassium, magnesium, manganese and zinc, thus affecting the utilisation of carbohydrates and proteins and causing deficiencies in the essential amino acids lysine, tryptophan and methionine (Holzapfel 2002, Waters et al. 2015). LAB fermentation can overcome these problems by conferring to the final product, a higher protein quality and nutritional value, by inactivating protein inhibitors or improving mineral solubility. In the case of cereal fermentation, the steps before fermentation, soaking and germination, enhance the reduction of those anti-nutritive compounds. LAB may also have a detoxifying action against external antinutrients that can take the form of filamentous fungal mycotoxins (Dalié et al. 2010).

2.2.4 Functionality to improve organoleptic properties

LAB fermented food product can also promote hedonistic aspects such as the improvement of taste, aroma, texture and mouthfeel, supporting a mood-enhancing functionality. Depending on the type of starter culture used and on the substrate being fermented, considerable changes in the flavor (taste and aroma) and texture profiles may be experienced (Waters et al. 2015).

All three main constituents of food materials (i.e., proteins, carbohydrates, and lipids) deliver precursors that, once processed, can affect the organoleptic characteristics of the final product. The first organoleptic change in fermentation is accomplished when LAB produce organic acids by metabolizing sugars; the pH decreases with the simultaneous decrease in sweetness and increase in sourness (McFeeters 2004). Then, a number of other changes can follow due to active functional metabolic pathways that process different compounds, such as simple organic acids (sourness), amino acids (sweetness, umami), and oligopeptides (bitterness) (Smid and Kleerebezem 2014), revealing the importance of proteolysis and amino acid metabolism in food flavor quality. Flavor compounds can be enhanced by using selected strains, by changes in population dynamics in mixed cultures, among others. The strain specificity can be so unique as to promote the conversion of phenylalanine to phenylacetate or phenylacetaldehyde, the flavor compounds responsible for the 'flowery' and 'honey like' flavor impression (Gänzle 2009). The strong flavors of fermented food vegetables, such as pickles, *gundruk,* and sauerkraut, may be used as condiments to enhance the overall flavor of a meal when diet is dull, bland, and repetitive (Montet et al. 2014).

LAB fermentation also has an important role in texture, which is related to the production of exopolysaccharides (EPS). EPS increase viscosity and firmness, improve the texture, and contribute to the mouthfeel of the fermented product (Zamfir et al. 2014).

2.2.5 Functionality to diminish wastes

Fermentation increases the range of raw materials available as food by salvaging waste food, which otherwise would not be usable, as well as by changing the consistency of the product while making it digestible. The trend of reducing waste through biotechnology is a growing area of research in developed countries in which social conscience presses producers to find new ways of dealing with wastes. This trend is, however, crucial in developing countries to improve food security. In Sudan, for instance, a wide range of waste products, including bones, hides, and locusts, is fermented to produce edible food products. In Indonesia, a variety of waste products, like fresh coconut residue, left over from the production of coconut cream or milk, are fermented to produce nutritious food products (Steinkraus 1996). LAB starter cultures can have a role to play in these areas of starter culture utilization.

3. LAB Functional Starter Cultures: Market and Consumer Needs and Wants

Consumers of the 21st century constantly want new products and experiences in many areas of their lives, including in the food they buy and prepare. These consumers select new products, and in particular fermented ones, following very particular criteria. The trends of choice are as diverse as increasing the number of artisan, organic, sustainable, fair-trade, natural and overall healthy food products, along with ever-growing 'free-from' forms (dairy-, lactose-, wheat-, gluten-, meat-, egg-, soya-, GMO-, additive-, fat- or nut-free). Therefore, the market demands new functional starter cultures and new functionally fermented food products that can answer to many of those desires.

Consumers are looking for free-from food products for general health purposes, with an increasing number of consumers also requiring such products due to food allergies, intolerances, and sensitivities. One of the trends is the increasing demand for non-dairy beverages, like fermented vegetable and fruit smoothies, or vegetable, fruit and cereal drinks, with a high functional value, a consequence of the rising communities of vegetarianism and veganism, and the increasing prevalence of lactose non-persistence, malabsorption or intolerance. International data shows the prevalence of hypolactasia, including lactose intolerance, as being approximately 70% worldwide, with increased occurrence in certain populations (Lomer et al. 2008). These types of drinks must be made available locally at an affordable price, be versatile enough to be regularly consumed as required, be appealing to all age categories, and above all, be palatable. Another aim of industry is to target the market of people with diseases that can be associated with diet, such as cardiovascular disease or elevated blood pressure, which also leads to the development of non-dairy based functional beverage substitutes (Zannini et al. 2012). Therefore, new LAB starter cultures should and are being developed for cereal or even soya and fruit beverages (Coda et al. 2012, Rathore et al. 2012, Waters et al. 2015, Santos et al. 2014). Maize and cassava particularly, as they lack gluten, are promising substrates to produce high quality LAB fermented baked goods instead of wheat or other gluten-rich cereals, in order to be adapted for people with coeliac disease (Dufour et al. 1996, Di Cagno et al. 2002, Maldonado Alvarado et al. 2013).

Companies are also exploring almond, rice and coconut milks and their fermented forms to produce lactose-free and lactose-reduced spreads, milks, and yogurt ice creams. With these new alternatives, dairy- or cereal-based products can be replaced to avoid food allergies and intolerances (e.g., lactose and gluten). The perception of the advantages for these new types of diets is, therefore, growing.

The need for free-from food products, and their functionally fermented counterparts, has provided an opportunity to promote a strong over-lap with healthy eating, "natural", and environmentally friendly trends which ensure they can maintain a multi-layered appeal to the consumer. In the production of table olives, the debittering step with alkali can be replaced by glucosidase active yeasts and bacteria able to hydrolyse oleuropein (Bevilacqua et al. 2015), a phenolic compound mainly responsible for the product's bitterness. The increasing demand for green products brought the focus on bioremediation as a way to reduce waste water and lye and design an eco-friendly approach to production processes in which LAB can participate. In the dairy industry a completely new range of products is appearing and occupying a new niche market, for instance Greek-style yogurts with high-protein levels advertised for men who work out, and high-protein and low-fat Greek-style yogurts for a more general public. EPS have to be present in this type of product to compensate for the change of texture and mouthfeel due to the lack of lipids and the mode of production. Producers had and have to look for different EPS producing LAB. So, in many of the already existing products, more than two LAB starters have to be added to raw material. In fact, there are products with up to five or more strains per starter culture, in order to have the necessary final texture and mouthfeel. These polysaccharides would also substitute for the ones from plants (starch, pectin, guar gum, alginate) or animals (gelatine), currently in use (Zamfir et al. 2014).

New savoury yogurts are flavorful but not sweet since unconventional flavours, like carrots, sweet potato, beet, parsnips, tomato or butternut squash are added, along with the milk and starter cultures. Starter cultures that can thrive in these new environments are also being tested and used.

Different, but also important, is the trend to find strains for different country dependent functionality needs. Studies with LAB and sourdough have been carried out especially in Germany, France, and Sweden where the sourdough fermentation process has been used, traditionally, for rye, rye-mixes and other flours which are difficult to bake without souring (Lönner et al. 1986, De Vuyst et al. 2002, Catzeddu et al. 2006, Valcheva et al. 2006). Further developments in starter cultures to overcome taste issues will reduce the barriers for these LAB fermented products to be purchased. This will lead to an increase in the number of products being positioned to a wide range of consumers for being healthier, and not just aimed at a free-from product demanding public.

Another criterion for the selection of starter cultures in general, and LAB starter cultures in particular, is their origin. In fact, LAB that are used as probiotics, for instance, were traditionally isolated from the GI tract of humans. In addition, many LAB strains that are used as starter cultures were isolated from sausages or other meat products. The search for LAB strains from plant origin, cereal, vegetable or fruit, is being done for some time now. These starter cultures are not only requested because of their different metabolic characteristics but, and most importantly, because there is growing market for consumers with specific needs, such as vegetarians or vegans.

4. The Search for New Lab Functional Starter Cultures

Triggered by the extensive studies performed on the beneficial effects of LAB fermented food products, and by market and consumer trends, producers of LAB starter cultures have to keep searching for new LAB isolates as well as new ways of changing already known isolates. The search for new LAB functional starter cultures must acknowledge that among all the LAB group genera, for any given specie, each strain can slightly differ from another, in gene content or solely in phenotypic characteristics. Screening for new strains is not a task of small numbers but a task of large numbers.

4.1 Finding New Strains

The main route for the discovery of new strains principally comprises continuously searching all over the world in academic culture collections and by continuously isolating LAB from all types of products and environments (meat, dairy, plant, fish, sea-food, GI tract, or even decaying materials) (Siezen et al. 2011, Bourdichon et al. 2012). LAB isolates from different sources, originating from different products or geographical regions, are subjected to metabolism screening programs and genomic studies.

Due to the economic importance of LAB, especially in food product fermentations, extensive genomic studies have been performed and have given significant information on the metabolism, physiology, and potential for new applications (Pfeiler and Klaenhammer 2007).

Around 50 LAB genome sequencings have been achieved and more are in the pipeline. To take full advantage of this data, bioinformatics tools were developed to perform data mining, diversity analysis, metabolic reconstruction, and comparative genomic analysis (Siezen et al. 2004). Full sequencing of LAB isolates enabled the development of whole genome microarrays technologies for the global analysis of gene expression (transcriptome analysis) and identification of genotype-phenotype associations (comparative genome hybridization) (Kuipers et al. 2002, Kok et al. 2005, Siezen et al. 2011, Gaspar et al. 2013). Comparative genomic analysis revealed that diversification of strains though the acquisition of new genes, and thus new functions, can be achieved by horizontal gene transfer, 'the non-genealogical transmission of genetic material from one organism to another' (Goldenfeld and Woese 2007).

Conjugation, transduction, and transformation are three independent gene transfer mechanisms associated with horizontal gene transfer. Rossi et al. (2014) published a review on this subject with examples of possibly exchanged genes between bacterial species in food, for niche adaptation, substrate utilization, as well as gene encoding for hazardous traits. Many genetic mobile elements were also found in LAB genomes, including plasmids, prophages, insertion sequence elements, transposons, and group II introns. These mobile elements contribute to genome plasticity, host competiveness, and environmental adaptation (Top and Springael 2003, Frost et al. 2005).

The selection of new strains with the use of an evolutionary engineering approach relying on techniques such as next-generation sequencing, yield selection, single-cell technologies, and metabolic crossfeeding, enables the discovery of mechanisms that otherwise would not have been evident (Bachmann et al. 2015). By combining functional genomics, proteomics, transcriptomics, and metabolomics, the knowledge of the composition and function of single and complex domesticated starter cultures has increased, which is of utmost importance for industrial uses (Erkus et al. 2013, Douillard and de Vos 2014).

During selection, when no single strain appears to have all the desirable functionalities, industrials may move to another option, that is selecting and using mixed starter cultures. This strategy mimics the fermentation of traditional food products, which have only a natural inoculum, and have their uniqueness related to mixed fermentation (Furukawa et al. 2013).

4.2 Changing Existing Strains

The other possibility that is being explored to obtain new functional starter cultures is to change existing strains by metabolic or inverse metabolic engineering, therefore constructing strains with a particularly desirable physiological phenotype (e.g., enhanced production of heterologous protein) (Bailey et al. 2002, Kern et al. 2007). Metabolic engineering was defined initially by Bailey as an "improvement of cellular activities by the manipulations of enzymatic, transport, and regulatory functions of the cell with use of recombinant DNA technology" (Bailey 1991). Later the concept was revised to "genetic modification of cellular biochemistry to introduce new properties or to modify existing ones" (Jacobsen and Khosla 1998). Functional genomics, proteomics, and metabolomics contribute towards metabolic engineering due to the data that is produced, allowing the identification of regulatory networks of known metabolic pathways.

Target of many research efforts in the last decades, the metabolic engineering of LAB has been dedicated to the bioproduction of chemicals (e.g., lactic acid, ethanol or butanol), food ingredients (e.g., alanine, diacetyl or acetaldehyde), low-calorie polyols, polysaccharides, vitamins, plant metabolites, and improving acid resistance, or engineering lactose and galactose utilization. LAB metabolism is simple and, as such, easy to manipulate by genetic modification; hence numerous genetic molecular tools for genetic engineering are now available (Zhu et al. 2009, Gaspar et al. 2013). Genetic engineering has been extensively done with *Lactococcus lactis*, with many different purposes like improving flavor, texture, and health applications (De Vos and

Hugenholtz 2004). Even the possibility of using a *L. lactis* in a LAB starter culture for the *in situ* production of vitamins, by overexpressing and/or disrupting relevant metabolic genes, has been explored (Ammor and Mayo 2007).

The improvement of robustness of strains for industrial purposes can also be performed by the manipulation of LAB genes (Bron and Kleerebezem 2011). In the case of not particularly robust LAB strains that have no spray-drying (low a_w and high temperatures), chill-, cryo- or baro-tolerance, patho-biotechnology methodologies can be used as a new approach for strain improvement. Patho-biotechnology seeks to exploit pathogenic stress survival factors for beneficial applications in medicine, biotechnology, and food. Among these adaptations is the ability to accumulate compatible solutes, such as betaine. Indeed, compatible solute uptake in the food-borne pathogen *Listeria monocytogenes* has previously been shown to protect the pathogen from the detrimental effects of low a_w, reduced temperature, and extremes of pressure, as well as prolonging GI persistence, all potentially desirable traits in selecting and designing versatile probiotic strains. One of the three compatible solute systems that are known to operate in *L. monocytogenes* is glycine betaine porter I (BetL), a glycine betaine-Na^+ symporter (Angelidis and Smith 2003). A nisin-controlled expression (NICE) system was inserted in *Lactobacillus salivarius* and it was demonstrated that heterologous expression of BetL can significantly improve betaine uptake rates, with a resulting increase in the stress tolerance profile of the probiotic strain *Lb. salivarius* UCC118 (Sheehan et al. 2006, Sleator and Hill 2006).

Most of the changes mentioned above are mainly metabolism oriented, but bioprocesses using industrial microbes need also to have good physiological performances. Since the entire microbial cell functions in actual bioprocesses, rather than only a specific pathway, it is crucial to globally modify the strain as opposed to only engineering a few pathways (Zhang et al. 2009, Zhu et al. 2012). Systems metabolic engineering, an upgrade of metabolic engineering with the aid of systems biology tools, provides systematic strategies for cellular and metabolic engineering while elucidating strain specific problems that can otherwise overlooked (Park et al. 2008, Park and Lee 2008, De Vos 2011).

To obtain ideal industrial microbes with enhanced and/or new features, natural microbes need to be engineered so that bioprocesses can efficiently compete with existing chemical processes. Synthetic biology offers opportunities to generate engineered microbial strains through the design and construction of finely controllable metabolic and regulatory pathways, circuits, and networks, as well as create new enzymes and pathways (Zhang et al. 2009, Zhu et al. 2012, Lee and Na 2013, Seo et al. 2013).

All methodologies of strain engineering imply the genetic modification of isolates. The problem is that in the European Union, consumers are not keen on having genetically modified organism/microorganisms (GMO/M) as a food product or being added to one. Nevertheless the potentialities of engineering strains as an alternative to conventional improvement have an immense potential and it is already used to produce food enzymes.

4.3 Choosing Safe Strains

A wide variety of bacterial and fungal species are used in food and feed production, either directly or as a source of additives or food enzymes. Some of these have a long history of apparent safe use, while others are less well understood and may represent a risk for consumers. In 2002, the International Dairy Federation (IDF) and the European Food and Feed Cultures Association (EFFCA) published the 'Inventory of Microorganisms with a Documented History of Use in Food' (IDFEFFCA 2002). This inventory contains the names of species that have a record of use in food production either as single strain, single species, or multiple species cultures, as defined by IDF Standard 149A:1997. The inventory was not designed to be comprehensive, but rather proposed as a 'live inventory' to which new entities could be added and where modifications could be made in accordance with the 'state of the art'. It consists of LAB, including bacterial species belonging to *Enterococcus* and *Streptococcus*, as well as yeasts and molds. These strains have a long history of use in the food industry without any adverse effects. In 2012, Bourdichon et al. published an updated list that included microorganisms from dairy origin, as well as from other matrices, and the numbers changed from 31 to 62 and from 133 to 264 in genera and species, respectively. Other parts of the world can have their own lists of authorized strains and criteria for the selection of safe strains to be used in food and feed. In the European Union, the European Food Safety Agency (EFSA) has a Scientific Committee that reviewed the range and number of microorganisms that can be used and published lists of microorganisms recommended for a Qualified Presumption of Safety (QPS) (EFSA 2013).

4.4 Technological Features

The selection of new LAB strains has to be matched with the industrial production and the desired qualities of the final food product they will ferment. Some technological features are already known as necessary for the selection of a LAB starter culture: rapid and adequate production of lactic acid, growth at different temperatures, salt concentrations, and pH, proteolytic and lypolytic enzyme production, production of gas for heterofermentative bacteria, and tolerance, or even synergy, with the other microbial components of the starter culture (Ammor and Mayo 2007). A starter culture has to be produced, tested, packaged, and sold in such conditions that guarantee that these characteristics emerge when it comes in contact with the food matrices. So it is difficult to separate the improvement of production of the starter culture and the improvement of the performance of the starter culture in the food product. Both aspects have to be considered when selecting new strains.

Strategies to improve productivity, and even functionality, of industrial starter cultures have been based on changes in the growth medium, sub-lethal pre-adaptation, and cross-protection (Bron and Kleerebezem 2011). In terms of improving the viability of starter cultures and probiotics in foods, as well as during the passage through the GI tract, producers of starter cultures are now testing, and in some cases using, microencapsulation techniques (Rouhi et al. 2013). Microencapsulation may also enhance microbial operating efficiency during fermentation. Other advantages include prevention of interfacial inactivation, stimulation of the production and excretion of secondary metabolites, and continuous utilization (Nazzaro et al. 2012).

Recently, the use of respiration in the metabolic sense, as opposed to fermentation, is being approached as an alternative way to improve LAB fermentation performances. In LAB, respiration is activated by exogenous haem, and for some species, haem and menaquinone. It is known that respiration metabolism increases growth yield and improves fitness. So, according to Pedersen et al. (2012) "the respiration capacity of numerous LAB opens the door for more efficient production of these industrial bacteria and for the development of novel uses in situations where good growth, prolonged survival, less oxygen, and a more neutral pH could constitute a benefit". In fact, when in dairy, cereal, vegetable, fruit, meat or fish fermentation, LAB starters try to use all potentialities from the fermentation environment, such as using haem and menaquinones from environmental sources, which could allow respiration. Respiration technology has not yet been applied to food products, but preparing fermented foods in respiration conditions could be a new way of improving the performance of LAB in some fermented food products (Kaneko et al. 1990, Kaneko et al. 1991, Pedersen et al. 2012).

Finally, producers have to face the problem of bacteriophage contamination that represents an important risk to any process requiring bacterial growth, particularly in the biotechnology and food industries. In the particular case of LAB, the phages that infect them are all members of the Caudovirales order, having double-stranded DNA genomes encapsidated in a head connected to a tail (Ackermann 1998), it being the head that hooks to the bacteria in the infection process. The selection of bacterial strains to be used as starter cultures have, therefore, to be screened for being refractory to phage infection, especially if they are to be used as a unique fermenting strain. When a food product is produced with multiple strains, the use of strains with different phage sensitivity profiles is common practice so that the fermentation reaches its end by avoiding complete lysis of starter cultures (Samson and Moineau 2013, Erkus et al. 2013). Phage-resistant derivatives can be selected naturally via exposure of the phage-sensitive wild-type strains to specific phages (Moineau 1999). The spontaneous resulting phage-resistant mutants are considered non-GMOs because they are produced by natural evolution (Coffey and Ross 2002). When considering the improvement of strains through genetic engineering, one of the options is to engineer phage resistance mechanisms for bacteria that can be tailor-made and used to protect economically important bacterial cultures. This can be efficiently done due to the increasing number of available phage genomes in public databases and the development of various genetic tools (Samson and Moineau 2013, Mahony et al. 2014). From the bacterial side also, the scenario is very promising since, due to recognized importance of cell-wall involvement in LAB functionality, growth and fitness, and sensitivity to bacteriophages, knowledge of the variations of structure of cell-wall glycopolymers of LAB has advanced immensely (Chapot-Chartier 2014).

5. Conclusion

A wide range of LAB fermented functional food products are on the market. Many of them rely on the use of LAB selected and studied strains that are commercialized and, as such, allow consistent productions. Traditional food products can be produced using starter cultures or not, depending on the product or on the country. Nevertheless, all the knowledge that was obtained from them along the years allowed the design of new or improved food products, while maintaining the traditional ones. However, market and consumers always push food and starter cultures producers to find and study new LAB functional starter cultures, to use LAB isolated from one matrix as starter cultures in another matrix or to try new mixtures of strains to be used as starter cultures. Through this chapter, some examples of those approaches have been summarized.

Overall, a wide variety of potential health benefits is being claimed, as also mentioned in the chapter. Although fermented functional foods offer considerable market potential, several issues still need to be addressed as most of the studies on functional fermented foods are of a rather descriptive and preliminary nature (Leroy and De Vuyst 2014). The support of these findings by mechanistic insights and the demonstration of effectiveness with well controlled *in-vivo* experiments, animal models, and clinical trials is, in many cases, still needed.

LAB have been used or have been present for centuries in a wide variety of fermented foods. They are very harsh bacteria that withstand many different and stressful situations during production conditions, industrial processes, and when in transit through the GI tract. Numerous studies have been dedicated to them, and many articles have been written and will be written about them. Nevertheless, their potentialities are so immense that any challenge that the industry has in terms of LAB functionality or productivity will, one day, certainly be covered by a new LAB strain.

Keywords: Lactic acid bacteria, functional starter culture, fermented food, probiotics, bacteriocins, market, consumer, metabolic engineering, physiological engineering, systems biology, bacteriophages, food safety, nutrition, health

References

Ackermann, H.W. 1998. Tailed bacteriophages: The order caudovirales. Advances in Virus Research 51: 135–201.

Ammor, M.S. and Mayo, B. 2007. Selection criteria for lactic acid bacteria to be used as functional starter cultures in dry sausage production: An update. Meat Science 76: 138–146.

Angelidis, A.S. and Smith, G.M. 2003. Role of the glycine betaine and carnitine transporters in adaptation of *Listeria monocytogenes* to chill stress in defined medium. Applied and Environmental Microbiology 69: 7492–7498.

Bachmann, H., Pronk, J.T., Kleerebezem, M. and Teusink, B. 2015. Evolutionary engineering to enhance starter culture performance in food fermentations. Current Opinion in Biotechnology 32: 1–7.

Bailey, J.E., Sburlati, A., Hatzimanikatis, V., Lee, K., Renner, W.A. and Tsai, P.S. 2002. Inverse metabolic engineering: A strategy for directed genetic engineering of useful phenotypes. Biotechnology and Bioengineering 79: 568–579.

Bailey, J.E. 1991. Toward a science of metabolic engineering. Science 252: 1668–1675.

Bevilacqua, A., de Stefano, F., Augello, S., Pignatiello, S., Sinigaglia, M. and Corbo, M.R. 2015. Biotechnological innovations for table olives. International Journal of Food Sciences and Nutrition 66: 127–131.

Bourdichon, F., Casaregola, S., Farrokh, C., Frisvad, J.C., Gerds, M.L., Hammes, W.P., Harnett, J., Huys, G., Laulund, S., Ouwehand, A. et al. 2012. Food fermentations: Microorganisms with technological beneficial use. International Journal of Food Microbiology 154: 87–97.

Bron, P.A. and Kleerebezem, M. 2011. Engineering lactic acid bacteria for increased industrial functionality. Bioengineered Bugs 2: 80–87.

Buckenhüskes, H.J. 1993. Selection criteria for lactic acid bacteria to be used as starter cultures for various food commodities. FEMS Microbiology Reviews 12: 253–271.

Catzeddu, P., Mura, E., Parente, E., Sanna, M. and Farris, G.A. 2006. Molecular characterization of lactic acid bacteria from sourdough breads produced in Sardinia (Italy) and multivariate statistical analyses of results. Systematic and Applied Microbiology 29: 138–144.

Chapot-Chartier, M.-P. 2014. Interactions of the cell-wall glycopolymers of lactic acid bacteria with their bacteriophages. Frontiers in Microbiology 5: 1–10.

Chaves-López, C., Serio, A., Paparella, A., Martuscelli, M., Corsetti, A., Tofalo, R. and Suzzi, G. 2014. Impact of microbial cultures on proteolysis and release of bioactive peptides in fermented milk. Food Microbiology 42: 117–121.

Chen, Y., Liu, W., Xue, J., Yang, J., Chen, X., Shao, Y., Kwok, L.-Y., Bilige, M., Mang, L. and Zhang, H. 2014. Angiotensin-converting enzyme inhibitory activity of *Lactobacillus helveticus* strains from traditional fermented dairy foods and antihypertensive effect of fermented milk of strain H9. Journal of Dairy Science 11: 6680–6692.

Coda, R., Lanera, A., Trani, A., Gobbetti, M. and Di Cagno, R. 2012. Yogurt-like beverages made of a mixture of cereals, soy and grape must: microbiology, texture, nutritional and sensory properties. International Journal of Food Microbiology 155: 120–127.

Coffey, A. and Ross, R.P. 2002. Bacteriophage-resistance systems in dairy starter strains: Molecular analysis to application. Antonie van Leeuwenhoek 82: 303–321.

Crowley, S., Mahony, J. and van Sinderen, D. 2013. Current perspectives on antifungal lactic acid bacteria as natural bio-preservatives. Trends in Food Science & Technology 33: 93–109.

Dalié, D., Deschamps, A.M. and Richard-Forget, F. 2010. Lactic acid bacteria—Potential for control of mould growth and mycotoxins: A review. Food Control 21: 370–380.

De Vos, W.M. and Hugenholtz, J. 2004. Engineering metabolic highways in Lactococci and other lactic acid bacteria. Trends in Biotechnology 22: 72–79.

De Vos, W.M. 2011. Systems solutions by lactic acid bacteria: From paradigms to practice. Microbial Cell Factories 10(Suppl.1): S2.doi: 10.1186/1475-2859-10-S1-S2.

De Vuyst, L., Schrijvers, V., Paramithiotis, S., Hoste, B., Vancanneyt, M., Swings, J., Kalantzopoulos, G., Tsakalidou, E. and Messens, W. 2002. The biodiversity of lactic acid bacteria in Greek traditional wheat sourdoughs is reflected in both composition and metabolite formation. Applied and Environmental Microbiology 68: 6059–6069.

Di Cagno, R., Coda, R., De Angelis, M. and Gobbetti, M. 2013. Exploitation of vegetables and fruits through lactic acid fermentation. Food Microbiology 33: 1–10.

Di Cagno, R., De Angelis, M., Lavermicocca, P., De Vincenzi, M., Giovannini, C., Faccia, M. and Gobbetti, M. 2002. Proteolysis by sourdough lactic acid bacteria: Effects on wheat flour protein fractions and gliadin peptides involved in human cereal intolerance. Applied and Environmental Microbiology 68: 623–633.

Douillard, F.P. and de Vos, W.M. 2014. Functional genomics of lactic acid bacteria: From food to health. Microbial Cell Factories 13(Suppl 1): S8.doi:1 0.1186/1475-2859-13-S1-S8.

Dufour, D., Larsonneur, S., Alarcón, F., Brabet, C. and Chuzel, G. 1996. Improving the bread-making potential of cassava sour starch. pp. 133–142. *In*: D. Dufour, G.M. O'Brien and R. Best (eds.). Cassava Flour and Starch: Progress in Research and Development [Internet]. Cali, Colombia: CIAT-CIRAD.

EFSA. 2013. Scientific Opinion on the maintenance of the list of QPS biological agents intentionally added to food and feed (2013 update). EFSA Journal [Internet] 11: 108.

Erkus, O., de Jager, V.C.L., Spus, M., van Alen-Boerrigter, I.J., van Rijswijck, I.M.H., Hazelwood, L., Janssen, P.W.M., van Hijum, S.A.F.T., Kleerebezem, M. and Smid, E.J. 2013. Multifactorial diversity sustains microbial community stability. The ISME Journal 7: 2126–2136.

Frost, L.S., Leplae, R., Summers, A.O. and Toussaint, A. 2005. Mobile genetic elements: The agents of open source evolution. Nature Reviews Microbiology 3: 722–732.

Furukawa, S., Watanabe, T., Toyama, H. and Morinaga, Y. 2013. Significance of microbial symbiotic coexistence in traditional fermentation. Journal of Bioscience and Bioengineering 116: 533–539.

Gaspar, P., Carvalho, A.L., Vinga, S., Santos, H. and Neves, A.R. 2013. From physiology to systems metabolic engineering for the production of biochemicals by lactic acid bacteria. Biotechnology Advances 31: 764–788.

Gänzle, M.G., Höltzel, A., Walter, J., Jung, G. and Hammes, W.P. 2000. Characterization of reutericyclin produced by *Lactobacillus reuteri* LTH2584. Applied and Environmental Microbiology 66: 4325–4333.

Gänzle, M.G. 2009. From gene to function: Metabolic traits of starter cultures for improved quality of cereal foods. International Journal of Food Microbiology 134: 29–36.

Gerez, C.L., Torres, M.J., Font de Valdez, G. and Rollán, G. 2013. Control of spoilage fungi by lactic acid bacteria. Biological Control 64: 231–237.

Gest, H. 2004. The discovery of microorganisms by Robert Hooke and Antoni Van Leeuwenhoek, Fellows of the Royal Society, London.

Goldenfeld, N. and Woese, C. 2007. Biology's next revolution. Nature 445: 369.

González, A., Vázquez-Baeza, Y. and Knight, R. 2014. SnapShot: The Human Microbiome. Cell 158: 690–690.e1.

Granier, A., Goulet, O. and Hoarau, C. 2013. Fermentation products: Immunological effects on human and animal models. Pediatric Research 74: 238–244.

Hammes, W.P. 1990. Bacterial starter cultures in food production. Food Biotechnology 4: 383–397.

Hansen, E.B. 2002. Commercial bacterial starter cultures for fermented foods of the future. International Journal of Food Microbiology 78: 119–131.

Heller, K.J. 2001. Probiotic bacteria in fermented foods: Product characteristics and starter organisms. The American Journal of Clinical Nutrition 73: 374S–379S.

Holzapfel, W.H. 2002. Appropriate starter culture technologies for small-scale fermentation in developing countries. International Journal of Food Microbiology 75: 197–212.

Hurtado, A., Reguant, C., Bordons, A. and Rozès, N. 2012. Lactic acid bacteria from fermented table olives. Food Microbiology 31: 1–8.

IDFEFFCA. 2002. Inventory of Microorganisms with a Documented History of Use in Food [place unknown]: Bulletin of the IDF No. 377/2002.

Jacobsen, J.R. and Khosla, C. 1998. New directions in metabolic engineering. Current Opinion in Chemical Biology 2: 133–137.

Kaneko, T., Takahashi, M. and Suzuki, H. 1990. Acetoin fermentation by citrate-positive *Lactococcus lactis* subsp. *lactis* 3022 grown aerobically in the presence of Hemin or Cu. Applied and Environmental Microbiology 56: 2644–2649.

Kaneko, T., Watanabe, Y. and Suzuki, H. 1991. Differences between *Lactobacillus casei* subsp. *casei* 2206 and citrate-positive *Lactococcus lactis* subsp. *lactis* 3022 in the characteristics of diacetyl production. Applied and Environmental Microbiology 57: 3040–3042.

Kau, A.L., Ahern, P.P., Griffin, N.W., Goodman, A.L. and Gordon, J.I. 2011. Human nutrition, the gut microbiome and the immune system. Nature 474: 327–336.

Kern, A., Tilley, E., Hunter, I.S., Legisa, M. and Glieder, A. 2007. Engineering primary metabolic pathways of industrial micro-organisms. Journal of Biotechnology 129: 6–29.

Kok, J., Buist, G., Zomer, A.L., van Hijum, S.A.F.T. and Kuipers, O.P. 2005. Comparative and functional genomics of lactococci. FEMS Microbiology Reviews 29: 411–433.

Kovač, B. and Raspor, P. 1997. The use of the mould *Rhizopus oligosporus* in food production. Food Technology and Biotechnology 35: 69–73.

Kuipers, O.P., de Jong, A., Baerends, R.J.S., van Hijum, S.A.F.T., Zomer, A.L., Karsens, H.A., Hengst, den C.D., Kramer, N.E., Buist, G. and Kok, J. 2002. Transcriptome analysis and related databases of *Lactococcus lactis*. Antonie van Leeuwenhoek 82: 113–122.

Lavermicocca, P., Valerio, F., Evidente, A., Lazzaroni, S., Corsetti, A. and Gobbetti, M. 2000. Purification and characterization of novel antifungal compounds from the sourdough *Lactobacillus plantarum* Strain 21B. Applied and Environmental Microbiology 66: 4084–4090.

Lee, G.N. and Na, J. 2013. The impact of synthetic biology. ACS Synthetic Biology 2: 210–212.

Leroi, F. 2010. Occurrence and role of lactic acid bacteria in seafood products. Food Microbiology 27: 698–709.

Leroy, F. and De Vuyst, L. 2004. Lactic acid bacteria as functional starter cultures for the food fermentation industry. Trends in Food Science & Technology 15: 67–78.

Leroy, F. and De Vuyst, L. 2014. Fermented food in the context of a healthy diet: How to produce novel functional foods? Current Opinion in Clinical Nutrition and Metabolic Care 17: 574–581.

Logan, A.C. and Katzman, M. 2005. Major depressive disorder: Probiotics may be an adjuvant therapy. Medical Hypotheses 64: 533–538.

Logan, A.C., Venket Rao, A. and Irani, D. 2003. Chronic fatigue syndrome: Lactic acid bacteria may be of therapeutic value. Medical Hypotheses 60: 915–923.

Lomer, M.C.E., Parkes, G.C. and Sanderson, J.D. 2008. Review article: Lactose intolerance in clinical practice—Myths and realities. Alimentary Pharmacology & Therapeutics 27: 93–103.

Lönner, C., Welander, T., Molin, N., Dostálek, M. and Blickstad, E. 1986. The microflora in a sour dough started spontaneously on typical Swedish rye meal. Food Microbiology 3: 3–12.

Mahony, J., Bottacini, F., van Sinderen, D. and Fitzgerald, G.F. 2014. Progress in lactic acid bacterial phage research. Microbial Cell Factories 13 Suppl. 1: S1.

Maldonado Alvarado, P., Grosmaire, L., Dufour, D., Toro, A.G., Sánchez, T., Calle, F., Santander, M.A.M., Ceballos, H., Delarbre, J.L. and Tran, T. 2013. Combined effect of fermentation, sun-drying and genotype on bread making ability of sour cassava starch. Carbohydrate Polymers 98: 1137–1146.

Marshall, E. and Mejia, D. 2012. Traditional Fermented Food and Beverages for Improved Livelihoods. FAO, Rome.

Martinussen, J., Solem, C., Holm, A.K. and Jensen, P.R. 2013. Engineering strategies aimed at control of acidification rate of lactic acid bacteria. Current Opinion in Biotechnology 24: 124–129.

McFeeters, R.F. 2004. Fermentation microorganisms and flavor changes in fermented foods. Journal of Food Science 69: FMS35–FMS37.

Moineau, S. 1999. Applications of phage resistance in lactic acid bacteria. Antonie van Leeuwenhoek 76: 377–382.

Montet, D., Ray, R.C. and Zakhia-Rozis, N. 2014. Lactic acid fermentation of vegetables and fruits. pp. 108–140. *In*: R.C. Ray and D. Montet (eds.). Microorganisms and Fermentation of Traditional Foods. CRC Press, Boca Raton, FL.

Nazzaro, F., Orlando, P., Fratianni, F. and Coppola, R. 2012. Microencapsulation in food science and biotechnology. Current Opinion in Biotechnology 23: 182–186.

Ogawa, J., Kishino, S., Ando, A., Sugimoto, S., Mihara, K. and Shimizu, S. 2005. Production of conjugated fatty acids by lactic acid bacteria. Journal of Bioscience and Bioengineering 100: 355–364.

Oliveira, P.M., Zannini, E. and Arendt, E.K. 2014. Cereal fungal infection, mycotoxins, and lactic acid bacteria mediated bioprotection: From crop farming to cereal products. Food Microbiology 37: 78–95.

Onyango, C., Noetzold, H., Ziems, A., Hofmann, T., Bley, T. and Henle, T. 2005. Digestibility and antinutrient properties of acidified and extruded maize–finger millet blend in the production of *uji*. LWT-Food Science and Technology 38: 697–707.

Paredes López, O. 1992. Nutrition and safety considerations. In: Office of International Affairs, National Research Council, editors. National Academy Press. Washington, D.C.: Applications of Biotechnology to Traditional Fermented Foods, pp. 143–158.

Park, J.-E., Oh, S.-H. and Cha, Y.-S. 2014. *Lactobacillus brevis* OPK-3 isolated from kimchi inhibits adipogenesis and exerts anti-inflammation in 3T3-L1 adipocyte. Journal of the Science of Food and Agriculture 94: 2514–2520.

Park, J.H. and Lee, S.Y. 2008. Towards systems metabolic engineering of microorganisms for amino acid production. Current Opinion in Biotechnology 19: 454–460.

Park, J.H., Lee, S.Y., Kim, T.Y. and Kim, H.U. 2008. Application of systems biology for bioprocess development. Trends in Biotechnology 26: 404–412.

Pawlowska, A.M., Zannini, E., Coffey, A. and Arendt, E.K. 2012. "Green Preservatives": Combating fungi in the food and feed industry by applying antifungal lactic acid bacteria. pp. 217–238. *In*: J. Henry (ed.). Advances in Food and Nutrition Research. Vol. 66. Academic Press. Waltham, MA.

Pedersen, M.B., Gaudu, P., Lechardeur, D., Petit, M,-A. and Gruss, A. 2012. Aerobic respiration metabolism in lactic acid bacteria and uses in biotechnology. Annual Review of Food Science and Technology 3: 37–58.

Pessione, E. 2012. Lactic acid bacteria contribution to gut microbiota complexity: Lights and shadows. Frontiers in Cellular and Infection Microbiology 2: 86. doi: 10.3389/fcimb.2012.00086.

Pfeiler, E.A. and Klaenhammer, T.R. 2007. The genomics of lactic acid bacteria. Trends in Microbiology 15: 546–553.

Pophaly, S.D., Poonam Singh, P., Kumar, H., Tomar, S.K. and Singh, R. 2014. Selenium enrichment of lactic acid bacteria and bifidobacteria: A functional food perspective. Trends in Food Science & Technology 39: 135–145.

Prema, P., Smila, D., Palavesam, A. and Immanuel, G. 2010. Production and characterization of an antifungal compound (3-phenyllactic acid) produced by *Lactobacillus plantarum* strain. Food Technology and Biotechnology 3: 379–386.

Rathore, S., Salmerón, I. and Pandiella, S.S. 2012. Production of potentially probiotic beverages using single and mixed cereal substrates fermented with lactic acid bacteria cultures. Food Microbiology 30: 239–244.

Rattanachaikunsopon, P. and Phumkhachorn, P. 2010. Lactic acid bacteria: Their antimicrobial compounds and their uses in food production. Annals of Biological Research 1: 218–228.

Reddy, N.R. and Pierson, M.D. 1994. Reduction in antinutritional and toxic components in plant foods by fermentation. Food Research International 27: 281–290.

Rodríguez, H., Curiel, J.A., Landete, J.M., las Rivas de, B., López de, Felipe F., Gómez-Cordovés, C., Mancheño, J.M. and Muñoz, R. 2009. Food phenolics and lactic acid bacteria. International Journal of Food Microbiology 132: 79–90.

Rossi, F., Rizzotti, L., Felis, G.E. and Torriani, S. 2014. Horizontal gene transfer among microorganisms in food: current knowledge and future perspectives. Food Microbiology 42: 232–243.

Rouhi, M., Sohrabvandi, S. and Mortazavian, A.M. 2013. Probiotic fermented sausage: Viability of probiotic microorganisms and sensory characteristics. Critical Reviews in Food Science and Nutrition 53: 331–348.

Ruas-Madiedo, P., Sánchez, B., Hidalgo-Cantabrana, C., Margolles, A. and Laws, A. 2012. Exopolysaccharides from lactic acid bacteria and bifidobacteria. pp. 125–152. *In*: Y.H. Hui and E. Ozgul Evranuz (eds.). 2nd ed. Handbook of Animal-Based Fermented Food and Beverage Technology, Boca Raton, FL.

Samson, J.E. and Moineau, S. 2013. Bacteriophages in food fermentations: New frontiers in a continuous arms race. Annual Review of Food Science and Technology 4: 347–368.

Santos, C.C.A.D.A., Libeck, B.D.S. and Schwan, R.F. 2014. Co-culture fermentation of peanut-soy milk for the development of a novel functional beverage. International Journal of Food Microbiology 186: 32–41.

Selhub, E.M., Logan, A.C. and Bested, A.C. 2014. Fermented foods, microbiota, and mental health: Ancient practice meets nutritional psychiatry. Journal of Physiological Anthropology 33: 2. doi: 10.1186/1880-6805-33-2.

Seo, D.J., Lee, M.H., Seo, J., Ha, S.-D. and Choi, C. 2014. Inactivation of murine norovirus and feline calicivirus during oyster fermentation. Food Microbiology 44: 81–86.

Seo, S.W., Yang, J., Min, B.E., Jang, S., Lim, J.H., Lim, H.G., Kim, S.C., Kim, S.Y., Jeong, J.H. and Jung, G.Y. 2013. Synthetic biology: Tools to design microbes for the production of chemicals and fuels. Biotechnology Advances 31: 811–817.

Sheehan, V.M., Sleator, R.D., Fitzgerald, G.F. and Hill, C. 2006. Heterologous expression of BetL, a betaine uptake system, enhances the stress tolerance of *Lactobacillus salivarius* UCC118. Applied and Environmental Microbiology 72: 2170–2177.

Siezen, R.J., Bayjanov, J.R., Felis, G.E., van der Sijde, M.R., Starrenburg, M., Molenaar, D., Wels, M., van Hijum, S.A.F.T. and van Hylckama Vlieg, J.E.T. 2011. Genome-scale diversity and niche adaptation analysis of *Lactococcus lactis* by comparative genome hybridization using multi-strain arrays. Microbial Biotechnology 4: 383–402.

Siezen, R.J., van Enckevort, F.H.J., Kleerebezem, M. and Teusink, B. 2004. Genome data mining of lactic acid bacteria: The impact of bioinformatics. Current Opinion in Biotechnology 15: 105–115.

Sleator, R.D. and Hill, C. 2006. Patho-biotechnology: Using bad bugs to do good things. Current Opinion in Biotechnology 17: 211–216.

Smid, E.J. and Kleerebezem, M. 2014. Production of aroma compounds in lactic fermentations. Annual Review of Food Science and Technology 5: 313–326.

Steinkraus, K.H. 1996. Handbook of Indigenous Fermented Foods. Marcel Dekker, New York.

Steinkraus, K.H. 1992. Lactic Acid Fermentations. In: Office of International Affairs, National Research Council, editors. National Academy Press. Washington, D.C.: Applications of Biotechnology in Traditional Fermented Foods, pp. 43–51.

Steinkraus, K.H. 2002. Fermentations in world food processing. Comprehensive Reviews in Food Science and Food Safety 1: 23–32.

Ström, K., Sjögren, J., Broberg, A. and Schnürer, J. 2002. *Lactobacillus plantarum* MiLAB 393 produces the antifungal cyclic dipeptides cyclo(L-Phe-L-Pro) and cyclo(L-Phe-*trans*-4-OH-L-Pro) and 3-phenyllactic acid. Applied and Environmental Microbiology 68: 4322–4327.

Talarico, T.L., Casas, I.A., Chung, T.C. and Dobrogosz, W.J. 1988. Production and isolation of reuterin, a growth inhibitor produced by *Lactobacillus reuteri*. Antimicrobial Agents and Chemotherapy 32: 1854–1858.

Tamime, A.Y. 2002. Fermented milks: A historical food with modern applications—a review. European Journal of Clinical Nutrition 56: S2–S15.

Top, E.M. and Springael, D. 2003. The role of mobile genetic elements in bacterial adaptation to xenobiotic organic compounds. Current Opinion in Biotechnology 14: 262–269.

Turnbaugh, P.J., Ley, R.E., Hamady, M., Fraser-Liggett, C.M., Knight, R. and Gordon, J.I. 2007. The human microbiome project. Nature 449: 804–810.

Valcheva, R., Ferchichi, M.F., Korakli, M., Ivanova, I., Gänzle, M.G., Vogel, R.F., Prévost, H., Onno, B. and Dousset, X. 2006. *Lactobacillus nantensis* sp. nov., isolated from French wheat sourdough. International Journal of Systematic and Evolutionary Microbiology 56: 587–591.

Waters, D.M., Mauch, A., Coffey, A., Arendt, E.K. and Zannini, E. 2015. Lactic acid bacteria as a cell factory for the delivery of functional biomolecules and ingredients in cereal based beverages: A review. Critical Reviews in Food Science and Nutrition 55: 503–520.

Yang, B., Chen, H., Gu, Z., Tian, F., Ross, R.P., Stanton, C., Chen, Y.Q., Chen, W. and Zhang, H. 2014. Synthesis of conjugated linoleic acid by the linoleate isomerase complex in food-derived lactobacilli. Journal of Applied Microbiology 117: 430–439.

Zamfir, M., Cornea, C.P. and Vuyst, L. 2014. Biodiversity and biotechnological potential of lactic acid bacteria. AgroLife Scientific Journal 3: 169–176.

Zannini, E., Pontonio, E., Waters, D.M. and Arendt, E.K. 2012. Applications of microbial fermentations for production of gluten-free products and perspectives. Applied Microbiology and Biotechnology 93: 473–485.

Zendo, T. 2013. Screening and characterization of novel bacteriocins from lactic acid bacteria. Bioscience, Biotechnology, and Biochemistry 77: 893–899.

Zhang, Y., Zhu, Y., Zhu, Y. and Li, Y. 2009. The importance of engineering physiological functionality into microbes. Trends in Biochemical Sciences 27: 664–672.

Zhu, L., Zhu, Y., Zhang, Y. and Li, Y. 2012. Engineering the robustness of industrial microbes through synthetic biology. Trends in Microbiology 20: 94–101.

Zhu, Y., Zhang, Y. and Li, Y. 2009. Understanding the industrial application potential of lactic acid bacteria through genomics. Applied Microbiology and Biotechnology 83: 597–610.

9

Microencapsulation of Probiotics and Applications in Food Fermentation

A.M. Mortazavian,[1,]* M. Moslemi[2] and S. Sohrabvandi[3]

1. Introduction

Probiotic encapsulation favors the cells in various aspects such as increasing their resistance to bacteriophage infection, chemical poisons, and genetic mutation, improved productivity in metabolite production and a denser biomass production (Mortazavian et al. 2007). Furthermore, the main factor is increased cell tolerance against the harsh environment of fermented and other acidic products, resulting in increased shelf-life (Mortazavian and Sohrabvandi 2006, Mortazavian et al. 2006a, 2006b, Mortazavian et al. 2007). It is recommended that each person should consume about 1×10^9 probiotic viable cells per day (approximately 100 g/day of probiotic products with 10^7 cfu (colony forming units)/g viability) (Kurmann and Robinson 1991, Shah 2000, Lourens-Hattingh and Viljoen 2001, Shah et al. 2011, Hickey 2005, Parracho et al. 2007). But it is not applicable in practice because of the susceptibility of probiotics to environmental stresses and their poor viability during fermentation and storage time (Shah 2000, Korbekandi et al. 2011). In general, a large number of internal and external parameters have significant effects on probiotics' viability, especially in fermented products. The most important are medium pH and acidity, presence of

[1] Department of Food Science and Technology, National Nutrition and Food Technology Research Institute, Faculty of Nutrition Sciences, Food Science and Technology, Shahid Beheshti University of Medical Sciences, P.O. Box 19395-4741, Tehran, Iran.
[2] Food and Drug Administration, Ministry of Health and Medical Education, P.O. Box 13147-15311, Tehran, Iran.
[3] Department of Food Technology Research, National Nutrition and Food Technology Research Institute, Shahid Beheshti University of Medical Sciences, P.O. Box 19395-4741, Tehran, Iran.
* Corresponding author: mortazvn@sbmu.ac.ir; mortazvn@yahoo.com

molecular oxygen and redox potential, inhibitors such as hydrogen peroxide, bacterial metabolites such as bacteriocins and short chain fatty acids, some flavors, bacterial competition, packaging, inoculation condition, fermentation process, encapsulation technology, nutritional supplements, thermal processing, time and temperature of storage, addition of osmo-regulator agents, carbonation, buffering capacity of the media, homogenization, rate of product cooling and quantity of production (De Vuyst 2000, Shah 2000, Tamime et al. 2005, Mortazavian et al. 2006a, Kosin and Rakshit 2006, Mortazavian et al. 2008).

Microencapsulation of probiotics provides a promising approach on how to overcome their loss under detrimental environmental conditions (Hansen et al. 2002). An appropriate encapsulation of probiotics provides proper protection of entrapped cells while it does not negatively affect the sensory properties of the final product. Apart from cell protection as the main function, microencapsulation could offer other advantages to probiotic products. For instance, acetic acid is one of the main metabolites especially produced by *Bifidobacterium* spp. which gives the vinegar off-flavor to fermented products. Bacterial encapsulation, by decreasing the amount of acid production by enclosed microorganisms, considerably covers the off-flavor produced during fermentation (Adhikari et al. 2000). Also, if 'prebiotics' (mostly, non-digestible carbohydrates) are used as encapsulating materials, they would have some protective effects on the probiotic bacterial cells. It is found that prebiotics lead to higher resistance to bile salts in viable cells (Saarela et al. 2003). In this chapter, different aspects of probiotic microencapsulation are briefly explained.

2. Probiotics and their Viability in Food Products

The term 'probiotic' refers to the live microorganisms with various health benefits to the host when administered in adequate numbers (WHO 2006). The most important health benefits that are attributed to probiotics include anti-mutagenic and anti-carcinogenic effects, immuno-modulatory and immuno-stimulatory roles, anti-infection properties, reduction of serum cholesterol and lactose intolerance symptoms, and nutritional benefits (Mortazavian et al. 2006a, Korbekandi et al. 2011). Generally, the bacterial species of lactobacilli and bifidobacteria are the most common species of probiotics used in food products, either as fermenting bacteria (starter probiotic, added before fermentation for proliferation to high viable counts, acidification, aroma production and nutritional consequences) or non-fermenting bacteria (non-starter probiotic, added after fermentation at proper viable numbers for retention of as much cell stability as possible during storage although this method may lead to an expensive product). Lactobacilli belong to the lactic acid bacteria group (LAB) that has good native adaptation to milk-based matrices (Korbekandi et al. 2011). Their high potential in acid production favors food in resistance to spoilages and helps enhance the desirable organoleptic attributes of products.

The main probiotic strains in preparation of fermented milk products or non-dairy probiotic products are *Bifidobacterium bifidum, B. adolescentis, B. breve, B. infantis, B. longum, B. lactis, Lactobacillus acidophilus, Lb. johnsonii, Lb. reuteri, Lb. rhamnosus, Lb. paracasei, Lb. casei, Lb. brevis, Enterococcus faecium, E. faecalis,* and *Pediococcus*. Recently, *Bacillus coagulans* and *B. subtilis* are increasingly used in food precuts as probiotics, especially in those exposed to elevated temperatures (Tamime et al. 1995, Tuohy et al. 2003, Korbekandi et al. 2011). Commercially, many probiotic strains are available in lyophilized (freeze dried) form with very high viable cell concentration.

Nutraceutical effectiveness of probiotics, their most important qualitative parameter, is determined by the rate of probiotics viability in the final products at the time of consumption. There is no fixed agreement on acceptable levels of probiotic viability in probiotic products; however, generally, the levels of 10^6 cfu/ml, 10^7 cfu/ml and $> 10^8$ cfu/ml are recognized as the minimum amount, acceptable level and satisfactory level, respectively (Korbekandi et al. 2011). Naturally, probiotics (not all) show relatively poor viability in fermented milks and fruit juices due to their slow growth and sensitivity to detrimental factors during production and storage. Thus, it is expected that their viable count in final products be lower than the acceptable level at the time of consumption. Furthermore, if a high number of viable cells has been protected in products by any method, it is not guaranteed that the same numbers would be reached to the target site of body owing to the exposure to very low stomach pH, severe pH changes from acid to alkaline condition and the presence of bile salts (Tamime et al. 2005, Korbekandi et al. 2011).

3. General Concepts and Definitions Considering Probiotic Microencapsulation

There are numerous compositional or process factors that have been practiced to improve the viability of probiotics in food products. However, it has now been proved that the best way is selecting originally resistant strains of probiotics or environmentally-adapted resistant ones (Korbekandi et al. 2011). The 'stepwise stress adaptation method' has been suggested as an applicable approach for increasing the probiotics' resistance to deteriorative factors in fermented milks. By this method, microorganisms are gradually introduced to the harsher situations step-by-step (Shah 2000). For example, it is observed that the adapted *Lb. acidophilus* cells could tolerate and grow at pH = 3.5 in the presence of 0.3% bile salts (Chou and Weimer 1999) and bifido bacteria tolerate pH values less than 4.0 (Korbekandi et al. 2011). An interesting result was observed in the survivability of free *Lb. acidophilus* CSCC 2401 and *B. infantis* CSCC 1912 in ice-cream; they showed higher survivability than the encapsulated ones (Korbekandi et al. 2011). However, it is not always possible to use isolated resistant strains with good fermentability and confer pleasant sensory properties to food products, as some of them might generate off-flavors. Therefore, microencapsulation of probiotic cells might be used as a suitable method in added-value products (justified price) with special medicinal claims, in which high viable counts of probiotics are required till their end of shelf life. Interestingly, the evolution of encapsulation solves some difficulties related to viable cell application (Korbekandi et al. 2011).

Encapsulation technology has a wide spectrum of usage in different fields of study including medicine, pharmaceutics, biotechnology, and the food industry (Anal and Singh 2007). Entrapment of living probiotics within polymeric materials by limiting their interaction with the surrounding medium is one of the effective means of protection of cells against disruptive parameters in fermented milks. The process is named as encapsulation. In other words, microencapsulation of probiotics could be defined as enclosing desired probiotics with efficient shells in order substantially to retain their viability (by shielding them from detrimental environmental factors) in food products (during processing and storage) until the time of consumption and subsequently in the gastrointestinal tract (GIT) (during transit through it), with proper release in the target GIT place. Therefore, the controlled release property (no release in products and non-target GIT places, but proper release in target site of GIT) of microcapsules is as important as their preservation property. As can be seen, while there are many affecting factors in food products, the two main factors in the GIT are stomach acid/pH and bile salts. Depending on the size of encapsulates, it is called as micro-encapsulation or macro-encapsulation. In other words, because the size of most probiotics is in the range of micrometers, the terms 'micro-' or 'macro-encapsulation' are routinely used for the coating of living cells (Bouwmeester et al. 2007, Hsieh and Ofori 2007, Sekhon 2010, Ijabadeniyi 2012).

Although encapsulation of micro-particles is considered a new technology, the root of this concept can be found naturally in capsule-generating bacteria. In nature, some bacteria are able to secrete exo-polysaccharide around their cells, protecting them against mechanical damage or any other extrinsic factors (Kailasapathy et al. 2008). It is considered natural encapsulation. These secretions, that are absolutely cell-friendly components, act as a supplementary layer or coating and protect the maternal cells from stress factors.

A microcapsule (micro-bead) consists basically of two parts: the core (that is entrapped within the encapsulation material, containing probiotic cells as well as enclosed solutions) and the shell (that encompasses the core). In some cases, microcapsules could be covered by a second layer of material that is usually named as 'support' or 'coat'. It leads to increased encapsulation efficiency (Mortazavian et al. 2007). Each bead might consist of one or several cells, or be empty. The interstitial liquid from the solution fills the free spaces of the micro-bead. Normally, the surface of micro-beads consist of different micro-cracks. These micro-cracks allows very slow metabolism of bacterial cells. However, their extension could lead to pore or void formation; thereby, reducing capsulation efficiency (Mortazavian et al. 2007).

Microcapsules shall not be digested in harsh environments in food products or in the stomach, but must be digested in the small and/or large intestines to release the viable bacterial cells. From another approach, microcapsules could possess sudden or sustained release. In sustained release method, the capsules are gradually digested and the probiotic cells are continuously released. Various factors affect the encapsulation core release, among the factors are harsh pH (Sun and Griffiths 2000), bile salts (Lee and Heo 2000), thermal shocks such as deep freezing and freeze drying (Shah and Ravula 2000) or high temperatures such as in spray drying, the presence of molecular oxygen as a inhibitory effect of obligatory anaerobes (Ohno et al. 1995), mechanical processes, enzymatic activities, osmotic pressure, bacteriophages (Steenson and Swaisgood 1987), moisture infusion along with the capsules layers, chemical interaction, antimicrobial perfusion (Sultana et al. 2000) and time of storage. For instance, microcapsules can be chemically designed to be resistant to stomach acid and pH and not be digested and release the probiotics, but be easily dissolved in the small intestine at pH ~ 7.

Microcapsules must meet following requirements for food applications (Lee and Salminen 2009):

- Must be food grade and compatible with food vehicles (safety considerations)
- Must be functionally efficient (protection of probiotics in food products during processing and storage and in GIT until the release in target location)
- Must be inexpensive and cost-effective
- Should not display undesirable impact (at least, any unacceptable effect) on organoleptic attributes of products
- Must have high encapsulation efficiency (high probiotic loading)
- Must have proper release of enclosed cells at target sites of gut

Probiotic encapsulation includes two distinct stages. At first, probiotic bacteria or cells slurry are added into the polymeric solution. Secondly, the final solution is dehydrated or dried to achieve the encapsulate suspension or powder (Gomes and Malcata 1999).

4. The Functional Properties of Probiotic Microencapsulation

Although it is evident that the most important function of microencapsulation is the increase of probiotics viability in food products until the time of consumption and during the delivery through the GIT, there might also be other intended functions such as intentional changes in starter culture metabolic activity (improving fermentation rate and profile) during product manufacture, producing encapsulated starter cultures, improving sensory characteristics of food products and probiotic cells immobilization (Krasaekoopt et al. 2003). These aspects are described below. Figure 1 indicates the functional properties and drawbacks of probiotic microencapsulation.

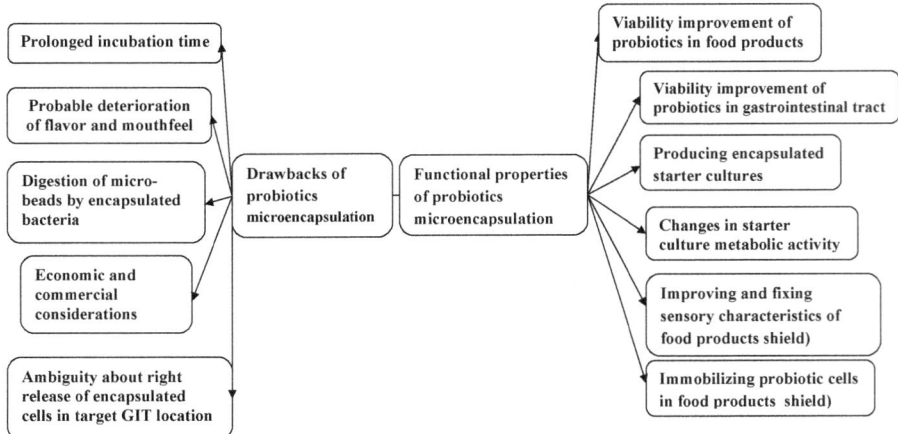

Figure 1. Functional properties and drawbacks of probiotic microencapsulation.

4.1 Viability Improvement of Probiotics in Food Products

Viability improvement of probiotics in food products is important from two points of view:

- increasing the therapeutic efficiency of the probiotic product
- product shelf life extension, which comprises economical and marketing merits.

High volumes of research have revealed that microencapsulation of different probiotic strains could significantly enhance their stability in food products during the production and shelf life. This is due to the protective effects of microcapsules against detrimental environmental factors in product matrices such as high acidity, low pH, molecular oxygen (for bifidobacteria), chemical and biochemical toxic agents generated during the process (especially heat-derived compounds and antibiotics), toxic food additives, digestive enzymes, bacteriophages, hydrogen peroxide, short-chain fatty acids, carbonyl-aromatic compounds (the last three cases could be produced by starter cultures during fermentation) and heat processing (e.g., drying) (Mortazavian et al. 2007). In addition, because microencapsulation of probiotic strains considerably decreases their metabolic activity, viability of the cells would increase due to the slower metabolic activity, acid production rate and ageing. The mild gradient in acid production could be considered a promising effect during storage periods because it can lead to a higher viability of probiotic cells (Mortazavian et al. 2006b). Regarding the production of organic acids in the fermentation process, acetic acid is one of the weakest ones that shows a high mortality effect on producing bacteria because of facilitated penetration of un-dissociated organic acids into bacterial cells. Therefore, coating of probiotics with polymers relatively permeable to the bacterial metabolites protects them against inhibitory effects (Mortazavian and Sohrabvandi 2006). Table 1 indicates selected publications on viability increase of encapsulated probiotics in food products.

High acidity and low pH of fermented products are the main factors that cause viability loss of probiotics, especially during refrigerated storage (Korbekandi et al. 2011). Most research has proven that in media with organic acids, microcapsules possess appropriate protective effects on probiotic strains; however, this effect might noticeably decrease in media with mineral acid (HCl) (Korbekandi et al. 2011). A mixture of alginate-HACS (high amylose corn starch) or alginate-RS (resistant starch) when compared with calcium alginate alone, considerably improves the viability of probiotics during eight weeks of refrigerated storage in yogurt (Sultana et al. 2000). According to the study carried out on shelf-life of *Lb. rhamnosus* VTT E-97800, the free bacteria kept at ambient temperature and high relative humidity had one-third shelf-life compared to the deep frozen encapsulated cells (6 months *vs.* 18 months) (Mattila-Sandholm et al. 2002). Encapsulation of *B. bifidum* in calcium alginate and its incorporation to yogurt improved its viability during three weeks refrigerated storage at 4°C. In this condition, the viable cell count did not reach below 10^7 cfu/ml and no undesirable sensory attributes were found in the product. The same results were observed at the frozen storage of the product (Sultana et al. 2000).

The tolerance of encapsulated probiotic cells towards stress factors differs from one microbial species to another and it is observed that resistance to acidic condition is case-sensitive. For example, bifidobacteria tolerated bile salts of the small

Table 1. Selected publications on viability increase of encapsulated probiotics in food products.

Encapsulating polymer(s)	Product	Probiotic	Reference
Ca-alginate	Yogurt	*Lb. acidophilus* and bifidobacteria	Sultana et al. (2000)
	Frozen ice milk	bifidobacteria	Kebary et al. (1998)
	Milk and fermented milk	*B. longum*	Hansen et al. (2002)
	Mayonnaise with pH 4.4	*Bifidobacterium* spp.	Khalil and Mansour (1998)
	Frozen ice milk	Probiotic lactobacilli	Sheu and Marshall (1993)
	Fermented frozen dairy dessert	*Lb. acidophilus* and bifidobacteria	Shah and Ravula (2000), Sultana et al. (2000), Godward and Kailasapathy (2003)
	Yogurt	*B. bifidum*	Sultana et al. (2000)
	Cheddar cheese	*B. longum*	Dinakar and Mistry (1994)
	Doogh	*Lb. acidophilus* and *B. lactis* BB-12	Mortazavian et al. (2008)
	Fermented milk	*Lb. acidophilus* and bifidobacteria	Hansen et al. (2002), Krasaekoopt et al. (2004)
Alginate/HACS and alginate/RS	Yogurt	*Lb. acidophilus* and bifidobacteria	Sultana et al. (2000)
Alginate-starch mixture	Yogurt	*Lb. acidophilus* and bifidobacteria	Sultana et al. (2000)
Xanthan-gelan mixture	Yogurt	*B. infantis*	Sun and Griffiths (2000)
k-carrageenan-locust bean mixture	Fermented milks	Different probiotics	Audet et al. (1988), Arnaud et al. (1992), Paquin et al. (1990), Sun and Griffiths (2000)
Alginate/chitosan and gelatin/ chitosan (coating)	Fermented milks	Different probiotics	Zhou et al. (1998), Krasaekoopt et al. (2003)

intestine rather than acidic conditions of gastric juice. In contrast, *Lb. acidophilus* showed a different response and is sensitive to bile salts (Krasaekoopt et al. 2004, Mortazavian et al. 2008). In another study, the viable count of *Lb. acidophilus* at pH values less than 4.2 reached to lower than 10^6 cfu/ml after seven days in fermented milk drinks; on the other hand, *B. lactis* showed the same results after 14 days of storage. Comparatively, viable counts of encapsulated *Lb. acidophilus* and bifidobacteria were more than 10^5 and 10^6 cfu/ml at the end of 42 days refrigerated storage, respectively. However, their free cell counts for the two bacteria were non-detectable and 10^2 cfu/ml, respectively (Mortazavian et al. 2008). In comparison, survivability of mentioned probiotics in simulated gastrointestinal conditions was studied in another simulated medium. In a harsh condition (pH = 1.5, 90 min; 2% bile salts, 90 min), encapsulation resulted in the viability of 18% and 9.5% compared to 0.6% and 0.2% for free cells of *Lb. acidophilus* and bifidobacteria, respectively. In a milder condition (pH = 2, 30 min; 0.6% bile salts, 60 min), cells survivability was 16% and 21% for free *Lb. acidophilus* and bifidobacteria in comparison to the 26% and 34% for encapsulated ones. Micro-beads were sized in the range of 300–500 μm with a high frequency in the size of around 340 μm (Mortazavian et al. 2008). Good efficiency after the encapsulation of *B. infantis* with xanthan-gelan mixture in yogurt with pH 4.0 during the six weeks of storage period at 4°C has been reported. The mentioned cells showed higher survivability during the pasteurization process (Sun and Griffiths 2000).

4.2 Viability Improvement of Probiotics in Gastrointestinal Tract

Although in a lesser extent when compared to food products, microencapsulation of probiotics can significantly improve their viability when passing through acidic-enzymatic-bile conditions of the GIT (Mortazavian et al. 2007). Most of the investigations have been carried out in simulated gastrointestinal conditions rather than *in vivo* trials. In this method, the cell suspension or product containing probiotics is exposed to simulated gastric juice (pH 1.5–2 with gastric enzymes) and intestinal juice (bile salt from 0.2–2% with digestive enzyme mixture as well as pH 6–7 for small intestine and 9–10 for large intestine) for defined holding times (e.g., 20–90 minutes, depending on the passing matrix) (Mainville et al. 2005). In a simulation study, *B. infantis* showed a significant loss from 1.23×10^9 to < 10 cfu/ml after 30 min storage at simulated gastric condition (pH = 1.5), whereas, its viability loss did not exceed 0.67% for encapsulated cells at the same condition (Sun and Griffiths 2000). Selected publications on viability increase of encapsulated probiotics in GIT are shown in Table 2.

4.3 Producing Encapsulated Starter Cultures

Most probiotic starter cultures are commercially launched in the free form rather than the encapsulated, in order to allow fermentability of probiotics during fermentation as well as due to economic considerations. However, in special cases based on especial market orders and products, it is justifiable, and even necessary from the cell stability point of view, to produce encapsulated starter cultures. Microencapsulation of starters can increase their viability in three stages: during storage until used, in food products

Table 2. Selected publications on viability increase of encapsulated probiotics in Gastro Intestinal Tract.

Encapsulating polymer(s)	GIT conditions	Probiotic	Reference
Ca-alginate	Gastric juice (pH 1.5) Gastric juice (pH 1.5, 30 min) GIT juices	*Bifidobacterium longum B. infantis Lactobacillus acidophilus* and *B. lactis* BB-12	Lee and Heo (2000) Sun and Griffiths (2000) Mortazavian et al. (2008)
Alginate/CaCl$_2$	(pH2; bile: 1%)	*Lb. acidophilus*	Chandramouli et al. (2004)
Cellulose acetate phthalate (CAP)	GIT juices	*B. pseudolongum*	Rao et al. (1989), Groboillot et al. (1993)
Resistant starch	Gastric juice	Different probiotics	Shah et al. (2011)

after being used (during production and storage) and during delivery through GIT. It has been shown that the shelf life of encapsulated *Lb. rhamnosus* VTT E-97800 which is kept under room temperature and relatively high relative humidity is at least six months. This shelf life was successfully increased to at least 18 months when the encapsulated cells were deep-frozen in liquid nitrogen. Only 10% deterioration of such beads was observed after passing through simulated gastrointestinal conditions (Mattila-Sandholm et al. 2002, Mortazavian et al. 2007).

4.4 Changes in Starter Culture Metabolic Activity

By the encapsulation of starter cultures metabolic activity, including acidification rate and aroma profile of the final product (especially in fermented milks), could be altered and controlled. Acidification rate highly influences viability of probiotics, texture development and quality, flavor quality and fermentation time. Specific encapsulation of probiotic (even traditional yogurt bacteria) cells can result in more desirable rates of cellular metabolic activity. For example, yogurt production with encapsulated traditional yogurt bacteria (*Streptococcus salivarius* ssp. *thermophilus* and *Streptococcus delbrueckii* ssp. *bulgaricus*) resulted in a product with relatively fixed sensory properties as well as greater viability of bacteria (Krasaekoopt et al. 2003). In this regard, size and thickness of capsules considerably affect the lactose absorption rate through the capsules and subsequently, acid production rate by the starters. Excessive fermentation times could be uneconomical.

4.5 Improving and Fixing Sensory Characteristics of Food Products

Microencapsulation of probiotics could results in improvement and/or fixation of sensory properties of the final food product in following ways:

- Because acidification rate and pH drop in fermented products (such as yogurt) produced by encapsulated starters is much slower than in those produced by free ones, products with controlled acidification during fermentation and storage can be produced. Therefore, using encapsulated starters would result in milder taste and inhibiting sharp- and over acidification. Also, post-acidification during the storage time, especially for those products kept over refrigeration temperatures, would be considerably avoided. Therefore, microencapsulation of starter cultures leads to flavor fixation of fermented products. For instance, no significant change in sensory properties of yogurt containing encapsulated *B. bifidum* was observed after three weeks of refrigerated storage at 4°C (Ohno et al. 1995). In a method for the production of yogurt with fixed flavor, first, the pH of yogurt milk is declined to the desired amount; then, the yogurt bacteria is inactivated by sufficient heat treatment and finally, the encapsulated probiotic cells are added (Kurmann and Robinson 1991).

- Acetic acid produced by *Bifidobacterium* spp. gives a vinegar taint to the fermented probiotic products such as yogurt (Kurmann and Robinson 1991). Therefore, microencapsulation of bifidobacteria would substantially solve this problem (Tamim et al. 2005, Mortazavian et al. 2006a). Other probiotics might also generate different types of off flavors in food products, especially when the moisture and the storage temperature are high enough to allow the cells to be metabolically active. Therefore, the addition of probiotic cells in such products in the encapsulated form, apart from its suitable impact on cell viability, could overcome the mentioned problem.

5. Major Components Used in Probiotic Microencapsulation

Different polymers might be used for microencapsulation of probiotics. In most cases, the combination of polymers (either in the form of polymer mixture or polymer coating) produces more appropriate results. Table 3 represents types of polymer used for probiotic microencapsulation along with their advantages and disadvantages.

5.1 Alginate and its Combinations

Alginate is the most popular polymer used for microencapsulation purposes because of its biocompatibility, cost-effectiveness for large scale production, simple structure and the proper digestive characteristics (Shah et al. 2011). Alginate is an algal originated polymer that is widely used in this purpose. It consists of two monomer units consisting of D-mannuronic and L-guluronic acids. Calcium incorporation into the polymer backbone induces a rigid egg-box structure for encapsulation of probiotic bacteria, especially in the concentration of 0.5–4% (Sheu and Marshall 1993, Jankowski et al. 1997, Khalil and Mansour 1998, Hansen et al. 2002, Krasaekoopt et al. 2004, Mortazavian et al. 2007). Calcium alginate gel structure is formed *via* interfacial polymerization process. By this method, calcium alginate precipitates will be formed by the infusion of calcium ions within the alginate backbone (Anal and Singh 2007). One of the popular approaches in probiotic encapsulation is the application of cross-linked alginate owing to its poly-anionic nature. Combination of alginate with other compounds resulted in higher density and strength compared to the polymer alone, simultaneously better controlled release of probiotics at target sites (Marx 1989). One of the main advantages of calcium alginate micro-beads is their surface relative porosity which facilitates proper diffusion. By this property, the extrinsic nutrients can

Table 3. Types of polymers used for probiotic microencapsulation; their advantages and disadvantages.

Type of polymers	Advantages	Disadvantages	Reference
Alginate	Biocompatibility Cost-effectiveness for large scale production Simple structure Proper digestive characteristics Maintaining the high integrity of calcium alginate beads in milk-based products	Low physical stability and leakage of calcium at highly acidity environment or in the presence of chelators	Lakkis (2007)
Resistant starch	Enhances cells viability through upper gastrointestinal tract Absorbing site for probiotic adherence to capsules' shell and digestive tract Prebiotic role for colonic bacteria	It can be used as a nutrient source for probiotic bacteria (due to its prebiotic role)	Mortazavian et al. (2007)
Xanthan-gelan mixture	Acid resistance	High gel-setting temperature of gelan and thermal injuries to viable cells	Lakkis (2007)
k-carrageenan	Its combination with locust bean gum shows negligible susceptibility to organic acids and is an appropriate candidate for probiotic encapsulation in fermented foods	Acid sensitive Inhibitory effect of monovalent used for polymer hardening on traditional fermenting bacteria Brittle gels which are susceptible to stresses of bacterial growth and agitation shear forces Large amount of potassium ions are not recommended in human diet	Lakkis (2007), Necas and Bartosikova (2013)
Cellulose acetate phthalate	Safety record for pharmaceutical and dietary supplements Insoluble at acidic pH (less than 5)	Beads' formation cannot be achieved by ionotropic gelation Capsules are formed by emulsification	Lakkis (2007), Gbassi and Vandamme (2012)
Chitosan	Biocompatibility Biodegradability Non-toxicity	Acid sensitive and non-efficient effect in viability improvement of bacterial cells other than in combination with other polymers such as alginate Inhibitory effects on bacteria because of its poly-cationic nature and the possibility of free cations binding to negative charge of bacterial surface	Groboillot et al. (1993), de Vos et al. (2010)
Whey protein	Acid resistant Amphoteric character depending on the pH of medium	Thermal sensitivity	Gbassi and Vandamme (2012), Dong et al. (2013)
Casein	Thermal resistance	Acid solubility at pH below 4.6 (P_I of casein)	Dhanasingh and Nallaperumal (2010)

enter the beads where they can be used by core probiotics, and metabolites produced by core cells can be released to the external medium across the capsule layer (Anal and Singh 2007, Kailasapathy et al. 2008).

Alginate beads possess some disadvantages. They are susceptible to acidic environments and undergo crackling and loss of mechanical stability in lactic acid-containing environments (Eikmeier and Rehm 1987, Roy et al. 1987, Audet et al. 1988, Ellenton 1998). Their integrity deteriorates when subjected to monovalent ions or chelating agents which absorb calcium ions such as phosphates, lactates and citrates (Roy et al. 1987, Smidsrod and Skjak-Braek 1990, Ellenton 1998). Other disadvantages include difficulties in industrial scale up due to their high expenses and presence of cracks on the surface of micro-beads (Gouin 2004). The defects mentioned can be modified by blending alginate with other polymer compounds, coating other compounds on its capsules and using various additives for its structural modifications (Krasaekoopt et al. 2003). Blending alginate with starch is a common practice and apart from good protection from bacterial cells, alginate-starch blends render the advantage of diffusion of micronutrients and metabolites through the capsules, inside and outside of the entrapped cells, thus allowing them to be metabolically active (Jankowski et al. 1997, Sultana et al. 2000, Sun and Griffiths 2000, Hansen et al. 2002, Krasaekoopt et al. 2003). Blending calcium alginate with Hi-maize starch produces capsules with high cell viability (good integrated structure in capsules) as well as the prebiotic effect of the latter compound (Sultana et al. 2000). Alginate-glycerol blends improved survivability of the cells deep frozen with liquid nitrogen and kept at $-20°C$ due to the cryogenic effect of glycerol (Hansen et al. 2002). Coating semi-permeable layers of chitosan polymer (as a poly-cationic compound) around the alginate capsules (which have negative charges) resulted in beads with improved physical and chemical stability with good tolerance against the deteriorative effects of calcium chelating and anti-gelling agents (Zhou et al. 1998, Krasaekoopt et al. 2003). The low molecular weight chitosan diffuses faster into the alginate matrix compared to the high molecular weight one, resulting in the formation of capsules with higher density and strength. Coating of calcium chloride on the alginate capsules caused the generation of more stable beads with a higher protective effect on the probiotic cells (Chandramouli et al. 2004). Similar to chitosan, poly-amino acids such as poly-*L*-lysine (PLL), can form other poly-cationic polymers coated on the alginate capsules and make strong complexes with the alginate matrix (Smidsrod and Skjak-Braek 1990, Champagne et al. 1992, Larisch et al. 1994). In a study, alternative multilayer shells of alginate/PLL were investigated: the first layer of PLL on the capsule surface produces positive charge and the second alginate coat gives the beads' surface a negative charge. This trend can be repeated several times (Marx 1989, Champagne et al. 1992, Larisch et al. 1994). Other similar investigations have been coatings of poly-ethylene-amine and glutaraldehyde (as other types of poly-cationic polymers) on the alginate capsules and cross-linked alginate matrices (produced at low pHs) obtained from modified alginate structures. Although the latter kind of matrix had more density and strength compared with the alginate matrix alone, it was able to successfully release the bacterial cells into the intestine (Marx 1989).

5.2 Resistant Starch

Resistant starch is a retrograded non-digestible form of native starch which is not degraded by amylase and reaches the large intestine in the intact form (Englyst et al. 1992, Morais et al. 1996, Mortazavian et al. 2007). This characteristic, in addition to its functional prebiotic role in probiotic bacteria consumption, makes the biopolymer an appropriate candidate in colon delivery of micro-beads (Situ et al. 2014). Preparation of resistant starch can be achieved by applying the thermal and retrogradation processes on high amylose corn starch (Dimantov et al. 2004, Dundar and Gocmen 2013). Its degradation takes place after fermentation by colonic microorganisms such as bifidobacteria, lactobacilli, streptococci and entrobacteriaceae, producing short chain fatty acids resulting in pH decline (Macfarlane and Cummings 1991, Kleessen et al. 1997, Le Blay et al. 1999).

Resistant starch, as one of the popular polymer and nutrient-rich sources of probiotics, enhances cell viability in the colon. Furthermore, it provides an absorbing site for probiotic adherence to encapsulates' shells and the GIT (Mattila-Sandholm et al. 2002). Moreover, resistant starch is a 'prebiotic' compound (Section 5.6) and confers an additional advantage to micro-capsules from the viability protection point of view. Incorporation of high amylose corn starch (HACS) to alginate capsules resulted in the improved viability of *Lb. acidophilus* and *Bifidobacterium* spp. as compared to the control group without prebiotics (Sultana et al. 2000).

5.3 Xanthan-Gelan Mixture and K-Carrageenan

Xanthan-gelan mixture is a proposed combination for probiotic encapsulation (Paquin et al. 1990, Sanderson 1990, Sultana et al. 2000), with an optimum mixing ratio of 1:0.75 for xanthan:gelan, respectively. Compared to some susceptible polymers, the mixture is acid resistant, but gelan gum cannot be used for microorganism encapsulation because of its high gel-setting temperature (approximately 85°C for 1 h). Therefore, the high temperature will result in thermal injuries of probiotics (Sun and Griffiths 2000).

In contrast to xanthan-gelan mixtures, *k*-carrageenan is a neutral polymer and requires relatively high temperatures (about 60°C or more) to dissolve, especially when high concentrations (2–5%) are used (Klein et al. 1983). Although it has different gelation temperatures depending on its cation content and gelation, the temperature will increase by incorporation of cations especially K^+ to the polymer structure (Kim et al. 2011). Microbial encapsulation is carried out when probiotic slurry at a mild temperature (40–45°C) is added to the polymer solution. Then, further cooling to ambient temperature leads to encapsulation. Micro-capsule hardening takes place by adding monovalent ions such as potassium (Krasaekoopt et al. 2003). KCl is one of the frequently used monovalent forms which show the inhibitory effect on traditional yogurt bacteria including *Streptococcus salivarius* ssp. *thermophilus* and *Lactobacillus delbrueckii* ssp. *bulgaricus* (Audet et al. 1988). Other monovalent ions, such as Rb^+, Cs^+ and NH_4^+, without probiotic inhibitory effects, are introduced for *k*-carrageenan beads preparation. The latter cations produce the stronger microcapsules compared to potassium. Mixture of *k*-carrageenan-locust bean showed negligible susceptibility to organic acids and has an appropriate efficiency in lactic fermented products. Therefore, the mentioned mixture has been widely used for probiotic encapsulation in fermented products (Audet et al. 1988, Arnaud et al. 1992).

5.4 Cellulose Acetate Phthalate (CAP)

CAP has a negative charge owing to the phthalate groups. The presence of ionizable groups makes it soluble at pHs ≥ 6 (Malm et al. 1951). Because the polymer is considered safe for oral consumption, it is widely used for encapsulation purposes (Rao et al. 1989, Krasaekoopt et al. 2003). Studying the freeze dried capsules of *B. pseudolangum* prepared by wax coated cellulose acetate phthalate showed a significant higher survivability passing through the stomach (Rao et al. 1989).

5.5 Chitosan

Chitosan is a crustacean and insect shell originated polymer with cationic nature arising from the amine groups (Kean and Thanou 2010). Depending on its degree of deacetylation and pK, it is soluble at acidic conditions and pHs lower than its pK (Bowman 2006). Further polymerization is done by chemical anionic cross-linkers (Klein et al. 1983, Zhang 2009). It cannot be used for the encapsulation of probiotics because of its non-efficient effect in viability improvement of bacterial cells. In contrast, in most cases, it is used as a coating layer at low concentrations (e.g., 0.4%) (Zhou et al. 1998). For example, using chitosan alone resulted in less stability than in its combination with other co-polymers like starch and any additives (Guerin et al. 2003).

5.6 Prebiotic Compounds

A number of benefits are associated with the incorporation of prebiotics in microencapsulation ingredients, including the ability to enhance the survival of microencapsulated probiotics. As the role and benefits associated with probiotics and prebiotics are further explored and reported, the synergistic relationship between the two has been shown to have a greater impact on the health benefits of probiotic microorganisms, hence conveying health promotion to the hosts. The technological developments made for prebiotic incorporation into microencapsulation techniques have allowed for the integration of synbiotics with the benefits of specific targeting of probiotic microorganisms to sites of interest within the GIT, enhanced survival and growth promotion of probiotic cells.

6. Microencapsulation Effectiveness of Probiotics

Regarding the main functions of probiotic microencapsulation (Section 4), the effectiveness of this process includes three aspects:

- Greatest protection from viability of cells during production and storage as well as during their transition through the GIT till the target location
- Complete digestion of the microcapsules at GIT target place (but not any other places until, which is also called 'controlled release')
- Improvement of the sensory properties of the food product, or at least, avoidance of negative side effect on the mentioned attributes.

The critical point is that the design of micro-capsules must face both stability (in product and GIT before the target site) and instability (at target GIT location). If the micro-beads possess a high level of stability but are not digested at their destination, the encapsulation concept would be totally meaningless and the probiotics cells will eliminated from the intestine. Therefore, for engineering microcapsules, it is highly recommend that simulated GIT tests to assess the digestibility of micro-beads at the target site be carried out. Not only digestibility, but digestion time is also very critical. For instance, it has been reported that the release of probiotic cells from calcium alginate and chitosan-coated alginate-starch (CCAS) capsules in simulated small intestine conditions (in porcine GIT contents, at 37°C) occurred within 1 hour, whereas nearly 8 hours were required in simulated colon conditions and partial release in the duodenal content was found after 10 hours of incubation (Iyer et al. 2004). Time release of *Lb. casei* strain Shirota (LCS) from CCAS capsules was also investigated in different regions of the GIT including under simulated ileum, colon, duodenal, jejunal and stomach conditions. Most of the cells were released in the ileum at 8 hours; 12 hours were required for complete release in the colon and partial release in the duodenal and jejunal, while no significant release occurred in stomach even after 24 hours (Iyer et al. 2005). Some key factors affect the efficiency of probiotic encapsulation, which are explained below.

6.1 Compatibility of Encapsulating Materials with Surrounding Environment

Using the proper materials for encapsulation purposes with regard to the interfering parameters such as surrounding environment (food matrix or GIT conditions) should not be avoided. For example, fixing the calcium ions within the alginate backbone is a fundamental principal. Therefore, using calcium alginate in high acidity environments or a medium containing chelating agents can lead to the leakage of cations. On the contrary, milk-based products (due to the high levels of calcium) maintain the micro-beads' integrity (shape and structure) and inhibit calcium leaching from the initial structure (Hansen et al. 2002). In parallel, incorporation of a second layer on alginate capsules can improve its physicochemical properties. By this method, unwanted calcium releases, following its interactions, can be inhibited. Moreover, the second layer increases complex mechanical strength (Martinsen et al. 1989).

It is evident that the lower acidic condition of fermented foods such as yogurt has much lower detrimental effects on micro-beads than gastric juice (Sultana et al. 2000, Hansen et al. 2002). For example, it is reported that alginate micro-beads with a mean diameter of 100 μm are relatively stable in various fermented products, but they usually cannot tolerate gastric environments (Cui et al. 2000). Bacteria could grow or become active within the micro-bead gel structure and secrete their metabolites to the internal space of micro-beads. It is reported that the bacteria loading can be more than 10^{11} cfu/g in some cases such as pectin and alginate coating or even less for starch based" (Champagne and Kailasapathy 2008). A contradictory result in Cheddar cheese was reported: encapsulated probiotics showed a decreased survivability compared to free cells. Nature of matrix has a significant role in probiotics survivability. If the matrix has a soft tissue, byproducts of the bacteria can penetrate through the matrix. Otherwise, metabolites such as organic acids entrap within the micro-beads would make them weaker and can cause to their loss compared to uncoated (free) cells; because encapsulated bacteria face double barriers, coating as well as the rigid matrix. Therefore, high acid concentration led to the inhibition of entrapped cells (Kailasapathy 2003, Shah et al. 2011).

For *in vivo* effectiveness, the encapsulating materials should not be digested either in the food products or in the GIT; they should be well digested only once situated in the target position. Such a characteristic is usually difficult to achieve and needs fine engineering in design. The designers must consider all the chemical characteristics of the encapsulating materials and their interactions with each other as well as with surrounding environmental compounds. For instance, resistant starch is an efficient component for the purpose of probiotics encapsulation, because it is not dissolved or decomposed in gastric acid, by neutral pH or by the enzymatic activity of pancreas, but releases its cells when the beads enter the intestine (Englyst et al. 1992).

6.2 Size and Thickness of Micro-Beads

The bead's diameter (especially, shell or coat) is an effective parameter in viability improvement of a core probiotic. It has been experienced that excessive reduction of bead diameters could lead to a lower or non-protective effect on the core materials (Sultana et al. 2000). For example, it has been reported that survivability of encapsulated probiotics with alginate capsules under acidic-bile conditions showed no significant difference when the diameter of gel-beads were 20 and 70 μm, compared with the bigger sizes (Sultana et al. 2000). The diameter of the capsules determines the lactose absorption rate through the capsules and subsequently, acid production rate by the starters. The mentioned factor also has an important effect on the cell-release rate from micro-beads. It has been reported, that for the beads with a small diameter (0.5–1.0 mm), cell release and acid production rates are carried out with higher speed (Krasaekoopt et al. 2003). It has also been reported that reduced oxygen permeation to microencapsulated cells reduced the loss of viability for oxygen sensitive strains (Talwalkar and Kailasapathy 2003).

Regarding polymers used for encapsulation and type of microorganisms, increasing the bead's diameter more than the specific level had no significant effect on the cell's viability (Hansen et al. 2002). The capsule's diameter plays a significant role in substrate absorption and subsequent metabolite release, and as a result, the viability of cells. Gas permeability of shells is a determinative point. Reduced oxygen permeation favors the higher viability of oxygen sensitive microorganisms (Talwalkar and Kailasapathy 2003) and helps to improve tolerance of bacteria such as bifidobacteria to environmental oxygen (McMaster et al. 2005). It has been reported that bacterial cell release across the micro-beads in their target place depends relatively on the capsule's size; capsules with the smaller diameters induced cell release and acid production at higher speeds (Mortazavian et al. 2007).

Micro-capsule size indirectly affects cell load (the number of entrapped cells in micro-beads) in them. By increase in cell load, quantitative efficiency of encapsulation would increase. It should be underlined that very small beads might result in declined viability of contained cells due to lack of internal space for them to absorb nutrients and secrete metabolites. Very large beads are also not very effective, because on one hand, softening and un-integrity might be the result, and on the other hand, by any probable damage, high cell load would be subject to detrimental surrounding factors. Moreover, excessive increase in capsule diameter leads to the emergence of roughness in the mouthfeel of food (Mortazavian et al. 2007).

6.3 Integrity of Micro-Beads

The integrity and strength of microcapsules considerably influence their effectiveness. Loss of the bead's integrity due to deteriorative parameters leads to diffusion of moisture and other fluids which reduces its barrier character against environmental factors (Gouin 2004). The application of alginate alone has some functional drawbacks which can be alleviated in the presence of additional layers (support) or by its structural modification using various additives (Krasaekoopt et al. 2003). For instance, one of disadvantages in alginate microcapsules is their susceptibility to acidic conditions following their cracking, increased porosity and deformation in the presence of organic acids (Eikmeier and Rehm 1987, Roy et al. 1987, Audet et al. 1988, Ellenton 1998). One alternative is coating the core compounds by alginate together with starch. It is reported that the encapsulation efficiency of LAB has been improved by several methods (Jankowski et al. 1997, Sultana et al. 2000, Sun and Griffiths 2000, Hansen et al. 2002, Krasaekoopt et al. 2003). One alternative is coating the core compounds by alginate together with starch. A promising approach is the blending of calcium alginate with high amylose corn starch which result in high cell viability because of inducing microcapsules with an optimum integrity parallel to the prebiotic role of high amylose corn starch which could be used by probiotic bacteria. Although, it is preferred to design the starch layer as an external surface. Because if the prebiotic layer consumed by probiotics, the whole integrity of micro-beads would be disrupted before accessing to target site (Mortazavian et al. 2007). A mixture of alginate-glycerol also improved survivability of liquid nitrogen deep frozen cells stored at –20°C; the cryogenic effect of glycerol made the greatest impact (Hansen et al. 2002). On the other hand, incorporation of other polymers such as chitosan, by using electrostatic interactions, can also improve alginate physicochemical characteristics. It is shown that the coating of poly-anionic alginate by chitosan (with a semi-permeabe and poly-cationic nature) led to improved physical and chemical characters. Inaccessibility of calcium ions in the first layer (interior), out of reach of chelators and anti-gelling agents, is one of the main factors. Similar results could be achieved in the presence of other poly-cationic polymers such as poly-*L*-lysine (PLL) as a poly-amino acid. Followed by ionic bonds, strong interactions form, and the mentioned advantages would appear (Martinsen et al. 1989, Mortazavian et al. 2007). In this regard, coating of alginate micro-beads by poly-ethylene-amine (poly-cation), cross-linked by glutaraldehyde, is another example.

High precision and paying attention to the processing factors during microencapsulation such as freezing (cryogenic freezing or freeze drying), spray drying, mixing, micronization, and to storage conditions are necessary in order to avoid injuries to the beads and contained cells. Process factors can also affect some important parameters related to bead effectiveness such as bead diameter (Mortazavian et al. 2007).

7. Evaluation of Microencapsulation Effectiveness of Probiotics

Microencapsulation effectiveness of probiotics can be evaluated by assessing different quality parameters such as the following.

7.1 Viability of Probiotic Cells

Encapsulation efficiency with regard to cell viability against deteriorative environmental factors has been measured in practice. The method is based on the calculation of cell loss kinetic in the product and its further changes through the GI tract at simulated conditions over time. Fermented milk contains viable microorganisms with metabolic activity during storage. Therefore, the product's condition is not stable over time; the logarithmic loss of cells does not show a linear trend and it needs more sophisticated calculations than the usual cases in thermal bacteriology. In this situation, D-value has to be calculated by the rule of logarithmic loss at different conditions (as defined in thermal bacteriology), so that the beads are introduced to the artificial static conditions. For example, D-value must be calculated at three distinct periods (in product, simulated gastric medium and simulated intestinal tract). For product preparation, lactic acid is used and for simulated gastric conditions, hydrochloric acid is added to the medium at the base level to supply pH of 1.5–2.0. A simulated intestinal tract is prepared by the addition of a phosphate buffer, pancreatic digestive enzymes and bile salts. It is recommended that the temperature be regulated between 4°C and 37°C (normal body temperature) for simulation of the product and GI tract, respectively (Sun and Griffiths 2000, Dimantov et al. 2004). By this method, kinetic of cell loss is assessable. For assessing the viability protection effectiveness of micro-beads, the beads are digested after being subjected to defined environmental conditions and then, the classic enumeration (cultivation in culture media) is applied.

7.2 Micro-Geometrical Properties of the Beads

Micro-geometrical properties of the beads include bead size (usually reported as an average bead size), shape, integrity and uniformity in shape and size. Beads size can be measured using a laser diffractometer or the light scattering technique (Mortazavian et al. 2007). Direct observation, using microscopy technique (light microscopy or scanning electron microscopy, SEM), sieving or membrane filtration of the encapsulated mix can also be used. For sieving, sequential sieving (150 μm, 500 μm and 1 mm) is recommended. The beads are mainly spherical or elliptical in shape. However, these can be observed directly by the two microscopic techniques. To determine integrity and uniformity of the beads, studying the pores, cracks and voids of the bead surface by SEM or light scattering technique is important. Uneven surfaces have a higher light scattering index. Apart from the SEM method, the uniformity of the beads from the size point of view could be considered by sieving, as mentioned above (Sultana et al. 2000, Mortazavian et al. 2007). In the evaluation of micro-geometrical properties of the beads, self-aggregated particles that do not carry any cells should not be mistaken for the real micro-beads. This can be evident, particularly for starch particles and small barely granules (Sultana et al. 2000, Hansen et al. 2002, Mortazavian et al. 2007).

7.3 Cell Load of Micro-Beads and their Dispersibility within the Product

To evaluate mean of the beads' cell load (mean number of cells in each bead), direct observation of the beads from different samples by the SEM method is recommended. SEM and light scattering methods can also be used to evaluate the homogeneity of bead distribution throughout the product (Mortazavian et al. 2007).

Recently, it has been claimed that confocal microscopy has provided the opportunity to 'view' the internal capsular environment with detailed imaginary of probiotic cellular distribution, and to obtain a cell quantification using imaginary analysis software (Moore and Kailasapathy 2011).

8. Techniques Used for Microencapsulation of Probiotics

A number of methods exist for the encapsulation of materials including physical approaches such as nozzle/centrifugal extrusion, air suspension and rotating disk/spray drying, and chemical approaches such as interfacial polymerization and phase separation/coacervation (Moore and Kailasapathy 2011). From another view, two basic methods for probiotics encapsulation are: extrusion (droplet) and emulsion (two-phase system) techniques (Krasaekoopt et al. 2003). There are various other methodsin addition to the two mentioned, including lyophlisation, spray drying, spray coating, spray chilling, and freeze drying. A brief description of these methods follows.

8.1 Extrusion Technique

Extrusion is the older and more common technique for microencapsulation of probiotics (King 1995). It is a simple, cheap and non-violent method, in which the cell injuries are minimal at the end of the process and can protect probiotics at the highest level of viability. This method is also biocompatible and flexible in practice (Klein et al. 1983, Gbassi and Vandamme 2012). However, it has an important drawback in that there is no possibility of large scale production. In fact, it cannot be scaled up to an industrial level because of the long time needed for micro-beads preparation. Compared to the emulsion method, this technology produces larger beads in the range of 2–5 mm in diameter. The process consists of multiple stages including: preparation of a polymeric solution, adding probiotic cells/slurry into the medium to obtain cells suspension and extrusion of mixture *via* syringe needle so that the droplets directly fall into the hardening solution. Depending on polymer type, in most cases, the hardening solution contains multivalent cations. In the case of alginate or any other anionic polymers, a three-dimensional network has been formed around cells immediately after dripping of intermediate mixture into the hardening solution by the act of calcium as a cross-linker (Krasaekoopt et al. 2003). Bead size and diameter are affected by various factors such as type of polymer, the concentration and viscosity of the polymer solution, distance of syringe needle to hardening solution and diameter of syringe needle (Martinsen et al. 1989).

8.2 Emulsion Technique

The emulsion method has been successfully used for LAB encapsulation (Audet et al. 1988, Lacroix et al. 1990). Compared to the extrusion method, emulsion techniques can be used with large quantities and the produced beads have the mean diameter of 25 μm–2 mm. In contrast, it is an expensive method due to the requirement of large volumes of vegetable oil for the formation of emulsion (Krasaekoopt et al. 2003). The method is based on the addition of probiotic cells and polymers as a disperse phase to high volumes of a vegetable oil as a continuous phase (Martin et al. 2014). Then, the mixture is homogenized by mechanical forces in the presence of emulsifiers till the water-in-oil emulsion is formed. By the addition of a cross-linker, the water soluble polymer, being insoluble, starts to form a gel structure surrounding the probiotic cells in the oil phase. The smaller particles also form in the water phase. Emulsifiers help the optimum interaction of the two immiscible phases by decreasing the interfacial tension between phases (Mortazavian et al. 2007). The micro-beads formed by this method are usually recovered by membrane filtration (Krasaekoopt et al. 2003). In the case of alginate, a fat soluble organic acid, such as acetic acid, is usually added to the solution to facilitate the gelation of polymer in the presence of calcium via decreasing pH (Mortazavian et al. 2007).

8.3 Other Techniques

'Lyophilization' (freeze drying) is a protective technology of cells during prolonged storage time. Using cryo-protectant agents such as glycerol can improve the survivability of encapsulated cells during the freezing stages (Kearney et al. 1990) due to their effect of decreasing the formation of water crystals through their binding to water molecules (Guerin et al. 2003). Freeze drying showed the least deteriorative effects on cell stability, but is commercially expensive and cannot be easily used at industrial scale. 'Spray drying' is a widespread encapsulation technique for heat resistant cores. Due to the high temperature and osmotic pressure of feeds as well as dehydration stress, it is not so appropriate for probiotic encapsulation, although is a relatively cheap technology and could be scaled up to industrial levels (Fu and Etzel 1995). Thermal processing results in a relative decomposition of bead structure, release of probiotic cells and reduced viability of microorganisms. The cell membrane and cell associated protein damages reduce microbial activity during further processing and during the storage period (Kailasapathy et al. 2008, Anal and Singh 2007). 'Spray coating' is based on the spraying of a coating material solution on the core dispersion and consequent formation of bead structure (Champagne and Fustier 2007). It is an appropriate tool in the formation of multiple shells around cells. 'Spray cooling' is rarely used in probiotic encapsulation; this may be due to the fact that other feasible approaches are more applicable especially at laboratory scale, although it is introduced as an efficient method for encapsulation of other agents (Champagne and Fustier 2007). 'Freeze drying' is a non-thermal and safe technology avoiding core loss through the water phase removal. The system has operated under vacuum conditions and reduces oxidative damages. However, it is an expensive method and is not suitable for any all components. Generally, the low temperature and vacuum line give the opportunity to encapsulate valuable cores by this method (Kailasapathy 2009). By freeze drying, spray drying and fluidized bed drying techniques, the preparation of microencapsulated cell powders or granules is possible (Dimantov et al. 2004).

9. Microencapsulation Drawbacks

Although microencapsulation of probiotics provides different advantages, as mentioned in Section 4, there are also some drawbacks.

9.1 Prolonged Incubation Time

One drawback relating to encapsulated probiotics is the prolonged incubation time in fermented food products. The metabolic activity of enclosed cells considerably falls and their acid production rate slows down in comparison to free cells. It is observed that incubation time of yogurt made with *Lb. casei* and *Lb. acidophilus,* up to the pH point of 5, was estimated at 6 hours for free cells compared to 30 hours for encapsulated cells (Sultana et al. 2000). In fact, the final product has a higher price in relation to the traditional type, due to the larger quantity of cells needed for inoculation because bacterial multiplication within the beads would be limited.

9.2 Deterioration of Flavor and Mouthfeel

Although microencapsulation of probiotics and even traditional yogurt starter bacteria can improve the sensory attributes of the probiotic products, especially fermented types (Section 4.5), its unsuitable usage might lead to the off flavor (changes in the metabolic pathways) and/or off texture of the final product, especially to defects in mouthfeel. For instance, encapsulated *B. longum* and *B. lactis* in milk possessed an especial off flavor, which was absent in the product with the same free cells. This was attributed to the production of small bitter peptides caused by changes in the metabolic pathways of encapsulated cells (Hansen et al. 2002, Krasaekoopt et al. 2006, Khosrokhavar and Mortazavian 2010). Micro-beads with diameters more than the special limit (100–400 µm, particularly more than 1 mm, depending on the type of micro-capsules and type of product) can deteriorate the mouthfeel of products such as liquid milk, yogurt and sour cream due to the appearance of the special sense of coarseness/roughness (Chandramouli et al. 2004).

9.3 Digestion of Micro-Beads by Encapsulated Bacteria

The utilization of some microcapsule compounds (e.g., starch or prebiotic compounds) by some bacteria is an aspect which should also be considered. Bacterial digestion of such molecules resulted in the uncontrolled and fast release of core probiotics (Mortazavian et al. 2007). There are no beneficial results in this regard.

9.4 Economic and Commercialization Considerations

As discussed in Section 8, commercialization and scale-up of microencapsulation techniques is difficult from technological and economical points of view. Most of the investigations are laboratory trials and have not been attempted in industrial scale and space. In the present time, the industrial and commercial production of encapsulated probiotics is not widespread and even universally renowned starter culture producing companies do not normally display these types of probiotic bacteria.

9.5 Ambiguity about Right Release of Encapsulated Cells in Target GIT Location

One of the most important drawbacks of probiotic microencapsulation is the ambiguity about the real fate of encapsulated cells after entry to the GIT. Most of the investigations are performed in simulated GIT instruction rather than real *in vivo* trials. Therefore, the results do not necessarily guarantee the reality. Furthermore, even according to some results of simulated GIT trials, as mentioned in Section 6, the micro-beads might not well-digest or digest in a prolonged time that causes excretion of cells from body (before digestion) rather than their settlement in the intestine.

This aspect can be deemed the biggest question mark on the technique of probiotic microencapsulation.

10. Conclusion

Encapsulation technology of probiotics has drawn great attention in food science in the recent years. Regarding the sensitivity of probiotic cells to deteriorative parameters such as thermal, chemical and mechanical stresses, coating of cells by resistant polymers could alleviate the stressful conditions. According to the available data, functional foods containing probiotics have to consist of more than 10^7 CFU/ml viable cells. Therefore, probiotic encapsulation helps to improve the cells viability and has a drastic advantage in controlled and target release of the bacterial cells. Furthermore, using prebiotics as coating layer result in additional health promotion in human body after their consumption by intestinal flora. However, some aspects such as low cost effectiveness, probability of shell digestion by probiotics as well as fast or delayed release of cells due to type of polymers and digestive potency of the medium, organoleptic attributes, and extended incubation time are some of the factors that have to be researched for better outcome. Overall, if all these aspects are to be considered in micro-beads preparation, the use of microencapsulated probiotics will have a promising result in the field of food technology.

Keywords: Controlled release, encapsulation, probiotics, prebiotic, prolonged shelf life, sensory, viability

References

Adhikari, K., Mustapha, A., Grun, I. and Fernando, L. 2000. Viability of microencapsulated bifidobacteria in set yogurt during refrigerated storage. Journal of Dairy Science 83: 1946–1951.
Anal, A.K. and Singh, H. 2007. Recent advances in microencapsulation of probiotics for industrial applications and targeted delivery. Trends in Food Science & Technology 18: 240–251.
Arnaud, J.P., Lacroix, C. and Choplin, L. 1992. Effect of agitation rate on cell release rate and metabolism during continuous fermentation with entrapped growing. Biotechnology Techniques 6: 265–270.

Audet, P., Paquin, C. and Lacroix, C. 1988. Immobilized growing lactic acid bacteria with κ-carrageenan-locust bean gum gel. Applied Microbiology and Biotechnology 29: 11–18.

Bouwmeester, H., Dekkers, S., Noordam, M., Hagens, W., Bulder, A., De Heer, C., Ten Voorde, S., Wijnhoven, S. and Sips, A. 2007. Health impact of nanotechnologies in food production. Wageningen: Rikilt, 2007 (Rapport/Rikilt 2007.0.14) pp. 1–91.

Bowman, K. and Leong, K.W. 2006. Chitosan nanoparticles for oral drug and gene delivery. International Journal of Nanomedicine 1: 117–128.

Champagne, C.P. and Kailasapathy, K. 2008. Encapsulation of probiotics. pp. 344–369. *In*: Garti, N. (ed.). Delivery and Controlled Release of Bioactives in Foods and Nutraceuticals. 1st edition. Woodhead Publishing Limited. England.

Champagne, C.P. and Fustier, P. 2007. Microencapsulation for the improved delivery of bioactive compounds into foods. Current Opinion in Biotechnology 18: 184–190.

Champagne, C.P., Gaudy, C., Poncelet, D. and Neufeld, R. 1992. *Lactococcus lactis* release from calcium alginate beads. Applied and Environmental Microbiology 58: 1429–1434.

Chandramouli, V., Kailasapathy, K., Peiris, P. and Jones, M. 2004. An improved method of microencapsulation and its evaluation to protect *Lactobacillus* spp. in simulated gastric conditions. Journal of Microbiological Methods 56: 27–35.

Chou, L.S. and Weimer, B. 1999. Isolation and characterization of acid- and bile-tolerant isolates from strains of *Lactobacillus acidophilus*. Journal of Dairy Science 82: 23–31.

Cui, J.H., Goh, J.S., Kim, P.H., Choi, S.H. and Lee, B.J. 2000. Survival and stability of bifidobacteria loaded in alginate poly-L-lysine microparticles. International Journal of Pharmaceutics 210: 51–59.

De Vos, P., Faas, M.M., Spasojevic, M. and Sikkema, J. 2010. Encapsulation for preservation of functionality and targeted delivery of bioactive food components. International Dairy Journal 20: 292–302.

De Vuyst, L. 2000. Technology aspects related to the application of functional starter cultures. Food Technology and Biotechnology 38: 105–112.

Dhanasingh, S. and Nallaperumal, S.K. 2010. Chitosan/casein microparticles: preparation, characterization and drug release studies. International Journal of Engineering and Applied Sciences 6: 234–238.

Dimantov, A., Greenberg, M., Kesselman, E. and Shimoni, E. 2004. Study of high amylose corn starch as food grade enteric coating in a microcapsule model system. Innovative Food Science and Emerging Technologies 5: 93–100.

Dinakar, P. and Mistry, V. 1994. Growth and viability of *Bifidobacterium bifidum* in cheddar cheese. Journal of Dairy Science 77: 2854–2864.

Dong, Q.Y., Chen, M.Y., Xin, Y., Qin, X.Y., Cheng, Z., Shi, L.E. and Tang, Z.X. 2013. Alginate-based and protein-based materials for probiotics encapsulation: A review. International Journal of Food Science and Technology 48: 1339–1351.

Dundar, A.N. and Gocmen, D. 2013. Effects of autoclaving temperature and storing time on resistant starch formation and its functional and physicochemical properties. Carbohydrate Polymers 97: 764–771.

Eikmeier, H. and Rehm, H. 1987. Stability of calcium-alginate during citric acid production of immobilized *Aspergillus niger*. Applied Microbiology and Biotechnology 26: 105–111.

Ellenton, J. 1998. Encapsulation bifidobacteria. M.S. Thesis, University of Guelph, Ontario, Canada.

Englyst, H.N., Kingman, S. and Cummings, J. 1992. Classification and measurement of nutritionally important starch fractions. European Journal of Clinical Nutrition 46: 33–50.

Fu, W.Y. and Etzel, M.R. 1995. Spray drying of *Lactococcus lactis* ssp. *lactis* C2 and cellular injury. Journal of Food Science 60: 195–200.

Gbassi, G.K. and Vandamme, T. 2012. Probiotic encapsulation technology: From microencapsulation to release into the gut. Pharmaceutics 4: 149–163.

Godward, G. and Kailasapathy, K. 2003. Viability and survival of free and encapsulated probiotic bacteria in cheddar cheese. Milchwissenschaft 58: 624–627.

Gomes, A.M. and Malcata, F.X. 1999. *Bifidobacterium* spp. and *Lactobacillus acidophilus*: Biological, biochemical, technological and therapeutical properties relevant for use as probiotics. Trends in Food Science and Technology 10: 139–157.

Gouin, S. 2004. Microencapsulation: Industrial appraisal of existing technologies and trends. Trends in Food Science and Technology 15: 330–347.

Groboillot, A., Champagne, C., Darling, G., Poncelet, D. and Neufeld, R. 1993. Membrane formation by interfacial cross-linking of chitosan for microencapsulation of *Lactococcus lactis*. Biotechnology and Bioengineering 42: 1157–1163.

Guerin, D., Vuillemard, J.C. and Subirade, M. 2003. Protection of bifidobacteria encapsulated in polysaccharide-protein gel beads against gastric juice and bile. Journal of Food Protection 66: 2076–2084.

Hansen, L.T., Allan-Wojtas, P., Jin, Y.L. and Paulson, A. 2002. Survival of Ca-alginate microencapsulated *Bifidobacterium* spp. in milk and simulated gastrointestinal conditions. Food Microbiology 19: 35–45.

Hickey, M. 2005. Current legislation on probiotic products. pp. 73–97. *In*: A. Tamime (ed.). Probiotic Dairy Products. Blackwell Publishing, Oxford.

Hsieh, Y.H.P. and Ofori, J.A. 2007. Innovations in food technology for health. Asia Pacific Journal of Clinical Nutrition 16: 65–73.

Ijabadeniyi, O.A. 2012. Safety of nanofood: A review. African Journal of Biotechnology 11: 15258–15263.

Iyer, C., Kailasapathy, K. and Peiris, P. 2004. Evaluation of survival and release of encapsulated bacteria in *ex vivo* porcine gastrointestinal contents using a green fluorescent protein gene-labelled *E. coli*. LWT-Food Science and Technology 37: 639–642.

Iyer, C., Phillips, M. and Kailasapathy, K. 2005. Release studies of *Lactobacillus casei* strain shirota from chitosan-coated alginate-starch microcapsules in *ex vivo* porcine gastrointestinal contents. Letters in Applied Microbiology 41: 493–497.

Jankowski, T., Zielinska, M. and Wysakowska, A. 1997. Encapsulation of lactic acid bacteria with alginate/starch capsules. Biotechnology Techniques 11: 31–34.

Kailasapathy, K. 2003. Protecting probiotics by microencapsulation. Microbiology Australia 24: 30–31.

Kailasapathy, K. 2009. Encapsulation technologies for functional foods and nutraceutical product development. CAB Reviews: Perspectives in Agriculture, Veterinary Science, Nutrition and Natural Resources 4: 1–19.

Kailasapathy, K., Versalovic, J. and Wilson, M. 2008. Formulation, administration, and delivery of probiotics. pp. 97–118. *In*: J. Versalovic and M. Wilson (eds.). Therapeutic Microbiology: Probiotics and Related Strategies, ASM Press, Washington DC.

Kean, T. and Thanou, M. 2010. Biodegradation, biodistribution and toxicity of chitosan. Advanced Drug Delivery Reviews 62: 3–11.

Kearney, L., Upton, M. and Mc Loughlin, A. 1990. Enhancing the viability of *Lactobacillus plantarum* inoculum by immobilizing the cells in calcium-alginate beads incorporating cryoprotectants. Applied and Environmental Microbiology 56: 3112–3116.

Kebary, K., Hussein, S. and Badawi, R. 1998. Improving viability of *Bifidobacterium* and their effect on frozen ice milk. Journal of Dairy Science 26: 319–337.

Khalil, A.H. and Mansour, E.H. 1998. Alginate encapsulated bifidobacteria survival in mayonnaise. Journal of Food Science 63: 702–705.

Khosrokhavar, R. and Mortazavian, A. 2010. Effects probiotic-containing microcapsules on viscosity, phase separation and sensory attributes of drink based on fermented milk. Milchwissenschaft 65: 177–179.

Kim, I.Y., Iwatsuki, R., Kikuta, K., Morita, Y., Miyazaki, T. and Ohtsuki, C. 2011. Thermoreversible behavior of κ-carrageenan and its apatite-forming ability in simulated body fluid. Materials Science and Engineering: c 31: 1472–1476.

King, A.H. 1995. Encapsulation of Food Ingredients. pp. 26–39. *In*: Sara J. Risch and Gary A. Reineccius (eds.). Encapsulation and Controlled Release of Food Ingredients.

Kleessen, B., Stoof, G., Proll, J., Schmiedl, D., Noack, J. and Blaut, M. 1997. Feeding resistant starch affects fecal and cecal microflora and short chain fatty acids in rats. Journal of Animal Science 75: 2453–2462.

Klein, J., Stock, J. and Vorlop, K.D. 1983. Pore size and properties of spherical Ca-alginate biocatalysts. European Journal of Applied Microbiology and Biotechnology 18: 86–91.

Korbekandi, H., Mortazavian, A.M. and Iravani, S. 2011. Technology and stability of probiotic in fermented milks containing probiotics and prebiotics. pp. 131–168. *In*: N.P. Shah, A.G.D. Cruz and J.A.F. Faria (eds.). Probiotic and Prebiotic Foods: Technology, Stability and Benefits to Human Health. Nova Science, New York.

Kosin, B. and Rakshit, S.K. 2006. Microbial and processing criteria for production of probiotics: A review. Food Technology and Biotechnology 44: 371–379.

Krasaekoopt, W., Bhandari, B. and Deeth, H. 2003. Evaluation of encapsulation techniques of probiotics for yoghurt. International Dairy Journal 13: 3–13.

Krasaekoopt, W., Bhandari, B. and Deeth, H. 2004. The influence of coating materials on some properties of alginate beads and survivability of microencapsulated probiotic bacteria. International Dairy Journal 14: 737–743.

Krasaekoopt, W., Bhandari, B. and Deeth, H.C. 2006. Survival of probiotics encapsulated in chitosan-coated alginate beads in yoghurt from UHT and conventionally treated milk during storage. LWT-Food Science and Technology 39: 177–183.

Kurmann, J. and Robinson, R. 1991. The health potential of products containing bifidobacteria. pp. 117–157. *In*: R.K. Robinson (ed.). Therapeutic Properties of Fermented Milks, Elsevier, Amsterdam.

Lacroix, C., Paquin, C. and Arnaud, J.P. 1990. Batch fermentation with entrapped growing cells of *Lactobacillus casei*. Applied Microbiology and Biotechnology 32: 403–408.

Lakkis, J.M. 2007. Encapsulation and Controlled Release Technologies in Food Systems, Blackwell Publishing, Minneapolis, 256pp.

Larisch, B., Poncelet, D., Champagne, C. and Neufeld, R. 1994. Microencapsulation of *Lactococcus lactis* subsp. *Cremoris*. Journal of Microencapsulation 11: 189–195.

Le Blay, G., Michel, C., Blottiere, H.M. and Cherbut, C. 1999. Enhancement of butyrate production in the rat caecocolonic tract by long-term ingestion of resistant potato starch. British Journal of Nutrition 82: 419–426.

Lee, K.Y. and Heo, T.R. 2000. Survival of *Bifidobacterium longum* immobilized in calcium alginate beads in simulated gastric juices and bile salt solution. Applied and Environmental Microbiology 66: 869–873.

Lee, Y.K. and Salminen, S. 2009. Handbook of Probiotics and Prebiotics, John Wiley & Sons Publication, New Jersey, 596pp.

Lourens-Hattingh, A. and Viljoen, B.C. 2001. Yogurt as probiotic carrier food. International Dairy Journal 11: 1–17.

Macfarlane, G. and Cummings, J. 1991. The colonic flora, fermentation, and large bowel digestive function. pp. 51–92. *In*: S.F. Phillips, J.H. Pemberton and R.G. Shoeter (eds.). The Large Intestine: Physiology, Pathophysiology and Disease. Raven Press, New York.

Mainville, I., Arcand, Y. and Farnworth, E. 2005. A dynamic model that simulates the human upper gastrointestinal tract for the study of probiotics. International Journal of Food Microbiology 99: 287–296.

Malm, C., Emerson, J. and Hiait, G. 1951. Cellulose acetate phthalate as an enteric coating material. Journal of the American Pharmaceutical Association 40: 520–525.

Martin, M.J., Lara-Villoslada, F., Ruiz, M.A. and Morales, M.E. 2014. Microencapsulation of bacteria: A review of different technologies and their impact on the probiotic effects. Innovative Food Science and Emerging Technologies (In Press), Available online 16 October 2014.

Martinsen, A., Skjak-braek, G. and Smidsrod, O. 1989. Alginate as immobilization material: Correlation between chemical and physical properties of alginate gel beads. Biotechnology and Bioengineering 33: 79–89.

Marx, J.L. 1989. A Revolution in Biotechnology. Cambridge University Press, UK.

Mattila-Sandholm, T., Myllarinen, P., Crittenden, R., Mogensen, G., Fonden, R. and Saarela, M. 2002. Technological challenges for future probiotic foods. International Dairy Journal 12: 173–182.

Mcmaster, L., Kokott, S., Reid, S. and Abratt, V. 2005. Use of traditional African fermented beverages as delivery vehicles for *Bifidobacterium lactis* DSM 10140. International Journal of Food Microbiology 102: 231–237.

Moore, S. and Kailasapathy, K. 2011. Microencapsulation: Science and technologies for probiotics and prebiotics. pp. 435–466. *In*: N.P. Shah, A.G.D. Cruz and J.A.F. Faria (eds.). Probiotic and Prebiotic Foods: Technology, Stability and Benefits to Human Health. Nova Science, New York.

Morais, M.B., Feste, A., Miller, R.G. and Lifschitz, C.H. 1996. Effect of resistant and digestible starch on intestinal absorption of calcium, iron, and zinc in infant pigs. Pediatric Research 39: 872–876.

Mortazavian, A., Ehsani, M., Azizi, A., Razavi, S., Mousavi, S., Sohrabvandi, S. and Reinheimer, J. 2008. Viability of calcium-alginate-microencapsulated probiotic bacteria in Iranian yogurt drink (doogh) during refrigerated storage and under simulated gastrointestinal conditions. Australian Journal of Dairy Technology 63: 25–30.

Mortazavian, A., Ehsani, M., Mousavi, S., Reinheimer, J., Emamdjomeh, Z., Sohrabvandi, S. and Rezaei, K. 2006a. Preliminary investigation of the combined effect of heat treatment and incubation temperature on the viability of the probiotic microorganisms in freshly made yogurt. International Journal of Dairy Technology 59: 8–11.

Mortazavian, A., Ehsani, M., Mousavi, S., Sohrabvandi, S. and Reinheimer, J. 2006b. Combined effects of temperature-related variables on the viability of probiotic microorganisms in yogurt. Australian Journal of Dairy Technology 61: 248–252.

Mortazavian, A., Razavi, S.H., Ehsani, M.R. and Sohrabvandi, S. 2007. Principles and methods of microencapsulation of probiotic microorganisms. Iranian Journal of Biotechnology 5: 1–18.

Mortazavian, A. and Sohrabvandi, S. 2006. Probiotics and food probiotic products; Based on dairy probiotic products. Eta Publication, Tehran.

Necas, J. and Bartosikova, L. 2013. Carrageenan: A review. Veterinarni Medicina 58: 187–205.

Ohno, T., Seki, K., Shibata, N. and Sunohara, H. 1995. Process for producing capsule and capsule obtained thereby. U.S. Patents 5478570 A.

Paquin, C., Lerog, M. and Lacroix, C. 1990. *Bifidobacterium longum* ATCC 15707 production using free and immobilized cell fermentation in whey permeate based medium. Proceeding of the 23rd International Dairy Federation, Brussels, Belgium, p. 321.

Parracho, H., Mccartney, A.L. and Gibson, G.R. 2007. Probiotics and prebiotics in infant nutrition. Proceedings of the Nutrition Society 66: 405–411.

Rao, A., Shiwnarain, N. and Maharaj, I. 1989. Survival of microencapsulated *Bifidobacterium pseudolongumin* simulated gastric and intestinal juices. Canadian Institute of Food Science and Technology Journal 22: 345–349.

Roy, D., Goulet, J. and Le Duy, A. 1987. Continuous production of lactic acid from whey perméate by free and calcium alginate entrapped *Lactobacillus helveticus*. Journal of Dairy Science 70: 506–513.

Saarela, M., Hallamaa, K., Mattila-Sandholm, T. and Matto, J. 2003. The effect of lactose derivatives lactulose, lactitol and lactobionic acid on the functional and technological properties of potentially probiotic *Lactobacillus* strains. International Dairy Journal 13: 291–302.

Sanderson, G. 1990. Gellan gum. pp. 201–232. *In*: P. Harris (ed.). Food Gels. Elsevier Science Publishers, London.

Sekhon, B.S. 2010. Food nanotechnology—An overview. Nanotechnology, Science and Applications 3: 1–15.

Shah, N. 2000. Probiotic bacteria: Selective enumeration and survival in dairy foods. Journal of Dairy Science 83: 894–907.

Shah, N. and Ravula, R. 2000. Microencapsulation of probiotic bacteria and their survival in frozen fermented dairy desserts. Australian Journal of Dairy Technology 55: 139–144.

Shah, N.P., Da Cruz, A.G. and Faria, J.A.F. 2011. Probiotic and Prebiotic Foods: Technology, Stability and Benefits to Human Health. Nova Science Publishers, New York, 545pp.

Sheu, T. and Marshall, R. 1993. Microentrapment of lactobacilli in calcium alginate gels. Journal of Food Science 58: 557–561.

Smidsrod, O. and Skjak-braek, G. 1990. Alginate as immobilization matrix for cells. Trends in Biotechnology 8: 71–78.

Steenson, L.R. and Swaisgood, H.E. 1987. Calcium alginate-immobilized cultures of lactic streptococci are protected from bacteriophages. Journal of Dairy Science 70: 1121–1127.

Sultana, K., Godward, G., Reynolds, N., Arumugaswamy, R., Peiris, P. and Kailasapathy, K. 2000. Encapsulation of probiotic bacteria with alginate–starch and evaluation of survival in simulated gastrointestinal conditions and in yoghurt. International Journal of Food Microbiology 62: 47–55.

Sun, W. and Griffiths, M.W. 2000. Survival of bifidobacteria in yogurt and simulated gastric juice following immobilization in gellan–xanthan beads. International Journal of Food Microbiology 61: 17–25.

Talwalkar, A. and Kailasapathy, K. 2003. Effect of microencapsulation on oxygen toxicity in probiotic bacteria. Australian Journal of Dairy Technology 58: 36–39.

Tamime, A., Saarela, M., Korslund Sondergaard, A., Mistry, V. and Shah, N. 2005. Production and maintenance of viability of probiotic microorganisms in dairy products. pp. 39–72. *In*: A.Y. Tamime (ed.). Probiotic Dairy Products, Blackwell Publishing, Oxford.

Tamime, A.Y., Marshall, V.M. and Robinson, R.K. 1995. Microbiological and technological aspects of milks fermented by bifidobacteria. Journal of Dairy Research 62: 151–187.

Tuohy, K.M., Probert, H.M., Smejkal, C.W. and Gibson, G.R. 2003. Using probiotics and prebiotics to improve gut health. Drug Discovery Today 8: 692–700.

World Health Organization. 2006. Probiotics in food: Health and nutritional properties and guidelines for evaluation. Food and Agriculture Organization of the United Nations, Rome, pp. 1–48.

Zhang, L., Dudhani, A., Lundin, L. and Kosaraju, S.L. 2009. Macromolecular conjugate based particulates: preparation, characterisation and evaluation of controlled release properties. European Polymer Journal 45: 1960–1969.

Zhou, Y., Martins, E., Groboillot, A., Champagne, C. and Neufeld, R. 1998. Spectrophotometric quantification of lactic bacteria in alginate and control of cell release with chitosan coating. Journal of Applied Microbiology 84: 342–348.

10

Probiotic Cereal-Based Fermented Functional Foods

Sultan Arslan[1,2],* and *Mustafa Erbas*[1]

1. Introduction

Functional foods are defined as those that are consumed in food form with the diet, do not contain synthetic compounds and reduce illness risks with their bioactive nutrients. In this way, they help optimize physical and mental well-being (Hardy 2000, Kedia et al. 2007, Noonan and Noonan 2004, Roberfroid 2000, Stanton et al. 2005). Functional foods can be classified as those containing original bioactive compounds, and foods enriched with bioactive substances. Functional components in functional foods includes probiotics, prebiotics and bioactive compounds such as dietary fiber, essential amino acids, polyunsaturated fatty acids, antioxidants, vitamins and minerals (Grajek et al. 2005). Probiotic microorganisms are a group of functional foods that have a beneficial effect on the gastrointestinal system (Ray et al. 2014). The adequate number of probiotic microorganisms for these effects to occur is 10^6–10^7 cfu/g in 100 g of the food product (Rivera-Espinoza and Gallardo-Navarro 2010). The other most important group of functional foods are prebiotic substances. Prebiotics are mostly carbohydrate derivates that are not digested in the upper gastrointestinal system, and are fermented by colon microflora.

In recent years, fermentation is the most common method for production of functional foods. Since cereals are a good source of dietary fiber, as well as being high in prebiotic and mineral content, they are suitable for producing fermented symbiotic food products with probiotic microorganisms. After degradation of phytic acid, a chelating agent, they also have good economic properties and easy availability compared with dairy products. However, there are few probiotic products based on cereals, as most

[1] Akdeniz University, Faculty of Engineering, Department of Food Engineering, 07058, Antalya, Turkey.
 E-mail: erbas@akdeniz.edu.tr
[2] Fırat University, Faculty of Engineering, Department of Food Engineering, 23119, Elazığ, Turkey.
* Corresponding author: sultanarslan04@akdeniz.edu.tr

of these products are dairy based. Also, there is a trend towards non-dairy probiotic products because of vegetarian diets, lactose intolerance and milk-protein allergies.

2. Cereal Fermentation

Fermentation is defined as the degradation of carbohydrates by microorganisms. It is one of the oldest techniques, used by millions of people, for food preservation and increasing food availability in the developing world (Kalui et al. 2010). Fermentation is a process that converts food ingredients with enzymes and microorganisms, and thus improves shelf-life and nutritional value (Mensah 1997, Nout and Motarjemi 1997, Steinkraus 2002). Organoleptic properties of fermented foods are improved with the production of organic acids such as lactic and acetic acid those impart sour taste to foods ensuring consumers' acceptability. Lactic acid, acetic acid, and propionic acid are increased during fermentation of tarhana dough (Erbas et al. 2006). An appropriate selection of substrate and starter culture is necessary to control the metabolic end-products (De Vuyst 2000, Rathore et al. 2012).

Fermentation is a complex process involving biochemical reactions in raw materials. The changes that occur during cereal fermentation mainly result from the enzyme activity of microorganisms (Steinkraus 2002). It is known that fermentation leads to the production of micro-nutrients including volatile and non-volatile aromatic compounds, biomass molecules and respiratory/biosynthetic products, such as lactic acid, carbon dioxide, ethanol and acetyl aldehydes (Annan et al. 2003, Beaumont 2002, Kalui et al. 2010, Steinkraus 2002, Ray and Joshi 2014). Fermentation is generally classified as alcoholic, lactic acid, acetic acid and alkali fermentation (McKay and Baldwin 1990). Cereal fermentation is mostly carried out by lactic acid fermentation.

Recently, consumer demand has necessitated the development of alternative food processes and preservation techniques. Fermentation of cereals decreases malnutrition in people by way of production of microbiological lipids, amino acids, organic acids and vitamins (Achi 1992, Caplice and Fitzgerald 1999, Paredeslopez and Harry 1988, Sanni et al. 1999). It was reported that the available content of lysine, methionine, and tryptophan increased during corn meal fermentation (Blandino et al. 2003). It was observed that the increase in the content of total free amino acids and essential amino acids of fermented tarhana was 57 and 93 percent, respectively (Erbas et al. 2005). It was also observed that the free phenolic acid content of rye bran and peeled rye bran increased by 90 and 30 percent, respectively, after fermentation (Lamsal and Faubion 2009).

One of the substrates commonly used in food fermentation is cereals, which make up 73 percent of the total world harvested area (Salmeron et al. 2009). Cereal grains such as wheat, sorghum, maize, millet and rice are common raw materials for the production of fermented foods, which are known by different names in different regions. In most of these products, the fermentation is spontaneous and includes mixed cultures of yeasts, bacteria, and fungi (Gupta and Abu-Ghannam 2012). Additionally, cereals are rich with minerals such as calcium, phosphorous, potassium and iron, and vitamin B (except B_{12}) such as thiamin, riboflavin and folate, when they are consumed after minimal processing (Capozzi et al. 2012, Ho and Cordain 2000, Katan et al. 2009). It was reported that cereals alone provide 36–43 percent of folate, 16 percent of pyridoxine, 20 percent of riboflavin of the daily intake (Capozzi et al. 2012, Kariluoto et al. 2010), with variability of the cereal source.

On the other hand, phytic acid, which is present in the aleuronic layer of seeds, decreases absorption of minerals. The phytase enzyme, which is produced by lactic acid bacteria and yeast, degrades phytic acid and increases mineral absorption. It was noted that during fermentation lactic acid, bacteria decreased the pH of the media to 5.5 and thus enhanced the activity of phytase (Blandino et al. 2003, Reale et al. 2007). In addition, mixed fermentation of cereals increased water-soluble vitamin content, such as riboflavin, thiamin and niacin (Blandino et al. 2003, Charalampopoulos et al. 2002b, Poutanen et al. 2009). It was determined that the thiamin, niacin and pantothenic acid content of a wheat and maize mixture was increased in fermentation by using probiotic microorganisms (Arslan et al. 2014). In another research, it was confirmed that the thiamin, riboflavin, niacin, vitamin B6 and folic acid content of tarhana dough doubled during fermentation (Certel et al. 2007). Moreover, the amount of folate increased, while the amount of pentosan decreased, during cereal fermentation (Lamsal and Faubion 2009).

Cereals are staple foods worldwide, such that 60 percent of world food production is provided from cereals (Salmeron et al. 2009). Fermentation is the main technology for the processing of cereals. The development of attractive flavor (diacetyl acetic and butyric acid) (Blandino et al. 2003), extended shelf-life and improved digestibility are good reasons for choosing this technique for food processing. In Africa and Asia, cereals commonly used in fermentation are maize, sorghum, rice and millets (Nout 2009). There are numerous of traditional fermented cereal products, depending on the raw material and region. Some of them classified as follows: maize products 'Ogi', 'Togwa', 'Pozol', 'Boza', 'Mawe' and 'Kenkey'; sorghum products 'Tchoukoutou', 'Uji', 'Bushera' and 'Jiu'; millet products 'Ben-Saalga' and 'Jnard', rice products 'Idli', 'Dosa', 'Dhokla' and 'Mifen' and wheat products 'Kiskh', 'Tarhana' (Gupta and Abu-Ghannam 2012, Nout 2009). The traditional fermented cereal products are listed in Table 1. Some of these fermented foods are used as weaning foods for infants and young children in various regions of Africa (Kalui et al. 2010). The production methods of these foods are nearly the same, cleaning the cereal grains, soaking them in water and fermentation. In such fermentations, endogenous grain amylases generate fermentable sugars as an energy source for the fermentation bacteria (Blandino et al. 2003). In nearly all of these products, fermentation is carried out simultaneously and spontaneously by yeast and lactic acid bacteria. *Lactobacillus fermentum, Lb. plantarum, Lb. salivarius, Lb. casei, Lb. acidophilus, Pediococcus pentosaceus* and *Leuconostoc* spp. are some species that can be isolated from cereal-based fermented foods (Kalui et al. 2010, Lei and Jakobsen 2004).

3. Functional Foods

The term 'functional foods' was first used in a project that studied food components and their effects on human health in 1984 in Japan (Ohama et al. 2006) and gained legal status in 1991, when they were named FOSHU (Foods for Specified Health Use) (Sanders 1998). The first applications of functional foods started with the fortification

Table 1. Most common traditional cereal based fermented foods and beverages (Blandino et al. 2003).

Product	Substrate	Microorganism	Nature of use	Regions
Adai	Cereal, legume	*Pediococcus, Streptococcus, Leuconostoc*	Breakfast, snack food	India
Anarshe	Rice	LAB	Breakfast, snack food	India
Ang-kak	Rice	*Monascus purpureus*	Colorant	China, Southeast Asia, Syria
Atole	Maize	LAB	Porridge	Southern Mexico
Bagni	Millet	Unknown	Beverage	Caucasus
Banku	Maize, cassava	LAB, Molds	Dough	Ghana
Bogobe	Sorghum	Unknown	Porridge	Botswana
Bhattejaanr	Rice	*Hansenula anomala, Mucor rouxianus*	Alcoholic paste	India
Boza	Wheat, millet	LAB, *Saccharomyces cerevisiae, Leuconostoc*	Thick, sweet, sour beverage	Albania, Turkey
Braga	Millet	Unknown	Beverage	Romania
Brem	Rice	Unknown	Cake	Indonesia
Brembali	Rice	*Mucorindicus, Candida*	Alcoholic beverage	Indonesia
Burukutu	Sorghum	*S. cerevisiae, S. chavalieri, Leuc. mesenteroides, Candida, Acetobacter*	Alcoholic beverage like vinegar Flavor	Nigeria, Benin, Ghana
Chee-fan	Soybean, wheat	*Mucor, Aspergillus glaucus*	Cheese-like product	China
Chicha	Maize	*Aspergillus, Penicillium,* yeasts, bacteria	Spongy solid	Peru
Chikokivana	Maize, millet	*S. cerevisiae*	Alcoholic beverage	Zimbabwe
Chongju	Rice	*S. cerevisiae*	Alcoholic beverage	Korea
Chinese yeast	Soybeans	*Mucor aceous,* yeasts	Solid	China
Dalaki	Millet	Unknown	Thick porridge	Nigeria
Darassum	Millet	Unknown	Liquid drink	Mongolia
Dhokla	Rice, wheat, bengal gram	*Leuc. mesenteroides, Streptococcus faecalis, Torulopsis candida, T. pullulans*	Steamed cake for breakfast or snack food	Northern India

Table 1. contd....

Table 1. contd.

Product	Substrate	Microorganism	Nature of use	Regions
Doro	Finger millet malt	Yeasts, bacteria	Alcoholic drink	Zimbabwe
Dosa	Rice, bengal gram	*Leuc. mesenteroides, Streptococcus faecalis, Torulopsis candida, T. pullulans*	Griddled cake for breakfast or snack food	India
Hamanatto	Wheat, soybeans	*Aspergillus oryzae, Streptococcus, Pediococcus*	Raisin-like, soft flavouring agent	Japan
Idli	Rice, black gram	*Leuc. mesenteroides, Entorococcus, Torulopsis*	Steamed cake for breakfast food	South India, Sri Lanka
Ilambazi	Maize	LAB, yeast, molds	Porridge as weaning food	Zimbabwe
Injera	Sorghum, maize	*Candida guilliermondii*	Bread-like staple	Ethiopia
Jaanr	Millet	*Hansenula anomala, Mucor rouxianus*	Alcoholic paste mixed with water	India, Himalaya
Jalebies	Wheat flour	*Saccharomyces bayanus*	Syrup-filled confection	India, Nepal, Pakistan
Jaminbang	Maize	Yeasts, bacteria	Bread, cake-like	Brazil
Kaanga	Maize	Bacteria, yeasts	Soft, slimy	New Zealand
Kachasu	Maize	Yeasts	Alcoholic beverage	Zimbabwe
Kaffir bira	Kaffir corn	LAB, yeasts	Alcoholic drink	South Africa
Kanji	Rice and carrots	*Hansenula anomala*	Liquid added to vegetables	India
Khanomjen	Rice	*Lactobacillus, Streptococcus*	Noodle	Thailand
Khaomak	Rice	*Rhizopus, Mucor, Saccharomyces*	Alcoholic sweet Beverage	Thailand
Kichudok	Rice	*Saccharomyces*	Steamed cake	Korea
Kishk	Wheat, milk	*Lb. plantarum, Lb. brevis, Lb. casei, Bacillus subtilis,* yeast	Solid, dried balls, dispersed rapidly in water	Egypt, Syria, Arabian countries
Kisra	Sorghum	Unknown	Staple as bread	Sudan
Kwunu	Millet	LAB, yeast	Paste used as breakfast dish	Nigeria
Kurdi	Wheat	Unknown	Solid, fried crisp	India

Name	Substrate	Microorganisms	Use/Form	Country
Lao chao	Rice	*Rhizopus oryzae, R. chinensis*	Paste, soft, juicy	China, Indonesia
Mahewu	Maize	*Streptococcus lactis*	Solid staple	South Africa
Mawe	Maize	LAB, yeast	Basis for preparation of dishes	South Africa
Mangisi	Millet	Unknown	Sweet-sour non-alcoholic drink	Zimbabwe
Mantou	Wheat flour	*Saccharomyces*	Steamed cake	China
Me	Rice	*Lactobacillus*	Sour food ingredient	Vietnam
Merissa	Sorghum, millet	*Saccharomyces*	Alcoholic drink	Sudan
Minchin	Wheat gluten	*Aspergillus, Cladosporium, Fusarium*	Solid as condiment	China
Mirin	Rice, alcohol	*Aspergillus oryzae, A. usamii*	Alcoholic liquid seasoning	Japan
Miso	Rice, soybeans	*Aspergillus oryzae, Torulopsis, Lactobacillus*	Paste used as seasoning	Japan, China
Mutwiwa	Maise	LAB, bacteria, moulds	Porridge	Zimbabwe
Munkoyo	Kaffir corn, millet	Unknown	Liquid drink	Africa
Nan	Wheat flour	*S. cerevisiae*, LAB	Solid as snack	India, Pakistan, Afghanistan, Iran
Nasha	Sorghum	*Streptococcus, Lactobacillus, Candida, S. cerevisiae*	Porridge as a snack	Sudan
Papadam	Black gram	*Saccharomyces*	Breakfast or snack food	India
Pito	Maize, sorghum	*Geotrichum candidum, Lactobacillus, Candida*	Alcoholic dark brown drink	Nigeria, Ghana
Pozol	Maise	Molds, yeasts, bacteria	Spongy dough formed into balls	Southeasters Mexico
Puto	Rice	*Leuconostoc mesenteroides, Strepromyces faecalis*, yeasts	Solid paste as seasoning agent, snack	Philippines
Rabdi	Maize, buttermilk	*Pediococcus acidilactici, Bacillus, Micrococcus*	Semi-solid mash	India
Sake	Rice	*Saccharomyces sake*	Alcoholic beverages	Japan

Table 1. contd....

Table 1. contd.

Product	Substrate	Microorganism	Nature of use	Regions
Sekete	Maize	*S. cerevisiae, S. chevalieri, S. elegans, Lb. plantarum, Lb. lactis. Bacillussubtilis, Aspergillus niger, A. flavus, Mucor rouxii*	Alcoholic beverage	Nigeria
Shaosinghjiu	Rice	*S. cerevisiae*	Alcoholic clear beverage	China
Shoyu	Wheat, soybeans	*Lactobacillus, Aspergillus, Zygosaccharomyces rouxi*	Liquid seasoning	Japan, China, Taiwan
Sierra rice	Rough rice	*Aspergillus flavus, A. candidus, Bacillus subtilis*	Brownish-yellow dry rice	Ecuador
Sorghum beer	Sorghum, maize	*Lactobacillus*, yeast	Liquid, acidic, weakly alcoholic	South Africa
Soybean milk	Soybeans	*Lactobacillus*	Drink	China, Japan
Takju	Rice, wheat	*Lactobacillus, S. cerevisiae*	Alcoholic turbid drink	Korea
Talla	Sorghum	Unknown	Alcoholic drink	Ethiopia
Tao-si	Wheat, soybean	*Aspergillus oryzae*	Seasoning	Philippines
Tapaipulut	Rice	*Chlamydomucor, Endomycopsis, Hansenula*	Alcoholic dense drink	Malaysia
Tapuy	Rice	*Saccharomyces, Mucor, Rhizopus, Aspergillus*	Sour sweet alcoholic drink	Philippines
Tarhana	Wheat, yoghurt	LAB, *S. cerevisiae*	Solid powder, dried seasoning	Turkey
Tauco	Cereals, soybeans	*Rhizopus oligosporus, Aspergillus oryzae*	Seasoning	West Java (Indonesia)
Tesgüino	Maize	Bacteria, yeasts, molds	Alcoholic beverage	North Western Mexico
Thumba	Millet	*Endomycops infibuliger*	Liquid drink	Eastern India
Tobwa	Maize	*Lactobacillus*	Non-alcoholic drink	Zimbabwe
Uji	Maize, sorghum, millet	*Leuc. mesenteriodes, Lb. plantarum*	Porridge as a staple	Kenia, Uganda, Tanganyika

LAB: Lactic acid bacteria.

of foods with minor nutrients such as ascorbic acid, tocopherol, water-soluble vitamins, folic acid, zinc, iron and calcium (Gupta and Abu-Ghannam 2012, Sloan 2002).

According to The European Commission Concerted Action on Functional Food Science (FUFOSE), functional foods must be appropriate for daily diet in the form of food rather than as tablets and have the effect of improving health and reducing the risk of disease, besides their nutritious effects; these functions must be proven scientifically (Roberfroid 2007). Functional foods and nutrition have important effects on the regulation of body functions. Recently, food scientist and technologists have focused on the production of functional foods including bioactive compounds such as vitamins, minerals, dietary fiber, omega-3 fatty acids and sterols (Angelov et al. 2006, Betoret et al. 2003, Blandino et al. 2003, Rivera-Espinoza and Gallardo-Navarro 2010, Saarela et al. 2000).

3.1 Probiotic Microorganisms

Probiotics are defined as "live microorganisms which, when administered in adequate amounts confer a health benefit on the host" according to the FAO (Food and Agriculture Organization) and WHO (World Health Organization). Probiotics are also defined as microorganisms that improve microflora of host and thus show beneficial effects (Betoret et al. 2003, Charalampopoulos et al. 2002a, FAO/WHO 2001, Prado et al. 2008, Rivera-Espinoza and Gallardo-Navarro 2010). The first few studies related to the positive effects of bacteria were by Eli Metchnikoff, who suggested that "the dependence of the intestinal microbes in the food makes it possible to adopt measures to modify the flora in human bodies and to replace the harmful microbes by useful microbes" (FAO/WHO 2001).

The number of scientific studies related to probiotics has increased due to the improvement of identification techniques (Prado et al. 2008). The most documented effects of probiotics are related to the treatment of intestinal disorder and diarrhea (Lamsal and Faubion 2009). Probiotic bacteria have multiple and simultaneous effects on the host and they regulate the digestive system with their lipolytic, proteolytic and β-galactose activity (Prado et al. 2008). The antimicrobial activity of the probiotic microorganisms arises from reducing the pH in the colon by the production of organic acids such as acetic and lactic acids, bacteriocins and other antimicrobial compounds. Thus, they block bacterial adhesion to the epithelial cells and decrease the pathogen population in the gut (Ng et al. 2009).

Nowadays specific strains of *Lactobacillus acidophilus, Lb. fermentum, Lb. plantarum, Lb. rhamnosus, Lb. casei, Bifidobacterium longum, B. bifidum, B. breve* and *Saccharomyces boulardii* are certified as probiotic and they are commonly used in the production of functional foods (Ray et al. 2014). Most of them were originally isolated from the human gastrointestinal system (Rivera-Espinoza and Gallardo-Navarro 2010). Microorganisms considered as probiotics are mentioned in Table 2 (Holzapfel et al. 2001). Probiotics must be of sufficient quantities (10^6–10^7 cfu/g) and consumed on a regular basis to show their beneficial effects on the gastrointestinal system (Prado et al. 2008, Rivera-Espinoza and Gallardo-Navarro 2010,

Table 2. Microorganisms considered as probiotics (Holzapfel et al. 2001).

Lactobacillus spp.	*Bifidobacterium* spp.	Other lactic acid bacteria	Nonlactic acid bacteria
Lb. acidophilus	*B. adeloscentis*	*Enterococcus faecalis*[1]	*Bacillus cereus* var. *Toyoi*[1,2]
Lb. amylovorus	*B. animalis*	*Enterococcus faecium*	*Eschericia coli* Nissle
Lb. casei	*B. bifidum*	*Lactococcus lactis*[3]	*Propionibacterium freudenreichii*[1,2]
Lb. crispatus	*B. breve*	*Leuconostoc mesenteroides*	*Saccharomyes boulardii*[2]
Lb. delbrueckii ssp. *bulgaricus*[3]	*B. infantis*	*Pediococcus acidilactici*[3]	*Saccharomyces cerevisiae*[2]
Lb. gallinarum[1]	*B. lactis*[4]	*Sporolactobacillus inulinus*[1]	
Lb. gasseri	*B. longum*	*Streptococcus thermophilus*[3]	
Lb. johnsonii			
Lb. paracasei			
Lb. plantarum			
Lb. reuteri			
Lb. rhamnosus			

[1] Main application for animals.
[2] Applied mainly as pharmaceutical preparations.
[3] Either the microorganism is nonprobiotic or little is known about the probiotic properties.
[4] Probably synonymous with *B. animalis*.

Salmeron et al. 2014). Furthermore, it was determined that dead cells of these microorganisms improve the immune system (Rivera-Espinoza and Gallardo-Navarro 2010, Saarela et al. 2000).

The adult gut is a stable microbial system in normal conditions and provides a living space for beneficial and pathogenic microbes. However, the protective microflora can be affected negatively by environmental factors such as aging, stress, diet, consumption of protein-enriched food and using antibiotics and they can lose their activity over a period of time. For these reasons, the viability of beneficial microorganism must be supplemented and increased for a healthy and robust body system (Gordon 2002, Lamsal and Faubion 2009, Laparra and Sanz 2010).

3.2 Prebiotic Compounds

Prebiotics are short-chain indigestible carbohydrates that are fermented by probiotic microorganisms and ensure the required energy for them in the gut. Thus, they support the selective growth of health-promoting probiotic bacteria such as *Lactobacillaceae* and *Bifidobacteriacea* (Holzapfel and Schillinger 2002, Laparra and Sanz 2010, Lee 2009, Roberfroid 2007). Nowadays, fructooligosaccharides (FOS), galacto-oligosaccharides (GOS), lactulose and inulin are the most often used prebiotics for this purpose. Cereals are rich in prebiotics in comparison with vegetables and fruits.

Cereal bran fibers are fermented much slower than inulin, and so they remain longer in the colon. As a result, the total production of short-chain fatty acids is slightly higher with oat bran than with inulin. Studies showed that cereal bran ensures a more balanced dietary fiber supplement because of lower gas production in the gut (Lamsal and Faubion 2009).

The number of studies based on prebiotics has increased with the growing trend of consumption of functional food ingredients. As a result of these studies, several criteria have been considered for identifying a compound as a prebiotic. Prebiotics must have the characteristics of protecting its chemical properties during the process of digestion, resistance to low acidity and absorption, fermentability by gastrointestinal microflora and supporting selective probiotics against pathogen microorganisms (Laparra and Sanz 2010, Manning and Gibson 2004, Roberfroid 2007).

Prebiotics are desired compounds in food systems due to their physiochemical and organoleptic properties. In many countries, the use of prebiotics is classified as food and food ingredients rather than as food additives. Prebiotic compounds are used in the food process for: improving texture and taste, replacing sugar and fat, increasing resistance to thermal processing, increasing viscosity and water absorption, depressing the freezing point and preventing crystal formation during freezing (Lee et al. 2009, Wang 2009).

4. Functional Fermented Foods and Health Effects

Probiotic microorganisms and prebiotic compounds, with their functional properties, are very important for human wellness and health. They are proposed for the treatment of many health problems alone or in conjunction with other methods (Roberfroid 2007). The summarized health benefit effects of functional foods are given in Table 3.

The most important effect of probiotic microorganisms and prebiotic compounds on health is the prevention of colon cancer. Some bacterial genera of the gastrointestinal microflora produce glycosidase, azoreductase and nitroreductase enzymes which activate carcinogenic compounds. These enzymes cause an accumulation of carcinogenic and mutagenic metabolites which are transferred to the intestine with bile salts. A high concentration of these harmful compounds in the gut also increases the risk of colon cancer. However, probiotic microorganisms living in the gut degrade these carcinogenic compounds and thereby reduce the risk of cancer (Laparra and Sanz 2010, Senok et al. 2005). Additionally, short chain fatty acids such as acetate, propionate and butyrate are produced as a result of probiotic-prebiotic interaction and these compounds inhibit carcinogenic cells. Furthermore, residence time of carcinogenic and mutagenic toxins in the intestine reduces because of the laxative effects of prebiotics.

It was reported that probiotic microorganisms repress gastrointestinal infections, reduce cholesterol and triglycerides, activate and support the immune system, produce vitamins and increase the absorption of minerals. Probiotics produce inhibitory substances (bacteriocins, organic acids, hydrogen peroxide, etc.) those compete against pathogenic microorganisms. Moreover, probiotics produce beneficial composites (bioactive compounds, immunoglobulin proteins, etc.) for the body and the immune system, ensure regeneration of damaged flora after antibiotic therapy and digest lactose in lactase intolerance disease. It has been proven that probiotic microorganisms and

Table 3. The health benefit effect of functional foods.

Agent	Health Effect	Mechanism	Reference
Probiotics and prebiotics	Prevention of colon cancer	Inhibition of carcinogenic and mutagenic compounds, production of short chain fatty acids, reduction of presence time of toxic substance in colon	(Laparra and Sanz 2010, Senok et al. 2005)
Probiotics and prebiotics	Treatment of gastrointestinal disease	Regulation of peristaltic waves of colon *via* production of organic acids and short chain fatty acids	(Lee and Salminen 2009, Saarela et al. 2000, Senok et al. 2005)
Probiotics and prebiotics	Regulation of immune system	Production of immunoglobulin proteins and stimulation of IgA and macrophage content	(Timmerman et al. 2007, Woodcock et al. 2004)
Probiotics and prebiotics	Increase of calcium absorption	Degradation of phytic acid and increase in solubility of calcium ions due to low pH	(Coudray et al. 1997, Greger 1999)
Probiotics	Inactivation of pathogen microorganisms	Decrease of pH, competition for nutrition and adhesiveness, production of antimicrobial components	(Manning and Gibson 2004, Saarela et al. 2000)
Probiotics	Digestion of lactose	Production of lactase enzyme	(Sanders 1998)
Probiotics	Decrease in cholesterol, triglyceride levels of blood and anti-obesity effect	Degradation of the precursor of bile acids and resultant decrease in lipid solubility and absorption	(Delzenne and Kok 1999, Laparra and Sanz 2010, Tahri et al. 1997)

prebiotics are more effective when they are used together because of synergistic effects (Charalampopoulos et al. 2002a, Manning and Gibson 2004, Rivera-Espinoza and Gallardo-Navarro 2010, Saarela et al. 2000, Sanders 1998, Stanton et al. 2005, Wang 2009).

5. Fermented Cereals as a Source of Functional Food

Nowadays, functional foods in markets are mostly milk based, and cereal based functional foods are very limited (Angelov et al. 2006). Yogurt is the main probiotic carrier food in European countries, while more than 53 different types of milk based probiotic foods are consumed in Japan (Lamsal and Faubion 2009). Increasing lactose intolerance, dyslipidaemia, vegetarianism and milk-protein allergy supports the development of non-dairy functional products such as cereals (Blandino et al. 2003, Prado et al. 2008, Ranadheera et al. 2010, Rivera-Espinoza and Gallardo-Navarro 2010, Saarela et al. 2000).

A large part of the daily diet of low-income countries consists of fermented grain products. For this reason, the consumption of fermented cereal products is assumed to be an impairment of social status and creates a negative public opinion. This situation is the most important challenge of fermented cereal foods. Apart from this, there has been little exploration of technologies and how they can be reproduced with the aim of adding value (Kalui et al. 2010). Therefore, cereals must be seriously considered as a functional food substrate with their rich content of micro- and macro-nutrients.

Recently, cereals have been extensively studied as probiotic food carriers. Cereals are a good source due to their content of dietary fiber and β-glucan when they are consumed as whole-grains (Kedia et al. 2007). They contribute about 50 percent of the fiber intake in Western countries (Lambo et al. 2005). It was determined that the resistant starch content of wheat, maize and wheat-maize mixed gruel was 12.03, 20.01 and 27.44 percent, respectively (Quintieri et al. 2012). Whole-grain consumption correlates with decreased risk of diabetes, cardiovascular disease, obesity and cancer. The bran and germ of cereal grains contains a significant amount of bioactive components, such as essential amino acids and vitamins (Lamsal and Faubion 2009).

Cereal-based probiotic carriers increase the resistance of probiotic microorganisms to low pH, gastric juice, digestive enzymes and bile salts while passing through the gastrointestinal system. *In vitro* and *in vivo* studies have shown that survivability rates of probiotics are affected by the food that is used for their delivery (Charalampopoulos et al. 2003, Patel et al. 2004). Food formulations with pH 3.5–4.5 and high buffering capacity increase the acidic conditions to which probiotics are exposed and enhance the stability of probiotics (Charalampopoulos et al. 2002b, Lamsal and Faubion 2009). The growth potential of probiotics in cereal based fermentation media has been reported in different researches. It was determined that the population of *Saccharomyces boulardii*, *Lb. acidophilus* and *Bifidobacterium bifidum*, in probiotic Boza was 6.84, 5.49 and 3.84 $\log_{10}$ cfu/mL respectively, after 24 hours of fermentation (Arslan et al. 2014). The cell population of *Lb. plantarum* probiotic strain reached 6.0 and 8.6 $\log_{10}$ cfu/ mL in barley-malt mixed and single malt media at the end of a 28 hour fermentation (Rathore et al. 2012). Another study showed that the maximum viability of *Lb. plantarum* was in 5.5 percent oats, 1.25 percent sugar and 5 percent fermentation inoculums, 10.4 log cfu/mL (Gupta et al. 2010). Michida et al. (2006) noted that the addition of malt and barley extracts into fermentation medium ensured acidity tolerance and high viability of *Lb. plantarum*. It was determined that oat and malt are the best substrates for lactic acid bacteria (Marklinder and Lonner 1992, Rathore et al. 2012). High survivability was ensured in malt extract medium under refrigeration conditions for 70 days. Among the strains used in these studies, *Lb. acidophilus* had the lowest viability due to its diverse nutritional requirements (Charalampopoulos and Pandiella 2010, Marklinder and Lonner 1992). Furthermore, reports are available that indicated that the growth of *Bifidobacteria* in cereal substrates is not possible without a growth promoter such as milk or yeast extract (Kabeir et al. 2005, Lee et al. 1999, Rozada-Sanchez et al. 2008). It can be concluded that cereals can be fermented by probiotic microorganisms in a sufficient number for acceptance as probiotic foods.

The symbiotic use of probiotics and prebiotics has an effect on wellness and health. Recently, there has been a trend of symbiotic foods, i.e., a combination of prebiotics and probiotics in a single product (Bielecka et al. 2002). It was reported that the feeding of probiotic oat gruel decreased abdominal bloating of irritable bowel disease in patients. Additionally, the concentration of carboxylic acid in faeces increased and the fibrinogen concentration of blood decreased (Lamsal and Faubion 2009). Similarly, adding amylases increased the proliferation of probiotic microorganisms during the fermentation of cereals (Lamsal and Faubion 2009). High-fiber cereal components provide the protection of probiotics against processing and storage (Charalampopoulos et al. 2002b). As a result, cereals are a good source for improving new functional foods due to their availability and economic properties and high content of prebiotic compounds.

Traditional fermented cereal products that contain natural probiotic microorganisms and prebiotic substances have been consumed since ancient times, mostly in Asia and Africa. Some probiotic strains can be isolated from non-dairy fermented products (Rivera-Espinoza and Gallardo-Navarro 2010). Todorov et al. (2008) determined that the *Lb. paracasei* ST284BZ strain, isolated from Boza (traditional cereal fermented food), showed good probiotic properties, such as antiviral and antibacterial activity. Several cereal-based functional foods have been shown at markets all around the world. Some examples of these commercial products include CornyActiv® cereal bars in Germany, Muesli® probiotic flakes in Portugal, Weetaflakes® whole-wheat breakfast cereals in France, wholegrain porridge in the UK, the JovitaProbiotisch® blend of cereals, fruit and probiotic yogurt in Germany, SOYosa® probiotic yogurt-like soy–oat product in Finland, Vita Biosa® Mixture of aromatic herbs in Denmark and the Goodness® snack bar in the UK (Dornblaser 2007, Gupta and Abu-Ghannam 2012, Lamsal and Faubion 2009, Prado et al. 2008).

Although several advances of cereals can be listed in the production of functional foods, there are important factors limiting their nutritional value, such as low protein content, lack of certain amino acids, bioavailability of micronutrients, anti-nutrients (phytic acids, tannins and polyphenols), low starch availability and the coarse nature of the grains (Blandino et al. 2003, Mouquet-Rivier et al. 2008). However, these problems can be overcome with some practical applications. Different cereals have complex nutrient compositions and when mixed, they can considerably modify the properties of the food (Rathore et al. 2012) in such a way that they can complement each other's nutrient requirements and increase bioavailability.

6. Conclusion

Nowadays, the expectation of quality of life, high healthcare costs and the trend for naturally preserving foods are driving factors for research into functional foods. As a result of this desire, functional foods that protect and improve wellness and reduce the risk of illness are preferred by the public. Functional foods are used in daily diet with numerous effects besides the nutritional functions, such as reducing illness risks and improving wealth and wellness. Increasing consumer awareness and changing eating habits due to urbanisation have resulted in a huge demand for functional foods and, for this reason, the functional food market has, day by day, grown rapidly all over the world. Cereals are the most planted raw material in agricultural areas of the world and have good availability and economic properties. Additionally, they supply the needs of prebiotics in the daily diet. For these reasons, cereals must be considered as a substrate for producing functional foods. Studies also indicate that cereals can be fermented by probiotic microorganisms and can increase viability of probiotics in the gastrointestinal system with their buffering capacity.

Keywords: Cereal, Fermentation, Functional foods, Probiotics, Prebiotics

References

Achi, O.K. 1992. Microorganisms associated with natural fermentation of prosopis-africana seeds for the production of okpiye. Plant Foods for Human Nutrition 42: 297–304.

Angelov, A., Gotcheva, V., Kuncheva, R. and Hristozova, T. 2006. Development of a new oat-based probiotic drink. International Journal of Food Microbiology 112: 75–80.

Annan, N.T., Poll, L., Sefa-Dedeh, S., Plahar, W.A. and Jakobsen, M. 2003. Volatile compounds produced by *Lactobacillus fermentum, Saccharomyces cerevisiae* and *Candida krusei* in single starter culture fermentations of Ghanaian maize dough. Journal of Applied Microbiology 94: 462–474.

Arslan, S., Durak, A.N., Erbas, M., Gulsum, U. and Tanriverdi, E. 2014. Determination of microbiological and chemical properties of probiotic boza and its consumer acceptability. Journal of American Colalge of Nutrition. DOI:10.1080/07315724.2014.880661.

Beaumont, M. 2002. Flavouring composition prepared by fermentation with *Bacillus* spp. International Journal of Food Microbiology 75: 189–196.

Betoret, N., Puente, L., Diaz, M.J., Pagan, M.J., Garcia, M.J., Gras, M.L., Martinez-Monzo, J. and Fito, P. 2003. Development of probiotic-enriched dried fruits by vacuum impregnation. Journal of Food Engineering 56: 273–277.

Bielecka, M., Biedrzycka, E. and Majkowska, A. 2002. Selection of probiotics and prebiotics for synbiotics and confirmation of their *in vivo* effectiveness. Food Research International 35: 125–131.

Blandino, A., Al-Aseeri, M.E., Pandiella, S.S., Cantero, D. and Webb, C. 2003. Cereal-based fermented foods and beverages. Food Research International 36: 527–543.

Caplice, E. and Fitzgerald, G.F. 1999. Food fermentations: Role of microorganisms in food production and preservation. International Journal of Food Microbiology 50: 131–149.

Capozzi, V., Russo, P., Duenas, M.T., Lopez, P. and Spano, G. 2012. Lactic acid bacteria producing B-group vitamins: A great potential for functional cereals products. Applied Microbiology Biotechnology 96: 1383–1394.

Certel, M., Erbas, M., Uslu, M.K. and Erbas, M.O. 2007. Effects of fermentation time and storage on the water-soluble vitamin contents of tarhana. Journal of the Science of Food and Agriculture 87: 1215–1218.

Charalampopoulos, D. and Pandiella, S.S. 2010. Survival of human derived *Lactobacillus plantarum* in fermented cereal extracts during refrigerated storage. LWT-Food Science and Technology 43: 431–435.

Charalampopoulos, D., Pandiella, S.S. and Webb, C. 2002a. Growth studies of potentially probiotic lactic acid bacteria in cereal-based substrates. Journal of Applied Microbiology 92: 851–859.

Charalampopoulos, D., Pandiella, S.S. and Webb, C. 2003. Evaluation of the effect of malt, wheat and barley extracts on the viability of potentially probiotic lactic acid bacteria under acidic conditions. International Journal of Food Microbiology 82: 133–141.

Charalampopoulos, D., Wang, R., Pandiella, S.S. and Webb, C. 2002b. Application of cereals and cereal components in functional foods: a review. International Journal of Food Microbiology 79: 131–141.

Coudray, C., Bellanger, J., CastigliaDelavaud, C., Remesy, C., Vermorel, M. and Rayssignuier, Y. 1997. Effect of soluble or partly soluble dietary fibers supplementation on absorption and balance of calcium, magnesium, iron and zinc in healthy young men. European Journal of Clinical Nutrition 51: 375–380.

De Vuyst, L. 2000. Technology aspects related to the application of functional starter cultures. Food Technology Biotechnology 38: 105–112.

Delzenne, N.M. and Kok, N.N. 1999. Biochemical basis of oligofructose-induced hypolipidemia in animal models. Journal of Nutrition 129: 1467S–1470S.

Dornblaser, L. 2007. Probiotics and prebiotics: What in the world is going on? Cereal Foods World 52: 20–21.

Erbas, M., Uslu, M.K., Erbas, M.O. and Certel, M. 2006. Effects of fermentation and storage on the organic and fatty acid contents of tarhana, a Turkish fermented cereal food. Journal of Food Composition and Analysis 19: 294–301.

Erbas, M., Ertugay, M.F., Erbas, M.O. and Certel, M. 2005. The effect of fermentation and storage on free amino acids of tarhana. International Journal of Food Science and Nutrition 56(5): 349–358.

FAO/WHO. 2001. Health and nutritional properties of probiotics in food including powder milk with live lactic acid bacteria, report of a joint FAO/WHO expert consultation on evaluation of health and nutritional properties of probiotics in food including powder milk with live lactic acid bacteria, Basel, Switzerland.

Gordon, D.T. 2002. Intestinal health through dietary fiber, prebiotics, and probiotics. Food Technol. 56: 23–23.

Grajek, W., Olejnik, A. and Sip, A. 2005. Probiotics, prebiotics and antioxidants as functional foods. Acta Biochimica Polonica 52: 665–671.

Greger, J.L. 1999. Nondigestible carbohydrates and mineral bioavailability. Journal of Nutrition 129: 1434S–1435S.

Gupta, S. and Abu-Ghannam, N. 2012. Probiotic fermentation of plant based products: possibilities and opportunities. Critical Reviews Food Science and Nutrition 52: 183–199.

Gupta, S., Cox, S. and Abu-Ghannam, N. 2010. Process optimization for the development of a functional beverage based on lactic acid fermentation of oats. Biochemical Engineering Journal 52: 199–204.

Hardy, G. 2000. Nutraceuticals and functional foods: Introduction and meaning. Nutrition 16: 688–689.

Ho, R.C. and Cordain, L. 2000. The potential role of biotin insufficiency on essential fatty acid metabolism and cardiovascular disease risk. Nutrition Research 20: 1201–1212.

Holzapfel, W.H., Haberer, P., Geisen, R., Bjorkroth, J. and Schillinger, U. 2001. Taxonomy and important features of probiotic microorganisms in food and nutrition. American Journal of Clinical Nutrition 73: 365S–373S.

Holzapfel, W.H. and Schillinger, U. 2002. Introduction to pre- and probiotics. Food Research International 35: 109–116.

Kabeir, B.M., Abd-Aziz, S., Muhammad, K., Shuhaimi, M. and Yazid, A.M. 2005. Growth of *Bifidobacterium longum* BB536 in medida (fermented cereal porridge) and their survival during refrigerated storage. Letters in Applied Microbiology 41: 125–131.

Kalui, C.M., Mathara, J.M. and Kutima, P.M. 2010. Probiotic potential of spontaneously fermented cereal based foods—A review. African Journal of Biotechnology 9: 2490–2498.

Kariluoto, S., Edelmann, M. and Piironen, V. 2010. Effects of environment and genotype on folate contents in wheat in the healthgrain diversity screen. Journal of Agricultural and Food Chemistry 58: 9324–9331.

Katan, M.B., Boekschoten, M.V., Connor, W.E., Mensink, R.P., Seidell, J., Vessby, B. and Willett, W. 2009. Which are the greatest recent discoveries and the greatest future challenges in nutrition? European Journal of Clinical Nutrition 63: 2–10.

Kedia, G., Wang, R.H., Patel, H. and Pandiella, S.S. 2007. Use of mixed cultures for the fermentation of cereal-based substrates with potential probiotic properties. Process Biochemistry 42: 65–70.

Lambo, A.M., Oste, R. and Nyman, M. 2005. Dietary fiber in fermented oat and barley beta-glucan rich concentrates. Food Chemistry 89: 283–293.

Lamsal, B.P. and Faubion, J.M. 2009. The beneficial use of cereal and cereal components in probiotic foods. Food Reviews International 25: 103–114.

Laparra, J.M. and Sanz, Y. 2010. Interactions of gut microbiota with functional food components and nutraceuticals. Pharmacology Research 61: 219–225.

Lee, J.H., Lee, S.K., Park, K.H., Hwang, I.K. and Ji, G.E. 1999. Fermentation of rice using amylolytic *Bifidobacterium*. International Journal of Food Microbiology 50: 155–161.

Lee, Y.K. and Salminen, S. 2009. Handbook of probiotics and prebiotics (2nd ed.). A John Wiley and Sons Inc. Publication, Canada.

Lei, V. and Jakobsen, M. 2004. Microbiological characterization and probiotic potential of koko and koko sour water, African spontaneous fermented millet porridge and drink. Journal of Applied Microbiology 96: 384–397.

Manning, T.S. and Gibson, G.R. 2004. Prebiotics. Best Practice & Research in Clinical Gastroenterology 18: 287–298.

Marklinder, I. and Lonner, C. 1992. Fermentation properties of intestinal strains of lactobacillus, of a sour dough and of a yogurt starter culture in an oat-based nutritive solution. Food Microbiology 9: 197–205.

McKay, L.L. and Baldwin, K.A. 1990. Applications for biotechnology—Present and future improvements in lactic-acid bacteria. FEMS Microbiology Letters 87: 3–14.

Mensah, P. 1997. Fermentation—The key to food safety assurance in Africa? Food Control 8: 271–278.

Michida, H., Tamalampudi, S., Pandiella, S.S., Webb, C., Fukuda, H. and Kondo, A. 2006. Effect of cereal extracts and cereal fiber on viability of *Lactobacillus plantarum* under gastrointestinal tract conditions. Biochemical Engineering Journal 28: 73–78.

Mouquet-Rivier, C., Icard-Verniere, C., Guyot, J.-P., Hassane Tou, E., Rochette, I. and Treche, S. 2008. Consumption pattern, biochemical composition and nutritional value of fermented pearl millet gruels in Burkina Faso. International Journal of Food Science and Nutrition 59: 716–729.

Ng, S.C., Hart, A.L., Kamm, M.A., Stagg, A.J. and Knight, S.C. 2009. Mechanisms of action of probiotics: recent advances. Inflammation Bowel Diseases 15: 300–310.

Noonan, W.P. and Noonan, C. 2004. Legal requirements for 'functional food' claims. Toxicology Letters 150: 19–24.

Nout, M.J.R. 2009. Rich nutrition from the poorest—Cereal fermentations in Africa and Asia. Food Microbiology 26: 685–692.

Nout, M.J.R. and Motarjemi, Y. 1997. Assessment of fermentation as a household technology for improving food safety: A joint FAO/WHO workshop. Food Control 8: 221–226.

Ohama, H., Ikeda, H. and Moriyama, H. 2006. Health foods and foods with health claims in Japan. Toxicology 221: 95–111.

Paredeslopez, O. and Harry, G.I. 1988. Food biotechnology review—Traditional solid-state fermentations of plant raw-materials—Application, nutritional significance, and future-prospects. Critical Reviews in Food Science and Nutrition 27: 159–187.

Patel, H.M., Pandiella, S.S., Wang, R.H. and Webb, C. 2004. Influence of malt, wheat, and barley extracts on the bile tolerance of selected strains of lactobacilli. Food Microbiology 21: 83–89.

Poutanen, K., Flander, L. and Katina, K. 2009. Sourdough and cereal fermentation in a nutritional perspective. Food Microbiology 26: 693–699.

Prado, F.C., Parada, J.L., Pandey, A. and Soccol, C.R. 2008. Trends in non-dairy probiotic beverages. Food Research International 41: 111–123.

Quintieri, L., Monteverde, A. and Caputo, L. 2012. Changes in prolamin and high resistant starch composition during the production process of Boza, a traditional cereal-based beverage. European Food Research Technology 235: 699–709.

Ranadheera, R.D.C.S., Baines, S.K. and Adams, M.C. 2010. Importance of food in probiotic efficacy. Food Research International 43: 1–7.

Rathore, S., Salmeron, I. and Pandiella, S.S. 2012. Production of potentially probiotic beverages using single and mixed cereal substrates fermented with lactic acid bacteria cultures. Food Microbiology 30: 239–244.

Ray, R.C. and Joshi, V.K. 2014. Fermented foods: Past, present and future scenario. pp. 1–36. *In*: R.C. Ray and D. Montet (eds.). Microorganisms and Fermentation of Traditional Foods, CRC Press, Boca Raton, Florida, USA.

Ray, R.C., El Sheikha, A.F. and Sashikumar, R. 2014. Oriental fermented functional (probiotics) foods. pp. 283–311. *In*: R.C. Ray and D. Montet (eds.). Microorganisms and Fermentation of Traditional Foods, CRC Press, Boca Raton, Florida, USA.

Reale, A., Konietzny, U., Coppola, R., Sorrentino, E. and Greiner, R. 2007. The importance of lactic acid bacteria for phytate degradation during cereal dough fermentation. Journal of Agriculture and Chemistry 55: 2993–2997.

Rivera-Espinoza, Y. and Gallardo-Navarro, Y. 2010. Non-dairy probiotic products. Food Microbiology 27: 1–11.

Roberfroid, M. 2007. Prebiotics: The concept revisited. Journal of Nutrition 137: 830S–837S.

Roberfroid, M.B. 2000. A European consensus of scientific concepts of functional foods. Nutrition 16: 689–691.

Rozada-Sanchez, R., Sattur, A.P., Thomas, K. and Pandiella, S.S. 2008. Evaluation of *Bifidobacterium* spp. for the production of a potentially probiotic malt-based beverage. Process Biochemistry 43: 848–854.

Saarela, M., Mogensen, G., Fonden, R., Matto, J. and Mattila-Sandholm, T. 2000. Probiotic bacteria: safety, functional and technological properties. Journal of Biotechnology 84: 197–215.

Salmeron, I., Fucinos, P., Charalampopoulos, D. and Pandiella, S.S. 2009. Volatile compounds produced by the probiotic strain *Lactobacillus plantarum* NCIMB 8826 in cereal-based substrates. Food Chemistry 117: 265–271.

Salmeron, I., Thomas, K. and Pandiella, S.S. 2014. Effect of substrate composition and inoculum on the fermentation kinetics and flavour compound profiles of potentially non-dairy probiotic formulations. LWT-Food Science and Technology 55: 240–247.

Sanders, M.E. 1998. Overview of functional foods: Emphasis on probiotic bacteria. International Dairy Journal 8: 341–347.

Sanni, A.I., Onilude, A.A. and Ibidapo, O.T. 1999. Biochemical composition of infant weaning food fabricated from fermented blends of cereal and soybean. Food Chemistry 65: 35–39.

Senok, A.C., Ismaeel, A.Y. and Botta, G.A. 2005. Probiotics: Facts and myths. Clinical Microbiology Infection 11: 958–966.

Sloan, A.E. 2002. The top 10 functional food trends: The next generation. Food Technology 56: 32–57.

Stanton, C., Ross, R.P., Fitzgerald, G.F. and Van Sinderen, D. 2005. Fermented functional foods based on probiotics and their biogenic metabolites. Current Opinion in Biotechnology 16: 198–203.

Steinkraus, K.H. 2002. Fermentations in world food processing. Comprehensive Reviews in Food Science and Food Safety 1: 23–32.

Tahri, K., Grill, J.P. and Schneider, F. 1997. Involvement of trihydroxyconjugated bile salts in cholesterol assimilation by bifidobacteria. Current Microbiology 34: 79–84.

Timmerman, H.M., Niers, L.E.M., Ridwan, B.U., Koning, C.J.M., Mulder, L., Akkermans, L.M.A., Rombouts, F.M. and Rijkers, G.T. 2007. Design of a multispecies probiotic mixture to prevent infectious complications in critically ill patients. Clinical Nutrition 26: 450–459.

Todorov, S.D., Botes, M., Guigas, C., Schillinger, U., Wiid, I., Wachsman, M.B., Holzapfel, W.H. and Dicks, L.M.T. 2008. Boza, a natural source of probiotic lactic acid bacteria. Journal of Applied Microbiology 104: 465–477.

Wang, Y.B. 2009. Prebiotics: Present and future in food science and technology. Food Research International 42: 8–12.

Woodcock, N.P., McNaught, C.E., Morgan, D.R., Gregg, K.L. and MacFie, J. 2004. An investigation into the effect of a probiotic on gut immune function in surgical patients. Clinical Nutrition 23: 1069–1073.

11

The Microbiota of Spontaneous Vegetable Fermentations

Spiros Paramithiotis[1,*] and *Eleftherios H. Drosinos*[1]

1. Introduction

A wide variety of fruits and vegetables has been traditionally utilized as a substrate for spontaneous lactic acid fermentation. The type and availability of raw materials, as well as climatic conditions, have led to the production of a wide range of spontaneously fermented products, the majority of which are nowadays recognized as characteristic for each region (Drosinos and Paramithiotis 2007).

Generally, spontaneous vegetable fermentations are characterized by a microbial succession at both species and sub-species level. Bacteria are considered to be the dominant epiphytic microorganisms of the phyllosphere, with archaea, filamentous fungi and yeasts playing a considerable role in that micro-ecosystem (Whipps et al. 2008). Upon preparation of the raw materials for spontaneous fermentation and once the environmental conditions, i.e., salinity and temperature are appropriately adjusted, the epiphytic microbiota is replaced by the fermentative one. Within the latter, a succession at species level is possible, depending on both salinity as well as the initial and accumulated acidity. A succession at sub-species level has recently been reported in such fermentations (Paramithiotis et al. 2014a,b) similar to the one that has been reported during spontaneous must fermentations (Egli et al. 1998, Sabate et al. 1998).

The advent of modern molecular techniques offered an alternative approach to the assessment of microbial dynamics during spontaneous vegetable fermentations. By extracting DNA directly from the sample, the non-culturable part of the micro-ecosystem can be studied. Many different techniques have been developed and evaluated, offering a glance in an otherwise hidden part of a micro-community.

[1] Laboratory of Food Quality Control and Hygiene, Department of Food Science and Human Nutrition, Agricultural University of Athens, Iera Odos 75, GR-11855 Athens, Greece.
* Corresponding author

Fermented products, in general, have enjoyed a very healthy reputation regarding safety despite some important outbreaks that have taken place, resulting mainly from the consumption of fermented products of animal origin (Adams and Mitchell 2002, Sartz et al. 2008). Infections associated with unpasteurized apple and orange juices, products with pH value close to the one of fermented vegetables, have raised some concerns regarding their safety (CDC 1996, 1999). More accurately, these concerns referred mainly to the ability of human pathogens to adapt and survive in extreme micro-environments, such as brined pickles, for extended periods of time (Castanie-Cornet et al. 1999, Tetteh et al. 2001, Mazzotta 2001, Kanellou et al. 2013). These factors, combined with the diversity of the ecological niches that are formed due to substantial differences in surface morphology, tissue composition and metabolic activities (Beuchat 2002, Klerks et al. 2007, Mitra et al. 2009, Burnett et al. 2000) of the plant parts used as raw materials along with their interaction with the pathogenic microorganisms (Brandl 2006, Raybaudi-Massilia et al. 2009, Critzer and Doyle 2010, Cooley et al. 2006, Kroupitski et al. 2011) renders at least uncertain any generalization schemes regarding the safety of these products.

In the following paragraphs, all available information on the above mentioned aspects is integrated and critically reviewed.

2. Techniques used for Micro-Ecosystem Assessment

The fundamental aspects that should be assessed when studying a micro-ecosystem are the identity of the microorganisms participating and their relative quantity. The former will give a measure of biodiversity at the desired level (genus/species/sub-species) and the latter, a measure of dominance, i.e., which taxonomic unit prevailed in which environment.

The approaches used to study microbial ecosystems may be divided into two categories: the culture-dependent and the culture-independent. The former requires cultivation of the microorganisms in specific substrates for enumeration and isolation. Then, the isolates are assigned to a taxonomic affiliation that may be reached by a variety of ways. The classical phenotypic identification procedure has been employed for many years; the ability of an isolate to grow under strictly defined conditions was used to create a dichotomous classification system. More recently, the phenotype was assessed by techniques such as FT-IR and MALDI-TOF MS that could depict in a spectrum, the chemical composition of a culture. Both phenotypic approaches suffer from severe drawbacks: the former cannot possibly describe the biodiversity at its full extent, while the more sophisticated latter one requires absolute accuracy in adjusting the analysis parameters and the results obtained are only as good as the respective databases that have been built. Finally, a consensus was reached regarding the need for a polyphasic approach in systematics (Vandamme et al. 1996); by using information from different levels of the microbial entity (genotype, proteome, phenotype), it was possible to overcome the limitations of each single method and to improve reliability (Paramithiotis et al. 2000). DNA and RNA have, very early, been recognized as molecules through which information about evolutionary history can be gained (Zuckerkandl and Pauling 1965). Ribosomal RNAs, in particular, have been used to divide the living world into three distinct primary groups, i.e., Archaea, Bacteria and Eucarya (Woese et al. 1990) since they are omnipresent and different positions in their sequences exhibit different mutation rates (Woese 1987). Regarding bacterial systematics, the comparison of 16S rRNA gene sequences has been rapidly adopted for the assessment of phylogenetic relationships as it consists of many domains that are considered independent from an evolutionary point of view, thus offering relative accuracy in that depiction (Woese 1987). However, there are certain drawbacks that should be taken into consideration when using 16S rRNA gene sequence analysis for classification purposes: it is not possible to differentiate between closely related species, horizontal transfer and recombination have been suggested as possible within this gene and the gene is present in multiple but not identical copies (Klappenbach et al. 2000, Schouls et al. 2003). Therefore, alternative options such as the single copy genes, *rpoB* (Case et al. 2007) and *gyrB* (Watanabe et al. 2001), have been studied for their comparative phylogenetic resolution, having so far no indication of recombination events through horizontal transfer.

The fact that many microorganisms do not grow under the conditions routinely used in a laboratory has also been documented very early (Akkermans et al. 1994) and therefore, only a fraction of a micro-ecosystem under study is revealed by culture-dependent techniques. An alternative approach has been provided by the advent of modern molecular techniques; DNA is extracted directly from the samples, separated and assigned to the microbial taxonomic unit from which it may have originated. Techniques such as Amplified Ribosomal DNA Restriction Analysis (ARDRA) (Gich et al. 2000, Park et al. 2014), Denaturing Gradient Gel Electrophoresis (DGGE) (Cocolin et al. 2004, Chang et al. 2008), Temperature Gradient Gel Electrophoresis (TGGE) (Gonzalez-Martinez et al. 2014, Pohlon et al. 2013) and Terminal-Restriction Length Polymorphism (T-RFLP) (Osborn et al. 2000, Tiquia 2010) have offered a glance into the non-culturable microbial community composition. However, there are two major aspects that should be taken into account when applying these techniques. Firstly, in most of the cases, the non-culturable fraction of a micro-ecosystem is assessed through differences in a partial sequence of 16S-rRNA gene. Since this gene is not only used for classification purposes, but also for detection as well, to the disadvantages mentioned before, another one should be added: the universal primers very often used may not hybridize to all potential targets due to differences in the sequence of even the conserved regions of small subunit rRNA genes (Baker et al. 2003). Therefore, the biodiversity of a sample may be underestimated. Secondly, it is very rare that an analytical step capable of differentiating DNA from cells that have lost their viability, such as treatment with ethidium monoazide bromide (EMA), is included in such protocols. Since DNA may persist relatively long after cell death (Josephson et al. 1993), failure to differentiate between living and non-living cells may offer a distorted view of the ecosystem, both qualitatively and quantitatively. Indeed, when a mature mixed-population microbial drinking-water biofilm was analyzed, the EMA-treated and untreated aliquots were significantly different (Nocker and Camper 2006). Despite those disadvantages, the application of the culture-independent approach revealed a remarkable biodiversity in several micro-communities. It has been reported that less than 1% of the bacterial species present in soil may be culturable (Malik et al. 2008); this percentage may increase to more than 10% when nutrient availability is improved, e.g., in the rhizosphere (Hirsch et al. 2010). However, this is the case mainly of environmental micro-ecosystems and rarely of food-fermentation ones. Regarding the latter, the fact that, at least vegetable fermentations are reproducible in terms of quality characteristics as well as physicochemical properties of the final products, even without any prior knowledge of the non-culturable fraction of the micro-ecosystem, may lead to the conclusion that this is rather restricted compared to the culturable one. Despite the size or relative importance from the technological point of view, it is of outmost importance to unravel the trophic relationships that lead to the development of a given microbial consortium with specific effects on specific raw materials. It is therefore important to gain as much integrated knowledge on an ecosystem as possible; this can be achieved by combining the two approaches and, at the same time, keeping in mind the respective strengths and weaknesses.

3. Lactic Acid Bacteria Population Dynamics of Spontaneously Fermented Vegetables

The lactic acid bacteria that participate in spontaneous vegetable fermentations are generally not specific to this habitat but are characterized by their omnipresence in food fermentations. The microorganisms that are more often isolated are *Leuconostoc mesenteroides* and *Lactobacillus plantarum*. The former is mostly associated with the onset of spontaneous vegetable fermentations, provided that the acidity is low; it is characterized by sensitivity to acidic conditions, ability to tolerate a wide range of salt concentrations and shorter generation time. On the other hand, *Lb. plantarum* is more acid tolerant and is generally distinguished by a large metabolic capacity that enables growth on a wide range of carbon sources; therefore it is mostly associated with the final stages of fermentation (Daeschel et al. 1987). Both species are very often isolated from fish, dairy, fruit, sausage and sourdough fermentations (Corsetti et al. 2001, Gobbetti et al. 1994, Ohhira et al. 1990, Marshall 1987, Leisner et al. 2001, Rantsiou et al. 2005, Drosinos et al. 2007). Many other lactic acid bacteria have been reported in spontaneous vegetable fermentations, mainly due to their ability to grow under stressful conditions, such as *Lb. sakei*, *Lb. brevis*, *Pediococcus pentosaceus* and *Weissella* spp., that can tolerate acidic conditions, *Ln. gelidum* that can grow at temperatures as low as 10°C, *Ln. gasicomitatum* that has more acid tolerance when compared to *Ln. mesenteroides* and *Ln. citreum*, and *W. koreensis* that has a psychrophilic character and can grow at even lower temperatures (Daeschel et al. 1987, Cho et al. 2006, Jung et al. 2013, Jeong et al. 2013a).

In the following paragraphs, the lactic acid bacteria dynamics of several spontaneously fermented vegetables is summarized and discussed.

3.1 Kimchi

Kimchi is a generic name referring to a group of brine-fermented vegetable-based products. It is prepared mainly by Chinese cabbage (*Brassica pekinensis*) that is accompanied by an extensive variety of other ingredients including vegetables, seafoods, spices, fruits, seasonings and cereals (Harris 1998). According to the raw materials used, the processing method, season or locality, a variety of more specific names currently exists. Representative studies on the structure and the dynamics of the lactic acid bacteria micro-community during fermentation are summarized in Table 1. *Leuconostoc* spp., *Weissella* spp. and *Lactobacillus* spp. are present throughout fermentation and no particular trend regarding dominance by a single species is apparent; fermentation seems to be driven by a microbial consortium consisting of species belonging to the aforementioned genera.

In a unique study by Chang et al. (2008) the yeast and archaeal population dynamics during spontaneous kimchi fermentation were assessed. Regarding the former, persistence of *Ladderomyces elongisporum*, *Trichosporon brassicae*, *Tr. middelhovenii*, *Candida sake*, *Saccharomyces castellii* and *Kluyveromyces marxianus* throughout fermentation was attributed to their possible suppression and therefore, no effect on fermentation was assumed. Regarding the haloarchaeal consortium that has been detected, *Halococcus dombrowskii* and *Hc. morrhuae* persisted throughout fermentation; *Haloarcheon trapanicum* and *Kaloterrigena thermotolerans* were detected only at the early stages of fermentation. Regarding their possible role in fermentation, it was assumed that the low pH value restricted their metabolic activities.

The variability in the raw materials used and the effect on the development of the micro-ecosystem has also been the subject of some study. Hong et al. (2014) studied the lactic acid bacteria biota of the main kimchi ingredient, namely salted Chinese cabbage, derived from three companies, by both culture-dependent and -independent approaches. It was reported that *Lb. curvatus*, *Lb. sakei*, *Ln. carnosum* and *Ln. mesenteroides* were detected by both approaches; members of the *Lb. plantarum*-group, *Ln. citreum*, *Ln. lactis*, *W. cibaria*, *W. paramesenteroides* and *Lactococcus lactis* were only detected by the culture-dependent approach while *Vibrio splendidus*, *V. mytili*, *Ln. gasicomitatum*, *Psychromonas arctica* and *Agrobacterium tumefaciens* by the culture-independent one. *Ln. mesenteroides* seemed to prevail among the samples originating from two companies and *Lb. curvatus* among the samples from the other. Occurrence of *Weissella* spp. in the ingredients of kimchi including Chinese cabbage, radish, red pepper powder, green onion, garlic, pickled shrimps, and ginger was studied by Kim et al. (2004). Green onion was reported as the major origin of *W. kimchii*. The presence of *Weissella* spp., *Leuconostoc* spp. and *Lactococcus* spp. in green onion as well as *Lactobacillus* spp., *Leuconostoc* spp. and *Weissella* spp. in garlic, was also reported by Jung et al. (2012). Regarding the latter, the addition of garlic resulted in a significant increase of the LAB population and a decrease of the other aerobic bacteria (Cho et al. 1988, Lee et al. 2008). As far as the effect of red pepper powder was concerned, slowing of the fermentation process, especially during the early period, as well as maintenance of a higher abundance of *Weissella* spp. and a lower quantity of *Leuconostoc* spp. and *Lactobacillus* spp. has been reported (Jeong et al. 2013b).

3.2 Cucumber

Cucumber fermentation is another typical brine-salted fermentation. Due to its high economic importance, it has been extensively studied. Initially, the pH of the brine is approximately 5.5 and fermentable carbohydrates, mainly glucose

Table 1. Structure and dynamics of the lactic acid bacteria micro-community during spontaneous kimchi fermentation.

Recipe	LAB Ecosystem Composition			Reference
	Early[1]	Middle	Final	
Chinese cabbage (74.5%); radish (13.5%); garlic (2.0%); ginger (0.5%); onion (2.0%); green onion (1.0%); red pepper powder (3.0%); leek (0.5%); shrimp paste (1.5%); anchovy paste (0.5%); sucrose (1.0%)	*Ln. carnosum; Ln. citreum; Ln. gelidum; Ln. gasicomitatum; Ln. inhae; Ln. kimchii; Ln. lactis; Ln. mesenteroides; W. cibaria; W. confusa; W. koreensis; Lb. curvatus; Lb. pentosus; Lb. plantarum; Lb. sakei*	*Ln. citreum; Ln. gelidum; Ln. gasicomitatum; Ln. kimchii; Ln. lactis; Ln. mesenteroides; W. cibaria; W. koreensis; Lb. curvatus; Lb. pentosus, Lb. sakei*	*Ln. gasicomitatum; Ln. mesenteroides; W. koreensis; Lb. pentosus; Lb. sakei*	Cho et al. (2006)
Chinese cabbage (89.8%); garlic (1.8%); ginger (0.9%); green onion (3.6%); red pepper powder (1.8%); fermented anchovy sauce (1.2%); sugar (0.9%)	*W. confusa;[2] Ln. citreum;[2] Lb. brevis;[3] Lb. sakei;[3] Lb. curvatus;[3] Lc. lactis lactis[3]*	*W. confusa;[2] Ln. citreum;[2] Lb. sakei;[2] Lc. lactis lactis;[3] Lb. brevis;[3] Lb. sakei;[3] Lb. curvatus[2,3]*	*W. confusa;[2] Ln. citreum;[2] Lb. sakei;[2] Lc. lactis lactis;[3] Ln. gelidum;[2] Se. marcescens;[2] Lb. brevis;[3] Lb. sakei;[3] Lb. curvatus[2,3]*	Lee et al. (2005)
Cabbage, red pepper powder, garlic, ginger, onion, radish, jeotgal	*B. mycoides; B. pseudomycoides; W. kandleri; Lb. sakei subsp. sakei; W. koreensis*	*B. mycoides; B. pseudomycoides; W. kandleri; Lb. sakei subsp. sakei; W. koreensis; Ln. gasicomitatum*	*Lb. sakei subsp. sakei; W. koreensis; Ln. gasicomitatum*	Chang et al. (2008)
Chinese cabbage, red pepper powder, radish, garlic, ginger, green onion, sugar, jeotgal	*Ln. mesenteroides; W. cibaria; Ln. citreum; Ln. lactis; Lb. sakei*	*Lb. sakei; W. koreensis; Ln. mesenteroides; Ln. gelidum; Ln. carnosum*	*Lb. sakei; W. koreensis; Ln. gelidum; Ln. mesenteroides; Ln. carnosum*	Jung et al. (2013)
Chinese cabbage, red pepper powder, Korean leek, garlic, ginger	*Ln. citreum; Ln. holzapfelii; Ln. gasicomitatum; W. soli; Lc. lactis*	*Lb. sakei; Ln. gasicomitatum; W. koreensis; Ln. gelidum*	*Lb. sakei; Ln. gasicomitatum; W. koreensis; Ln. gelidum*	Jeong et al. (2013c)
Radish, Korean leek, garlic, ginger	*Ln. citreum; Ln. gasicomitatum; W. cibaria; Ln. holzapfelii; Ln. lactis*	*Ln. gasicomitatum; Ln. gelidum; Lb. sakei*	*Ln. gasicomitatum; Ln. gelidum; Lb. sakei*	Jeong et al. (2013a)
Salted cabbage, dycon, red pepper powder, garlic, green onion, ginger, onion, salted shrimp, fish sauce, sugar	*W. confusa;[4,5] Ln. citreum;[4,5] Ln. mesenteroides;[4,5] Lc. lactis;[4] Lb. sakei;[5] Lb. curvatus;[4,5] Lc. gelidum;[4,5] Lc. carnosum;[4,5] S. salivarius;[4,5] B. subtilis[4,5]*	*W. confusa;[4,5] Ln. citreum;[5] Lb. sakei;[4,5] Lb. curvatus;[4,5] Lc. gelidum;[4,5] Lc. carnosum[4,5]*	*W. confusa;[4,5] W. koreensis;[5] Ln. citreum;[5] Lb. parabrevis;[5] Lb. sakei;[4,5] Lb. curvatus;[5] Lc. gelidum;[4] Lb. plantarum;[5] Lc. carnosum;[4,5] Lb. spicheri[5]*	Yeun et al. (2013)

[1] early, middle, final: These designate the respective stage of spontaneous kimchi fermentation.

Lb: Lactobacillus; Lc: lactococcus; Ln: leuconostoc; W: Weissella; Se: Serratia; B: Bacillus.

[2] fermentation at 10°C; [3] fermentation at 20°C.

[4] fermentation at 4°C; [5] fermentation at 10°C.

and fructose, are present. As a result, a rapid growth of several Gram-positive and Gram-negative bacteria, as well as yeasts, takes place. Lactic acid bacteria prevail mainly due to acidification and the fermentation stage is usually completed, primarily by *Lb. plantarum* and secondarily by *Pd. pentosaceus* and *Lb. brevis*. Regarding *Ln. mesenteroides* strains, their growth is most of the times inhibited. Depending on NaCl concentration (5–8%) and temperature (15–32°C), fermentation may be completed in 2–3 days or it may take as long as 7 days. At low salt concentrations and 20–27°C, fermentation is rapidly dominated by LAB and fermentable sugars are mostly converted to lactic acid. On the other hand, at elevated NaCl concentrations, lactic acid production is reduced and growth of both fermentative and oxidative yeasts is enhanced (Fleming et al. 1995, Singh and Ramesh 2008).

The dominating microbiota of spontaneously fermented cucumbers from the eastern Himalayas (Tamang et al. 2005) as well as four different counties of Taiwan was reported (Chen et al. 2012). Regarding the former, the pH value of the samples ranged from 3.5 to 4.2; *Lb. brevis* and *Lb. plantarum* dominated the lactic acid bacteria microbiota while presence of *Ln. fallax* was also mentioned in a sample. As far as the latter were concerned, each sample was characterized by different pH values and NaCl content and dominated by different lactic acid bacteria species. More accurately, *W. cibaria* was found dominant at pH 5.1 and 5 g/L NaCl, *Ln. lactis* at pH 4.9 and 14 g/L NaCl, *Lb. pentosus* at pH 4.3 and 22 g/L NaCl and a consortium consisting of *Lb. pentosus* and *Lb. plantarum* at pH 4.2 and 11 g/L NaCl.

Secondary cucumber fermentation has been the epicenter of intensive study due to the resulting spoilage of the product and the concomitant economic losses. During this secondary fermentation, lactic acid is depleted, pH value rises and an increase in the concentration of n-propanol, as well as acetic, butyric and propionic acids, takes place (Fleming et al. 1989). A wide variety of microorganisms have been isolated from commercial spoilage samples as well as experimental trials. Among them, the lactic acid bacteria, *Lb. buchneri* and *Pd. ethanolidurans*, the yeasts, *Issatchenkia occidentalis* and *Pichia manshurica*, in addition to *Enterobacter cloacae* and *Clostridium bifermentans*, have been repeatedly observed (Franco and Perez-Diaz 2012a). In a study conducted by Johanningsmeier et al. (2012), lactic acid bacteria isolated from fermented cucumber spoilage have been evaluated for their ability to degrade lactic acid under anaerobic conditions and various combinations of pH value and NaCl concentrations and, therefore, initiate spoilage-associated secondary cucumber fermentation. Cucumbers fermented with 6% NaCl to a final pH 3.2 exhibited no anaerobic lactic acid fermentation. Concerning the other combinations, only *Lb. buchneri* was able to metabolize lactic acid with a concurrent increase in acetic acid and 1, 2-propanediol. Moreover the ability of *Lb. buchneri,* as well as *Pi. manshurica* and *I. occidentalis,* to utilize lactic acid under aerobic storage of the fermented product has been reported (Franco and Perez-Diaz 2012b) with the metabolic activity of the yeasts being more rapid than that of the lactic acid bacterium. As far as *Cl. bifermentans* and *Eb. cloacae* were concerned, it seemed that an increase in the pH value is required before they can produce butyric and propionic acids. Based on this data, a scheme regarding the secondary cucumber fermentation has been proposed by Franco and Perez-Diaz (2012b). According to that, the oxidative yeasts *Pi. manshurica* and *I. occidentalis,* as well as the lactic acid bacterium, *Lb. buchneri*, initiate secondary fermentation by consuming lactic acid, causing a decrease in the redox potential and an increase in the pH value. As a result, the conditions become favorable for *Eb. cloacae* and *Cl. bifermentans* growth and for the production of butyric and propionic acids.

3.3 Sauerkraut

Sauerkraut fermentation is a typical dry-salted fermentation; it is characterized by an initial hetero-fermentative stage that is followed by a homo-fermentative one. Typically, fermentation is initiated by *Ln. mesenteroides* that can tolerate a wide range of salt concentrations and is characterized by a shorter generation time. Then, as fermentation proceeds, the environment becomes anaerobic and acidity is accumulated, leading to the substitution of the sensitive to acidic conditions *Ln. mesenteroides* sequentially by *Lb. brevis* and, at the final stage of fermentation, by the less sensitive *Lb. plantarum*. During this final stage, large quantities of lactic acid are formed by the remaining carbohydrates, leading to further lowering of the pH value. Sauerkraut production at 18–20°C normally leads to an acidity of 1.6–2.3% (calculated as lactic acid) and a pH value of 3.5 or less.

Apart from the dominant microbiota, in spontaneous sauerkraut fermentations, a secondary microbiota consisting of species such as *Enterococcus faecalis*, *Lb. confusus*, *Lb. curvatus*, *Lb. sakei*, *Lc. lactis lactis*, *Ln. fallax* and *Pd. pentosaceus* have also been described although their roles remain unclear (Harris et al. 1992, Barrangou et al. 2002).

3.4 Other Products

Almagro eggplant spontaneous fermentation is very interesting not only due to the utilization of a local eggplant variety (*Solanum melongena* L. var. *esculetum*) in the province of Ciudad Real (Spain) but also due to the very low initial pH value compared to the majority of spontaneously fermented vegetables. A rather restricted LAB diversity has been revealed with *Lb. brevis* and *Lb. fermentum* initiating the fermentation and *Lb. plantarum* dominating at the final stages. The low biodiversity was assigned to the low initial pH value (3.8) and the relatively high temperatures at which the fermentations take place because of the time of the year when this product is manufactured. The former inhibits leuconostocs and pediococci that tolerate a minimum pH for growths of 4.0 and 4.5, respectively, and offers lactobacilli an ecological advantage due to their comparatively enhanced acid tolerance. Regarding the latter, it is an additional factor that prevents leuconostocs from dominating, since they seem to prefer temperatures between 8 and 18°C (Sanchez et al. 2000). In a study by Sesena and Palop (2007), differences in the initial pH value and brine NaCl concentration resulted in differences in the composition of the lactic acid bacteria microbiota. When the initial pH value was 3.7 and NaCl content 4%, fermentation was almost exclusively driven by *Lb. fermentum*, whereas when the initial pH value was 3.3 and NaCl content 6%, fermentation was driven by a microbial consortium dominated by *Lb. plantarum* and *Lb. brevis* with *Lb. acidophilus* present as a secondary biota. Similar micro-ecosystem composition was reported for fermented eggplant of Vietnamese origin; *Lb. plantarum*, *Lb. fermentum* and *Lb. pentosus* were reported as dominating species with *Lb. paracasei, Lb. pantheris* and *Pd. pentosaceus* forming an occasional secondary biota (Nguyen et al. 2013).

Spontaneously fermented mustard and beet were also studied by Nguyen et al. (2013). The final pH values of both types of products were rather high; the former ranged from 4.2 to 5.5 and the latter from 4.2 to 4.9. In both cases, the dominating species was *Lb. fermentum,* with *Lb. pentosus* and *Lb. plantarum* forming a secondary biota. Spontaneous mustard fermentation was also studied by Chao et al. (2009), Chen et al. (2006) and Tamang et al. (2005). The differences in the production procedure were reflected in the composition of the micro-community. More accurately, the pH value of inziangsang (a traditional fermented leafy vegetable product of Nagaland and Manipur of India) ranged from 4.6 to 4.9 and fermentation was dominated by *Lb. brevis* and *Lb. plantarum* (Tamang et al. 2005). The suan-tsai studied by Chen et al. (2006) had a final pH value of approximately 5.0, an initial NaCl content of approximately 10% and final of 13% and the fermentation was driven by *Pd. pentosaceus* during the first days and *Tetragenococcus halophilus* until the end of fermentation (Chen et al. 2006). On the other hand, the suan-tsai studied by Chao et al. (2009) had a final pH value of 3.8, an initial NaCl content of 4% and fermentation was dominated in the initial stages by a consortium consisting of *Lb. alimentarius, Lb. brevis, Lb. coryniformis, Lb. farciminis, Lb. plantarum, Ln. citreum, Ln. mesenteroides, Ln. pseudomesenteroides, Pd. pentosaceus, W. cibaria* and *W. paramesenteroides* and at the final stages by *Lb. plantarum* and *Lb. brevis.* Fu-tsai is a product obtained by removing the partly fermented mustard, and after a series of sun-drying and fermentation steps, the NaCl concentration is increased to 12% and allowed to ferment for at least another 3 months. The final pH of the product is 3.6–3.7 and the dominating microbiota consists of *Lb. versmoldensis* and *Lb. alimentarius,* with *Pd. pentosaceus* being present at much lower population.

A combination of culture-dependent and -independent techniques was used to study spontaneous fermentation of caper berries (Pulido et al. 2005). During the first day of fermentation, a rapid pH value decrease from 7.5 to 4.37 was observed; it then gradually decreased to a final value of 3.55. *Lb. plantarum* was the dominant species throughout fermentation with *Lb. paraplantarum, Lb. pentosus* and *Lb. brevis* having a much lower relative abundance. *Lb. fermentum, Ec. faecium, Pd. pentosaceus* and *Pd. acidilactici* were also occasionally detected. Application of a culture-independent technique seemed to confirm the results obtained by the culture-dependent approach and, furthermore, detected *Ec. casseliflavus.*

An integrated study of spontaneous leek fermentation was performed by Wouters et al. (2013a). Both culture-dependent and independent approaches were used for population dynamics assessment and HPLC analysis for metabolite kinetics study. The effect of salt concentration, fermentation duration, as well as harvesting season, on the micro-ecosystem composition and community dynamics were also assessed. It was concluded that the fermentation conditions tested had no effect on the micro-ecosystem development and dynamics. A remarkable biodiversity at species level was revealed. *Ln. mesenteroides, Lb. plantarum, Lb. sakei, Lb. sakei/curvatus, Lb. parabrevis* and *Lb. brevis* were reported as present in all spontaneous leek fermentations while the occasional appearance of *Lb. hammessii/parabrevis, Ln. kimchii, Ln. lactis/garlicum, Ln. lactis/citreum, Ln. gasicomitatum, Ln. gelidum, Lc. lactis, Lc. raffinolactis,*

Lb. crustorum, *Carnobacterium maltaromaticum*, *Lb. nodensis* and *W. soli* were reported. Regarding metabolites, glucose and fructose were the main carbohydrates, with sucrose present as a minor one. Growth of lactic acid bacteria resulted in the production of lactic and acetic acids, ethanol and mannitol.

The dominant microbiota of a variety of spontaneously fermented vegetables by both culture-dependent and -independent techniques was studied by Wouters et al. (2013b). The final pH value ranged from 3.3 to 4.2. The majority of the products were dominated by a microbial consortium consisting of *Lb. brevis* and *Lb. plantarum* and only a few of them by either of the aforementioned species. *Ln. citreum* and *Ln. mesenteroides,* as well as *Lb. sakei,* appeared only occasionally. Application of the culture-independent approach also revealed the presence of *W. confusa/cibaria*, *Lc. lactis*, *Pd. pentosaceus* as well as *Ln. pseudomesenteroides* in some samples, although it failed to detect the dominant microbiota in some cases.

A non-salted fermented product, indigenous to the Himalayas, namely Gundruk, was studied by Tamang et al. (2005). This product was made of fresh leaves of local vegetables, mustard leaves, cauliflower leaves and cabbages. Final pH values ranged from 4.8 to 5.0 and *Lb. plantarum* was reported to dominate the micro-ecosystem.

Development of the microbial community during spontaneous cauliflower fermentation was studied by Paramithiotis et al. (2010) and Wouters et al. (2013b). In the former study, a lactic acid bacteria succession similar to the one reported in sauerkraut fermentation was reported. More accurately, an initial hetero-fermentative stage, driven by strains belonging to *Ln. mesenteroides*-group, was followed by a homo-fermentative one dominated by strains belonging to *Lb. plantarum*-group. Strains belonging to *Ec. faecium*-group and *Ec. faecalis*-group were also present at the early stages of fermentation. The final pH value of the brine was 3.78. In the latter study, *W. kimchii* was found dominant in the early stages of the fermentation with *Lb. plantarum*, *Lb. sakei/curvatus* and *W. viridescens* present as a secondary microbiota. Then, fermentation was driven by a consortium consisting of *Enterococcus* sp., *Lb. sakei/curvatus*, *Ln. mesenteroides* and *Pd. pentosaceus* and finally terminated by *Lb. brevis*. The final pH value of the brine was 3.5. These differences between the two studies may be assigned primarily to the differences in the production procedure, namely brine formulation and temperature of fermentation. In the first study, brine consisted of 8% NaCl and fermentation took place at 20°C whereas in the second study, brine consisted of 3.5% NaCl and 2% sucrose and fermentation took place at 25°C during the first three days and 16°C until the end.

The effect of the ripening stage on the development of the microbial community during spontaneous fermentation of green tomatoes was studied by Paramithiotis et al. (2014b). Tomatoes belonging to 'immature green' and 'mature green' ripening stages, as defined by Gautier et al. (2008) and having gradual pH values, i.e., 3.8–4.2, 4.3–4.8 and 4.9–5.4, were selected. Spontaneous fermentation was driven by lactic acid bacteria with yeasts present as a secondary microbiota. In all cases, *Ln. mesenteroides* initiated the spontaneous fermentation and dominated when initial pH values were below 4.8. On the other hand, when initial pH value was above 4.9, *Lb. curvatus*, *Lb. casei* and *Ln. citreum* also emerged, with the first appearing only transiently at the early stages of fermentation and the rest actually co-dominating the fermentation with *Ln. mesenteroides*. The approach used in the current study, i.e., combined clustering of the isolates with RAPD-PCR and rep-PCR, led to their complete differentiation at subspecies level. Thus, a microbial succession at that level was suggested, occurring throughout spontaneous fermentation. Such a succession was previously reported during spontaneous must fermentations (Egli et al. 1998, Sabate et al. 1998).

The bacterial community dynamics of spontaneous fermentation of a mixture of green tomatoes, cauliflower and carrot was studied by Wouters et al. (2013b). It was reported that *W. kimchii, W. paramesenteroides, Lb. plantarum, Ln. mesenteroides* and *Ln. pseudomesenteroides* were present during the early fermentation stages with none of them being prevalent. Then, as fermentation proceeded, only *Ln. citreum* and *Lb. brevis* were detected and upon further fermentation, *Lb. brevis* became the prevailing species.

Paramithiotis et al. (2014a) reported on the microbial population dynamics during spontaneous fermentation of *Asparagus officinalis* L. young sprouts. Both culture-dependent and -independent approaches were used in order to obtain a more integrated view of the microbiota dynamics. *Lb. sakei* and strains belonging to *Ec. faecium*-group prevailed until the 5th day of fermentation, while *W. viridescens* and *W. cibaria* dominated from the 7th day until the end of fermentation. Combination of SDS-PAGE with rep-PCR, apart from the effective clustering at species level, resulted in an equally effective differentiation at sub-species level, revealing that the succession at species level was accompanied by a succession at sub-species level. On the other hand, application of PCR-DGGE allowed characterization of the lactic acid bacteria biota only at species level. Dominance of *Weissella* spp. was verified and occasional presence of *Lb. sakei, Ec. faecium* and *Bacillus licheniformis* was also reported. Interestingly, the latter was only detected by PCR-DGGE and not by the culture-dependent approach. However, the dominant LAB biota, as identified by the culture-dependent approach, was only partially revealed by PCR-DGGE.

Brovada, an ancient traditional product of northeastern Italy, made by the spontaneous fermentation of turnips, was studied by Maifreni et al. (2004). The initial pH value of 6.4 was decreased within 24 hours to 3.7 and remained stable until the end of the fermentation. The lactic acid bacteria microbiota was initially dominated by *Lb. coryniformis*. As the fermentation proceeded, *Lb. hilgardii* and *Pd. parvulus* emerged, with the latter dominating until the final day of the fermentation, in which it co-dominated with *Lb. plantarum. Lb. viridescens* and *Lb. maltaromicus* were only detected during the initial stages while *Lb. plantarum* and *Lb. higardii* were present throughout fermentation.

A unique product of the Himalayas, namely sinki, was also described by Tamang et al. (2005). This product is made of radish tap-root prepared by pit fermentation, an ancient method that is still applied in some places of the world. The pH value of the product ranged from 4.0 to 4.2 and fermentation was dominated primarily by *Lb. brevis* and secondarily by *Ln. fallax*.

4. Fate of Pathogens During Spontaneous Vegetable Fermentation

Generally, fermented foods are a less likely vehicle for food-borne disease than the respective fresh produce; the antagonistic effect of the lactic acid bacteria, the low pH value and the high acidity have been recognized as the most significant inhibitory agents along with the important contribution of the type of the organic acid produced, not to mention carbon dioxide, hydrogen peroxide, ethanol, diacetyl and bacteriocins production (Adams and Nicolaides 1997). However, in the case of fermented vegetables, generalization schemes regarding their safety are, at the very least, uncertain due to the extended diversity in the phyllosphere niche, i.e., the differences in surface morphology, tissue composition and metabolic activities and the ability of the human pathogenic microorganisms to adapt (Beuchat 2002, Klerks et al. 2007, Mitra et al. 2009, Burnett et al. 2000, Takeuchi and Frank 2000, 2001, Critzer and Doyle 2010, Warriner and Namvar 2010, Kroupitski et al. 2009, 2011). Therefore, many studies have been conducted considering the fate of human pathogens during spontaneous vegetable fermentation and many interesting results have been reported.

Survival of *Escherichia coli* O157:H7, *Salmonella* Enteritidis, *Staphylococcus aureus* and *Listeria monocytogenes* in commercial and laboratory prepared kimchi was studied by Inatsu et al. (2004). Both types of kimchi were inoculated with the pathogens at 5–6 log CFU/g and incubated at 10°C for 7 days. All pathogens could survive during this period; their population decreased to the enumeration limit only upon prolongation of the incubation period. More accurately, *St. aureus* decreased to the enumeration level after 12 days incubation, in both types of kimchi; *S.* Enteritidis after 16 days and only in the commercially prepared products and *L. monocytogenes* after 16 and 20 days in the commercially and laboratory made products, respectively. Population of *E. coli* O157:H7 remained at high levels in both products throughout this incubation period, as well as counts of *S.* Enteritidis in the laboratory prepared sample. However, no further details regarding pH values, acidity of NaCl concentration were provided.

The effect of fermentation temperature (18 & 22°C) and salt concentration (1.8, 2.25 & 3%) on the survival of *E. coli* O157:H7 and *L. monocytogenes* during the production of whole-head and shredded sauerkraut was studied by Niksic et al. (2005). The final pH value of both sauerkrauts ranged from 3.5 to 4.0 but the differences in the acidity were more pronounced; acidity of whole-head sauerkraut reached 7 g/L at 22°C while that of shredded sauerkraut was recorded at 18 g/L at 22°C. Both pathogens persisted in the fermentation brines for most of the fermentation but their population decreased below enumeration limit at the end of it; *E. coli* O157:H7 population decreased more rapidly in the shredded sauerkraut, most probably due to the elevated acidity.

Kim et al. (2008) evaluated the relationship between lactic acid bacteria population or pH value and growth of *Bacillus cereus*, *L. monocytogenes* and *St. aureus* during kimchi fermentation under two treatments: heat treatment (85°C for 15 min) or neutralization treatment (pH 7) taking place at day 0 or day 3. It was observed that heat treatment at day 0 had no effect on *B. cereus* population, resulted in the reduction of *L. monocytogenes* population and a marginal reduction of *St. aureus* population. On the other hand, heat treatment on day 3 resulted in the reduction of the counts of all pathogens. More accurately, significant and complete inactivation was observed for *B. cereus* and *L. monocytogenes*, respectively, but it was only marginal for *St. aureus*. Neutralization treatment on day 0 resulted in slight population decrease of *B. cereus*, significant decrease of *L. monocytogenes* population and complete inactivation of *St. aureus*. On the other hand, neutralization treatment on day 3 resulted in marginal population decrease of *St. aureus*, significant decrease of *B. cereus* counts and complete inactivation of the *L. monocytogenes* population. All these effects have been assigned to lactic acid bacteria growth and concomitant pH decrease.

Survival of *E. coli* O157:H7 in cucumber fermentation brines at different stages of fermentation that were obtained from commercial plants was studied by Breidt, Jr. and Caldwell (2011). The pH value of the brines ranged from 3.2 to 4.5, the lactic acid concentration from less than 1 mM to 150 mM (all brines contained acetic acid) and NaCl from 5.5 to 8.7%. It was reported that the time required to obtain the 5-log reduction standard by Food and Drug Administration for *E. coli* O157:H7 was positively correlated with the pH value and the NaCl composition of the brines. Moreover, *E. coli* O157:H7 was not able to compete with either *Lb. plantarum* or *Ln. mesenteroides* at 30°C resulting in a decrease of the cell counts of the pathogen to below detection limit within 2–3 days.

The fate of *E. coli* O157:H7 and *L. monocytogenes* during spontaneous cabbage and radish kimchi fermentation was assessed by Cho et al. (2011). The pathogens were inoculated at 5 and 7 log CFU/mL and the fermentation took place at 4°C for 15 days. Survival of the pathogens was correlated with the final pH value obtained; when final pH value was below 4.0, the pathogens were not recovered, whereas when the final pH value was above 4.0, the pathogens population was significantly reduced but not completely inhibited.

In a study by Paramithiotis et al. (2012), *L. monocytogenes* and *S.* Typhimurium were inoculated from the beginning of spontaneous cauliflower fermentation at 2, 4 and 6 log CFU/mL. The initial NaCl concentration was 8% and was reduced to 6% after equilibration. Initial pH value and acidity of the brine were recorded at 6.19 and 0.11% lactic acid, respectively, while the final pH value and acidity were 3.8 and 0.6% lactic acid, respectively. Population of both pathogens initially increased, then decreased but remained at high levels until the end of fermentation. This has been attributed either to the gradual decrease of pH and increase of total titratable acidity that may trigger adaptive responses or to the development of cross-protection against low pH triggered by high osmolarity.

5. Conclusion and Future Perspectives

Besides the fermented vegetables that have met commercial significance, i.e., sauerkraut, kimchi and fermented cucumbers, there is a still unexplored variety of traditional recipes that lead to products with surprising organoleptic properties and great potential. A glimpse of these regional products enriched our knowledge regarding the microbial consortia that drive spontaneous vegetable fermentations and the succession that takes place at both species and subspecies level due to the dynamically changing environment. The advent of modern molecular techniques and their widespread use is not expected to improve our comprehension regarding the composition of either these micro-ecosystems or their dynamics. On the other hand, it is extremely likely that it will improve our understanding of the trophic relationships that lead to the development of these micro-communities.

Keywords: Lactic acid fermentation, spontaneous, fruits, vegetables, lactic acid bacteria

References

Adams, M. and Mitchell, R. 2002. Fermentation and pathogen control: A risk assessment approach. International Journal of Food Microbiology 79: 75–83.

Adams, M.R. and Nicolaides, L. 1997. Review of the sensitivity of different pathogens of fermentation. Food Control 8: 227–239.

Akkermans, A.D.L., Mirza, M.S., Harmse, H.J.M., Blok, H.J., Herron, P.R., Sessitsch, A. and Akkermans, W.M. 1994. Molecular ecology of microbes: A review of promises, pitfalls and true progress. FEMS Microbiology Reviews 15: 185–194.

Baker, G.C., Smith, J.J. and Cowan, D.A. 2003. Review and re-analysis of domain-specific 16S primers. Journal of Microbiological Methods 55: 541–555.

Barrangou, R., Yoon, S.-S., Breidt, Jr., F., Fleming, H.P. and Klaenhammer, T.R. 2002. Identification and characterization of *Leuconostoc fallax* strains isolated from an industrial sauerkraut fermentation. Applied and Environmental Microbiology 68: 2877–2884.

Beuchat, L.R. 2002. Ecological factors influencing survival and growth of human pathogens on raw fruits and vegetables. Microbes and Infection 4: 413–423.

Brandl, M.T. 2006. Fitness of human enteric pathogens on plants and implications for food safety. Annual Review of Phytopathology 44: 367–392.

Breidt, F. Jr, and Caldwell, J.M. 2011. Survival of *Escherichia coli* O157:H7 in cucumber fermentation brines. Journal of Food Science 76: M198–M203.

Burnett, S.L., Chen, J. and Beuchat, L.R. 2000. Attachment of *Escherichia coli* O157:H7 to the surfaces and internal structures of apples as detected by confocal scanning laser microscopy. Applied and Environmental Microbiology 66: 4679–4687.

Case, R.J., Boucher, Y., Dahllof, I., Holmstrom, C., Doolittle, W.F. and Kjelleberg, S. 2007. Use of 16S *rRNA* and *rpoB* genes as molecular markers for microbial ecology studies. Applied and Environmental Microbiology 73: 278–288.

Castanie-Cornet, M.P., Penfound, T.A., Smith, D., Elliptt, J.F. and Foster, J.W. 1999. Control of acid resistance in *Escherichia coli*. Journal of Bacteriology 181: 3525–3535.

CDC. 1996. Outbreak of *Escherichia coli* O157:H7 infections associated with drinking unpasteurized commercial apple juice: British Columbia, California, Colorado, and Washington. Morbidity and Mortality Weekly Report 45: 975.

CDC. 1999. Outbreak of *Salmonella* serotype Muenchen infections associated with unpasteurized orange juice: United States and Canada. Morbidity and Mortality Weekly Report 48: 582–585.

Chang, H.-W., Kim, K.-H., Nam, Y.-D., Roh, S.W., Kim, M.-S., Jeon, C.O., Oh, H.-M. and Bae, J.-W. 2008. Analysis of yeast and archaeal population dynamics in kimchi using denaturing gradient gel electrophoresis. International Journal of Food Microbiology 126: 159–166.

Chao, S.-H., Wu, R.-J., Watanabe, K. and Tsai, Y.-C. 2009. Diversity of lactic acid bacteria in suan-tsai and fu-tsai, traditional fermented mustard products of Taiwan. International Journal of Food Microbiology 135: 203–210.

Chen, Y.-S., Wu, H.-C., Lo, H.-Y., Lin, W.-C., Hsu, W.-H., Lin, C.-W., Lin, P.-Y. and Yanagida, F. 2012. Isolation and characterization of lactic acid bacteria from jiang-gua (fermented cucumbers), a traditional fermented food in Taiwan. Journal of the Science of Food and Agriculture 92: 2069–2075.

Chen, Y.-S., Yanagida, F. and Hsu, J.-S. 2006. Isolation and characterization of lactic acid bacteria from suan-tsai (fermented mustard), a traditional fermented food in Taiwan. Journal of Applied Microbiology 101: 125–130.

Cho, G.-Y., Lee, M.H. and Choi, C. 2011. Survival of *Escherichia coli* O157:H7 and *Listeria monocytogenes* during kimchi fermentation supplemented with raw pork meat. Food Control 22: 1253–1260.

Cho, J., Lee, D., Yang, C., Jeon, J., Kim, J. and Han, H. 2006. Microbial population dynamics of kimchi, a fermented cabbage product. FEMS Microbiology Letters 257: 262–267.

Cho, N.C., Jhon, D.Y., Shin, M.S., Hong, Y.H. and Lim, H.S. 1988. Effect of garlic concentrations on growth of microorganisms during kimchi fermentation. Korean Journal of Food Science and Technology 20: 231–235.

Cocolin, L., Rantsiou, K., Iacumin, L., Urso, R., Cantoni, C. and Comi, G. 2004. Study of the ecology of fresh sausages and characterization of populations of lactic acid bacteria by molecular methods. Applied and Environmental Microbiology 70: 1883–1894.

Cooley, M.B., Chao, D. and Mandrell, R.E. 2006. *Escherichia coli* O157:H7 survival and growth on lettuce is altered by the presence of epiphytic bacteria. Journal of Food Protection 69: 2329–2335.

Corsetti, A., Lavermicocca, P., Morea, M., Baruzzi, F., Tosti, N. and Gobbetti, M. 2001. Phenotypic and molecular identification and clustering of lactic acid bacteria and yeasts from wheat (species *Triticum durum* and *Triticum aestivum*) sourdoughs of southern Italy. International Journal of Food Microbiology 64: 95–104.

Critzer, F.J. and Doyle, M.P. 2010. Microbial ecology of foodborne pathogens associated with produce. Current Opinion in Biotechnology 21: 125–130.

Daeschel, M.A., Anderson, R.E. and Fleming, H.P. 1987. Microbial ecology of fermenting plant materials. FEMS Microbiology Reviews 46: 357–367.

Drosinos, E.H. and Paramithiotis, S. 2007. Trends in lactic acid fermentation. pp. 39–92. In: M.V. Palino (ed.). Food Microbiology Research Trends. Nova Publishers, Hauppauge.

Drosinos, E.H., Paramithiotis, S., Kolovos, G., Tsikouras, I. and Metaxopoulos, I. 2007. Phenotypic and technological diversity of lactic acid bacteria and staphylococci isolated from traditionally fermented sausages in Southern Greece. Food Microbiology 24: 260–270.

Egli, C.M., Edinger, W.D., Mitrakul, C.M. and Henick-Kling, T. 1998. Dynamics of indigenous and inoculated yeast populations and their effects on the sensory character of Riesling and Chardonnay wines. Journal of Applied Microbiology 85: 779–789.

Fleming, H.P., Daeschel, M.A., McFeeters, R.F. and Pierson, M.D. 1989. Butyric acid spoilage of fermented cucumbers. Journal of Food Science 54: 636–639.

Fleming, H.P., Kyung, K.H. and Breidt, F. 1995. Vegetable fermentations. pp. 629–662. In: G. Reed and T.W. Nagodawithana (eds.). Biotechnology Second, Completely Revised Edition, vol. 9 Enzymes, Biomass, Food and Feed. VCH, Weinheim.

Franco, W. and Perez-Diaz, I.M. 2012a. Role of selected oxidative yeasts and bacteria in cucumber secondary fermentation associated with spoilage of the fermented fruit. Food Microbiology 32: 338–344.

Franco, W. and Perez-Diaz, I.M. 2012b. Microbial interactions associated with secondary cucumber fermentation. Journal of Applied Microbiology 114: 161–172.

Gautier, H., Diakou-Verdin, V., Benard, C., Reich, M., Buret, M., Bourgaud, F., Poessel, J.L., Caris-Veyrat, C. and Genard, M. 2008. How does tomato quality (sugar, acid, and nutritional quality) vary with ripening stage, temperature, and irradiance? Journal of Agricultural and Food Chemistry 56: 1241–1250.

Gich, F.B., Amer, E., Figueras, J.B., Abella, C.A., Balaguer, M.D. and Poch, M. 2000. Assessment of microbial community structure changes by amplified ribosomal DNA restriction analysis (ARDRA). International Microbiology 3: 103–106.

Gobbetti, M., Corsetti, A., Rossi, J., La Rosa, F. and De Vincenzi, S. 1994. Identification and clustering of lactic acid bacteria and yeasts from wheat sourdoughs of central Italy. Italian Journal of Food Science 6: 85–94.

Gonzalez-Martinez, A., Pesciaroli, C., Martinez-Toledo, M.V., Hontoria, E., Gonzalez-Lopez, J. and Osorio, F. 2014. Study of nitrifying microbial communities in a partial-nitritation bioreactor. Ecological Engineering 64: 443–450.

Harris, L.J. 1998. The microbiology of vegetable fermentations. pp. 45–72. In: B.J.B. Wood (ed.). Microbiology of Fermented Foods. Blackie Academic and Professional, London.

Harris, L.J., Fleming, H.P. and Klaenhammer, T.R. 1992. Characterisation of two nisin-producing *Lactococcus lactis* subsp. *lactis* strains isolated from a commercial sauerkraut fermentation. Applied and Environmental Microbiology 58: 1484–1489.

Hirsch, P.R., Mauchline, T.H. and Clark, I.M. 2010. Culture-independent molecular techniques for soil microbial ecology. Soil Biology and Biochemistry 42: 878–887.

Hong, Y., Yang, H.-S., Li, J., Han, S.-K., Chang, H.-C. and Kim, H.-Y. 2014. Identification of lactic acid bacteria in salted Chinese cabbage by SDS-PAGE and PCR-DGGE. Journal of the Science of Food and Agriculture 94: 296–300.

Inatsu, Y., Bari, M.L., Kawasaki, S. and Isshiki, K. 2004. Survival of *Escherichia coli* O157:H7, *Salmonella enteritidis*, *Staphylococcus aureus*, and *Listeria monocytogenes* in Kimchi. Journal of Food Protection 67: 1497–1500.

Jeong, S.H., Jung, J.Y., Lee, S.H., Jin, H.M. and Jeon, C.O. 2013a. Microbial succession and metabolite changes during fermentation of dongchimi, traditional Korean watery kimchi. International Journal of Food Microbiology 164: 46–53.

Jeong, S.H., Lee, H.J., Jung, J.Y., Lee, S.H., Seo, H.-Y., Park, W.-S. and Jeon, C.O. 2013b. Effects of red pepper powder on microbial communities and metabolites during kimchi fermentation. International Journal of Food Microbiology 160: 252–259.

Jeong, S.H., Lee, S.H., Jung, J.Y., Choi, E.J. and Jeon, C.O. 2013c. Microbial succession and metabolite changes during long-term storage of kimchi. Journal of Food Science 78: M763–M769.

Johanningsmeier, S.D., Franco, W., Perez-Diaz, I. and McFeeters, R.F. 2012. Influence of sodium chloride, pH, and lactic acid bacteria on anaerobic lactic acid utilization during fermented cucumber spoilage. Journal of Food Science 77: M397–M404.

Josephson, K.L., Gerba, C.P. and Pepper, I.L. 1993. Polymerase chain reaction detection of nonviable bacterial pathogens. Applied and Environmental Microbiology 59: 3513–3515.

Jung, H.J., Hong, Y., Yang, H.S., Chang, H.C. and Kim, H.Y. 2012. Distribution of lactic acid bacteria in garlic (*Allium sativum*) and green onion (*Allium fistulosum*) using SDS-PAGE whole cell protein pattern comparison and 16S rRNA gene sequence analysis. Food Science and Biotechnology 21: 1457–1462.

Jung, J.Y., Lee, S.H., Jin, H.M., Hahn, Y., Madsen, E.L. and Jeon, C.O. 2013. Metatranscriptomic analysis of lactic acid bacterial gene expression during kimchi fermentation. International Journal of Food Microbiology 163: 171–179.

Kanellou, G., Paramithiotis, S., Mataragas, M. and Drosinos, E.H. 2013. Field study on the microbiological quality of pickles in brine and survival of *Salmonella typhimurium* and *Listeria monocytogenes* during storage at 4°C. European Food Research and Technology 236: 391–396.

Kim, T.-W., Lee, J.-Y., Song, H.-S., Park, J.-H., JI, G.-E. and Kim, H.-Y. 2004. Isolation and identification of *Weissella kimchii* from green onion by cell protein pattern analysis. Journal of Microbiology and Biotechnology 14: 105–109.

Kim, Y.-S., Zheng, Z.-B. and Shin, D.-H. 2008. Growth inhibitory effects of kimchi (Korean traditional fermented vegetable product) against *Bacillus cereus*, *Listeria monocytogenes*, and *Staphylococcus aureus*. Journal of Food Protection 71: 325–332.

Klappenbach, J.A., Dunbar, J.M. and Schmidt, T.M. 2000. rRNA operon copy number reflects ecological strategies of bacteria. Applied and Environmental Microbiology 66: 1328–1333.

Klerks, M.M., Franz, E., van Gent-Pelzer, M., Zijlstra, C. and van Bruggen, A.H.C. 2007. Differential interaction of *Salmonella enterica* serovars with lettuce cultivars and plant-microbe factors influencing the colonization efficiency. The ISME Journal 1: 620–631.

Kroupitski, Y., Golberg, D., Belausov, E., Pinto, R., Swartzberg, D., Granot, D. and Sela, S. 2009. Internalization of *Salmonella enterica* in leaves is induced by light and involves chemotaxis and penetration through open stomata. Applied and Environmental Microbiology 75: 6076–6086.

Kroupitski, Y., Pinto, R., Belausov, E. and Sela, S. 2011. Distribution of *Salmonella* Typhimurium in romaine lettuce leaves. Food Microbiology 28: 990–997.

Lee, J.S., Heo, G.Y., Lee, J.W., Oh, Y.J., Park, J.A., Park, Y.H., Pyun, Y.R. and Ahn, J.S. 2005. Analysis of kimchi microflora using denaturing gradient gel electrophoresis. International Journal of Food Microbiology 102: 143–150.

Lee, J.Y., Choi, M.K. and Kyung, K.H. 2008. Reappraisal of stimulatory effect of garlic on kimchi fermentation. Korean Journal of Food Science and Technology 40: 479–484.

Leisner, J.J., Vancanneyt, M., Rusul, G., Pot, B., Lefebvre, K., Fresi, A. and Tee, L.K. 2001. Identification of lactic acid bacteria constituting the predominating microflora in an acid-fermented condiment (tempoyak) popular in Malaysia. International Journal of Food Microbiology 63: 149–157.

Maifreni, M., Marino, M. and Conte, L. 2004. Lactic acid fermentation of *Brassica rapa*: Chemical and microbial evaluation of a typical Italian product (brovada). European Food Research and Technology 218: 469–473.

Malik, S., Beer, M., Megharaj, M. and Naidu, R. 2008. The use of molecular techniques to characterize the microbial communities in contaminated soil and water. Environmental International 34: 265–276.

Marshall, V.M.E. 1987. Fermented milks and their future trends: I. Microbiological aspects. Journal of Dairy Research 54: 559–574.

Mazzotta, A.S. 2001. Thermal inactivation of stationary-phase and acid-adapted *Escherichia coli* O157:H7, *Salmonella*, and *Listeria monocytogenes* in fruit juices. Journal of Food Protection 64: 315–320.

Mitra, R., Cuesta-Alonso, E., Wayadande, A., Talley, J., Gilliland, S. and Fletcher, J. 2009. Effect of route of introduction and host cultivar on the colonization, internalization, and movement of the human pathogen *Escherichia coli* O157:H7 in spinach. Journal of Food Protection 72: 1521–1530.

Nguyen, D.T.L., Van Hoorde, K., Cnockaert, M., De Brandt, E., Aerts, M., Thanh, L. and Vandamme, P. 2013. A description of the lactic acid bacteria microbiota associated with the production of traditional fermented vegetables in Vietnam. International Journal of Food Microbiology 163: 19–27.

Niksic, M., Niebuhr, S.E., Dickson, J.S., Mendonca, A.F., Koziczkowski, J.J. and Ellingson, J.L.E. 2005. Survival of *Listeria monocytogenes* and *Escherichia coli* O157:H7 during sauerkraut fermentation. Journal of Food Protection 68: 1367–1374.

Nocker, A. and Camper, A.K. 2006. Selective removal of DNA from dead cells of mixed bacterial communities by use of ethidium monoazide. Applied and Environmental Microbiology 72: 1997–2004.

Ohhira, I., Jeong, C.M., Miyamoto, T. and Kataoka, K. 1990. Isolation and identification of lactic acid bacteria from traditional fermented sauce in Southeast Asia. Journal of Dairy and Food Science 39: 175–182.

Osborn, A.M., Moore, E.R. and Timmis, K.N. 2000. An evaluation of terminal-restriction fragment length polymorphism (T-RFLP) analysis for the study of microbial community structure and dynamics. Environmental Microbiology 2: 39–50.

Paramithiotis, S., Doulgeraki, A.I., Karahasani, A. and Drosinos, E.H. 2014a. Microbial population dynamics during spontaneous fermentation of *Asparagus officinalis* L. young sprouts. European Food Research and Technology 239: 297–304.

Paramithiotis, S., Doulgeraki, A.I., Tsilikidis, I., Nychas, G.J.E. and Drosinos, E.H. 2012. Fate of *Listeria monocytogenes* and *Salmonella* sp. during spontaneous cauliflower fermentation. Food Control 27: 178–183.

Paramithiotis, S., Hondrodimou, O.L. and Drosinos, E.H. 2010. Development of the microbial community during spontaneous cauliflower fermentation. Food Research International 43: 1098–1103.

Paramithiotis, S., Kouretas, K. and Drosinos, E.H. 2014b. Effect of ripening stage on the development of the microbial community during spontaneous fermentation. Journal of the Science of Food and Agriculture 94: 1600–1606.

Paramithiotis, S., Mueller, M.R.A., Ehrmann, M.A., Tsakalidou, E., Seiler, H., Vogel, R. and Kalantzopoulos, G. 2000. Polyphasic identification of wild yeast strains isolated from Greek sourdoughs. Systematic and Applied Microbiology 23: 156–164.

Park, J.M., Yang, C.Y., Park, H. and Kim, J.M. 2014. Development of a genus-specific PCR combined with ARDRA for the identification of *Leuconostoc* species in kimchi. Food Science and Biotechnology 23: 511–516.

Pohlon, E., Ochoa Fandino, A. and Marxsen, J. 2013. Bacterial community composition and extracellular enzyme activity in temperate streambed sediment during drying and rewetting. PLoS ONE 8: e83365.

Pulido, R.P., Ben Omar, N., Abriouel, H., Lopez, R.L., Martinez Canamero, M. and Galvez, A. 2005. Microbiological study of lactic acid fermentation of caper berries by molecular and culture-dependent methods. Applied and Environmental Microbiology 71: 7872–7879.

Rantsiou, K., Drosinos, E.H., Gialitaki, M., Urso, R., Krommer, J., Gasparik-Reichardt, J., Toth, S., Metaxopoulos, I., Comi, G. and Cocolin, L. 2005. Molecular characterization of *Lactobacillus* species isolated from naturally fermented sausages produced in Greece, Hungary and Italy. Food Microbiology 22: 19–28.

Raybaudi-Massilia, R.M., Mosqueda-Melgar, J., Soliva-Fortuny, R. and Martin-Belloso, O. 2009. Control of pathogenic and spoilage microorganisms in fresh-cut fruits and fruit juices by traditional and alternative natural antimicrobials. Comprehensive Reviews in Food Science and Food Safety 8: 157–180.

Sabate, J., Cano, J., Querol, A. and Guillamon, J.M. 1998. Diversity of *Saccharomyces cerevisiae* strains in wine fermentations: Analysis for two consecutive years. Letters in Applied Microbiology 26: 452–455.

Sanchez, I., Palop, L. and Ballesteros, C. 2000. Biochemical characterization of lactic acid bacteria isolated from spontaneous fermentation of 'Almagro' eggplants. International Journal of Food Microbiology 59: 9–17.

Sartz, L., De Jong, B., Hjertqvist, M., Plym-Forshell, L., Alsterlund, R., Lofdahl, S., Osterman, B., Stahl, A., Eriksson, E., Hansson, H.-B. and Karpman, D. 2008. An outbreak of *Escherichia coli* O157:H7 infection in southern Sweden associated with consumption of fermented sausage; aspects of sausage production that increase the risk of contamination. Epidemiology and Infection 136: 370–380.

Schouls, L.M., Schot, C.S. and Jacobs, J.A. 2003. Horizontal transfer of segments of the 16S *rRNA* genes between species of the *Streptococcus anginosus* group. Journal of Bacteriology 185: 7241–7246.

Sesena, S. and Palop, M.Ll. 2007. An ecological study of lactic acid bacteria from Almagro eggplant fermentation brines. Journal of Applied Microbiology 103: 1553–1561.

Singh, A.K. and Ramesh, A. 2008. Succession of dominant and antagonistic lactic acid bacteria in fermented cucumber: Insights from a PCR-based approach. Food Microbiology 25: 278–287.

Takeuchi, K. and Frank, J.F. 2000. Penetration of *Escherichia coli* O157:H7 into lettuce tissues as affected by inoculum size and temperature and the effect of chlorine treatment on cell viability. Journal of Food Protection 63: 434–440.

Takeuchi, K. and Frank, J.F. 2001. Quantitative determination of the role of lettuce leaf structures in protecting *Escherichia coli* O157:H7 from chlorine disinfection. Journal of Food Protection 64: 147–151.

Tamang, J.P., Tamang, B., Schillinger, U., Franz, C.M.A.P., Gores, M. and Holzapfel, W.H. 2005. Identification of predominant lactic acid bacteria isolated from traditionally fermented vegetable products of the Eastern Himalayas. International Journal of Food Microbiology 105: 347–356.

Tetteh, G.L. and Beuchat, L.R. 2001. Sensitivity of acid-adapted and acid-shocked *Shigella flexneri* to reduced pH achieved with acetic, lactic, and propionic acids. Journal of Food Protection 64: 975–981.

Tiquia, S.M. 2010. Using terminal restriction fragment length polymorphism (T-RFLP) analysis to assess microbial community structure in compost systems. Methods in Molecular Biology 599: 89–102.

Vandamme, P., Pot, B., De Vos, P., Kersters, K. and Swings, J. 1996. Polyphasic taxonomy, a consensus approach to bacterial systematics. Microbiology Reviews 60: 407–438.

Warriner, K. and Namvar, A. 2010. The tricks learnt by human enteric pathogens from phytopathogens to persist within the plant environment. Current Opinion in Biotechnology 21: 131–136.

Watanabe, K., Nelson, J., Harayama, S. and Kasai, H. 2001. ICB database: The *gyrB* database for identification and classification of bacteria. Nucleic Acids Research 29: 344–345.

Whipps, J.M., Hand, P., Pink, D. and Bending, G.D. 2008. Phyllosphere microbiology with special reference to diversity and plant genotype. Journal of Applied Microbiology 105: 1744–1755.

Woese, C.R. 1987. Bacterial Evolution. Microbiology Reviews 51: 221–271.

Woese, C.R., Kandler, O. and Wheelis, M.L. 1990. Towards a natural system of organisms: Proposal for the domains Archaea, Bacteria and Eucarya. Proceedings of the National Academy of Sciences USA 87: 4576–4579.

Wouters, D., Bernaert, N., Conjaerts, W., Van Droogenbroeck, B., De Loose, M. and De Vuyst, L. 2013a. Species diversity, community dynamics and metabolite kinetics of spontaneous leek fermentations. Food Microbiology 33: 185–196.

Wouters, D., Grosu-Tudor, S., Zamfir, M. and De Vuyst, L. 2013b. Bacterial community dynamics, lactic acid bacteria species diversity and metabolite kinetics of traditional Romanian vegetable fermentations. Journal of the Science of Food and Agriculture 93: 749–760.

Yeun, H., Yang, H.-S., Chang, H.-C. and Kim, H.-Y. 2013. Comparison of bacterial community changes in fermenting kimchi at two different temperatures using a denaturing gradient gel electrophoresis analysis. Journal of Microbiology and Biotechnology 23: 76–84.

Zuckerkandl, E. and Pauling, L. 1965. Molecules as documents of evolutionary history. Journal of Theoretical Biology 8: 357–366.

12

Functional Fermented Food and Feed from Seaweed

Izabela Michalak[1] and *Katarzyna Chojnacka*[1,]*

1. Introduction

Fermented foods are food substrates that are processed by edible microorganisms (bacteria, fungi: yeast and mold) whose enzymes (e.g., amylases, proteases and lipases) hydrolyze the polysaccharides, proteins and lipids to non-toxic products with flavors, aromas and textures pleasant and attractive to the human consumer (Steinkraus 1997). Fermentation is based on the production of lactic acids due to the naturally present lactic acid bacteria in the fermented material or by the bacteria that are added (Ennouali et al. 2006).

Food fermentations can be classified in a number of ways; for example, by classes: (1) beverages, (2) cereal products: bread, idli, dosa, (3) dairy products: dahi, kefir, kumys, yogurt (yoghurt), cheese, (4) fish products, (5) fruit and vegetable products (sauerkraut, kimchi, pickled vegetables, olives), plant root products (gari, fufu), (6) legumes: soy sauce, miso, tempeh, natto, sufu and (7) meat products (Steinkraus 1997, Tamang and Kailasapathy 2010, Prajapati and Nair 2008). Today, there are about 5,000 varieties of fermented foods and beverages prepared and consumed worldwide, which account for contributions of between 5 and 40% of daily meals (Tamang and Kailasapathy 2010). The above mentioned fermented foods and beverages are described in detail in a number of papers (e.g., Steinkraus 1997, Shiby and Mishra 2013) and books (e.g., Prajapati and Nair 2008, Tamang and Kailasapathy 2010). Therefore, this chapter brings the focus on another group of organisms—algae—which can constitute a raw material for the production of fermented foods. This group comprises

---
[1] Department of Advanced Material Technologies, Faculty of Chemistry, Wrocław University of Technology, Smoluchowskiego 25, 50-372 Wrocław, Poland.
* Corresponding author: katarzyna.chojnacka@pwr.edu.pl

of macroalgae (called also 'seaweed') and microalgae and includes cyanobacteria (called 'blue-green algae'). Macroalgae are divided into three groups: green (*Chlorophyta*), red (*Rhodophyta*) and brown (*Phaeophyta*). This chapter is devoted mainly to macroalgae. Production of algae based fermented products is not common and the literature data are quite limited. However, it is underlined that fermented products produced from algae need, due to their abundance in marine and freshwater environments and to rich chemical composition, to be developed (Uchida and Miyoshi 2013).

Fermented products from algae can be ingredients of functional food, which is known to have several health-specific advantages. In addition to the basic nutritive value, functional food contains a proper balance of ingredients, which help in the prevention and treatment of illnesses and diseases. Within this category, products containing lactic acid bacteria (LAB) or probiotics are gaining importance (Gupta and Abu-Ghannam 2011); for example: fermented dairy product (functional food) with probiotics (functional ingredient) supports intestinal tract health and boosts immunity (Plaza et al. 2008). Therefore, algae can constitute a raw material for production of fermented products for human and animals (Fig. 1).

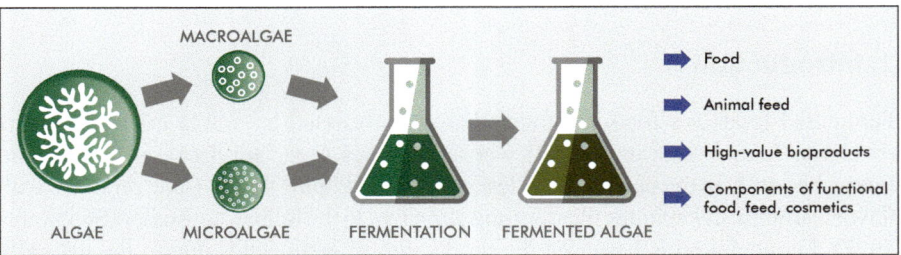

Figure 1. Algae as a raw material for production of fermented products for human and animals.

2. Seaweeds and their Functional Ingredients

Seaweeds are a natural source of new compounds with biological activity; they have the potential to be used as functional ingredients (Plaza et al. 2008). The types and abundance of these compounds vary strongly between algal species. Algae contain, for example, polyphenols, lipids (e.g., polyunsaturated fatty acid—PUFA), vitamins, minerals, dietary fiber, polysaccharides, proteins and pigments (Dawczynski et al. 2007, Plaza et al. 2008, Michalak and Chojnacka 2015). Some of these compounds are found only in algae, mainly polysaccharides: agar-agar, carrageenan or alginate, sulphated polysaccharides (fucoidan, laminaran). Hydrocolloids have already found applications in food, dairy, pharmaceutical industries because of their gelling, thickening and stabilizing properties (Mayakrishnan et al. 2013). It was confirmed that algae possess anti-diabetic, anti-allergic, anti-coagulant, anti-cancer, anti-inflammatory, anti-oxidant, immune-protective, cardiovascular, immune-enhancing properties, which are owed to the biologically active compounds found in algal cells (Mišurcová et al. 2012, Kim 2013, Michalak and Chojnacka 2015).

The most known functional ingredients from algae are: sterols and soluble fiber which reduce total and LDL cholesterol, PUFAs which reduce risk of certain heart diseases, polysaccharides that have antiviral and apoptotic activity and fucoxanthin which has a preventive effect on cerebrovascular diseases and activates the metabolism (Plaza et al. 2008). These biologically active compounds are usually extracted from the raw algal biomass by traditional and/or novel extraction techniques (e.g., supercritical fluid extraction, microwave assisted extraction, enzyme assisted extraction) (Michalak and Chojnacka 2014). However, they can be isolated also from the fermented algae, as was shown in the work of Nikapitiya et al. (2007). From fermented red alga *Grateloupia filicina*, polysaccharides were extracted. For this purpose, the broth was filtered and then extracted with ethanol. The precipitate collected by centrifugation was purified by fast protein liquid chromatography. The obtained polysaccharides were characterised by anticoagulant activity. This indicates that seaweed fermentation can be useful for the pharmaceutical industries.

3. Seaweeds as a Raw Material for Fermentation

Seaweeds can be used in the fermentation processes for many reasons. First of all, they are a rich source of many bioactive compounds, and saccharides are the key substrate for this process. Seaweeds are also one of the cheapest available sources of protein, but the utilization is limited by the presence of crude fiber, which can be eliminated by fermentation (Felix and Brindo 2008).

The second reason for algal biomass utilization is the necessity to manage waste algae that cause nuisance blooms in many coastal waters (Uchida and Miyoshi 2010) or are a solid waste after the extraction of compounds from algae (e.g., alginate, iodine and mannitol) (Zhang et al. 2012). There are many problems in disposal of waste biomass, mainly the small scale and the high utilization cost. Therefore it causes environmental pollution but, on the other hand, it is a valuable resource (Zhang et al. 2012). This algal biomass still seems to contain useful components, which can be recycled after improving the quality by fermentation (Wijesinghe et al. 2012). Fermentation is an efficient method that enables the stabilization and the transformation of biomass of diverse origins to a more sanitized quality product (Ennouali et al. 2006).

Fermentation of algal biomass has practical (production of functional products), environmental (coping with the effects of eutrophication) and economic implications (processing instead of landfilling). For example, Ennouali et al. (2006) fermented red algae, *Gelidium sesquipedale*, from a factory after extraction of agar-agar; it would have normally been discharged into the environment. The product obtained from fermentation is rich in minerals, proteins, sugars and a small amount of lipids and can be used as fertilizer and/or integrated in animal feed.

Fermentation of seaweeds is not a simple process. The main problem is the rigid algal cell wall. The cell wall of *Chlorophyta* is composed of cellulose, hydroxyproline glucosides, xylans and mannans. The main components in the cell wall of *Phaeophyta* are: cellulose, alginic acid and sulfated mucopolysaccharides (fucoidan). *Rhodophyta* in the cell wall contain cellulose, xylans, several sulfated polysaccharides (galactans) and alginate in corallinaceae (Davis et al. 2003). The solution to this problem can be pretreatment of the biomass, during which the cells are disrupted, making biologically active compounds more bioavailable (Huang et al. 2014). There are different pretreatment procedures of algae, for example: mechanical, thermal, physical, chemical and enzymatic (Michalak and Chojnacka 2014). Gupta et al. (2011) found that heat treated brown macroalgae, *Laminaria digitata* and *Laminaria saccharina* were suitable substrates for lactic acid fermentation by *Lactobacillus plantarum*.

4. Fermentation of Seaweeds

Seaweeds could be used as a substrate for lactic acid or ethanol fermentation in order to obtain such products as may be used as functional foods or components of diets and fertilizers (Uchida and Miyoshi 2013). Microbial fermentation of algae may result in the production of a wide range of primary and secondary metabolites that may constitute an ingredient of functional food (Wijesinghe et al. 2012, Eom et al. 2013). Fermentation of seaweeds with lactic acid bacteria and yeast enhances their nutritive value by enriching the protein, vitamin, mineral, essential amino acid and essential fatty acid content. Additionally, this process improves the digestibility of seaweed-based feeds (Uchida and Murata 2002).

In the work of Felix and Brindo (2014), proximate compositions of raw and fermented *Padinatetrastomatica* were compared. It was shown that after fermentation, the content of proteins and lipids increased in fermented algae (from 10.5 in raw to 15.9% in fermented seaweed and from 1.14 to 3.23%, respectively). The content of ash and fiber decreased after fermentation (from 27.0 to 24.0% and from 24.0 to 3.60%, respectively). These findings were also confirmed in the work of Choi et al. (2014). In the fermented *Undaria pinnatifida*, the content of crude proteins, ether extract and carbohydrates increased and crude ash decreased when compared to raw seaweed. Additionally, it was shown that the fermented seaweed released minerals (essential for the proper functioning of living organisms) through cell walls which made them more soluble and easily absorbed; for example: Ca (before fermentation 0.93%, after fermentation 1.27%), P (before 0.28%, after 0.37%), Fe (before 0.59%, after 0.98%) and Zn (before 0.24%, after 0.31%).

Fermentation of algae is generally a two-step process. First of all, saccharification must be conducted. Seaweeds are mainly composed of polysaccharides, which need to be degraded to monosaccharides. In the process of saccharification, many enzymes can be used, mainly cellulase, then agarase, alginate lyase, hemicellulase and mannase (Uchida 2014). In the work of Uchida and Miyoshi (2010), it was shown that the combinational use of macerozyme and lactase with cellulase caused an increase of the lactic acid content in the algal cultures. The use of cellulase to obtain fermented products from algae is considered as a basis for creation of a new branch of industry based on algal fermentation technology (Uchida 2014). For seaweeds characterised by high protein content (e.g., *Porphyra yezoensis*), the application of protease is recommended for effectively conducting lactic acid fermentation (Tsuchiya et al. 2007).

The second process is fermentation using microorganisms, usually lactic acid bacteria (*Lactobacillus* species: *Lb. brevis*, *Lb. casei* and *Lb. plantarum*) or yeast (of food grade *Saccharomyces cerevisiae*) to initiate the process (Felix and Pradeepa

2012, Uchida and Miyoshi 2013). It was found that the inoculation of LAB alone or with yeast strain for lactic acid fermentation facilitated the process. Inoculation of the yeast strains alone gave unsatisfactory results—the growth of contaminant bacteria was observed (Uchida and Miyoshi 2013). Seaweed products fermented with listed microorganisms are known to act as growth promoters, immune enhancers and probiotics in cultivable organisms (Felix and Brindo 2014).

The current research is focused on the development of a starter culture of halophilic lactic acid bacteria (HLAB) that allow the preparation of algal fermented products containing a high (> 10%) salt content and having long-term preservation. The most commonly used in fermentation, *Lactobacillus* group bacteria are not halotolerant; their growth is significantly restricted under a salt concentration exceeding 5% (Uchida and Miyoshi 2013). In the work of Uchida et al. (2014) it was indicated that for the production of highly salted products (e.g., sauces) by fermentation process, halophilic lactic acid bacteria such as *Tetragenococcus* and *Pediococcus* species should be used.

Another important factor in the fermentation of seaweeds is the addition of an adequate source of nitrogen (e.g., ammonium chloride), which is necessary for microbial propagation and to regulate their metabolism. Microorganisms synthesize many compounds, for example amino acids and nucleic acids. The addition of inorganic nitrogen can also promote microbial metabolism to improve the protein content in the algae residue (Han et al. 2012).

A general scheme of the fermentation process of seaweeds is presented in Fig. 2. These two processes (saccharification and fermentation) can be performed simultaneously by adding polysaccharide decomposing enzymes and starter microorganisms (Uchida 2007). The fermented algae collected from the fermenter can be in the form of (1) liquid fermentation (Hayisama-ae et al. 2014, Prachyakij et al. 2008), (2) fermented seaweed powder dried in a hot air oven (Felix and Brindo 2014), dried at the natural temperature or at the temperature of 65°C in the blast drier (Liu et al. 2012) and milled or (3) a raw material for the extraction of biologically active compounds; for example, sugars and oligosaccharides (Hayisama-ae et al. 2014), sulphated polysaccharides (Nikapitiya et al. 2007), phlorotannins (Eom et al. 2013) and polyphenols (Wijesinghe et al. 2012). Exemplary procedures of extraction of biologically active compounds from fermented algae are contained in Table 1.

The degree to which fermentation has proceeded can be confirmed by microscopic observations. Uchida and Miyoshi (2010) found that after 35 days of fermentation, microalgal cells were changed—they were partially damaged after enzymatic treatment. However, the decomposition was not complete; cell wall residues remained. After fermentation, the color of the algal cells was also changed to brown. This indicates chlorophyll decomposition under acidic culture conditions.

The examples of fermentation of various seaweed species (mainly: brown *Laminaria* sp. and *Undaria* sp., green *Ulva* sp., red *Gracilaria* sp.) with different starter cultures and produced fermented products are presented in Table 2. Nowadays, fermented algal products dominate in aquaculture. The actual and potential applications of algal fermented products are presented in Fig. 3.

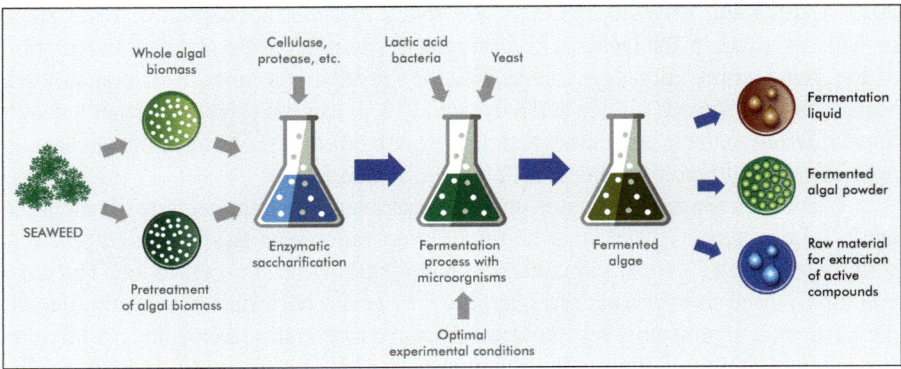

Figure 2. Fermentation process of seaweed—general scheme.

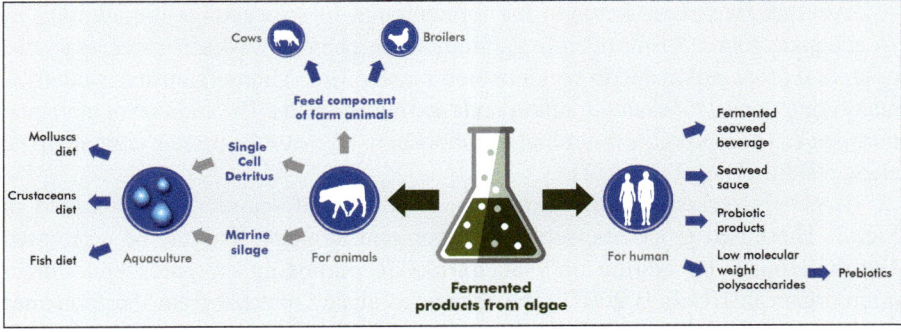

Figure 3. The actual and potential applications of fermented products from algae.

5. Fermented Algal Products for Animals

Macroalgae, rich in bioactive compounds, could potentially be exploited as functional ingredients for animal health applications. Anaerobic fermentation of seaweed proved to be a method to increase the nutritional value of animal feed (e.g., the content of proteins and polysaccharides). The algal biomass is a high-quality animal feed ingredient and can effectively enhance the immunity of animals and increase economic benefits (Han et al. 2012). Some functional effects were observed when fish and animals were fed with fermented products from algae (Uchida and Miyoshi 2013).

5.1 Fermented Products From Algae in Aquaculture

Algal fermented products can be applied in aquaculture (aqua-farming) in the cultivation of aquatic organisms such as fish, crustaceans (e.g., shrimp, freshwater and farm prawns), molluscs (e.g., bivalves: oysters, mussels, clams). This approach was proposed by several authors: Uchida et al. (1997), Uchida and Murata (2002), Pérez Camacho et al. (2007), Felix and Pradeepa (2011, 2012). Seaweeds decomposed to a cellular level (particle sizes of 5 to 12 μm), called Single Cell Detritus (SCD), are

Table 1. Examples of isolation steps of biologically active compounds from fermented seaweeds.

Eom et al. 2013	Fermentation of *Eisenia bicyclis*	Collection of fermented samples	Extraction with methanol	Rotary evaporation of methanolic extract	Fractionation of antibacterial substance (phlorotannins) by liquid–liquid solvent partition procedure
Hayisama-ae et al. 2014	Fermented red seaweed beverage (FSB)	Freeze drying of FSB after 60 days of fermentation	Extraction of freeze dried FSB with hot water at 80°C	Filtration	Analysis of sugars and oligosaccharides by HPLC
Nikapitiya et al. 2007	Fermentation of red alga *Grateloupia filicina*	Filtration of the fermented sample	Extraction with ethanol	Sample centrifugation	Purification of sulphated polysaccharides by fast protein liquid chromatography
Wijesinghe et al. 2012	Fermentation of *Ecklonia cava* processing by-product (ECPB)	Freeze drying and homogenization of fermented ECPB	Extraction with distilled water and 80% ethanol, separately in a shaking incubator	Sample centrifugation Filtration of supernatant Evaporation of filtrate under vacuum	Determination of total phenolic content (with Folin–Ciocalteu reagent)
Prachyakij et al. 2008	Fermentation of *Gracilaria fisheri*	Analysis of total sugar as glucose, total acidity as lactic acid (by gas chromatography), Cu, Zn, Fe, As and Pb by inductively coupled plasma-atomic emission spectroscopy (ICP-AES) and antibacterial activity by the agar well diffusion technique in fermentation liquid (FSB)			

Table 2. Review of microbial fermentation of seaweeds.

Fermented product	Seaweed	Microorganism	Fermentation equipment	Reference
single cell detritus (SCD)—a dietary material for brine shrimp (*Artemia nauplii*)	*Laminaria japonica*	*Alteromonas espejiuna* AR06	not available	Uchida et al. 1997
single cell products	*Ulva* spp.	*Lactobacillus brevis* strain B5201, *Debaryomyces hansenii* var. *hansenii* strain Y5201, *Candida zeylanoides*-related strain Y5206	not available	Uchida 2005
SCD for young pearl oysters (*Pinctada fucata martensii*)	*Undaria pinnatifida*	*Lactobacillus brevis* strain B5201; yeasts: *Debaryomyces hansenii* strain Y5201 and *Candida* sp. strain Y5206	500 ml polycarbonate bottles	Uchida and Murata 2002
SCD for clam (*Ruditapes decussatus*)	*Laminaria saccharina*	bacteria *Pseudoalteromonas espejiana* CECT 5255 and marine vibrio (species unknown)	6 L polycarbonate vessels	Pérez Camacho et al. 2007
SCD (for shrimp larval feeding to replace algae and *Artemia* in the marine hatchery sector)	*Ulva reticulata*	*Lactobacillus plantarum* and *Saccharomyces ceriviceae*	10 L fermentors *in situ*	Felix and Pradeepa 2012
marine silage (MS)—a dietary material for pearl oysters (*Pinctada fucata martensii*)	*Undaria pinnatifida*	*Lactobacillus brevis* strain B5201, *Debaryomyces hansenii* var. *hansenii* strain Y5201, *Candida* sp. strain Y5206	10 L polycarbonate bottles	Uchida et al. 2004
MS for fish red sea bream (*Pagrus major*)	*Undaria pinnatifida*	*Lactobacillus brevis* strain B5201; yeasts: *Debaryomyces hansenii* strain Y5201 and *Candida* sp. strain Y5206	sterile 500 mL polycarbonate bottles with screw caps	Uchida 2007
MS for aquaculture (fishmeal)	*Ulva* spp.	*Lactobacillus casei*	20 L polycarbonate tank	Uchida and Miyoshi 2010
fermented seaweed powder for giant freshwater prawn (*Macrobrachium rosenbergii*)	*Padinatetrastomatica*	*Lactobacillus* spp. and *Saccharomyces cerevisiae*	fermenter vessel	Felix and Brindo 2014

animal feed	Seaweed residue (not specified)	EM inoculants (culture of photosynthetic bacteria, lactic acid bacteria, yeasts, actinomycetes bacteria, *Acetobacter*)	biochemical incubator	Han et al. 2012
dietary fermented seaweed for broilers	*Undaria pinnatifida Hizikia fusiformis*	*Bacillus subtilis, Aspergillus oryzae*	not available	Choi et al. 2014
fertilizer and/or animal feed	*Gelidium sesquipedale*	mixture of yeasts and lactic bacteria	barrels	Ennouali et al. 2006
methanol extract from fermented algae (antimicrobial and antioxidant properties)	*Eisenia bicyclis*	*Candia utilis YM-1* isolated from meju (fermented soybeans)	1 L flask	Eom et al. 2013
fermented seaweed beverage (FSB)	*Gracilaria fisheri*	*Lactobacillus plantarum*	15 L plastic bucket	Prachyakij et al. 2008
fermented seaweed beverage	*Gracilaria fisheri*	*Lactobacillus plantarum* DW12	not available	Hayisama-ae et al. 2014
fermented seaweed beverage	*Gracilaria fisheri*	*Lactobacillus plantarum* DW12	15 L plastic bucket	Ratanaburee et al. 2011

expected to be utilized in aquaculture in place of unicellular algae. Their cultivation is laborious and is accompanied by various technical difficulties in large scale production. Felix and Pradeepa (2012) proposed to use marine SCD in the shrimp larval feeding. This product is also ideally utilized as a fish diet and can replace algae and *Artemia* (aquatic crustaceans known as brine shrimp, used to feed fish and crustacean larvae) in the marine hatchery sector (a place for artificial breeding, hatching and rearing through the early life stages of animals, finfish and shellfish (Crespi and Coche 2008)).

A considerable reduction in the size of the particles of macroalgae in the fermentation process makes them accessible for digestion by shellfish (e.g., clams) and also increases their digestibility (Pérez Camacho et al. 2007). Aquaculture based on algal fermented materials is eco-friendly (Uchida 2007). Moreover, the simplicity of the production technique enables SCD to be obtained on both a laboratory and an industrial scale. Additionally, the production costs of SCD organic matter are much lower than the production costs for live or freeze-dried phytoplankton (Pérez Camacho et al. 2007).

Pioneering research in this area was carried out by Dr. Motoharu Uchida and his team. Experiments on the use of seaweeds for the production of SCD by fermentation, from *Undaria pinnatifida* (Uchida and Murata 2002) and from *Laminaria japonica* (Uchida et al. 1997), for example, were carried out. For the production of SCD, the optimal conditions should be established: content of alga (5% on a dry basis), salts (e.g., NaCl 2.5–3.5%), enzymes (cellulose 0.5% w/v), microbial mixture, duration of incubation (6–14 days), incubation temperature (5–50°C, optimal 20°C) (Uchida and Murata 2002). An important factor for fermentation is salt concentration. Other bacteria can contaminate the cultures prepared without salt. The advantage of the obtained fermented product is the acidic reaction, which allows for long-shelf life at room temperature without excessive storage energy costs (Uchida and Miyoshi 2010). SCD technology is relatively economical with the final product that can be stored up to a year at room temperature (Felix and Pradeepa 2011).

Marine silage (MS) is a new dietary product prepared by the decomposition of seaweeds to the cellular level and by carrying out lactic acid fermentation (Uchida et al. 2004). The marine silage production process is simple, practical and economical (low cost). The process of hydrolysis requires neither expensive equipment nor procedures. The product can be stored for a long time; for example, MS obtained from *Undaria pinnatifida* has long-term (18 months) shelf-life at 20°C (Uchida et al. 2004). In the available literature, there are some examples of MS production from seaweeds: *Undaria pinnatifida* (Uchida et al. 2004, Uchida 2007), *Ulva* spp. (Uchida and Miyoshi 2010), *Undaria pinnatifida* and *Ecklonia maxima* (Uchida 2007).

The SCD and MS dietary materials for aquaculture are expected to have useful functions that contribute to fish health (Uchida and Miyoshi 2010). The diet based on SCD and MS has the potential to be used as a probiotic agent since it was treated by bacteria (bacteria are attached to the surface of the detritus). This approach has some useful functions, such as an anti-pathogenic activity and a vitamin-producing ability (Uchida et al. 1997, Uchida and Miyoshi 2010). In the work of Uchida (2007), it was shown that marine silage (obtained from *Ecklonia maxima*), when used as a supplement to the diet of fish—red sea bream—at 10% (w/w) promoted the survival rate of the fish by protecting it against viral disease (iridovirus) when compared with those fed with a control diet. Pérez Camacho et al. (2007) used single cell detritus produced from *Laminaria saccharina* in the feeding of clam (*Ruditapes decussatus*). The growth of clam was equivalent to, or even greater than that obtained with exclusively live phytoplankton diets. Additionally, the absence of proliferation of pathogenic bacteria in the culture vessels was observed (possibly as a result of the control of the bacteria used in the production of the SCD).

5.2 Fermented Products from Algae in the Feeding of Farm Animals

Researchers are looking for unconventional feed for farm animals, such as seaweed-based feed, which will replace the conventional feed (e.g., corn, wheat, soybean). Compared with the conventional, the seaweed feed is much richer in crude protein, crude fiber and minerals (especially iodine), which are beneficial to the growth and reproduction of domestic and farm animals (Zhang et al. 2012, Uchida and Miyoshi 2013). Hong et al. (2010) determined the effect of fermented brown seaweed waste (FBSW, addition of 1% and 2%) as functional feed on milk production and composition and physiological responses in Holstein dairy cows. It was found that daily milk yield and composition (fat, protein) were not affected by FBSW supplementation, but Ca level in milk significantly increased. Plasma thyroxine (T_4) concentration in the cows' diet increased with FBSW when compared to the control group (iodine is necessary for the production of T_4). Choi et al. (2014) examined the effect of the supplementation of dietary fermented seaweed (*Undaria pinnatifida* and *Hizikia fusiformis*) on growth performance and blood profile of broilers. The results suggested that the addition of fermented algae to the diet may provide positive effects on growth performance (increase of body weight gain, decrease of mortality rate) and immune response (increase of the level of serum immunoglobulin).

6. Fermented Algal Products for Humans

Until now, fermented products have not been produced from aquatic plant materials (e.g., algae) (Uchida and Miyoshi 2013). The interest in the use of 'natural green' extracts in the food industry (e.g., in various food and beverages) is increasing. Several products of microbial fermentation (e.g., anti-oxidants, flavors, colorants, preservatives and sweeteners) are incorporated to food as additives and supplements (Couto and Sanroman 2006). Seaweed can be used as a raw material rich in valuable nutraceutical compounds in the production of functional food (Gupta and Abu-Ghannam 2011). Wijesinghe et al. (2012) suggested finding the use for low-value algal biomass (e.g., *Ecklonia cava*) as a bio-resource for the development of functional food and cosmetics. In this work, the by-product of *Ecklonia cava* was fermented by yeast—*Candida utilis*. In the next step, ethanol extract was obtained which exhibited antioxidant activity. It was shown that fermentation could offer a tool for further increase in the bioactive potential of waste algal biomass.

Another area of research is focused on the concept of probiotic microbes and dietary prebiotics from seaweeds as functional ingredients for gut health (O'Sullivan et al. 2010). Food products that contain either probiotic microbes or prebiotics have been considered to be one of the functional foods that can promote health and prevent diseases (Qiang et al. 2009). Gupta et al. (2011) showed that the use of lactic acid bacteria (*Lactobacillus plantarum*) in the fermentation of seaweeds provides the possibility of developing a range of functional foods; for example, probiotic products. Strains of several *Lactobacillus* species proved to bring about a range of health promoting activities such as immunomodulation, enhancement of resistance against pathogens and reduction of blood cholesterol levels (de Vries et al. 2006).

The use of algal materials for the production of edible fermented products has been limited due to several reasons (Gupta et al. 2011). The human gastrointestinal tract does not digest the typical algal carbohydrates because they are dietary fibres (Dawczynski et al. 2007). These fibers have an inert nature and low fermentability by gut microbiota, probably due to high molecular weight polymers that pass through the gut too rapidly to allow bacterial utilization (Warrand 2006). Polysaccharides from marine macroalgae, such as fucoidan, laminarin, alginate and their derivatives, may offer the potential to be used as prebiotics (O'Sullivan et al. 2010). Ramnani et al. (2012) evaluated the fermentation and prebiotic properties of low molecular weight polysaccharides derived from agar and alginate present in seaweeds (*Gracilaria*, *Gelidium* and *Ascophyllum* spp.). These forms are more soluble and can be added at high concentrations to foods without affecting the sensory properties of the product. It is supposed that these compounds can stimulate the beneficial bacteria in the human gut. It was shown that low molecular weight polysaccharides were fermented by gut bacteria and exhibited the potential to be used as a novel source of prebiotics, enabling bifidobacterial populations and increasing acetate and propionate.

The application of the microbial fermentation of seaweeds can be one of the more interesting strategies to develop natural antibiotic or therapeutic substances. For example: fermentation broth of *Eisenia bicyclis* with *Candia utilis* YM-1 exhibited enhanced antimicrobial activity against methicillin-resistant *Staphylococcus aureus* and food-borne pathogenic bacteria (Eom et al. 2013).

Macroalgae can be also used for the production of a fermented plant beverage (FPB) (Prachyakij et al. 2008, Hayisama-ae et al. 2014). This non-alcoholic beverage is produced over a period of 2 to 6 months by lactic fermentation from different kinds of plants. The components required for making FPB are plant material (e.g., algae), sugar and potable water (3:1:10 (w/w/v)) (Prachyakij et al. 2008). FPBs are considered to be a novel functional food product because they are rich in nutrients, often containing antimicrobial compounds and potentially some human LAB probiotics (Kantachote et al. 2005, Prachyakij et al. 2008). Prachyakij et al. (2008) tested sensory quality (flavor, odor, color, clearness and overall acceptance) of the FSB obtained from *Gracilaria fisheri* (a fermented red seaweed beverage). The solution of FSB was a clear yellow color, tasted a little sweet with sour flavor and had a slight seaweed odor. Inoculation of seaweeds during fermentation with *Lb. plantarum* DW3 does not disturb the organoleptic properties of the beverage.

Hayisama-ae et al. (2014) produced a novel functional fermented seaweed beverage (FSB) from red seaweed—*Gracilaria fisheri*—using *Lb. plantarum* DW12 as the starter culture. In addition to the normal fermentation products, the bacteria (that is a potential probiotic) produced γ-aminobutyric acid (GABA—a non-protein amino acid) and prebiotic compounds (fructo-oligosaccharides) from sucrose. One of the most beneficial probiotic properties is their antibacterial activity; a cell free culture supernatant also inhibited the growth of potential food borne pathogens and spoilage bacteria (*Bacillus cereus*, *Staphylococcus aureus*, *Escherichia coli*, *Salmonella typhi* and *Vibrio parahaemolyticus*) and has antioxidant activity. In the work of Ratanaburee et al. (2011), *Lb. plantarum* DW12, a γ-aminobutyric acid producing strain, was used as a starter culture to obtain a functional fermented red seaweed beverage. The development of functional foods containing GABA has recently been intensively investigated. GABA has a variety of physiological functions; for example, it reduces hypertension, has diuretic effects and inhibits the proliferation of cancer cells (Park and Oh 2007).

Prachyakij et al. (2008) found that the fermented seaweed beverage produced from *Gracilaria fisheri* by lactic fermentation is a good means of providing more available essential elements such as Fe, Cu and Zn. The application of probiotic LAB for fermentation process is the best method to control contamination by yeast strains in fermented beverages. LAB produce several antimicrobial compounds such as: carbon dioxide, hydrogen peroxide and organic acids such as lactic and acetic acid which participate in the reduction of microbial contamination.

The development of lactic acid fermentation gives rise to the possibility of producing fermented food from algae. According to Uchida and Miyoshi (2013), seaweed sauce is a possible product. A major difficulty in the production of seaweed sauce of commercial value is the shortage of amino acid compounds found in the supernatant of fermented products prepared from seaweed. Another problem is the isolation of the appropriate bacteria strains that can grow in a highly-salted seaweed culture. In the work of Uchida et al. (2014), halophilic LAB (i.e., *Tetragenococcus halophilus*) that can be a starter culture for manufacturing fermented seaweed-sauce (a Nori–*Porphyra yezoensis*), were isolated from environmental samples.

7. Units Used for Fermentation of Seaweeds

Currently, the fermentation process of seaweeds is performed mainly on the laboratory scale, in order to characterize the obtained products in terms of physico-chemical properties and to check their utilitarian properties. There have been some attempts to increase the volume of fermenters used (e.g., 15 L—Prachyakij et al. 2008, Ratanaburee et al. (2011), 20 L—Uchida and Miyoshi 2010).

The process most developed is the fermentation of algae for aquaculture purposes. The production of bio-enriched feed (e.g., SCD) for aquaculture can be performed in large quantities in simple air-tight containers or in more sophisticated fermenters or bioreactors specially designed for the purpose. The advantage of the second solution is the time required to obtain the fermented product—only two to three days, whereas in air-tight containers two weeks would be required. Fermenters are also recommended to enhance the purity and quality of the final product (Felix and Pradeepa 2011). For example, Algal Scientific Corporation (Northville, Michigan, www.algalscientific. com, access 19th January 2015) produces algae based products in two processes: sterile fermentation where the only objective is the production of high value algal bio-products, and open fermentation where the primary objective is to reduce the treatment costs for high strength process water as is found in the food and beverage industries. After the fermentation process, the dense algal monoculture biomass is harvested from the reactor, dried and milled to produce a whole cell animal feed ingredient called Algamune™ AM. The principal compound in the product is beta-1,3-glucan that exhibits its health benefits in animals and humans. This is a versatile compound that has been incorporated into many products such as nutraceuticals, functional foods and beverages and animal feed.

However, in the case of fermentation of seaweeds, more trials are needed in order to optimize the process (the choice of physico-chemical conditions of the fermentation process, as well as the appropriate amount of algal biomass, enzymes, microbial starter cultures, etc.). Several technological aspects should be also considered in the design of a novel fermented food, such as the growth capability and productivity of the starter culture and the stability of the final product during storage (Gupta et al. 2011).

8. Conclusions

The advantage of the fermentation of algae is the possibility of executing this process on a large scale, especially in coastal areas; however, more in-depth experiments are needed. Seaweeds are recognized as a useful raw material for the production of functional fermented foods because they contain safe biologically-active substances. Fermentation of algal biomass enables to obtain product with defined characteristics and health-promoting properties. Fermentation improves the nutritive value of the obtained fermented products (the content of proteins, vitamins, minerals, essential amino acids and essential fatty acids), which can be in the form of liquid fermentation (e.g., beverages), fermented seaweed powder (e.g., animal feed) or a raw material for the extraction of biologically active compounds (functional ingredients of food, feed, cosmetics, etc.). These functional compounds (polysaccharides, PUFAs, polyphenols) have confirmed anti-diabetic, anti-allergic, anti-coagulant, anti-cancer, anti-inflammatory, anti-oxidant, immune-protective, cardiovascular, immune-enhancing properties, beneficial for humans and animals. Fermented algal products have also the potential to be used as probiotics (lactic acid bacteria) and prebiotics (polysaccharides) that can promote health and prevent diseases.

On the basis of presented examples of products, it can be concluded that there is a significant possibility of expanding the algal fermentation industry in the nearest future, for several reasons: practical (production of functional products), environmental (coping with the effects of eutrophication) and economic (processing instead of landfilling).

Acknowledgments

This work was supported by grants entitled 'Biologically Active Compounds In Extracts From Baltic Seaweeds' (2012/05/D/ST5/03379) attributed by The National Science Centre and 'Innovative Technology Of Seaweed Extracts—Components Of Fertilizers, Feed And Cosmetics' (PBS/1/A1/2/2012) attributed by The National Centre for Research and Development in Poland.

Keywords: Microalgae, macroalgae, fermentation, functional food, feed, high-value products

References

Choi, Y.J., Lee, S.R. and Oh, J-W. 2014. Effects of dietary fermented seaweed and seaweed Fusiforme on growth performance, carcass parameters and immunoglobulin concentration in broiler chicks. Asian Australasian Journal of Animal Sciences 27: 862–870.

Couto, S.R. and Sanroman, M.A. 2006. Application of solid-state fermentation to food industry: A review. Journal of Food Engineering 76: 291–302.

Crespi, V. and Coche, A. 2008. Food and Agriculture Organization of the United Nations (FAO) Glossary of Aquaculture.

Dawczynski, C., Schubert, R. and Jahreis, G. 2007. Amino acids, fatty acids, and dietary fibre in edible seaweed products. Food Chemistry 103: 891–899.

Davis, T.A., Volesky, B. and Mucci, A. 2003. A review of the biochemistry of heavy metal biosorption by brown algae. Water Research 37: 4311–4330.

deVries, M.C., Vaughan, E.E., Kleerebezem, M. and de Vos, W.M. 2006. *Lactobacillus plantarum*—Survival, functional and potential probiotic properties in the human intestinal tract. International Dairy Journal 16: 1018–1028.

Ennouali, M., Ouhssine, M., Ouhssine, K. and Elyachioui, M. 2006. Biotransformation of algal waste by biological fermentation. African Journal of Biotechnology 5: 1233–1237.

Eom, S.-H., Lee, D.-S., Kang, Y.M., Son, K.-T., Jeon, Y.-J. and Kim, Y.-M. 2013. Application of yeast *Candida utilis* to ferment *Eisenia bicyclis* for enhanced antibacterial effect. Applied Biochemistry and Biotechnology 171: 569–582.

Felix, N. and Brindo, R.A. 2008. Fermented feed ingredients as fish meal replacer in aqua feed production. Aquaculture Asia Magazine April–June: 33–34.

Felix, S. and Pradeepa, P. 2011. Single-cell detritus: Fermented, bioenriched feed for marine larvae. Global Aquaculture Advocate July/August: 72–73.

Felix, S. and Pradeepa, P. 2012. Lactic acid fermentation of seaweed (*Ulva reticulata*) for preparing Marine Single Cell Detritus (MSCD). Tamil Nadu J. Veterinary & Animal Sciences 8: 76–81.

Felix, N. and Brindo, R.A. 2014. Effects of raw and fermented seaweed, *Padinatetrastomatica* on the growth and food conversion of giant freshwater prawn *Macrobrachium rosenbergii*. International Journal of Fisheries and Aquatic Studies 1: 108–113.

Gupta, S. and Abu-Ghannam, N. 2011. Recent developments in the application of seaweeds or seaweed extracts as a means for enhancing the safety and quality attributes of foods. Innovative Food Science and Emerging Technologies 12: 600–609.

Gupta, S., Abu-Ghannam, N. and Scannell, A.G.M. 2011. Growth and kinetics of *Lactobacillus plantarum* in the fermentation of edible Irish brown seaweeds. Food and Bioproducts Processing 89: 346–355.

Han, L., Zhang, S., Ma, J. and Liu, X. 2012. Research and optimization of technological process based on fermentation for production of seaweed feed. Green and Sustainable Chemistry 2: 47–54.

Hayisama-ae, W., Kantachote, D., Bhongsuwan, D., Nokkaew, U. and Chaiyasut, C. 2014. A potential synbiotic beverage from fermented red seaweed (*Gracilaria fisheri*) using *Lactobacillus plantarum* DW12. International Food Research Journal 21: 1789–1796.

Hong, Z.-S., Lee, Z.-H., Xu, C.-X., Yin, J.-L., Jin, Y.-C., Lee, H.-J., Lee, S.-B., Choi, Y.-J. and Lee, H.-G. 2010. Effect of Fermented Brown Seaweed Waste (FBSW) on milk production, composition and physiological responses in Holstein dairy cows. Journal of Animal Science and Technology 52: 287–296.

Huang, Y., Hong, A., Zhang, D. and Li, L. 2014. Comparison of cell rupturing by ozonation and ultrasonication for algal lipid extraction from *Chlorella vulgaris*. Environmental Technology 35: 931–937.

Kantachote, D., Charernjiratrakul, W. and Asavaroungpipop, N. 2005. Characteristics of fermented plant beverages in southern Thailand. Songklanakarin Journal of Science and Technology 27: 601–615.

Kim, S.K. 2013. Marine Nutraceuticals: Prospects and Perspectives. CRC Press, Boca Raton, Florida.

Liu, X., Zhang, S.-P., Han, L. and Li, Y. 2012. Influence of several fermentation on seaweed waste of feed. Journal of Sustainable Bioenergy Systems 2: 108–111.

Mayakrishnan, V., Kannappan, P., Abdullah, N. and Ahmed, A.B.A. 2013. Cardioprotective activity of polysaccharides derived from marine algae: An overview. Trends in Food Science & Technology 30: 98–104.

Michalak, I. and Chojnacka, K. 2014. Algal extracts: Technology and advances. Engineering in Life Sciences 14: 581–591.

Michalak, I. and Chojnacka, K. 2015. Algae as production systems of bioactive compounds. Engineering in Life Sciences DOI: 10.1002/elsc.201400191.

Mišurcová, L., Škrovánková, S., Samek, D., Ambrožová, J. and Machů, L. 2012. Health benefits of algal polysaccharides in human nutrition. Advances in Food and Nutrition Research 66: 75–145.

Nikapitiya, C., de Zoysa, M., Jeon, Y.-J. and Lee, J. 2007. Isolation of sulfated anti-coagulant compound from fermented red seaweed *Grateloupia filicina*. Journal of the World Aquaculture Society 38: 407–417.

O'Sullivan, L., Murphy, B., McLoughlin, P., Duggan, P., Lawlor, P.G., Hughes, H. and Gardiner, G.E. 2010. Prebiotics from marine macroalgae for human and animal health applications. Marine Drugs 8: 2038–2064.

Park, K.-B. and Oh, S.-H. 2007. Production of yogurt with enhanced levels of gamma-aminobutyric acid and valuable nutrients using lactic acid bacteria and germinated soybean extract. Bioresource Technology 98: 1675–1679.

Pérez Camacho, A., Salinas, J.M., Delgado, M. and Fuertes, C. 2007. Use of single cell detritus (SCD) produced from *Laminaria saccharina* in the feeding of the clam *Ruditapes decussatus* (Linnaeus, 1758). Aquaculture 266: 211–218.

Plaza, M., Cifuentes, A. and Ibáñez, E. 2008. In the search of new functional food ingredients from algae. Trends in Food Science & Technology 19: 31–39.

Prachyakij, P., Charernjiratrakul, W. and Kantachote, D. 2008. Improvement in the quality of a fermented seaweed beverage using an antiyeast starter of *Lactobacillus plantarum* DW3 and partial sterilization. World Journal of Microbiology & Biotechnology 24: 1713–1720.

Prajapati, J.B. and Nair, B.M. 2008. The history of fermented foods. pp. 1–24. *In*: E.R. Farnworth (ed.). Handbook of Fermented Functional Foods. CRC Press, Taylor and Francis Group, Boca Raton, Florida.

Qiang, X., YongLie, C. and QianBing, W. 2009. Health benefit application of functional oligosaccharides. Carbohydrate Polymers 77: 435–441.

Ramnani, P., Chitarrari, R., Tuohy, K., Grant, J., Hotchkiss, S., Philp, K., Campbell, R., Gill, C. and Rowland, I. 2012. *In vitro* fermentation and prebiotic potential of novel low molecular weight polysaccharides derived from agar and alginate seaweeds. Anaerobe 18: 1–6.

Ratanaburee, A., Kantachote, D., Charernjiratrakul, W., Penjamras, P. and Chaiyasut, C. 2011. Enhancement of γ-aminobutyric acid in a fermented red seaweed beverage by starter culture *Lactobacillus plantarum* DW12. Electronic Journal of Biotechnology 4: DOI: 10.2225/vol14-issue3-fulltext-2.

Shiby, V.K. and Mishra, H.N. 2013. Fermented milks and milk products as functional foods: A review. Critical Reviews in Food Science and Nutrition 53: 482–496.

Steinkraus, K.H. 1997. Classification of fermented foods: Worldwide review of household fermentation techniques. Food Control 8: 311–317.

Tamang, J.P. and Kailasapathy, K. 2010. Fermented foods and beverages of the world. CRC Press, Boca Raton, Florida.

Tsuchiya, K., Matsuda, S., Hirakawa, G., Shimada, O., Horio, R., Taniguchi, C., Fujii, T., Ishida, A. and Iwahara, M. 2007. GABA production from discolored laver by lactic acid fermentation and physiological function of fermented laver. Nihon Shokuhin Hozou Kagaku-kaishi 33: 121–125.

Uchida, M., Nakata, K. and Maeda, M. 1997. Introduction of detrital food webs into an aquaculture system by supplying single cell algal detritus produced from *Laminaria japonica* as a hatchery diet for *Artemia nauplii*. Aquaculture 154: 125–137.

Uchida, M. and Murata, M. 2002. Fermentative preparation of single cell detritus from seaweed, *Undaria pinnatifida*, suitable as a replacement hatchery diet for unicellular algae. Aquaculture 207: 345–357.

Uchida, M., Numaguchi, K. and Murata, M. 2004. Mass preparation of marine silage from *Undaria pinnatifida* and its dietary effect for young pearl oysters. Fisheries Science 70: 456–462.

Uchida, M. 2005. Studies on lactic acid fermentation of seaweed. Bulletin of Fisheries Research Agency 14: 21–85.

Uchida, M. 2007. Preparation of marine silage and its potential for industrial use. pp. 51–56. *In*: R. Stickney, R. Iwamoto and M. Rust (eds.). Aquaculture and Stock Enhancement of Finfish: Proceedings of the

Thirty-fourth U.S.-Japan Aquaculture Panel Symposium, San Diego, California, November 7–9, 2005. U.S. Dept. Commerce, NOAA Tech. Memo. NMFS-F/SPO-85.

Uchida, M. and Miyoshi, T. 2010. Development of a new dietary material from unutilized algal resources using fermentation skills. Bulletin of Fisheries Research Agency 31: 25–29.

Uchida, M. and Miyoshi, T. 2013. Review: Algal fermentation—the seed for a new fermentation industry of foods and related products. JARQ 47: 53–63.

Uchida, M. 2014. Fermentation of seaweeds and its applications. pp. 14–46. *In*: S.K. Kim (ed.). Seafood Science: Advances in Chemistry, Technology and Applications. CRC Press, Taylor and Francis Group, Boca Raton, Florida.

Uchida, M., Miyoshi, T., Yoshida, G., Niwa, K., Mori, M. and Wakabayashi, H. 2014. Isolation and characterization of halophilic lactic acid bacteria acting as a starter culture for sauce fermentation of the red alga Nori (*Porphyra yezoensis*). Journal of Applied Microbiology 116: 1506–1520.

Warrand, J. 2006. Healthy polysaccharides. Food Technology and Biotechnology 44: 355–370.

Wijesinghe, W.A.J.P., Lee, W.-W., Kim, Y.-M., Kim, Y.-T., Kim, S.-K., Jeon, B.-T., Kim, J.-S., Heu, M.-S., Jung, W.-K., Ginnae, A., Ki-Wan, L. and You-Jin, J. 2012. Value-added fermentation of *Ecklonia cava* processing by-product and its antioxidant effect. Journal of Applied Phycology 24: 201–209.

Zhang, S., Hu, X., Ma, J., Ma, Z., Liu, X. and Cui, L. 2012. Study on feed fermented from seaweed waste. African Journal of Microbiology Research 6: 7610–7615.

13

Antimicrobial Resistance of Fermented Food Bacteria

Nevijo Zdolec

1. Introduction

Lactic acid bacteria (LAB) are the most important bacterial group involved in fermented food production. Some fermented products include additional microbiological groups important for products' safety and quality, such as coagulase-negative staphylococci (CoNS), yeasts and molds. CoNS and molds are particularly important in fermented meat products, due to their contribution to proteolytic and lipolytic succession. Yeasts are more commonly employed in the production of cheese and fermented milks (Boekhout and Samson 2005). Some species or strains have been selected as technologically or hygienically superior to others, leading to raw material inoculation (e.g., milk, sausage mixture). Their implementation should upgrade the safety and quality of the final products. Such microbial species or strains are known as (protective) starter cultures which are recognized as safe in fermented food production (Lücke 2000). Recently, there has been expansion of research in the area of autochthonous starter cultures, i.e., the potential application of strains isolated from traditionally fermented foodstuffs in industrial conditions. The purpose of this approach is to produce industrial food products with autochthonous sensorial features with the help of wild strain(s) (Frece et al. 2014). The selection of potential starter cultures should include health-hazard testing, such as the possibility of strains producing biogenic amines, toxins or antibiotic resistance gene carriage.

Today, hundreds of substances are currently used as active principles in health products developed for the treatment of food producing animals. Therefore, many

University of Zagreb, Faculty of Veterinary Medicine, Department of Hygiene, Technology and Food Safety, Heinzelova 55, 10000 Zagreb, Croatia.
E-mail: nzdolec@vef.hr

animals are exposed to chemicals that may leave residues in the carcass at the time of slaughter. The possibility of resistant organisms of animal origin becoming directly pathogenic to man, or transferring their resistance genes to pathogens of medical importance, is of particular concern (Teuber 2001). Due to the intensive use of antibiotics in public health and animal husbandry, antibiotic resistance in pathogens has increasingly become a medical problem over the last decades.

The prudent use of antibiotics in veterinary medicine and agriculture is an important factor in spreading antimicrobial resistance through food production chain. In addition to the spread of resistant zoonotic food borne pathogens, there is also possibility that food-related commensal bacteria or opportunistic pathogens are carriers of resistance genes and therefore a potential hazard to consumers. Fermented food products could be an ideal vehicle for antimicrobial resistance transfer from animals, environment or humans through food chain, because these products are not thermally treated and are naturally rich in indigenous microbiota (Zdolec et al. 2011). During the last few decades, the hazard potential of natural microbiota from fermented food was assessed in relation to potential hazards, including antimicrobial resistance (Resch et al. 2008). In general, low occurrence of hazardous determinants was found in food CoNS and LAB, but the presence of resistance genes is still a reason for concern. Antimicrobial resistance is most important in clinically relevant zoonotic bacteria, since they directly threaten human life. The significance of antimicrobial resistance in food-borne non-pathogens is not recognized as a rising public health problem, but theoretically, it could be taken into account as a potential threat. In this respect, naturally present LAB or CoNS in different fermented food can be observed as potentially hazardous microbial groups as transmitters of antibiotic resistance determinants. In the present chapter, the antimicrobial resistance of CoNS and LAB (enterococci) is discussed in terms of fermented meat/milk products and raw materials.

2. Coagulase-Negative Staphylococci (CoNS) and Enterococci

Coagulase negative staphylococci (CoNS) are an important microbial group in meat fermentation, together with lactic acid bacteria (LAB). Their favorable actions are expressed in color and aroma development and in proteolytic and lipolytic succession, all of which contribute to the final product's sensorial characteristics. The degree of proteolitic changes during the fermented meat's maturation is related mainly to CoNS proteases (Hammes and Hertel 1998), in addition to some LAB (Casaburi et al. 2008). Further, CoNS influence the course of change and intensity of lipid degradation (Hadžiosmanović 1978). Products of lipolysis and proteolysis such as peptides, amino acids, carboniles and volatile compounds contribute to specific flavor and texture of fermented meat products (Hughes et al. 2002, Casaburi et al. 2008, Toldrá 2008). The importance of CoNS in fermented sausage technology is also manifested in catalase activity (peroxide degradation) and nitrite reduction (Cocconcelli and Fontana 2008). An understanding of the biochemical activity of CoNS in meat matrix lead to the selection of technologically acceptable, safe strains as autochthonous starter cultures in order to standardize sensorial features and overall product quality (Zdolec et al. 2013).

The selection criteria for appropriate starter cultures are now extended to the consideration of potential public health hazards like the ability to produce biogenic amines or to transfer resistant genes (Talon and Leroy 2011). Differences in production technology of traditionally fermented meat products, raw materials, additives, climate, etc. significantly influence the affirmation of specific CoNS species and strains. For example, huge differences are found in CoNS succession during the ripening of fermented sausages, reaching a final number of 3 to 9 $\log_{10}$ cfu (colony forming units)/g depending on sausage type, competitiveness with LAB, pH, salt and other factors (Samelis et al. 1998, Coppola et al. 2000, Cocolin et al. 2001, Comi et al. 2005, Urso et al. 2006, Zdolec et al. 2008). The most frequently isolated CoNS species in fermented meats are *Staphylococcus xylosus, Staphylococcus eqourum, Staphylococcus saprophyticus* and *Staphylococcus carnosus*. In addition, several studies reported the presence of other staphylococci, such as *Staphylococcus epidermidis* or *Staphylococcus warneri* (Blaiotta et al. 2004, Martin et al. 2006, Bonomo et al. 2009, Marty et al. 2012, Zdolec et al. 2013). These staphylococcal species are frequently isolated from different specimens in meat chain production, including final meat products, and harbor transmissible resistance determinants (Simeoni et al. 2008).

Enterococci, as representatives of LAB groups, are particularly important in meat and milk products as potential protective and probiotic cultures, respectively. Opinions on their hygienic and technological importance are quite controversial; they could promote favorable sensory properties of fermented products and suppress pathogens or spoilage bacteria (bacteriocins—enterocins) but also they could threat the consumer's health by producing biogenic amines or transfer antimicrobial resistance genes (Čanžek Majhenić 2006). Enterocin-producing strains of enterococci are frequently isolated from different fermented meat or milk products and proposed as bioprotective cultures for industrial applications since they are active against foodborne pathogens (Hugas et al. 2003). According to Giraffa (2002), enterococci survive and grow during meat and milk fermentation, especially in natural products without competitive starter cultures. Zdolec et al. (2008) demonstrated the continuous growth of enterococci in spontaneously fermented sausages during 28-day maturation (from 3.7 to 5.3 log cfu/g), while, in the presence of a bacteriocinogenic culture of *Lactobacillus sakei*, the population was 1 log lower. Similar findings were reported by Urso et al. (2006) in Italian dry fermented sausages. Our experiments showed significantly lower (P < 0.05) enterococci count in all fermentation phases of sausages by implementing *Lb. sakei* and bacteriocin mesenterocin Y (Zdolec et al. 2008). According to Bover-Cid (2001) and other authors, the majority of enterococci isolated from fermented sausages produce biogenic amines or harbor resistance genes. These potential risks could be prevented by the application of competitive starter cultures. As mentioned before, enterococci possess a strong probiotic potential, especially in fermented dairy products. Numerous dairy-related *Enterococcus faecalis* and *E. faecium* strains are now well characterized and tested for probiotic characteristics (Laukova 2012, Franz et al. 2011). At the same time, these species are known as potential pathogens and carriers of resistance genes (Franz et al. 2011). In general, enterococci possess intrinsic antibiotic resistance to cephalosporins, ß-lactams, sulphonamides, and to certain levels of clindamycin and aminoglycosides, while acquired resistance exists to chloramphenicol, erythromycin, clindamycin, aminoglycosides, tetracycline, ß-lactams, fluoroquinolones and glycopeptides (Gimenez Pereira 2005). During the selection of potential probiotic cultures, these risks must be excluded by phenotypic and genotypic evaluations. Another public health issue is enterococcal contamination of raw milk intended for the production of unpasteurized (traditional) products.

3. Antimicrobial Resistance of CoNS and Enterococci in Raw Material: Meat and Milk

Staphylococci of animal origin are carriers of antimicrobial resistance genes that confer resistance to veterinary antimicrobial agents such as penicillins, cephalosporins, tetracyclines, macrolides, lincosamides, phenicols, aminoglycosides, aminocyclitols, pleuromutilins, and diaminopyrimidines (Wendlandt et al. 2013). CoNS are a microbial group involved in several animal and human diseases and frequently present in the environment, food, food-producing facilities or hospitals. Their resistance to antimicrobials could lead to serious public-health problems, especially due to the transfer of *mec*A, the methicillin resistance gene, to pathogenic *Staphylococcus aureus* (Ito et al. 2009). Bhargava and Zhang (2014) have found different methicillin resistant CoNS (MRCoNS) in retail meat (beef, chicken, turkey), which can serve as reservoirs for *mec*A. In addition to the clonal transmission of MRCoNS in meat, horizontal occurrence of staphylococcal cassette chromosome *mec* (SCC*mec*) was observed in staphylococcal species (Bhargava and Zhang 2014). Simeoni et al. (2008) monitored the persistence of resistant CoNS in different production stages of pork, including sampling of faeces, feedstuffs, raw meat, processed meat and fermented sausages. The highest percentage of resistant strains was found in raw meat and processed meat, and the lowest in fermented sausages. Results observed show that the swine production chain is a source of antibiotic-resistant staphylococci, thus suggesting the importance of resistance surveillance in the food production environment. From the epidemiological point of view, the contamination routes of resistant bacteria (CoNS) in meat must be determined, starting from the farm level, and including slaughtering procedures and meat processing.

It is well known that CoNS are the most important bacteria involved in subclinical bovine mastitis, next to *Staphylococcus aureus* (Kalmus et al. 2011). Their resistance to antimicrobial agents is common due to a high antibiotic pressure in conventional dairy farming. Sampimon et al. (2011) stated that different CoNS species from bovine milk differ significantly in their phenotypic and genotypic antimicrobial resistance profile which is important for udder health management. They noticed the highest phenotypic penicillin resistance in *Staphylococcus epidermidis* strains which were also frequently multiresistant and harbored *mec*A genes. Kalmus et al. (2011) isolated 38.5% penicillin resistant CoNS from cows that had clinical or subclinical mastitis. However, it is also important to evaluate the presence of antimicrobial resistant staphylococci in regularly collected raw milk samples from clinically healthy animals in order to assess the potential spreading of resistant strains from raw material to dairy products (from unpasteurized milk).

3.1 Case Study: Antimicrobial Resistance of Milk Bacteria from Healthy and Drug-Treated Cow Udder

Recently, milk samples were examined from healthy cows without any signs of mastitis and drug-treated cows from the same farms which had mastitis (Zdolec et al. 2014). The aim of this study was to evaluate the composition and antimicrobial resistance of milk microbiota, considering udder health status. Milk was aseptically sampled from healthy cows without any signs of mastitis (n = 17) and drug-treated cows with mastitis (n = 19). Total viable count, psychrophilic bacteria, LAB, staphylococci, *Escherichia coli,* enterococci, enterobacteria, *Listeria* spp. and sulphite-reducing clostridia were determined using standard cultured methods, and results are shown in Table 1.

Presumptive colonies of enterococci (n = 56), enterobacteria (n = 30), *E. coli* (n = 24) and staphylococci (n = 94) were randomly selected for antimicrobial susceptibility testing by disk diffusion method (CLSI 2010). Depending on microbial group or species, the following antimicrobial disks (Bio-Rad, Marnes-la-Coquette, France) were used: ampicillin (10 μg), rifampin (5 μg), chloramphenicol (30 μg), tetracycline (30 μg), erythromycin (15 μg), nitrofurantoin (300 μg), vancomycin (30 μg), penicillin (10 IU), linezolid (30 μg), trimethoprim (5 μg), cefoperazone (75 μg), kanamycin (30 μg), trimethoprim/sulfamethoxazole (25 μg), nalidixic acid (30 μg), ciprofloxacin (5 μg), gentamicin (10/300 μg), teicoplanin (30 μg), sulfonamides (250 μg), levofloxacin (5 μg), clindamycin (2 μg) and amoxicillin/clavulanic acid (30 μg). Bacterial culture (0.5 McFarland) was streaked on Mueller-Hinton agar (Bio-Rad, France) and a maximum of 6 disks were used per plate by means of a disk dispenser (Bio-Rad, France). After incubation (35°C, 18–24 hours), zone diameter was measured and interpreted according to CLSI document M100-S20. Vancomycin resistance in enterococci was tested by both agar diffusion method and E-test (AB BIODISK, bioMérieux). Results of antimicrobial susceptibility test are presented in Table 2. A total of 94 staphylococci strains were isolated and subjected to susceptibility tests against 17 selected antimicrobial agents. Staphylococci isolated from drug-treated udder milk samples were most frequently resistant to clindamycin, penicillin, ampicillin, linezolid and erythromycin. A significantly lower share of resistant staphylococci was found in milk samples from healthy udders (Fig. 1) where the most common resistance was found toward penicillin (14.6%), erythromycin and kanamycin (12.2%). There is a lack of similar studies performed in raw milk from healthy cows; on the other hand, antimicrobial resistance surveys are done in milk from mastitic cows with *Staphylococcus aureus* or CoNS. Jurmanović et al. (2012) reported that *S. aureus* isolates from mastitic cows in Croatia are resistant to *β*-lactam antibiotics, aminoglycosides, lincosamides, oxytetracyclin, fluoroquinolones and sulfonamides. Results of their 9-year survey showed the increase of resistance of *S. aureus* (n = 2719) towards kanamycin, neomycin, enrofloxacin, lincomycin and penicillin. Frey et al. (2013) reported coagulase negative staphylococci (CoNS) from mastitic cows mostly resistant to oxacillin (47.0% of the isolates), fusidic acid (33.8%), tiamulin (31.9%), penicillin (23.3%), tetracycline (15.8%), streptomycin (9.6%), erythromycin (7.0%), sulfonamides (5%), trimethoprim (4.3%), clindamycin (3.4%), kanamycin (2.4%), and gentamicin (2.4%).

Table 1. Bacterial counts ($\log_{10}$ cfu/ml) in milk according to udder health status.

	Enterococci		Enterobacteria		E. coli		Staphylococci		Total viable count		Psychrophilic bacteria		Lactic acid bacteria	
	Drug-treated udder	Healthy udder	Drug-treated udder	Healthy udder	Drug-treated udder	Healthy udder	Drug-treated udder	Healthy udder	Drug-treated udder	Healthy udder	Drug-treated udder	Healthy udder	Drug-treated udder	Healthy udder
X	2.28	1.77	1.71	1.00	0.9	1.02	3.09	3.07	4.25a	3.18b	2.45	2.07	2.55c	0.81d
Max	5	3.6	4.69	3.00	3.3	3.23	4.0	4.0	5.39	3.77	5.17	3.95	4.77	3.65
Min	<1	<1	<1	<1	<1	<1	2	<1	3	3	<2	<2	<2	<2
SD	1.6	1.36	1.61	1.19	1.35	1.26	0.62	0.94	1.21	0.29	1.89	1.24	1.51	1.26

Mean values marked with different letters are significantly different ($P < 0.05$).
Source: Dobranić et al. (2014).

Table 2. Percentage of resistant strains depending on health status of cow udder.

Antimicrobial agent	Enterococci (n = 56)		Enterobacteria (n = 24)		E. coli (n = 30)		Staphylococci (n = 94)	
	Drug-treated udder	Healthy udder	Drug-treated udder	Healthy udder	Drug-treated udder	Healthy udder	Drug-treated udder	Healthy udder
Ampicillin	27	0	n.t.	n.t.	n.t.	n.t.	35.85	9.8
Rifampin	56	47	n.t.	n.t.	n.t.	n.t.	13.2	0
Chloramphenicol	81	68	14.3	0	50	0	30.18	4.8
Tetracycline	24	10	57.1	40	50	7.1	15.09	0
Erythromycin	78	63	n.t.	n.t.	n.t.	n.t.	35.85	12.2
Nitrofurantoin	86	63	35.7	10	18.8	7.1	n.t.	n.t.
Penicillin	81	53	n.t.	n.t.	n.t.	n.t.	47.16	14.6
Linezolid	0	0	n.t.	n.t.	n.t.	n.t.	35.85	0
Trimethoprim	n.t.	n.t.	50	10	68.8	0	n.t.	n.t.
Cefoperazone	n.t.	n.t.	0	0	31.3	0	1.9	2.4
Kanamycin	n.t.	n.t.	28.6	10	43.8	14.3	15.09	12.2
Trimethoprim/Sulfamethoxazole	n.t.	n.t.	35.7	20	50	50	0	2.4
Nalidixic acid	n.t.	n.t.	14.3	0	37.5	0	n.t.	n.t.
Ciprofloxacin	0	0	7.1	0	37.5	0	0	0
Gentamicin	0	0	0	30	12.5	7.1	3.8	2.4

Teicoplanin	n.t.	n.t.	n.t.	n.t.	n.t.	n.t.	30.18	0
Sulfonamides	n.t.	n.t.	35.7	10	50	0	13.2	4.8
Levofloxacin	0	0	n.t.	n.t.	n.t.	n.t.	0	0
Clindamycin	n.t.	n.t.	n.t.	n.t.	n.t.	n.t.	49.05	4.8
Amoxicillin/clavulanic acid	n.t.	n.t.	n.t.	n.t.	n.t.	n.t.	1.9	0
Vancomycin*	35	32	n.t.	n.t.	n.t.	n.t.	n.t.	n.t.
Streptomycin	n.t.	n.t.	42.9	20	62.5	0	n.t.	n.t.

*intermediate zones in agar diffusion test.
n.t.: not tested.
Source: Zdolec et al. (2014).

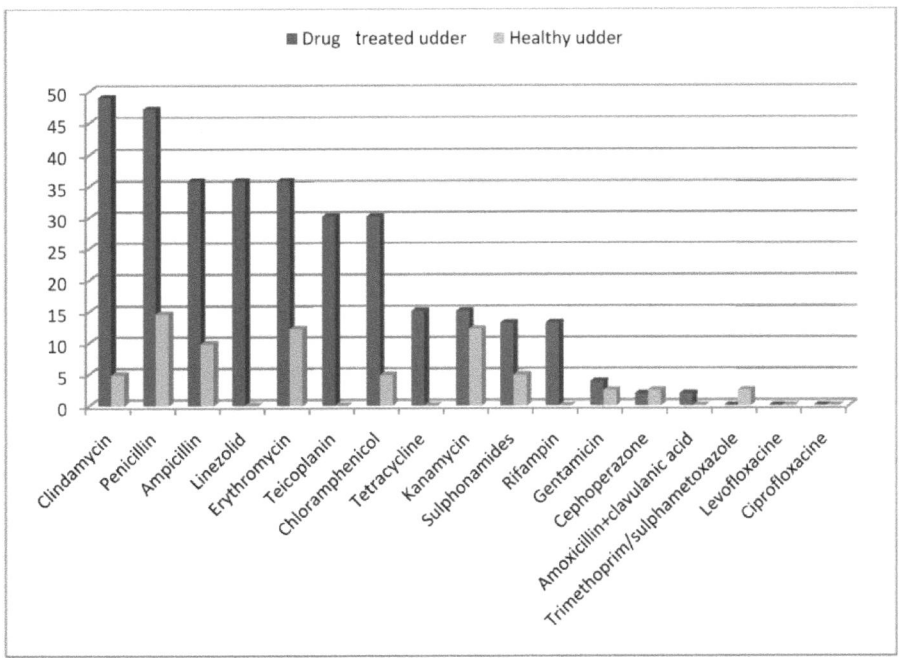

Figure 1. Percentage of resistant CoNS strains (n = 94) depending on health status of cow udder. Source: Zdolec et al. (2014).

The prevalence of multiresistant bacteria was significantly higher in the milk of drug-treated cows (Table 3). The percentage ratio (drug-treated: healthy udders) of multiresistant staphylococci was 56.5:4.8 and enterococci 87.2:73.7. Three enterococci strains and five staphylococci strains were resistant to 7 and 8 antimicrobial agents, respectively. Enterococci were resistant to agents from antimicrobial groups of Aminopenicillins, Rifampin, Chloramphenicol, Tetracyclines, Macrolides, Nitrofurans, Penicillins G and V. More than 80% of enterococci strains from drug-treated udders were resistant to penicillin, chloramphenicol and nitrofurantoin. However, the share of resistant enterococci from healthy udders towards the same antimicrobials ranged from 50–60%. Such a high share of resistant enterococci in both cow populations could be the result of animal cohabitation and cross contamination. Using the agar disk diffusion method, 19 enterococci with intermediate vancomycin zones were found. However, E-tests did not show any MIC above 32 µg/ml, indicating the absence of vancomycin-resistant enterococci (VRE). Vancomycin is a highly important antibiotic used in the treatment of human bacterial infections, including enterococci or methicillin-resistant *S. aureus*. Linezolid is the second important agent in treatment of vancomycin-, oxacillin- or methicillin-resistant enterococci/staphylococci infections (Deresinski 2009). Jimenez et al. (2013) reported that all enterococci isolates from ovine, feline, canine, porcine and human milk were susceptible to vancomycin and linezolid. The most isolates were resistant to chloramphenicol and tetracycline, and frequently resistant to erythromycin, which is partially presented in our study. During

Table 3. Percentage of multiresistant strains related to udder health status and distribution of multiresistance toward antimicrobial agents.

Multiresistant strains	Drug-treated udder	Healthy udder	Number of resistant strains (healthy udder/drug-treated udder)										
			Number of antimicrobial agents										
			2	3	4	5	6	7	8	9	10	11
Enterococci	87.2	73.7	2/3	1/1	1/4	6/7	4/10	0/6	0/3	-	-	-
Enterobacteria	57.14	30	2/2	0/1	1/2	1/1	0/2	0/1	1/1	-	-	-
E. coli	56.3	0	2/2	-	0/1	0/1	0/1	-	-	-	0/2	0/4
Staphylococci	56.6	4.8	5/5	1/2	0/6	0/7	1/7	0/3	0/5	-	-	-

Source: Zdolec et al. (2014).

the last few years, a low occurrence of VRE in raw milk was reported by other authors (Cetinkaya et al. 2013, Kročko et al. 2011). Dairy products, primarily cheeses, are more frequently positive to VRE strains (Furlaneto-Maia 2014).

4. Antimicrobial Resistance of CoNS and Enterococci from Fermented Meat and Milk Products

Some CoNS species are considered opportunistic pathogens and a major cause of nosocomial infections in humans and animals (Piette and Verschraegen 2009). In the food industry, the specific reason for concern is biofilm-positive strains such as *S. epidermidis* (Schlegelová et al. 2008a). Considering fermented meats, Even et al. (2010) reported a low occurrence of safety hazards in CoNS population. However, possible risks are related to biogenic amine and enterotoxin producers (Dobranić et al. 2013) as well as resistance gene carriers (Marty et al. 2012). Generally, the most relevant issue in food-related CoNS is antimicrobial resistance, mostly to erythromycin, tetracycline and penicillin (Simeoni et al. 2008, Resch et al. 2008, Even et al. 2010). Some recent studies showed a significant share of resistant strains in CoNS population of spontaneously fermented sausages (Zdolec et al. 2013, Marty et al. 2012). Zdolec et al. (2013) reported a frequent presence of *tet*K and *tet*M genes in *S. epidermidis* as dominant species in naturally produced fermented sausages, 'Kulen' and 'Homemade Slavonian sausage' from Croatia. These results could be of public-health significance due to gene transfer between different intrinsic bacterial species during dry sausage fermentation (Gazzola et al. 2012).

Research in recent years showed that resistance to antibiotics, including resistance to some antibiotics of therapeutic importance, also occurs in strains of the important starter cultures, *Staphylococus xylosus*/*S. carnosus,* as well as in opportunistic pathogens isolated from fermented sausages, such as *S. epidermidis, S. warneri* and *S. saprophyticus* (Martin et al. 2006). *Staphylococcus epidermidis, S. equorum* and *S. saprophyticus* show the highest rate of antibiotic resistance among staphylococci isolated from food (Schlegelová et al. 2008b). Even et al. (2010) found that 69% of multiresistant strains isolated from food/clinical specimens belonged to *S. epidermidis* species with dominant resistance to erythromycin, tetracycline and penicillins which were traced back (for the last two agents) to the presence of *tet*K and *bla*Z genes. The *tet*K and *tet*M genes were also reported as the most prevalent antibiotic resistance genes in CoNS from the production chain of swine meat commodities where *S. epidermidis* was the dominant species (Simeoni et al. 2008).

Many studies showed a comparably low correlation between phenotypic and genotypic detection of antibiotic resistance. Resch et al. (2008) have found *tet*K in 93% of CoNS which were phenotypically resistant to tetracycline, but *tet*L and *tet*M could not be detected; *lnu*A gene in only 3% of 120 strains were phenotypically resistant to lincomycin; and finally the absence of *mec*A gene in strains which showed phenotypic resistance to oxacillin/methicillin. Zdolec et al. (2013) detected *tet*K or *tet*M in 8 strains from 12 phenotypically resistant strains, while *erm*B gene was not found in erythromycin resistant strains. Schlegelová et al. (2008b) reported a relatively low occurrence of erythromycin resistant staphylococci in the food chain, but isolates were carriers of *erm, msr* and *mph* genes. In our previous study (Zdolec et al. 2012a), antimicrobial susceptibility of CoNS isolates (n = 25) from spontaneously fermented wild boar sausages was evaluated, with expectation of lower frequency of resistant strains. All strains were susceptible to penicillin, oxacillin, ampicillin, amoxicillin, enrofloxacine, gentamycin, vancomycin and streptomycin, while resistance to tetracycline, erythromycin and lincomycin was detected in 3, 7 and 9 strains, respectively. Three strains were resistant to two antibiotics (TET-ERY; ERY-LIN phenotype). Recently, Marty et al. (2012) showed that *S. xylosus* strains from spontaneously fermented wildlife meat sausages possess lower antimicrobial resistance compared to commercial *S. xylosus* starter strains. Our results indicate that relatively high rates of CoNS isolates possess phenotypic antimicrobial resistance (16 from 25 strains; 64%); this was also reported in staphylococci from other wildlife meat and/ or products (Pipova et al. 2012). These preliminary results indicate the need for more extensive research to provide relevant data for risk assessment related to antimicrobial resistance in wildlife meat microbiota.

The results of a two-year survey of CoNS antimicrobial resistance from Croatian traditional and industrial meat and milk products are summarized in Tables 4 and 5. All isolates were susceptible to clindamycin, oxacillin, amoxicillin, enrofloxacine, vancomycin, streptomycin, amikacin, amoxicillin + clavulanic acid, chloramphenicol and ciprofloxacin. The majority of isolates were resistant to erythromycin (34.5%; n = 142), penicillin (17.47%; n = 103), lincomycin (14.06%; n = 64), tetracycline (13.38%; n = 142) and ampicillin (10.93%; n = 64). From household produced dry fermented sausages, resistance was recorded in 35% (n = 64) CoNS, while in industrially produced cheeses from pasteurized milk resistance even more, 55.2% (n = 58). Within the sausage isolates, a majority of the strains harbored resistance to erythromycin, tetracycline and lincomycin, while in dairy related CoNS, the resistance toward erythromycin and penicillin dominated. Fermented sausages and dairy products were free of penicillin and lincomycin-resistant CoNS, respectively.

Table 4. The number and percentage of resistant CoNS from milk and meat products from Croatian markets.

Antimicrobial agent	Number of isolates	Number of resistant isolates	Percentage (%) of resistant isolates
Erythromycin	142	49	34.5
Penicillin	103	18	17.47
Lincomycin	64	9	14.06
Tetracycline	142	19	13.38
Ampicillin	64	7	10.93
Gentamicin	64	2	3.1
Trimethoprim	78	2	2.56
Trimethoprim + sulfametoxazole	78	1	1.28
Tobramycin	78	1	1.28

Source: Zdolec et al. (2012b).

Table 5. The number and percentage of resistant CoNS according to food matrices.

Food matrices	Number of isolates	Number of resistant isolates	Percentage (%) of resistant isolates
Wild boar fermented sausage	25	16	64
Dry sausage kulen	25	15	60
Slavonian dry sausage	14	4	28.5
Raw cheese from pasteurized milk	58	32	55.2
Cheese brine	9	0	0
Butter	3	3	100
Hard cheeses	8	0	0
Total	**142**	**66**	**46.47**

Source: Zdolec et al. (2012b).

The occurrence of multi-resistant CoNS was common in both spontaneously fermented sausages and pasteurized-milk cheeses (Zdolec et al. 2013). Mikulašova et al. (2014) have also found high multiresistance in CoNS from Bryndza, a traditional Slovak cheese. Multidrug resistant staphylococci are frequently found in ready-to-eat food, including cheeses (Chajecka-Wierzchowska et al. 2014).

Within the food LAB group, antimicrobial resistance is most relevant in enterococci, particularly in fermented products such as cheeses or sausages. The results of many studies support the potential risk of acquired antimicrobial resistance in enterococci and transfer of mobile genetic material to other bacteria, even in the condition of low antimicrobial pressure (Cocconcelli et al. 2003). In general, as showed in Tables 6 and 7, the majority of enterococci from fermented milk and meat products harbor transmissible resistant determinants, but the resistance toward clinically relevant agents is rare.

5. Conclusion

The microbiota of cheeses or fermented meat products (ready-to-eat products) naturally consists of enterococci and/or staphylococci/micrococci. These microbial groups are technologically important, but they could harbor transmissible resistance genes. The possible contaminant sources are raw materials (milk, meat), animals, humans, the environment and food contact surfaces. Since ready-to-eat products are not usually treated before consumption, it is highly important to follow a good standard of manufacturing and hygiene practices along the whole food chain to minimize contamination with initially resistant strains.

Table 6. Examples of enterococci resistance studies in ready-to-eat dairy products.

Enterococci species	Milk products	Country	Dominant resistance	Methods used	Multiresistance	Reference
E. faecalis, E. faecium	Traditional dairy products	Nigeria	Chloramphenicol, Vancomycin	Disk diffusion	10 antimicrobials	Oguntoyinbo and Okueso (2013)
E. faecalis, E. faecium, E. casseliflavus, E. mundtii, E. durans	Milk and dairy products	Czech Republic	Cephalotin, Ofloxacin	Disk diffusion		Trivedi et al. (2011)
E. faecalis, E. faecium, E. casseliflavus, E. gallinarum	Soft cheese	Brasil	Tetracycline, Nalidixic acid, Amikacin, Erythromycin, Vancomycin, Cephalothin	Disk diffusion, PCR	66.6% of *E. faecium*, 58.3% of *E. faecalis*	Furlaneto-Maia et al. (2014)
E. faecalis	Mozzarella di Bufala Campana	Italy	Tetracycline	Broth microdilution, PCR	-	Devirgiliis et al. (2010)
E. faecalis	Cheese	Turkey	Erythromycin, Tetracycline, Kanamycin	Disk diffusion	9% of *E. faecium*, 11% of *E. faecalis*	Togay et al. (2010)
E. italicus	Cheeses	Italy	Tetracycline	PCR	-	Maietti et al. (2007)
E. faecalis, E. faecium, E. durans, E. avium	Soft cheeses, Mozzarella cheese	Italy	Erythromycin, Gentamicin, Ciprofloxacin	Disk diffusion	12.3% of *E. faecalis*, 11.4% of *E. faecium*	Pesavento et al. (2014)
E. faecalis	Soft, semi-hard, hard cheeses	France	Tetracycline, Minocycline, Erythromycin, Kanamycin, Chloramphenicol	Disk diffusion PCR, PFGE	29% of *E. faecalis*	Jamet et al. (2012)

Table 7. Examples of enterococci resistance studies in ready-to-eat meat products.

Enterococci species	Meat products	Country	Dominant resistance	Methods used	Multiresistance	Reference
E. faecalis, E. faecium, E. gallinarum	Dry fermented sausages, dry cured hams	Canada	Clindamycin, Tetracycline hydrochloride, Tylosin, Erythromycin	Broth microdilution, PCR	17 of 29 strains were resistant to three to 8 antibiotics	Jahan et al. (2013)
E. faecalis, E. faecium	Sausages, ham	Germany	Enrofloxacin, Erythromycin, Avilamycin, Quinupristin/Dalfopristin (*E. faecium*); Tetracycline, Erythromycin (*E. faecalis*)	Broth microdilution	-	Peters et al. (2003)
E. faecalis, E. faecium, E. durans, E. gallinarum	Ham	Italy	Tetracycline, Erythromycin	Disk diffusion	12.3% of *E. faecalis*, 11.4% of *E. faecium*	Pesavento et al. (2014)
Enterococcus faecalis, Enterococcus faecium, Enterococcus casseliflavus	Fermented sausages Alheira, Salpicãode Vinhais, Chouriça de Vinhais	Portugal	Rifampycin, Tetracycline, Erythromycin, Ciprofloxacin	Broth microdilution	8.3% isolates resistant to 3 or 4 agents	Barbosa et al. (2009)
E. faecium	Dry-cured sausages	Spain	Tetracycline, Rifampicin, Ciprofloxacin	Disk diffusion, PCR	All isolates (n = 19) resistant to three or more agents	Landeta et al. (2013)
E. faecalis, E. faecium	Sausages	Turkey	Erythromycin, Tetracycline, Kanamycin	Disk diffusion	9% of *E. faecium*, 11% of *E. faecalis*	Togay et al. (2010)

Keywords: ampicillin, antimicrobial resistance, bacteriocins, cefoperazone, coagulase-negative staphylococci, enterocins, enterococci, erythromycin, chloramphenicol, lactic acid bacteria, methicillin resistant CoNS, nitrofurantoin, rifampin, *Staphylococcus epidermidis, Staphylococcus saprophyticus, Staphylococcus warneri.* tetracycline, trimethoprim/sulfamethoxazole

References

Barbosa, J., Ferreira, V. and Teixeira, P. 2009. Antibiotic susceptibility of enterococci isolated from traditional fermented meat products. Food Microbiology 26: 527–532.

Bhargava, K. and Zhang, Y. 2014. Characterization of methicillin-resistant coagulase-negative staphylococci (MRCoNS) in retail meat. Food Microbiology 42: 56–60.

Blaiotta, G., Ercolini, D., Pennacchia, C., Fusco, V., Casaburi, A., Pepe, O. and Villani, F. 2004. PCR detection of staphylococcal enterotoxin genes in *Staphylococcus* spp. strains isolated from meat and dairy products. Evidence for new variants of seG and seI in *S. aureus* AB-8802. Journal of Applied Microbiology 97: 719–730.

Boekhout, T. and Samson, R. 2005. Fungal biodiversity and food. pp. 29–41. *In:* R.M.J. Nut, W.M. de Vos and M.H. Zwitering (eds.). Food Fermentation. Wageningen Academic Publishers, Wageningen.

Bonomo, M.G., Ricciardi, A., Zotta, T., Sico, M.A. and Salzano, G. 2009. Technological and safety characterization of coagulase-negative staphylococci from traditionally fermented sausages of Basilicata region (Southern Italy). Meat Science 83: 15–23.

Bover-Cid, S., Izquierdo-Pulido, M. and Vidal-Carou, M.C. 2001. Biogenic amineformation in ripened sausages formulated with and without sugar. Meat Science 57: 215–221.

Čanžek Majhenič, A. 2006. Enterococci: Yin-yang microbes. Mljekarstvo 56: 5–20.

Casaburi, A., Di Monaco, R., Cavella, S., Toldrá, F., Ercolini, D. and Villani, F. 2008. Proteolytic and lipolytic starter cultures and their effect on traditional fermented sausages ripening and sensory traits. Food Microbiology 25: 335–347.

Çetinkaya, F., Elal Muş, T., Soyutemiz, G.E. and Çibik, R. 2013. Prevalence and antibiotic resistance of vancomycin-resistant enterococci in animal originated foods. Turkish Journal of Veterinary and Animal Sciences 37: 588–593.

Chajęcka-Wierzchowska, W., Zadernowska, A., Nalepa, B., Sierpińska, M. and Laniewska-Trokenheim, L. 2014. Retail ready-to-eat food as a potential vehicle for *Staphylococcus* spp. harboring antibiotic resistance genes. Journal of Food Protection 77: 993–998.

Cocconceli, P.S. and Fontana, C. 2008. Characteristics and applications of microbial starters in meat fermentations. pp. 129–148. *In:* F. Toldrá (ed.). Meat Biotechnology. Springer, New York, USA.

Cocconcelli, P.S., Cattivelli, D. and Gazzola, S. 2003. Gene transfer of vancomycin and tetracycline resistances among *Enterococcus faecalis* during cheese and sausage fermentations. International Journal of Food Microbiology 88: 315–323.

Cocolin, L., Manzano, M., Aggio, D., Cantoni, C. and Comi, G. 2001. A novel polymerase chain reaction (PCR)—denaturing gradient gel electrophoresis (DGGE) for the identification of *Micrococcaceae* strains involved in meat fermentations. Its application to naturally fermented italian sausages. Meat Science 57: 59–64.

Comi, G., Urso, R., Iacumin, L., Rantsiou, K., Cattaneo, P., Cantoni, C. and Cocolin, L. 2005. Characterisation of naturally fermented sausages produced in the North East of Italy. Meat Science 69: 381–392.

Coppola, S., Mauriello, G., Aponte, M., Moschetti, G. and Villani, F. 2000. Microbial succession during ripening of Naples-type salami, a southern Italian fermented sausage. Meat Science 56: 321–329.

Deresinski, S. 2009. Vancomycin in Combination with other antibiotics for the treatment of serious Methicillin-Resistant *Staphylococcus aureus* infections. Clinical Infectious Diseases 49: 1072–1079.

Devirgiliis, C., Barile, S., Caravelli, A., Coppola, D. and Perozzi, G. 2010. Identification of tetracycline- and erythromycin-resistant Gram-positive cocci within the fermenting microflora of an Italian dairy food product. Journal of Applied Microbiology 109: 313–323.

Dobranić, V., Zdolec, N., Račić, I., Vujnović, A., Zdelar-Tuk, M., Filipović, I. and Špičić, S. 2013. Determination of enterotoxin genes in coagulase-negative staphylococci from autochthonous Croatian fermented sausages. Veterinarski Arhiv 83: 145–152.

Dobranić, V., Butković, I., Koturić, A., Medvid, V., Zdolec, N. 2014. The influence of udder health on milk microbiota. Book of abstracts, 41st Croatian symposium of dairy experts with international cooperation. Lovran, Croatia 77–78.

Even, S., Leroy, S., Charlier, C., Zakour, N.B., Chacornac, J.P., Lebert, I., Jamet, E., Desmonts, M.H., Coton, E., Pochet, S., Donnio, P.Y., Gautier, M., Talon, R. and Leloir, Y. 2010. Low occurrence of safety hazards in coagulase negative staphylococci isolated from fermented food stuffs. International Journal of Food Microbiology 139: 87–95.

Franz, C.M.A.P., Huch, M., Abriouel, H., Holzapfel, W. and Gálvez, A. 2011. Enterococci as probiotics and their implications in food safety. International Journal of Food Microbiology 151: 125–140.

Frece, J., Kovačević, D., Kazazić, S., Mrvčić, J., Vahčić, N., Ježek, D., Hruškar, M., Babić, I. and Markov, K. 2014. Comparison of sensory properties, shelf-life and microbiological safety of industrial sausages produced with autochthonous and commercial starter cultures. Food Technology and Biotechnology 52: 307–316.

Frey, Y., Rodriguez, J.P., Thomann, A., Schwendener, S. and Perreten, V. 2013. Genetic characterization of antimicrobial resistance in coagulase-negative staphylococci from bovine mastitis milk. Journal of Dairy Science 96: 2247–2257.

Furlaneto-Maia, L., Rocha, K.R., Henrique, F.C., Giazzi, A. and Furlaneto, M.C. 2014. Antimicrobial resistance in *Enterococcus* sp. isolated from soft cheese in Southern Brazil. Advances in Microbiology 4: 175–181.

Gazzola, S., Fontana, C., Bassi, D. and Cocconcelli, P.S. 2012. Assessment of tetracycline and erythromycin resistance transfer during sausage fermentation by culture-dependent and -independent methods. Food Microbiology 30: 348–354.

Giménez Pereira, M.L. 2005. Enterococci in Milk Products. M.S. Thesis, Massey University Palmerston North, New Zealand.

Giraffa, G. 2002. Enterococci from foods. FEMS Microbiology Review 744: 1–9.

Hadžiosmanović, M. 1978. Influence of micrococcaceae on lipolytic changes in fermented sausages. Ph.D. Thesis, Faculty of veterinary medicine, University of Zagreb, Zagreb (in Croatian).

Hammes, W.P. and Hertel, C. 1998. New developments in meat starter cultures. Meat Science 49: Suppl. 1: 125–138.

Hugas, M., Garriga, M. and Aymerich, M.T. 2003. Functionality of enterococci in meat products. International Journal of Food Microbiology 88: 223–233.

Hughes, M.C., Kerry, J.P., Arendt, E.K., Kenneally, P.M., McSweeney, P.L.H. and O'Neill, E.E. 2002. Characterization of proteolysis during the ripening of semi-dry fermented sausages. Meat Science 62: 205–216.

Ito, T., Hiramatsu, K., Oliveira, D., de Lencastre, H., Zhang, K., Westh, H., O'Brien, F., Giffard, P., Coleman, D., Tenover, F., Boyle-Vavra, S., Skov, R.L., Enright, M.C., Kreiswirth, B., Kwan, S.K., Grundmann, H., Laurent, F., Sollid, J.E., Kearns, A.M., Goering, R., John, J.F., Daum, R. and Soderquist, B. 2009. International working group on the classification of *Staphylococcal* cassette chromosome elements (IWG-SCC): Classification of *Staphylococcal* cassette chromosome mec (SCCmec): Guidelines for reporting novel SCCmec elements. Antimicrobial Agents Chemotherapy 53: 4961–4967.

Jahan, M., Krause, D.O. and Holley, R.A. 2013. Antimicrobial resistance of Enterococcus species from meat and fermented meat products isolated by a PCR-based rapid screening method. International Journal of Food Microbiology 163: 89–95.

Jamet, E., Akary, E., Poisson, M.A., Chamba, J.F., Bertrand, X. and Serror, P. 2012. Prevalence and characterization of antibiotic resistant *Enterococcus faecalis* in French cheeses Food Microbiology 31: 191–198.

Jiménez, E., Ladero, V., Chico, I., Maldonado-Barragán, A., López, M., Martín, V., Fernández, L., Fernández, M., Álvarez, M.A., Torres, C. and Rodríguez, J.M. 2013. Antibiotic resistance, virulence determinants and production of biogenic amines among enterococci from ovine, feline, canine, porcine and human milk. BMC Microbiology 13: 288.

Jurmanović, J., Bačanek, B., Pavljak, I., Sukalić, T., Jaki, V., Majnarić, D., Končurat, A. and Sokolović, J. 2012. Susceptibility of *Staphylococcus aureus* strains isolated from bovine intramammary infections to different antimicrobial agents. Rad Hrvatske Akademije Znanosti i Umjetnosti, Medicinske Znanosti, Rad 511, 37: 105–110.

Kalmus, P., Aasmäe, B., Kärssin, A., Orro, T. and Kask, K. 2011. Udder pathogens and their resistance to antimicrobial agents in dairy cows in Estonia. Acta Veterinaria Scandinavica 53: 4.

Kročko, M., Čanigová, M., Ducková, V., Artimová, A., Bezeková, J. and Poston, J. 2011. Antibiotic resistance of *Enterococcus* species isolated from raw foods of animal origin in South West part of Slovakia. Czech Journal of Food Science 29: 654–659.

Landeta, G., Curiel, J.A., Carrascosa, A.V., Muñoz, R. and de las Rivas, B. 2013. Technological and safety properties of lactic acid bacteria isolated from Spanish dry-cured sausages. Meat Science 95: 272–280.

Laukova, A. 2012. Potential applications of probiotic, bacteriocin-producing enterococci and their bacteriocins. pp. 39–61. *In*: S. Lahtinen, A.C. Ouwehand, S. Salminen and A. von Wright (eds.). Lactic Acid Bacteria: Microbiological and Functional Aspects. 4th ed. CRC Press, Taylor & Francis Group, Boca Raton.

Lücke, F.-K. 2000. Utilization of microbes to process and preserve meat. Meat Science 56: 105–115.

Maietti, L., Bonvini, B., Huys, G. and Giraffa, G. 2007. Incidence of antibiotic resistance and virulence determinants among *Enterococcus italicus* isolates from dairy products. Systematic and Applied Microbiology 30: 509–517.

Martin, B., Garriga, M., Hugas, M., Bover-Cid, S., Veciana-Nogues, M.T. and Aymerich, T. 2006. Molecular, technological and safety characterization of Gram-positive catalase-positive cocci from slightly fermented sausages. International Journal of Food Microbiology 107: 148–158.

Marty, E., Buchs, J., Eugster-Meier, E., Lacroix, C. and Meile, L. 2012. Identification of staphylococci and dominant lactic acid bacteria in spontaneously fermented Swiss meat products using PCR-RFLP. Food Microbiology 29: 157–166.

Mikulášová, M., Valáriková, J., Dušinský, R., Chovanová, R. and Belicová, A. 2014. Multiresistance of *Staphylococcus xylosus* and *Staphylococcus equorum* from Slovak Bryndza cheese. Folia Microbiologica 59: 223–227.

Oguntoyinbo, F.A. and Okueso, O. 2013. Prevalence, species distribution and antibiotic resistant enterococci species in two traditional fermented dairy foods. Annals of Microbiology 63: 755–761.

Pesavento, G., Calonico, C., Ducci, B., Magnanini, A. and Lo Nostro, A. 2014. Prevalence and antibiotic resistance of *Enterococcus* spp. isolated from retail cheese, ready-to-eat salads, ham, and raw meat. Food Microbiology 41: 1–7.

Peters, J., Mac, K., Wichmann-Schauer, H., Klein, G. and Ellerbroek, L. 2003. Species distribution and antibiotic resistance patterns of enterococci isolated from food of animal origin in Germany. International Journal of Food Microbiology 88: 311–314.

Piette, A. and Verschraegen, G. 2009. Role of coagulase-negative staphylococci in human disease. Veterinary Microbiology 134: 45–54.

Pipová, M., Jevinová, P., Kmet', V., Regecová, I. and Marušková, K. 2012. Antimicrobial resistance and species identification of staphylococci isolated from the meat of wild rabbits (*Oryctolagus cuniculus*) in Slovakia. European Journal of Wildlife Research 58: 157–165.

Resch, M., Nagel, V. and Hertel, C. 2008. Antibiotic resistance of coagulase-negative staphylococci associated with food and used in starter cultures. International Journal of Food Microbiology 127: 99–104.

Samelis, J., Metaxopoulos, J., Vlassi, M. and Pappa, A. 1998. Stability and safety of traditional Greek salami: A microbiological ecology study. International Journal of Food Microbiology 44: 69–82.

Sampimon, O.C., Lam, T.J.G.M., Mevius, D.J., Schukken, Y.H. and Zadoks, R.N. 2011. Antimicrobial susceptibility of coagulase-negative staphylococci isolated from bovine milk samples. Veterinary Microbiology 150: 173–179.

Schlegelová, J., Babák, V., Holasová, M. and Dendis, M. 2008a. The biofilm-positive *Staphylococcus epidermidis* isolates in raw materials, foodstuffs and on contact surfaces in processing plants. Folia Microbiologica 53: 500–504.

Schlegelová, J., Vlková, H., Babák, V., Holasová, M., Jaglic, Z., Stosová, T. and Sauer, P. 2008b. Resistance to erythromycin of *Staphylococcus* spp. isolates from the food chain. Veterinarni medicina 53: 307–314.

Simeoni, D., Rizzotti, L., Cocconcelli, P., Gazzola, S., Dellaglio, F. and Torriani, S. 2008. Antibiotic resistance genes and identification of staphylococci collected from the production chain of swine meat commodities. Food Microbiology 25: 196–201.

Talon, R. and Leroy, S. 2011. Diversity and safety hazards of bacteria involved in meat fermentations. Meat Science 89: 303–309.

Teuber, M. 2001. Veterinary use and antimicrobial resistance. Current Opinion in Microbiology 4: 493–499.

Toğay, S.O., Keskin, A.C., Açik, L. and Temiz, A. 2010. Virulence genes, antibiotic resistance and plasmid profiles of *Enterococcus faecalis* and *Enterococcus faecium* from naturally fermented Turkish foods. Journal of Applied Microbiology 109: 1084–1092.

Toldrá, F. 2008. Biotechnology of flavor generation in fermented meats. pp. 199–215. *In*: F. Toldrá (ed.). Meat Biotechnology. Springer, New York, USA.

Trivedi, K., Cupakova, S. and Karpiskova, R. 2011. Virulence factors and antibiotic resistance in enterococci isolated from food-stuffs. Veterinarni Medicina 56: 352–357.

Urso, R., Rantsiou, K., Cantoni, C., Comi, G. and Cocolin, L. 2006. Technological characterization of a bacteriocin-producing *Lactobacillus sakei* and its use in fermented sausages production. International Journal of Food Microbiology 110: 232–239.

Wendlandt, S., Feβler, A.T., Monecke, S., Ehricht, R., Swarz, S. and Kadlec, K. 2013. The diversity of antimicrobial resistance among staphylococci of animal origin. International Journal of Medical Microbiology 303: 338–349.

Zdolec, N., Dobranić, V., Butković, I., Koturić, A., Filipović, I. and Medvid, V. 2014. Antimicrobial resistance of milk bacteria from healthy and drug-treated udder. Book of abstracts, 41st Croatian symposium of dairy experts with international cooperation. Lovran, Croatia 97–98.

Zdolec, N., Račić, I., Vujnović, A., Zdelar-Tuk, M., Matanović, K., Filipović, I., Dobranić, V., Cvetnić, Ž. and Spičić, S. 2013. Antimicrobial resistance of coagulase-negative staphylococci isolated from spontaneously fermented sausages. Food Technology and Biotechnology 51: 240–246.

Zdolec, N., Dobranić, V., Filipović, I. and Marcincakova, D. 2012a. Antimicrobial resistance of coagulase-negative staphylococci isolated from spontaneously fermented wild boar sausages. Folia Veterinaria 56: 60–62.

Zdolec, N., Dobranić, V. and Filipović, I. 2012b. Occurrence of resistant coagulase-negative staphylococci in meat and dairy products. Proceedings of Fifth Croatian Veterinary Congress with international participation. Croatia 71–76.

Zdolec, N., Filipović, I., Cvrtila Fleck, Ž., Marić, A., Jankuloski, D., Kozačinski, L. and Njari, B. 2011. Antimicrobial susceptibility of lactic acid bacteria isolated from fermented sausages and raw cheese. Veterinarski Arhiv 81: 133–141.

Zdolec, N., Hadžiosmanović, M., Kozačinski, L., Cvrtila, Ž., Filipović, I., Škrivanko, M. and Leskovar, K. 2008. Microbial and physicochemical succession in fermented sausages produced with bacteriocinogenic culture of *Lactobacillus sakei* and semi-purified bacteriocin mesenterocin Y. Meat Science 80: 480–487.

14

Bioactive Components of Fermented Foods

Ami Patel[1],* *and Nihir Shah*[1]

1. Introduction

Fermented foods are vital food of choice due to their excellent functional and nutritional properties, aided by fermenting microorganisms. Novel bioactive compounds from fermented foods are an emerging area of research with great promise. Bioactive components may be described as 'food derived components that, in addition to their nutritional value, exert a physiological effect in the body' (Vermeirssen et al. 2004). These extra nutritional constituents occur naturally in small amounts in any food products *viz.* food grade pigments, plant growth factors and phenolic compounds such as flavonoids, phenolic acids, and tannins; secondary microbial metabolites such as antibiotics, mycotoxins and alkaloids are considered as most common bioactive compounds (Kris-Etherton et al. 2002, Nigam 2009). Furthermore, fermented foods also contain organic acids and flavor compounds that contribute to develop flavor, aroma and taste; antimicrobial components like hydrogen peroxide, short chain fatty acids (SCFA), bacteriocins; functional enzymes and vitamins; exopolysaccharides (EPS); and conjugated fatty acids and diverse microbes including bacteria, yeasts and moulds contribute to synthesize such bioactive components.

Traditionally, lactic acid bacteria (LAB) have been used as starter cultures to drive fermentations of diverse substrates involving milk, cereals-legumes, vegetables, meat and fish-based products. Their widespread human consumption and technological importance made them generally recognized as safe (GRAS) (Klaenhammer et al. 2005). LAB produce organic acids like lactate, acetate, propionate and butyrate

[1] Division of Dairy and Food Microbiology, Mansinhbhai Institute of Dairy & Food Technology-MIDFT, Dudhsagar Dairy Campus, Mehsana 384002, Gujarat, INDIA.
E-mail: nihirshah13@yahoo.co.in
* Corresponding author: amiamipatel@yahoo.co.in

which lower the pH of the food matrix and exert antimicrobial activity against the pathogenic and spoilage microorganisms. Moreover, LAB produce various low-molecular-carbon mass compounds such H_2O_2, carbon dioxide (CO_2), diacetyl (2,3-butanedione), acetaldehyde, SCFA, etc. and high-molecular-mass compounds like antimicrobial peptides and bacteriocins. Yeasts and molds are associated with biosynthesis of enzymes, several antioxidants and melatonin, a neurohormone involved in physiological activities like reproductive functions is produced into yeast fermented products like red wine, beer, and bread crump (Mas et al. 2014). An overview of bioactive compounds synthesized by diverse microorganisms in fermented foods is depicted in Fig. 1.

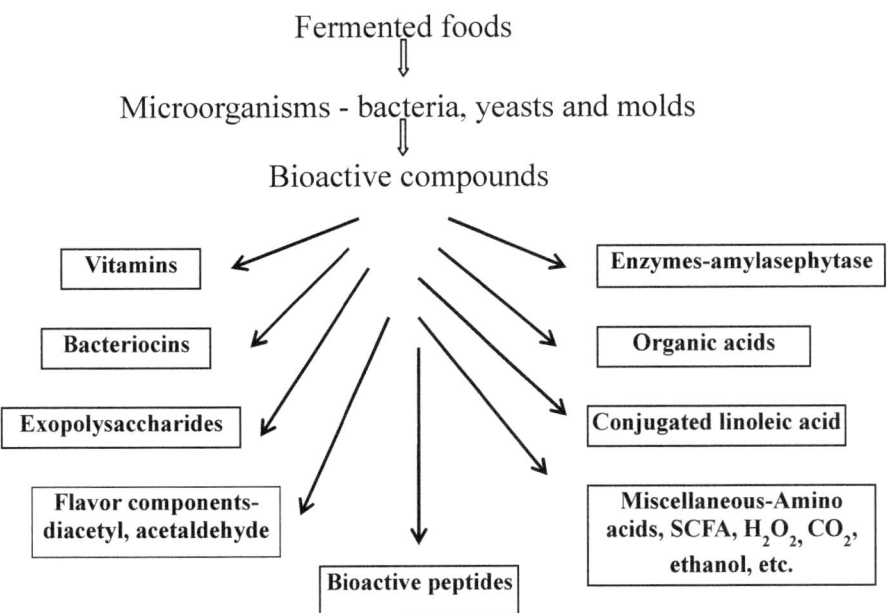

Figure 1. Overview of bioactive compounds synthesized by diverse microorganisms in fermented foods.

The fermenting microbiota of cereals improves food safety and quality through the destruction or biotransformation of toxic or anti-nutritional factors of native substrates like phytic acids, polyphenols, enzyme inhibitors and lectins (Steinkraus et al. 1993, Rose 2015). A lot of literature reveals the biosynthesis of essential amino acids, vitamins, SCFA, CLA, exopolysaccharides by different fermenting microorganisms *in vitro* and *in vivo*, which contribute to the well-being of the host (Marques et al. 2010).

Fermenting microflora were found to have antioxidant activities through scavenging free radicals (Osuntoki and Korie 2010). Fermented milks and soya based foods are also a source of physiologically important bioactive peptides that have a positive impact on the body's functions. Soybean isoflavones appear to be the major components responsible for bioactive functions such as lowering the risks of cancers of breast, prostate, and colon, cardiovascular diseases and osteoporosis (Yang et al. 2011).

Hence, due to presence of such biological active components, regular consumption of fermented foods is considered to reduce the risks of acquiring some of the chronic diseases and metabolic disorders that are associated with unbalanced diets.

2. Bioactive Components in Fermented Foods

The following bioactive compounds are produced in fermented foods:

2.1 Organic Acids

Most of the fermentation processes employing LAB or related food grade microorganisms are characterized by the accumulation of organic acids such as lactic acid, acetic acid, propionic acid and butyric acid, with simultaneous reduction in pH. The level and type of organic acids produced depend on the species of organisms, culture composition and growth conditions of the fermentation process (Patel and Shah 2014). In addition to imparting typical acidic flavor, organic acids are also responsible for preventing or controlling the growth of unwanted microorganisms in the food. Antimicrobial activity is believed to result from the action of organic acids on the cytoplasmic membrane of the bacterial cell, which causes solubilization of membrane lipids and their diffusion into the cytoplasm. It interferes with the maintenance of membrane potential and hampers active transport (Gottschalk 1988).

Bifidobacteria synthesize more amount of acetate than lactate (3:2 ratio). Unusually several bacteria found to produce phenyllactate and benzoic acid. Phenyllactic acid (PLA) has been recognized as the major factor responsible for antifungal activity and prolonging the shelf-life of food products. The inhibitory properties of PLA have been reported against molds isolated from bakery products, flours and cereals, including some mycotoxigenic species such as *Aspergillus ochraceus, Penicillium verrucosum* and *P. citrinum*; and some bacterial contaminants, namely *Listeria* spp., *Staphylococcus aureus* and *Enterococcus faecalis* (Dieuleveux and Gueguen 1998, Lavermicocca et al. 2003).

The antimicrobial action of each of the acids at given molar concentration is not equal. Acetic acid and propionic acids are claimed to be stronger antimicrobials than lactate and can inhibit yeasts, molds and bacteria. Food and Drug Administration (FDA)-approved Microgard[R] is a commercial food additive that is a growth extract of *Propionibacterium freudenreichii* subsp. *shermanii* containing propionic acid and is used in yoghurt, sour cream, cottage cheese, dairy desserts and chocolate confectionaries. In the U.S. another product, namely BioProfit, contains viable cells of *P. freudenreichii* subsp. *shermanii* strain JS, which is effective on bacteria and yeasts growth in dairy products and sourdough. It is also used to preserve grains and produce good quality silages.

2.2 Bacteriocins

Certain bacteria have the ability to synthesize bacteriocins, which are normally proteinaceous compounds in nature and have an inhibitory effect to themselves and closely related species (Servin 2004). Bacteriocin is produced by various species of *Lactobacillus, Pediococcus, Lactococcus, Weissella, Enterococcus* and *Bifidobacterium* (Servin 2004, Patel et al. 2013b). Their synthesis is ribosomally regulated in cells; however the precise mechanisms are still not clear. Most of the bacteriocins have narrow spectrum while some have wide spectrum activities. Bacteriocins produced by LAB are classified mainly into 4 classes as shown in Fig. 2 (O'Sullivan et al. 2002, Servin 2004). Class 1 bacteriocins—lantibiotics—are the most documented and industrially exploited. In that, nisin is produced by *Lactococcus lactis* subsp. *lactis* and the only fully characterized legally permitted to use as a biopreservative as (food additive E234) in cheese, canned products, salad dressings, bakery products and cooked meat sausages. Several bacteriocins produced by food grade bacteria with their activity spectrum are comprised in Table 1 and discussed in Chapter 7 in this book.

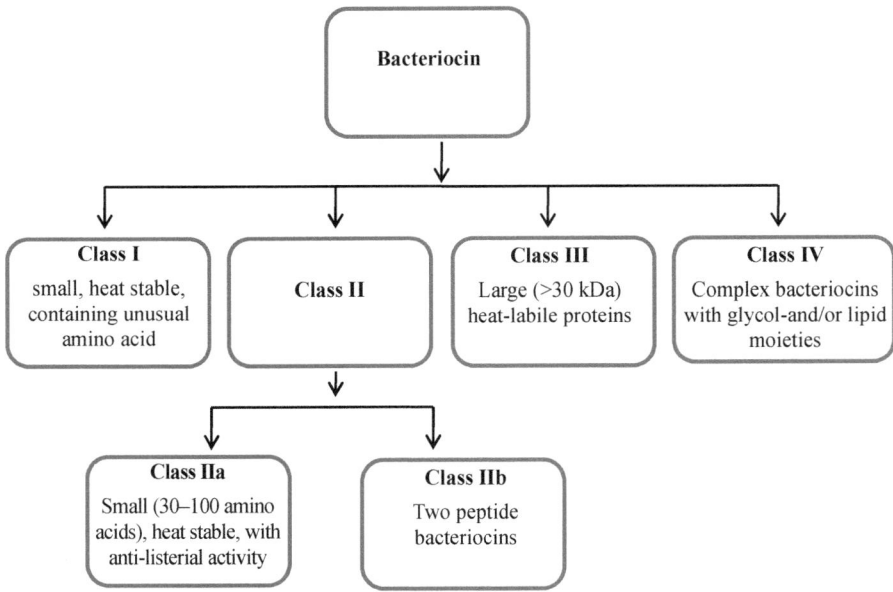

Figure 2. Classes of bacteriocins produced by lactic acid bacteria.

Application of nisin is somewhat limited by its comparatively narrow spectrum of activity and also because it is most effective at low pH only. The effectiveness of nisin can be improved if chelating agents like EDTA are present since such agents increase the permeability of the bacterial cell wall (Thomas et al. 2000). Pediocin PA-1/AcH obtained from *Pediococcus* spp. and enterocin AS-48 (class IIc) from *Enterococcus faecalis* have been the most likely bacteriocins to be used for bio-preservatives, next to nisin. Pediocin PA-1/AcH producing strain *Pd. pentosaceus* BCC 3772, when used as starter culture in traditional Thai fermented pork sausage Nham, was found to efficiently control the growth of *Listeria monocytogenes* (Kingcha et al. 2012). Other bacteriocins including diplococcin, plantaricin, acidophilin and bulgaricin have not been utilized yet commercially. It is notable that majority of bacteriocins

Table 1. Bacteriocins produced from food grade bacteria with their antimicrobial spectrum.
Lc. = Lactococcus; Lb. = Lactobacillus.

Producer	Bacteriocin	Inhibitory spectrum
Lc. lactis subsp. *lactis*	Nisin	Active against Lactococci, Clostridia, *S. aureus,* Micrococci
Lc. lactis subsp. *cremoris*	Diplococcin	*Lc. lactis* subsp. *lactis, Lc. lactis* subsp. *cremoris*
Lb. acidophilus	Lactacin B	
	Lactacidin	Lactobacilli Broad spectrum active against Gram +ve and Gram –ve bacteria
	Acidophilin	Lactic acid bacteria, spore formers, *Salmonella* spp., *E. coli, S. aureus, pseudomonas*
	Acidolin	Broad spectrum, spore formers, enteric pathogens
Lb. reuteri	Reuterin	Broad spectrum-*Salmonella, Shigella*, Listeria, Clostridia, yeasts-candida
Lb. plantarum	Plantaricin C	Active against different *Lactobacillus* spp., *Pediococcus*, Clostridia, *L. monocytogenes*
Lb. sake	Sakacin A	Active against different species of *Pediococcus, Lactobacillus, L. monocytogenes*
Pediococcus strains	Pediocin PA-1/AcH	Active against *Carnobacterium, Enterococcus*, LAB, *L. monocytogenes, B. cereus, Staphylococcus* spp. and *Clostridium* spp.
Enterococcus faecalis	Enterocin AS-48	Active against *Enterococcus, Lactobacillus, Pediococcus*, and *L. monocytogenes,*
Bifidobacterium spp.	Bifidocin B	Gram-positive bacteria—*Lactobacillus, L. monocytogenes*

produced by the strains of *Lb. acidophilus* have broad spectrums of activity against foodborne pathogens and spoilage bacteria. Reuterin, chemically identified as 3-hydroxy propionaldehyde, is an example of a non-protein bacteriocin produced by certain strains of *Lb. reuteri*. It is water-soluble, stable over a wide pH range and has a broad spectrum of activity against most bacteria, yeasts and molds, which makes it a prospective food preservative (Rattanachaikunsopon and Phumkhachorn 2010).

2.3 Vitamins

Vitamins play a key role in many essential functions of the body, as co-factors during biochemical reactions of metabolism and respiration, biosynthesis of nucleic acids and antioxidant activities. Essentially, many of the vitamins are not synthesized in humans and animals by natural means; however, it has long been known that a few bacteria, yeasts, fungi and algae have the ability to produce riboflavin, thiamine, folic acid, vitamin B_{12} (cobalamin), vitamin K_2 (menaguino) and other essential vitamins (LeBlanc et al. 2011). Studies carried out in fermented products like curd, yoghurt, cheeses, cereal based fermented foods and vegetable products making use of this important trait of microorganisms showed encouraging results for biosynthesis and the liberation of vitamins. Additionally, gut microflora have been recognized as a significant source of some water soluble vitamins to our body. Currently, dairy and food industries focus and approach to select and employ vitamin producing starters, especially probiotic strains, to produce fermented products with elevated amounts of 'natural' vitamins without uplifting manufacturing cost and with an inherent propensity to provide desired health benefits.

Folic acid, essential for the biosynthesis of DNA and RNA and the inter-conversions of amino acids, is found in legumes, fruits, leafy vegetables and in fermented milks. Wouters et al. (2002) suggested that in yoghurt and related fermented milks, the amount of folate can be increased by making use of proper starter cultures and storage conditions, to values above 200 µg/litre. Many LAB, including *Lb. plantarum, Lb. delbrueckii* subsp. *bulgaricus, Lc. lactis, Strept. thermophilus* and *Enterococcus* spp., have the ability to produce folate (LeBlanc et al. 2007). The production of folate from *Lc. lactis, Strept. thermophilus*, and *Leuconostoc* spp. and from most *Lactobacillus* spp., except *Lb. plantarum*, was reported by Sybesma et al. (2003).

Bifidobacterium spp. are generally found to produce folate, cobalamin, pyridoxine, riboflavin and thiamine which are strain specific. About 76 strains of 15 *Bifidobacterium* spp. of human and animal origin were tested for their ability to produce folate, in which a total of 17 strains belonging to 9 species showed folate biosynthesis, with two strains of *B. adolescentis* and one strain of *B. pseudocatenulatum* being the top three producers (Pompei et al. 2007). In another study, all tested *Bifidobacterium* strains were able to produce pyridoxine (vitamin B6) and small amounts of cobalamin, which are very crucial since they are responsible for the normal growth and development of the nervous system in neonates (Marques et al. 2010). Fermentation of reconstituted skim milk using folate-producing *Bifidobacterium* spp. in conjunction with yoghurt cultures demonstrated that it is possible to increase the concentration of folate in fermented milks through the appropriate selection of bacterial strains (Crittenden et al. 2003). Tempeh, a traditional Indonesian soybean based fermented food has shown elevated levels of B group vitamin from *Streptococcus* and *Enterococcus* species in addition to native microflora (Keuth and Bisping 1993). Goswami (2012) reported biosynthesis of folic acid and biotin from the probiotic strain *Lb. helveticus* MTCC 5463 in combination with *Lb. rhamnosus* MTCC 5462, a normal lactobacilli. In another *in vitro* investigation, increased production of folic acid was observed in human subjects after the administration of probiotic *Bifidobacterium* strains (Strozzi and Mogna 2008).

Vitamin B_{12}, cobalamin is the only vitamin that is not synthesized by mammals. Thus, it must be obtained from exogenous sources like fermented foods or the gut microbiota. Some members of the genus *Lactobacillus* such as *Lb. reuteri* have the ability to produce vitamin B_{12} and a few species of Propionibacteria, such as *Propionibacterium shermani*, are also responsible for the synthesizing of vitamin B_{12} in certain cheese varieties (Taranto et al. 2003, Burgess et al. 2004). Vitamin B_2, popularly known as riboflavin, plays a crucial role in cellular metabolism. In a convenient biotechnological application, riboflavin producing LAB were isolated and used for the preparation of fermented sourdough and pasta to enrich them naturally with vitamin B_2 (Spano 2011). Genetic analysis of the riboflavin biosynthetic (*rib*) operon in *Lc. lactis* subsp. *cremoris* strain NZ9000 showed improved riboflavin synthesis because of simultaneous over-expression of riboflavin biosynthetic genes (ribG, ribH, ribB and ribA) in *Lc. lactis* (Burgess et al. 2004). Sybesma et al. (2004) modified two complicated biosynthetic pathways through site directed mutagenesis followed by metabolic engineering in *Lc. lactis* that resulted in the simultaneous overproduction of both folate and riboflavin.

Vitamin K, which plays significant role during blood clotting and in the functions of bones and kidneys, is present as phylloquinone (Vitamin K_1) in green plants and is produced by some gut bacteria, especially by various species of *Lactococcus, Lactobacillus, Enterococcus, Leuconostoc* and *Streptococcus* as menaquinone (K_2) (O'Connor et al. 2005, Cooke et al. 2006). The production of various vitamins by different microorganisms has been compiled in Table 2.

Table 2. Vitamin biosynthesis by LAB and related genera.

Vitamin	Producing bacterium/strain	Product/ Subject	Reference(s)
Folic acid	*Streptococcus thermophilus, Lactobacillus delbrueckii* subsp. *bulgaricus*	Yoghurt	Wouters et al. 2002
	Bifidobacterium spp., *Strept.* thermophilus and/or *Lb. delbrueckii* subsp. *bulgaricus*	Reconstituted skim milk	Crittenden et al. 2003
	Lc. lactis, Strept. *thermophilus, Leuconostoc* spp., *Lactobacillus* spp.		Sybesma et al. 2003
Folic acid and biotin	*Bifidobacterium* spp.	Human volunteers	Strozzi and Mogna 2008
	Bifidobacterium spp.		Pompei et al. 2007
	Bifidobacterium spp.	Synthetic media	D'Aimmo et al. 2012
	Lb. helveticus MTCC 5463, *Lb. rhamnosus* MTCC 5462	Fermented milk	Goswami 2012
Folic acid, Thiamin and Vitamin B3 (nicotinic acid)	*B. bifidum, B. breve, B. longum longum, B. longum infantis, B. adolescentis*		Deguchi et al. 1985
Riboflavin	Lactic acid bacteria	Sourdough	Spano 2011
Vitamin B6 (pyridoxine) and Vitamin B12 (cobalamin)	*B. bifidum, B. breve, B. longum longum, B. infantis*		Marques et al. 2010
Vitamin B12 (cobalamin)	*Lb. reuteri* and other *lactobacilli*		Taranto et al. 2003
	Propionibacterium shermani		Burgess et al. 2004
Vitamin B complex	*Enterococcus* spp.	Tempeh	Keuth and Bisping 1993
	Bifidobacterium lactis	Fermented milk	Beitane and Ciprovica 2012
Vitamin K	*Lactococcus, Lactobacillus, Enterococcus, Leuconostoc* and *Streptococcus*		O'Connor et al. 2005, Cooke et al. 2006

The strain improvement methods also showed promising outcomes for vitamin biosynthesis in fermented foods. Burgess et al. (2004) carried out over-expression of riboflavin biosynthesis genes in *Lc. lactis* ssp. *cremoris* NZ9000, that led to the conversion of riboflavin utilizing strains into a vitamin B_2 producing factory. Similarly, metabolic engineering also been showed to improve folate production in *Lb. gasseri* and *Lb. reuteri* (Santos et al. 2008). In *Lc. lactis*, site directed mutagenesis, followed by selection and metabolic engineering, resulted in the simultaneous overproduction of folate and riboflavin (Sybesma et al. 2004). Random mutagenesis and genetic engineering lead to improved production of Vitamin B_{12} by *Propionibacterium freudenreichii* (Martins et al. 2002, Burgess et al. 2004). Moreover, the biosafety assessment of genetically modified LAB (*fol*C, *fol*KE, *fol*C + *fol*KE) demonstrated that they were as safe as the native strains from which they were derived (Le-Blanc et al. 2011).

2.4 Enzymes

Microorganisms synthesize various enzymes which may influence the compositional, processing and organoleptic properties, along with the overall quality of fermented products. Gut microflora releases hydrolytic enzymes into the gastrointestinal (GI) tract that may exert potential synergistic effects on digestion and alleviate symptoms of intestinal malabsorption. These beneficial organisms may serve as an alternate source for the preparation of enzyme extracts under the environmental conditions of fermentation (Tamang 2011). The enzymatic activity of the LAB isolated from wine, cheeses and yoghurt has been reported in numerous studies. Various species of *Lactobacillus*, *Lactococcus*, *Leuconostocs*, *Pediococcus*, *Weissella* and *Bifidobacterium* have been reported to produce carbohydrate degrading enzymes like glucosidases, amylases and xylanases (Patel et al. 2013a, Novik et al. 2007). LAB has been found to produce higher percentages of alpha and/or beta-galactosidase in comparison to bifidobacteria (Alazzeh et al. 2009). Several LAB produce the most stable kind of amylases which find application as starter cultures during sourdough preparations for the natural improvement of bread texture (Mogensen 1993). LAB that can produce β glycosidase are most likely found in the fermented foods obtained from plant material.

Fermentation leads to improve flavor and aroma of fermented products. For instance, some peptidases produced by starters like *Strept. thermophilus*, *Lc. lactis* subsp. *cremoris* and species of *Lactobacillus* improved the body texture and sensory quality of fermented milks, while proteolysis and lipolysis enhanced the flavor of most varieties of cheese (Guldfeldt et al. 2001, Gonzalez et al. 2010). LAB strains isolated from a traditional Spanish cheese were evaluated for their enzymatic activity and it was reported that high levels of dipeptidase activity was associated with *Lactococcus* spp. and enterolytic activity was detected for *Enterococcus* spp., while carboxypeptidase activity was very less or undetectable (Gonzalez et al. 2010). Enzymes of various LAB play an important role in wine-making; the typical wine flavor and aroma develop as a result of fermentation. These bacteria grow in wine during malolactic fermentation, following alcoholic fermentation, while a broad range of secondary modifications improve the taste and flavor of wine (Mtshali 2007).

2.4.1 Conversion or biotransformation of soy isoflavones through production of β-glucosidases

Several cereals, particularly fermented soybean products, contain the highest concentration of isoflavones; isoflavones have been reported to possess an anti-carcinogenic effect and anti-oxidative properties. Isoflavones exist in two basic categories, the aglycones (daidzein and genistein) and the glucosidic conjugates (daidzin and genistin). Among these, the isoflavones which are present in form of glucoside conjugates are not absorbed and require hydrolysis for bioavailability and subsequent metabolism. Some LAB have been shown to synthesize β-glucosidases, enhancing the bioavailability of isoflavones. Hydrolysis of isoflavones takes place in the GI tract by action of gut microflora. According to Pyo et al. (2005), the β-glucosidase enzymes produced by *Lb. acidophilus, Bifidobacterium lactis* and *Lb. casei* were responsible for the breakdown of the β-1-6 glucosidic bond which conjugates the pran ring of isoflavone and the sugar moieties. Hence, incorporation of such bacteria could improve the biological activity of cereals during fermentation.

2.4.2 Phytase breakdown of phytic acids

Phytic acid is the major anti-nutritional compound that blocks the availability of minerals in cereals and legumes, particularly, in soya beans (Anderson et al. 1995). Some fermenting microflora, including many LAB, are able to synthesize the phytase enzyme that degrades phytate into myo-inositol and phosphate during fermentation. Phytate is mainly present in cereals in the form of complexes with polyvalent cations like iron, zinc, calcium, magnesium and some proteins. Fermentation provides optimum pH for enzymatic degradation of phytic acid and thus, along with reduction of phytate content, the process may also lead to increase the amount of soluble iron, zinc and calcium levels (Chavan and Kadam 1989, Haard et al. 1999).

Similar to phytic acid, fermentation also diminishes polyphenols due to the activity of polyphenol oxidase present in the food grain or fermenting microflora. Microbial fermentation also help to destroy or reduce other anti-nutritional compounds present in substrates like oligosaccharides, tannins, lectins and saponins, either by producing specific degradation enzymes, or by biotransformation through metabolic reactions.

2.5 Hydrogen Peroxide

Some obligatory homofermentative LAB such as *Strept. thermophilus, Lb. acidophilus* and *Lb. casei* are found to produce hydrogen peroxide in fermented milk. Though it is produced in minute concentration, H_2O_2 oxidizes sulfhydryl groups of proteins causing denaturation of a number of enzymes, and the peroxidation of membrane lipids increases membrane permeability. Moreover, H_2O_2 may serve as a precursor for the production of bactericidal free radicals such as superoxide (O^{-2}) and hydroxyl (OH^-) radicals which can damage DNA (Yadav et al. 1993).

2.6 Flavor Compounds: Diacetyl and Acetaldehyde

A few strains of LAB are found to produce flavor compounds like diacetyl (2, 3-butanediol) and acetaldehyde that could exert antimicrobial activity. Diacetyl is formed during citrate metabolism and is responsible for the typical aroma and flavor of butter and other fermented milk products. Diacetyl is produced by different species of the genera *Leuconostocs, Lactococcus, Pediococcus*, and *Lactobacillus* while many of the lactobacilli are responsible for generating acetaldehyde during milk fermentation. These flavor compounds show an inhibitory effect against spoilage bacteria even though produced in very small amounts. Gram-negative bacteria, yeasts and molds are more sensitive to diacetyl and it is believed to interfere with the utilization of arginine or arginine binding proteins (Yadav et al. 1993, Rattanachaikunsopon and Phumkhachorn 2010). Acetaldehyde could get converted into H_2O_2 by xanthine oxidase enzyme. Nevertheless, the amount of flavor compounds is much lower than the level considered necessary to achieve the inhibition of microorganisms.

2.7 Short Chain Fatty Acids

The antimicrobial activity of short chain fatty acids (SCFA) has been recognized since many years. The SCFA such as formic acid, acetic acid, propionic acid, butyric acid and lactic acid are produced during the aerobic or anaerobic metabolism of carbohydrates by some of the lactobacilli and lactococci (Broekaert et al. 2011). Formation of SCFA leads to a decrease in the pH of the food or media and thus, creates an unfavourable environment for the growth of contaminating spoilage bacteria. The antimicrobial activity of SCFA is dependent on its concentration, chain length and pH of the medium (Yadav et al. 1993). The antimicrobial activity of SCFA has been thought to be due to the undissociated molecules, not the anions, since pH had profound effects on their activity, with a more rapid killing effect at lower pH levels.

Usually, formation of SCFA during the assimilation of prebiotic oligosaccharides is considered an important metabolic trait. Acidification can affect the balance of the bacterial species, bacterial metabolic activity and product formation. The presence of SCFA in the intestines contributes to lower pH, enhances bio-availability of minerals like calcium, zinc and magnesium and the inhibition of harmful bacteria (Teitelbaum and Walker 2002, Wong et al. 2006). Even the cancer-suppressing properties of a few dietary fibers are linked with their ability to generate butyric acid upon fermentation though colonic microbiota (Perrin et al. 2001). Acetate is mainly metabolized in human muscle, kidney, heart and brain, whereas propionate acts as a possible gluconeogenic precursor suppressing cholesterol synthesis (Gibson 1999).

2.8 Carbon Dioxide and Ethanol

Most yeasts and some hetero-fermentative LAB produce CO_2 and ethanol during growth. Formation of CO_2 gas is imperative during bread making and sourdough preparation for the leavening of the dough. Traditional methods employed to manufacture several traditional fermented milks like kefir and kumiss are characterized by an effervescent appearance accompanied by a tingling taste due to the development of CO_2 by inherent bacteria and yeasts. On the other hand, ethanol produced by lactobacilli in kefir and kumiss also gives the typical taste, aroma and flavor to the products. Ethanol producing attributes of yeasts have helped in the manufacturing of alcoholic beverages like beer, wine, rum and whisky since ancient times.

The precise mechanism of antimicrobial action exerted by CO_2 is still unknown; however it creates an anaerobic environment which may inhibit enzymatic decarboxylation reactions and the accumulation of CO_2 in the membrane lipid bilayer may affect membrane permeability. CO_2 has a strong antifungal activity and can effectively inhibit the growth of many food spoilage microorganisms, especially Gram-negative psychrotrophic bacteria (Yadav et al. 1993). The levels of ethanol produced in food systems by hetero-fermentative LAB are so low that their contribution to antimicrobial effects is negligible.

2.9 Exopolysaccharides

Microorganisms are able to produce principally two types of polysaccharides, intracellular polysaccharides and extracellular polysaccharides, classified according to their location in the cell. The intracellular polysaccharides are produced by plants (starch, inulin), animals (glycogen), and microorganisms (glycogen) while the extracellular polysaccharides, known as exopolysaccharides (EPS), are produced by a majority of bacteria and microalgae and less frequently by yeasts and fungi (De Vuyst et al. 2001). Most of the LAB producing EPS belong to the genera *Streptococcus*, *Lactococcus*, *Leuconostoc*, *Pediococcus*, and *Lactobacillus*, while EPS production is also reported from some Non Starter Lactic Acid Bacteria (NSLAB) like *Bifidobacteria* (Patel and Prajapati 2013). Microbial EPS occur as capsules (capsular exopolysaccharides: CPS) which are covalently bounded to the cell surface or secreted as ropy-slime forms in cell environments. The most frequently found monomers in the structure of EPS are glucose and galactose, followed by rhamnose, mannose, fructose, arabinose and xylose, the sugar derivatives, N-acetyl galactosamine and N-acetyl glucosamine. On the basis of their chemical composition and mechanism of biosynthesis, EPS can be classified as homopolysaccharides, which contain a single type of monosaccharide, and heteropolysaccharides, which comprise repeating units of different monosaccharides (De Vuyst et al. 2001).

EPS impart highly desirable rheological changes in the food matrix including increased viscosity, improved texture and reduced syneresis (Badel et al. 2011). EPS producing strains are very significant in the manufacturing of yoghurt, dahi, kefir and sourdough. In the dairy and food industries, microbial EPS is used in confectionary products to improve moisture retention, viscosity and inhibit sugar crystallization; in veterinary and human medicine as a blood plasma extender or blood flow improving agent and as a cholesterol lowering agent; in separation technology and in aqueous two phase systems, as a micro-carrier in tissue/cell culture (De Vuyst et al. 2001, Patel and Prajapati 2013).

EPS-producing LAB have a greater ability to withstand technological stresses (Stack et al. 2010) and survive the passage through the GI tract, compared to their non-producing bacteria (Lebeer et al. 2010). It is observed that bacterial strains with a relatively high EPS producing capacity were insensitive against bacteriophages and nisin, suggesting their beneficial physiological function in natural environments (Durlu-Ozkaya et al. 2007). Apart from industrial importance, EPS from various LAB claimed to possess physiological functions and health beneficiary effects such as prebiotic activity, cholesterol reducing activity, antigastritis, antiulcer, antitumour and anti-mutagenic properties as examined through some *in vitro* or *in vivo* studies (Korakli et al. 2002, Rodríguez et al. 2008, Tsuda et al. 2008).

2.10 Bioactive Peptides

Bioactive peptides are protein sequences that remain inactive in the primary structure of native protein, but when released by actions of proteolytic enzymes or processing steps, may regulate most of the body's physiological functions (Dziuba and Darewicz 2007). Usually, they contain 3 to 20 amino acid residues per molecule. Such biologically active peptides have been isolated from various foods such as soybean, fish, meat and maize; however, milk appears to be one of the most important sources of bioactive peptides identified so far (Farnworth 2003). Casein and whey proteins, which are latent and encrypted in their native form in milk, act as excellent sources of highly valuable proteins, accounting for approximately 80% and 20%, respectively of total milk proteins (Haque and Chand 2006, Hayes et al. 2007). In last few decades, due to their numerous health advantages, milk protein derived bioactive peptides have created commercial interests in the milieu of health improving functional foods (Pihlanto 2006).

Different methods may be employed to generate and stimulate biologically active peptides, such as (a) food processing (b) protein hydrolysis through digestive enzymes, and (c) proteolytic activity by means of fermenting microbial enzymes, especially LAB and related food grade microorganisms. Though LAB are not highly proteolytic in nature, some of the dairy starter cultures have evolved highly sophisticated proteolytic system. A number of fermented dairy products such as yoghurt, cheese and fermented milks came out as potential sources of bioactive peptides. Among LAB, *Lb. helveticus* and *Lb. delbrueckii* ssp. *bulgaricus* have shown the highest proteolytic activity during the manufacturing of fermented products, followed by *Lb. casei*, *Lb. plantarum*, *Lb. rhamnosus*, *Lb. acidophilus* and *Lc. lactis* ssp. *lactis*.

2.10.1 Physiological functions of bioactive peptides

Fermented dairy products serve as a source of physiologically important bioactive peptides that have a positive impact on the body's functions. The potential health benefits may be linked with the production of microbial metabolites *viz.* cell wall components, bacteriocins and the proteolysis of substrate proteins by cell-free extracts containing proteinase and peptidase (Nakajima et al. 1995, Hernández et al. 2005). The major means for producing biologically active peptides from milk is shown in Fig. 3. Initially, these peptides are inactive in the native protein structures; they are released in three ways including processing of foods, action of digestive enzymes during passage through the GI tract, and by microbial fermentation. Studies suggests that upon consumption, the bioactive peptides so generated may affect distinct physiological systems of humans, including the digestive, nervous, endocrine, immune

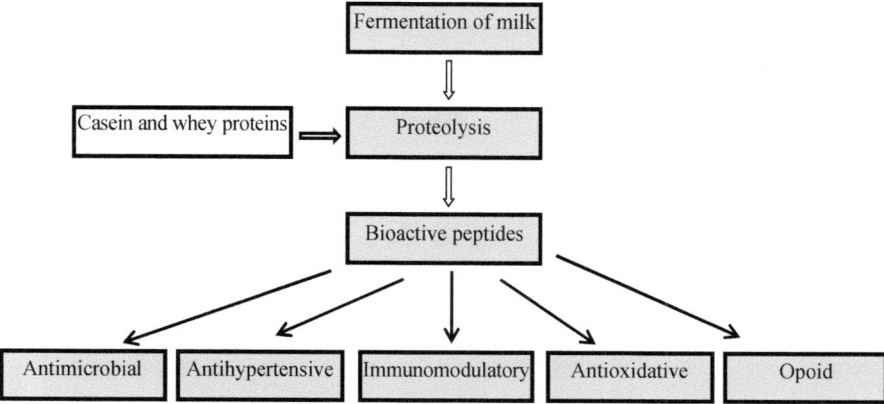

Figure 3. Bioactive peptides form during microbial fermentation of milk.

and cardiovascular systems (Korhonen 2009). Bioactive peptides have been found to have definite activities such as antimicrobial, anti-hypertensive, anti-oxidative, immunomodulatory, opiod or mineral-binding activities (Fitzgerald and Meisel 2003). Some milk-derived bioactive peptides have explored multifunctional properties, i.e., specific peptide sequences may exert two or more different biological activities together.

Fermented milks prepared by using proteolytic LAB were found to contain a high number of bioactive peptides including ACE-inhibitory and anti-hypertensive peptides (López-Fandiño et al. 2006). The blood pressure lowering effects of fermented dairy products or specific casein and whey protein hydrolysates offer compelling evidence of the role played by peptides to induce clinically significant reductions in both, systolic and diastolic blood pressure with no reported adverse effects (Huth et al. 2006). According to Migliore-Samour (1989), bioactive peptides derived from casein are involved in the stimulation of the immune system of neonates. Immunomodulatory peptides obtained from milk act on the immune system and trigger cell proliferation responses which, in turn, influence downstream immunological responses and cellular functions in the body. Bioactive peptides have been reported to possess antioxidant activity due to free radical scavenging, metal ion chelation and singlet oxygen quenching (Wong and Kitts 2003). Pihlanto (2006) suggested that caseins and whey proteins may possess possible free radical scavenging activity by amino acids like tyrosine and cysteine. Bioactive peptides are discussed in detail in Chapter 16 of this book.

2.11 Conjugated Linoleic Acid

Linoleic acid (LA) is a polyunsaturated omega-6 fatty acid and an essential fatty acid that must be consumed for proper health. Conjugated linoleic acids (CLA) are a family of at least 28 structurally similar forms of linoleic acid (cis-9, trans-11 octadecadienoic acid). They are chiefly found in meat and dairy products derived from ruminants. Fermented milks like yoghurt and cheese have shown higher concentration of CLA (Settani and Moschetti 2010).

Table 3. Overview of various bioactive peptides obtained from milk proteins.

Type of Bioactive peptide	Mechanism	Reference
Angiotensin converting enzyme (ACE) inhibitory action	Inhibition of ACE through a decrease of angiotensin II and an increase of bradykinin	Gobbetti et al. (2000), Donkor et al. (2007)
Anti-hypertensive peptides	Reduction of systolic and diastolic BP	Seppo et al. (2002), Tsai et al. (2008)
Antimicrobial peptides	Exact mechanism is not fully understood	Hernández et al. (2005)
Anti-oxidative peptides	Scavenging of free radical by amino acids such as tyrosine and cysteine	Pihlanto (2006)
Immunomodulatory peptides	Stimulation of phagocytosis, triggering of immunoglobulin production, maintenance of cytokines production and thus ability to influence cellular immunity	Otani and Hata (1995), Phelan et al. (2009), Otani et al. (2001)
Opiod peptides	Act as neuro-modulators and influence the release of various neurotransmitters	Clare and Swaisgood (2000), Teschemacher et al. (1997)

In addition to rumen microbiota, other microorganisms such as *Lb. acidophilus* (Lin et al. 1999, Ogawa et al. 2001, 2005), *Lb. plantarum* (Kishino et al. 2002); *Propionibacterium freudenreichii* (Jiang et al. 1998); and Bifidobacteria (Coakley et al. 2003) also reported formation of CLA from linoleic acid. Several strains of probiotic Lactobacilli are also capable of converting linoleic acid to CLA (Ogawa et al. 2005, Yadav et al. 2007). CLA have been reported to possess anti-carcinogenic (Whigham et al. 2000, Zu and Schut 1992), cholesterol reducing properties (Blankson et al. 2000), thus resulting in a reduced risk of cardiovascular diseases (Lee et al. 1994, Gavino et al. 2000) and immunostimulatory activities (Yu et al. 2002) as comprised in Table 4. CLA have also been reported to inhibit the *in vitro* proliferation of human malignant melanoma, particularly colorectal and breast cancer cells (Shultz et al. 1992).

2.12 Gamma-Aminobutyric Acid

GABA (γ-aminobutyric acid) is a four carbon non-protein amino acid that is abundantly present in plants, animals and microorganisms. GABA is a largely distributed neurotransmitter of the central nervous system. GABA deficiency, or decreased levels, in the brain is associated with psychiatric and neurological disorders, including depression, anxiety and insomnia (Dhakal et al. 2012). A few studies indicate that GABA can improve relaxation and sleep (Kim et al. 2010). Some microorganisms have the ability to produce glutamate decarboxylase enzyme which converts L-glutamate into GABA. Major GABA producing microflora involve LAB such as *Lb. hilgardii*, which make food spoilage pathogens unable to grow and act as probiotics in the GI tract. Thus, incorporation of such strains in fermentation processes can help increase GABA levels in fermented foods. Several strains of *Lactobacillus* and *Lactococcus* have been isolated from fermented foods, including fermented vegetable kimchi, paocai, soya extract, alcohol distillery lees and black raspberry juice (Huang et al. 2007, Kim et al. 2009, Servili et al. 2011). *Lb. delbrueckii* subsp.

Table 4. Biological attributes of CLA.

Biological attributes	Mechanism	Reference
Anti-carcinogenic activities	CLA might act by antioxidant mechanisms, inhibition of nucleotide synthesis or by inhibiting both DNA adduct formation and carcinogen activation	Ha et al. 1990, Shultz et al. 1992, Zu and Schut 1992
Fat reducing activity	-	Blankson et al. 2000, Baumgard et al. 2001
Anti atherogenic activity— cardiovascular disease	Decline in the levels of total plasma cholesterol, triacylglycerol, and the ratio of LDL to HDL cholesterol	Lee et al. 1994, Gavino et al. 2000
Reduced risk of diabetes	CLA can improve glucose homeostasis and inhibit body fat accretion as demonstrated in mice, rats, and pigs	Yadav et al. 2007
Immune function	Strengthening of the immune system, negatively regulated expression of certain pro-inflammatory genes	Yu et al. 2002

bulgaricus, *Lb. plantarum* and *Lb. paracasei* were isolated from Italian cheese (Siragusa et al. 2007) and Japanese traditional fermented fish (Komatsuzaki et al. 2005), respectively. The alternate way is to employ LAB with the potential to produce L-glutamic acid, which can facilitate the production of functional foods rich in GABA.

2.13 Antibiotics—Reutericyclin

It is the first antibiotic detected from LAB and the spectrum of inhibition of the antibiotic is restricted to gram-positive bacteria including *Lactobacillus* spp., *Bacillus subtilis*, *B. cereus*, *E. faecalis*, *S. aureus* and *Listeria innocua* (Rattanachaikunsopon and Phumkhachorn 2010). Unlike nisin, reutericyclin does not form pores, but selectively dissipates the transmembrane proton potential that led to disrupt cell membrane function.

2.14 Low-Calorie Sweeteners

In recent times, production of low calorie sweeteners like sorbitol, xylitol, manitol, tagatose, and thehalose by some LAB has attracted the interest of researchers. All these compounds are polyols, i.e., sugar alcohols, and can be produced by the fermentation of foods. Many species of *Leuconostoc* and *Lactobacillus* seem to be the most promising producers of low calorie sweeteners (Soetaert 1990, Kim 2004, Patra et al. 2009). These microbially synthesized compounds can be incorporated as vital food ingredients, particularly in diabetic food products like sugar-free candies, cookies and chewing gums. Low calorie sweeteners can be incorporated into foods after microbial production, or else may be produced directly in the foods by using LAB, where they may lead to synthesize such sweeteners (Patra et al. 2009). A number of health benefits

Table 5. Several bioactive components liberated in fermented foods.

Fermented food product	Fermenting microorganisms	Bioactive compounds	Reported health benefits
Yoghurt and fermented milks	*Lb. delbrueckii* subsp. *bulgaricus*, Strept. *thermophilus*	Folic acid, bioactive peptides	Antioxidant activity, antimicrobial activities
Kimchi	*Leuconostoc* spp., *Lb. brevis*, *Lb. curvatus*, *Lb. plantarum*	Beta carotene, chlorophyll	Anti-mutagenic activities, angiotensin converting activity
Sauerkraut	*Leuc. mesenteroides Lb. brevis*, *Lb. plantarum* and *Pd. Pentosaceous*	Vitamin C, dietary fibres	Anticancer property, antioxidant activity
Olives	*Leuc. mesenteroides, Saccharomyces cerevisiae* and *Candida boidinii*	Vitamin C, flavonoids	Antioxidant activity
Soya bean fermented products	LAB, yeasts and molds	Isoflavones, alpha linolenic acid	Antioxidant activity
Sourdough	*Lactobacillus* spp., *Leuconostoc* spp., *Lactococcus* spp., *Saccharomyces cerevisiae*	Riboflavin, exopolysaccharides	Co-factor in various biochemical reactions
Doenjang	*Bacillus subtilis*	Flavonoids, vitamins, minerals, plant hormones (phytoestrogens)	Anti-carcinogenic properties
Tempeh	*Aspergillus* spp. *Rhizopus* spp., *Streptococcus* and *Enterococcus* spp.	B group vitamin	

have been attributed to these microbially produced low calorie sweeteners, like low glycemic index which is an essential characteristic for diabetic people, followed by osmotic diuretic, weight control as well as antiplaque and prebiotic activities. Hence, these GRAS substances could be useful, especially to diabetic patients and weight watchers (Monedero et al. 2010, Patra et al. 2009).

3. Conclusion

Fermenting food grade microorganisms served as a unique source for developing novel products and applications, especially those that can satisfy the consumer's increasing demands for natural products and health benefits. The identification and application of LAB and related strains delivering health-promoting biologically active components is a very promising field and their ability to enrich the food matrix or human body with the aid of producing vitamins, enzymes and bioactive peptides further enhance the scope of utilizing them for medicinal and health applications. Such microorganisms could be a worthwhile alternative to fortification programmes and useful in the elaboration of novel functional foods. Adequate information regarding the application of microorganisms for the biosynthesis of organic acids, exopolysaccharides and bacteriocins is available. It is well established that certain LAB and *bifidobacteria* have an ability to synthesize vitamins, bioactive peptides, GABA or CLA in food matrixes; however, not much information is available regarding their amount of production as they are synthesized in very minute quantities. Recent advances dealing with fermenting microflora, particularly LAB and their functional ingredients, propose that we have yet to realize their full potential. Application of genetic engineering and gene technology methods to improve production of bioactive compounds by native strains is highly fascinating.

Keywords: Antimicrobials, beta-galactosidase, bioactive peptides, congugated linoleic acid, Gamma-aminobutyric acid, exopolysaccharides, vitamins

References

Anderson, J.W., Johnstone, B.M. and Cook-Newell, M.E. 1995. Meta-analysis of the effects of soy protein intake on serum lipids. The New England Journal of Medicine 333: 276–282.

Alazzeh, A.Y., Ibrahim, S.A., Song, D., Shahbazi, A. and AbuGhazaleh, A.A. 2009. Screening for alpha- and beta-galactosidases in *Lactobacillus reuteri* compared to different strains of Bifidobacteria, Milchwissenschaft 64: 434–437.

Badel, S., Bernardi, T. and Michaud, P. 2011. New perspectives for *Lactobacilli* exopolysaccharides. Biotechnology Advances 29: 54–66.

Baumgard, L.H., Sangster, J.K. and Bauman, D.E. 2001. Milk fat synthesis in dairy cows is progressively reduced by increasing supplemental amounts of trans-10, cis-12 conjugated linoleic acid (CLA). Journal of Nutrition 131: 1734–1769.

Beitane, I. and Ciprovica, I. 2012. The study of added prebiotics on b group vitamins concentration during milk fermentation. Romanian Biotechnological Letters 16: 92–96.

Blankson, H., Stakkestad, J.A., Fagertun, H., Thom, E., Wadstein, J. and Gudmundsen, O. 2000. Conjugated linoleic acid reduces body fat mass in overweight and obese humans. Journal of Nutrition 130: 2943–2948.

Broekaert, W.F., Courtin, C.M., Verbeke, K., Van De Wiele, T., Verstraete, W. and Delcour, J.A. 2011. Prebiotic and other health-related effects of cereal-derived arabinoxylans, arabino-xylanoligosaccharides and xylooligosaccharides. Critical Reviews in Food Science and Nutrition 51: 178–194.

Burgess, C., O'Connelll-Motherway, M., Sybesma, W., Hugenholtz, J. and van Sinderen, D. 2004. Riboflavin Production in *Lactococcus lactis*: Potential for *in situ* production of vitamin-enriched foods. Applied and Environmental Microbiology 70: 5769–5777.

Chavan, J.K. and Kadam, S.S. 1989. Critical reviews in food science and nutrition. Food Science 28: 348–400.

Clare, D.A. and Swaisgood, H.E. 2000. Bioactive milk peptides: A prospectus. Journal of Dairy Science 83: 1187–1195.

Coakley, M., Ross, R.P., Nordgren, M., Fitzgerald, G., Devery, D. and Stanton, C. 2003. Conjugated linoleic acid biosynthesis by human-derived *Bifidobacterium* species. Journal of Applied Microbiology 94: 138–145.

Cooke, G., Behan, J. and Costello, M. 2006. Newly identified vitamin K-producing bacteria isolated from the neonatal faecal flora. Microbial Ecology in Health and Disease 18: 133–138.

Crittenden, R.G., Martinez, N.R. and Playne, M.J. 2003. Synthesis and utilisation of folate by yoghurt starter cultures and probiotic bacteria. International Journal of Food Microbiology 80: 217–222.

D'Aimmo, M.R., Mattarelli, P., Biavati, B., Carlsson, N.G. and Andlid, T. 2012. The potential of bifidobacteria as a source of natural folate. Journal of Applied Microbiology 112: 975–984.

Deguchi, Y., Morishita, T. and Mutai, M. 1985. Comparative studies on synthesis of water-soluble vitamins among human species of bifidobacteria. Agricultural and Biological Chemistry 49: 13–19.

De Vuyst, L., De Vin, F., Vaningelgem, F. and Degeest, B. 2001. Recent developments in the biosynthesis and applications of heteropolysaccharides from lactic acid bacteria. International Dairy Journal 11: 687–707.

Dhakal, R., Bajpai, V.K. and Baek, K.-H. 2012. Production of gaba (γ-Aminobutyric acid) by microorganisms: A review. Brazilian Journal of Microbiology 43(4): 1230–1241.

Dieuleveux, V. and Gueguen, M. 1998. Anti-microbial effects of d-3-phenyllactic acid on *Listeria monocytogenes* in TSB-YE medium, milk, and cheese. Journal of Food Protection 61: 1281–1285.

Donkor, O.N., Henriksson, A., Vasiljevic, T. and Shah, N.P. 2007. Proteolytic activity of dairy lactic acid bacteria and probiotics as determinant of growth and *in vitro* angiotensin-converting enzyme inhibitory activity in fermented milk. Lait 87: 21–38.

Durlu-Ozkaya, F., Aslimb, B. and Ozkaya, M.T. 2007. Effect of exopolysaccharides (EPSs) produced by *Lactobacillus delbrueckii* subsp. *bulgaricus* strains to bacteriophage and nisin sensitivity of the bacteria. LWT-Food Science and Technology 40: 564–568.

Dziuba, M. and Darewicz, M. 2007. Food proteins as precursors of bioactive peptides classification into families. Food Science and Technology International 13: 393–404.

Farnworth, E.R. 2003. Handbook of Fermented Functional Foods, FL, CRC Press, Boca Raton.

Fitzgerald, R.J. and Meisel, H. 2003. Milk protein hydrolysates and bioactive peptides. *In*: P.F. Fox and P.L.H. Mcsweeney (eds.). Advanced Dairy Chemistry, Vol. 1: Proteins 3rd ed. New York, NY, USA: Kluwer Academic/Plenum Publishers.

Gavino, V.C., Gavino, G., Leblanc, M. and Tuchweber, B. 2000. An isomeric mixture of conjugated linoleic acids but not pure cis-9, trans-11-octadecadienoic acid affects body weight gain and plasma lipids in hamsters. Journal of Nutrition 130: 27–29.

Gibson, G.R. 1999. Dietary modulation of the human gut micro flora using the prebiotics oligofructose and inulin. Journal of Nutrition 129: 1438S–1441S.

Gobbetti, M., Lavermicocca, P., Minervini, F., De Angelis, M. and Corsetti, A. 2000. Arabinose fermentation by *Lactobacillus plantarum* in sourdough with added pentosans and αα-L-arabinofuranosidase: A tool to increase the production of acetic acid. Journal of Applied Microbiology 88: 317–324.

Gonzalez, L., Sacristan, N., Arenas, R., Fresno, J.M. and Tornadijo, M.E. 2010. Enzymatic activity of lactic acid bacteria (with antimicrobial properties) isolated from a traditional Spanish cheese. Food Microbiology 27: 592–597.

Goswami, R. 2012. Biosynthesis of folic acid and biotin from probiotic strain *L. helveticus* MTCC 5463. M.Sc. Thesis submitted to Dairy Microbiology Department, SMC College of Dairy Science, Anand Agricultural University, Anand, Gujarat, India.

Gottschalk, G. 1988. Bacterial Metabolism, 2nd ed. Springer-Verlag, New York.

Guldfeldt, L.U., Sorensen, K.I., Stroman, P., Behrndt, H., Williams, D. and Johansen, E. 2001. Effect of starter cultures with a genetically modified peptidolytic or lytic system on cheddar cheese ripening. International Dairy Journal 11: 373–382.

Ha, Y.L., Storkson, J. and Pariza, M.W. 1990. Inhibition of benz (a) pyrene induced mouse forestomach neoplasia by conjugated dienoic derivatives of linoleic acid. Cancer Research 50: 1097–1101.

Haard, N.F., Odunfa, S.A., Lee, C.-H., Quintero-Ramírez, R., Lorence-Quinones, A. and Wacher-Radarte, C. 1999. Fermented cereals. A global perspective. FAO Agricultural Services Bulletin, 138.

Haque, E. and Chand, R. 2006. Milk protein derived bioactive peptides [Online]. Dairy Science. Available: http://www.dairyscience.info/exploitation-of-anti-microbial proteins/111-milk-protein-derived-bioactive-peptides.html [Accessed November 2014].

Hayes, M., Stanton, C., Fitzgerald, G.F. and Ross, R.P. 2007. Putting microbes to work: Dairy fermentation, cell factories and bioactive peptides. Part II: Bioactive peptide functions. Biotechnology Journal 2: 435–449.

Hernández, D., Cardell, E. and Zárate, V. 2005. Anti-microbial activity of lactic acid bacteria isolated from Tenerife cheese: Initial characterization of plantaricin TF711, a bacteriocin-like substance produced by *Lactobacillus plantarum* TF711. Journal of Applied Microbiology 99: 77–84.

Huang, J., Mei, L., Sheng, Q., Yao, S. and Lin, D. 2007. Purification and characterization of glutamate decarboxylase of *Lactobacillus brevis* CGMCC 1306 isolated from fresh milk. Chinese Journal of Chemical Engineering 15: 157–161.

Huth, P.J., Dirienzo, D.B. and Miller, G.D. 2006. Major scientific advances with dairy foods in nutrition and health. Journal of Dairy Science 89: 1207–1221.

Jiang, J., Bjorck, L. and Fonden, R. 1998. Production of conjugated linoleic acid by dairy starter cultures. Journal of Applied Microbiology 85: 95–102.

Keuth, S. and Bisping, B. 1993. Formation of vitamins by pure cultures of tempeh moulds and bacteria during the tempeh solid substrate fermentation. Journal of Applied Biology 75: 427–434.

Kingcha, Y., Tosukhowong, A., Zendo, T., Roytrakul, S., Luxananil, P., Chareonpornsook, K., Valyaseyi, R., Sonomoto, K. and Visessanguan, W. 2012. Anti-listeria activity of *Pediococcus pentosaceus* BCC 3772 and application as starter culture for Nham, a traditional fermented pork sausage. Food Control 25: 190–196.

Kim, J.Y., Lee, M.Y., Ji, G.E., Lee, Y.S. and Hwang, K.T. 2009. Production of γ-aminobutyric acid in black raspberry juice during fermentation by *Lactobacillus brevis* GABA 100. International Journal of Food Microbiology 130: 12–16.

Kim, P. 2004. Current studies on biological tagatose production using L-arabinose isomerase: A review and future perspective. Applied Microbiology and Biotechnology 65: 243–249.

Kim, S.S., Oh, S.H., Jeong, M.H., Cho, S.C., Kook, M.C., Lee, S.H., Pyun, Y.R. and Lee, H.Y. 2010. Sleep-inductive effect of GABA on the fermentation of mono sodium glutamate (MSG). Korean Journal of Food Science and Technology 42: 142–146.

Kishino, S., Ogawa, J., Omura, Y., Matsumura, K. and Shimizu, S. 2002. Conjugated linoleic acid production from linoleic acid by lactic acid bacteria. Journal of the American Oil Chemists Society 79: 159–163.

Klaenhammer, T.R., Barragou, R., Buck, B.L., Azcarate-Peril, M.A. and Altermann, E. 2005. Genomic features of lactic acid bacteria effecting bioprocessing and health. FEMS Microbiology Reviews 29: 393–409.

Komatsuzaki, N., Shima, J., Kawamotoa, S., Momosed, H. and Kimurab, T. 2005. Production of g-aminobutyric acid (GABA) by *Lactobacillus paracasei* isolated from traditional fermented foods. Food Microbiology 22: 497–504.

Korakli, M., Ganzle, M.G. and Vogel, R.F. 2002. Metabolism by bifidobacteria and lactic acid bacteria of polysaccharides from wheat and rye, and exopolysaccharides produced by *Lactobacillus sanfranciscensis*. Journal of Applied Microbiology 92: 958–965.

Korhonen, H. 2009. Milk-derived bioactive peptides: From science to applications. Journal of Functional Foods 1: 177–187.

Kris-Etherton, P.M., Hecker, K.D., Bonanome, A., Coval, S.M., Binkoski, A.E. and Hilpert, K.F. 2002. Bioactive compounds in foods: Their role in the prevention of cardiovascular disease and cancer. American Journal of Medicines 113: 71S–88S.

Lebeer, S., Claes, I.J.J., Verhoeven, T.L.A., Vanderleyden, J. and De Keersmaecker, S.C.J. 2010. Exopolysaccharides of *Lactobacillus rhamnosus* GG form a protective shield against innate immune factors in the intestine. Microbial Biotechnology, pp. 1–7.

Lavermicocca, P., Valerio, F. and Visconti, A. 2003. Antifungal activity of phenyllactic acid against molds isolated from bakery products. Applied & Environmental Microbiology 69: 634–640.

LeBlanc, J.G., de Giori, G.S., Smid, E.J., Hugenholtz, J. and Sesma, F. 2007. Folate production by lactic acid bacteria and other food-grade microorganisms. Communicating Current Research and Educational Topics and Trends in Applied Microbiology 1: 329–339.

LeBlanc, J.G., Laino, J.E., Juarez del Valle, M., Vannini, V., van Sinderen, D., Taranto, M.P., Font de Valdez, G., Savoy de Giori, G. and Sesma, F. 2011. B-group vitamin production by lactic acid bacteria— Current knowledge and potential applications. Journal of Applied Microbiology 111: 1297–1309.

Lee, K.N., Kritchevsky, D. and Pariza, M.W. 1994. Conjugated linoleic acid and atherosclerosis in rabbits. Atherosclerosis 108: 19–25.

Li, H., Gao, D., Cao, Y. and Xu, H. 2008. A high γ-aminobutyric acid-producing *Lactobacillus brevis* isolated from Chinese traditional paocai. Annals of Microbiology 58(4): 649–653.

Lin, T.Y., Lin, C.W. and Lee, C.H. 1999. Conjugated linoleic acid concentration as affected by lactic cultures and added linoleic acid. Food Chemistry 67: 1–5.

López-Fandiño, R., Otte, J. and Van Camp, J. 2006. Physiological, chemical and technological aspects of milk-protein-derived peptides with anti-hypertensive and ACE-inhibitory activity. International Dairy Journal 16: 1277–1293.

Marques, T.M., Wall, R., Ross, P., Fitzgerald, G.F., Ryan, A. and Stanton, C. 2010. Programming infant gut microbiota: influence of dietary and environmental factors. Current Opinion in Biotechnology 21: 149–156.

Martins, J.H., Harg, H., Warren, M.J. and Jahn, D. 2002. Microbial production of vitamin B12. Applied Microbiology and Biotechnology 58: 275–285.

Mas, A., Guillamon, J.M., Torija, M.J., Beltran, G., Cerezo, A.B., Troncoso, M. and Garcia-Parrilla, M. 2014. Bioactive compounds derived from the yeast metabolism of aromatic amino acids during alcoholic fermentation. BioMed Research International Article ID 898045, 7 pages, http://dx.doi.org/10.1155/2014/898045.

Migliore-Samour, D., Floc'h, F. and Jollès, P. 1989. Biologically active casein peptides implicated in immunomodulation. Journal of Dairy Research 56: 357–362.

Mogensen, G. 1993. Starter cultures. pp. 1–25. *In*: Smith, J. (ed.). Technology of Reduced Additive Foods London: Blackie Academic & Professional, UK.

Monedero, V., Perez-Martínez, G. and Yebra, M.J. 2010. Perspectives of engineering lactic acid bacteria for biotechnological polyol production. Applied Microbiology and Biotechnology 86: 1003–1015.

Mtshali, P.S. 2007. Screening and characterisation of wine-related enzymes produced by wine-associated lactic acid bacteria, M.Sc. Thesis, Stellenbosch University, Matieland, South Africa.

Nakajima, K., Hata, Y., Osono, Y., Hamura, M., Kobayashi, S. and Watanuki, M. 1995. Anti-hypertensive effect of extracts of *Lactobacillus casei* in patients with hypertension. Journal of Clinical Biochemistry and Nutrition 18: 181–187.

Nigam, P.S. 2009. Production of bioactive secondary metabolites. pp. 129–45. *In*: P.S. Nigam and A. Pandey (eds.). Biotechnology for Agro-Industrial Residues Utilization. First ed. Netherlands: Springer.

Novik, G., Astapovich, N. and Ryabaya, N. 2007. Production of hydrolases by lactic acid bacteria and bifidobacteria and their antibiotic resistance. Applied Biochemistry and Microbiology 43: 164–169.

O'Connor, E.B., Barrett, E., Fitzgerald, G., Hill, C., Stanton, C. and Ross, R.P. 2005. Production of vitamins, exopolysaccharides and bacteriocins by probiotic bacteria. pp. 167–194. *In*: Tamime. A.Y. (ed.). Probiotic Dairy Products. Blackwell Publishing, Oxford, UK.

Ogawa, J., Kishino, S., Ando, A., Sugimoto, S., Nishara, K. and Shimizu, S. 2005. Production of conjugated linoleic acid by lactic acid bacteria. Journal of Bioscience and Bioengineering 100: 355–364.

Ogawa, J., Matsumura, K., Kishino, S., Omura, Y. and Shimizu, S. 2001. Conjugated linoleic acid accumulation via 10-hydroxy-12-octadecaenoic acid during microaerobic transformation of linoleic acid by *Lactobacillus acidophilus*. Applied and Environmental Microbiology 67: 1246–1252.

O'Sullivan, L., Ross, R.P. and Hill, C. 2002. Potential of bacteriocin-producing lactic acid bacteria for improvements in food safety and quality. Biochemistry 84: 593–604.

Osuntoki, A. and Korie, I. 2010. Antioxidant activity of whey from fermented milk with *Lactobacillus* species isolated from Nigerian fermented foods. Food Technology and Biotechnology 48: 505–511.

Otani, H. and Hata, I. 1995. Inhibition of proliferative responses of mouse spleen lymphocytes and rabbit Peyer's patch cells by bovine milk caseins and their digests. Journal of Dairy Research 62: 339–348.

Otani, H., Watanabe, T. and Tashiro, Y. 2001. Effects of bovine beta-casein (1-28) and its chemically synthesized partial fragments on proliferative responses and immunoglobulin production in mouse spleen cell cultures. Bioscience Biotechnology and Biochemistry 65: 2489–95.

Patel, A. and Prajapati, J.B. 2013. Food and health applications of exopolysaccharides produced by lactic acid bacteria. Advanced Dairy Research 1: 107. doi: 10.4172/2329-888X.1000107.

Patel, A. and Shah, N. 2014. Recent advances in anti-microbial compounds produced by food grade bacteria in relation to enhance food safety and quality. Journal of Innovative Biology 1(4): 189–194.

Patel, A., Falck, P., Shah, N., Immerzeel, P., Adlercreutz, P., Stålbrand, H., Prajapati, J.B., Holst, O. and Nordberg Karlsson, E. 2013a. Evidence for xylooligosaccharide utilization in *Weissella* strains isolated from Indian fermented foods and vegetables. FEMS Microbiology Letters 346: 20–28.

Patel, A., Shah, N., Ambalam, P., Prajapati, J.B., Holst, O. and Ljungh, A. 2013b. Anti-microbial profile of lactic acid bacteria isolated from vegetables and indigenous fermented foods of India against clinical pathogens using microdilution method. Biomedical and Environmental Sciences 26(9): 759–764.

Patra, F., Tomar, S.K. and Arora, S. 2009. Technological and functional applications of low-calorie sweeteners from lactic acid bacteria. Journal of Food Science 74: 16–21.

Perrin, P., Pierre, F., Patry, Y., Champ, M., Berreur, M., Pradal, G., Bornet, F., Meflah, K. and Menanteau, J. 2001. Only fibers promoting a stable butyrate producing colonic ecosystem decrease the rate of aberrant crypt foci in rats. Gut 48: 53–61.

Phelan, M., Aherne, A., Fitzgerald, R.J. and O'brien, N.M. 2009. Casein-derived bioactive peptides: Biological effects, industrial uses, safety aspects and regulatory status. International Dairy Journal 19: 643–654.

Pihlanto, A. 2006. Bioactive peptides: Functionality and production. Agro Food Industry Hi-Tech 17: 24–26.

Pompei, A., Cordisco, L., Amaretti, A., Zanoni, S., Raimondi, S., Matteuzzi, D. and Rossi, M. 2007. Folate production by bifidobacteria as a potential probiotic property. Applied and Environmental Microbiology 73: 179–185.

Pyo, Y.H., Lee, T.C. and Lee, Y.C. 2005. Effect of Lactic Acid Fermentation on Enrichment of Antioxidant Properties and Bioactive Isoflavones in Soybean. Journal of Food Science 70(3): S215–S220.

Rattanachaikunsopon, P. and Phumkhachorn, P. 2010. Lactic acid bacteria: Their anti-microbial compounds and their uses in food production. Annals Biological Research 14: 218–228.

Rodríguez, C., Medici, M., Rodríguez, A.V., Mozzi, F. and Font de Valdez, G. 2008. Prevention of chronic gastritis by fermented milks made with exopolysaccharide-producing *Streptococcus thermophilus* strains. Journal of Dairy Science 92: 2423–2434.

Rose, A. 2015. Soy and phytic acid: Stick with fermented tempeh and miso reducing phytic acid in your food: A visual analysis of the research on home kitchen remedies for phytic acid. Rebuild Market. Retrieved 20 January 2015.

Santos, F., Wegkamp, A., de Vos, W.M., Smid, E.J. and Hugenholtz, J. 2008. High-level folate production in fermented foods by the B12 producer *Lactobacillus reuteri* JCM1112. Applied and Environmental Microbiology 74: 3291–3294.

Seppo, L., Kerojoki, O., Suomalainen, T. and Korpela, R. 2002. The effect of a *Lactobacillus helveticus* LBK-16 H fermented milk on hypertension: A pilot study on humans. Ilchwissenschaft 57: 124–127.

Servili, M., Rizzello, C.G., Taticchi, A., Esposto, S., Urbani, S., Mazzacane, F., Di Maio, I., Selvaggini, R., Gobbetti, M. and Di Cagno, R. 2011. Functional milk beverage fortified with phenolic compounds extracted from olive vegetation water, and fermented with functional lactic acid bacteria. International Journal of Food Microbiology 147(1): 45–52.

Servin, A.L. 2004. Antagonistic activities of lactobacilli and bifidobacteria against microbial pathogens. FEMS Microbiology Review 28: 405–440.

Settani, L. and Moschetti, G. 2010. Non-starter lactic acid bacteria used to improve cheese quality and provide health benefits. Food Microbiology 27: 691–697.

Shultz, T.D., Chew, B.P. and Seaman, W.R. 1992. Differential stimulatory and inhibitory responses of human MCF-7 breast cancer cells to linoleic acid and conjugated linoleic acid in culture. Anticancer Research 12: 2143–2146.

Siragusa, S., Angelis, M.De., Cagno, R.Di., Rizzello, C.G., Coda, R. and Gobbetti, M. 2007. Synthesis of γ-aminobutyric acid by lactic acid bacteria isolated from a variety of Italian cheeses. Applied and Environmental Microbiology 81: 7283–7290.

Soetaert, W. 1990. Production of mannitol with *Leuconostoc mesenteroides*. Mededelingen Faculteit Landbouwwetenschappen Rijks universiteit Gent. 55: 1549–52.

Spano, G. 2011. Biotechnological production of vitamin B2-enriched bread and pasta. Journal of Agricultural Food Chemistry 59: 8013–8020.

Stack, H.M., Kearney, N., Stanton, C., Fitzgerald, G.F. and Ross, R.P. 2010. Association of beta-glucan endogenous production with increased stress tolerance of intestinal Lactobacilli. Applied and Environmental Microbiology 76(2): 500–507.

Steinkraus, K.H., Ayres, R., Olek, A. and Farr, D. 1993. Biochemistry of Saccharomyces. pp. 517–519. *In*: K.H. Steinkraus (ed.). Handbook of Indigenous Fermented Foods. New York: Marcel Dekker.

Strozzi, G.P. and Mogna, L. 2008. Quantification of folic acid in human feces after administration of *Bifidobacterium* probiotic strains. Journal of Clinical Gastroenterology 42: S179–184.

Sybesma, W., Burges, C., Starrenburg, M., van Sinderen, D. and Hugenholtz, J. 2004. Multivitamin production in *Lactococcus lactis* using metabolic engineering. Metabolic Engineering 6: 109–115.

Sybesma, W., Starrenburg, M., Kleerebezem, M., Mierau, I., de Vos, W.M. and Hugenholtz, J. 2003. Increased production of folate by metabolic engineering of *Lactococcus lactis*. Applied and Environmental Microbiology 69: 3069–3076.

Tamang, J.P. 2011. Prospects of Asian Fermented Foods in Global Markets. 11th ASEAN Food Conference, Bangkok, Thailand.

Taranto, M.P., Vera, J.L., Hugenholtz, J., De Valdez, G.F. and Sesma, F. 2003. *Lactobacillus reuteri* CRL1098 produces cobalamin. Journal of Bacteriology 185: 5653–5647.

Teitelbaum, J.E. and Walker, W.A. 2002. Nutritional impact of pre- and probiotics as protective gastrointestinal organisms. Annual Review of Nutrition 22: 107–138.

Teschemacher, H., Koch, G. and Brantl, V. 1997. Milk protein-derived opioid receptor ligands. Biopolymers 43: 99–117.

Thomas, L.V., Clarkson M.R. and Delves-Broughton, J. 2000. Nisin. pp. 463–524. *In*: A.S. Naidu (ed.). Natural Food Anti-microbial Systems, CRC Press, Boca Raton, FL.

Tsai, J.-S., Chen, T.-J., Pan, B.S., Gong, S.-D. and Chung, M.-Y. 2008. Anti-hypertensive effect of bioactive peptides produced by protease-facilitated lactic acid fermentation of milk. Food Chemistry 106: 552–558.

Tsuda, H., Hara, K. and Miyamoto, T. 2008. Binding of mutagens to exopolysaccharide produced by *Lactobacillus plantarum* mutant strain 301102S. Journal of Dairy Science 91: 2960–2966.

Vermeirssen, V., Van Camp, J. and Verstraete, W. 2004. Bioavailability of angiotensin I converting enzyme inhibitory peptides. British Journal of Nutrition 92: 357–366.

Wong, J.M.W., de Souza, R., Kendall, C.W.C., Emam, A. and Jenkins, D.J.A. 2006. Colonic health: Fermentation and short chain fatty acids. Journal of Clinical Gastroenterology 40: 235–243.

Wong, P.Y.Y. and Kitts, D.D. 2003. Chemistry of butter milk solid antioxidant activity. Journal of Dairy Science 86: 1541–1547.

Wouters, J.T.M., Ayad, E.H.E., Hugenholtz, J. and Smit, G. 2002. Microbes from raw milk for fermented dairy products. International Dairy Journal 12: 91–109.

Yadav, H., Jain, S. and Sinha, P.R. 2007. Antidiabetic effect of probiotic dahi containing *Lactobacillus acidophilus* and *Lactobacillus casei* in high fructose fed rats. Nutrition 23: 62–68.

Yadav, J.S., Grover, S. and Batish, V.K.A. 1993. Comprehensive *Dairy Microbiology*, Metropolitan Publisher, New Delhi, India, pp. 463–524.

Yang, H.J., Park, S., Pak, V., Chung, K.R. and Kwon, D.Y. 2011. Fermented soybean products and their bioactive compounds, soybean and health. *In*: Prof. Hany El-Shemy (ed.). ISBN: 978-953-307-535-8, InTech, Available from: http://www.intechopen.com/books/soybean-and-health/fermentedsoybean-products-and-their-bioactive-compounds.

Yu, Y., Correll, P.H. and VandenHeuvel, J.P. 2002. Conjugated linoleic acid decreases production of pro-inflammatory products in macrophages: Evidence for a PPAR gamma-dependent mechanism. Biochimica et Biophysica Acta 1581: 89–99.

Zu, H.X. and Schut, H.A.J. 1992. Inhibition of 2-amino-3-methylimidazo [4,5-f] quinoline (IQ)-DNA adduct formation in CDF1 mice by heat altered derivatives of linoleic acid. Food Chemistry and Toxicology 30: 9–16.

15

Nutritional and Therapeutic Significance of Protein-based Bioactive Compounds Liberated by Fermentation

Lata Ramchandran

1. Introduction

Lifestyle related diseases are currently taking the form of a global epidemic with unhealthy diet and social lifestyles contributing to chronic, non-communicable health disorders such as obesity, certain cancers, diabetes and cardiovascular (CV) diseases (Chaput et al. 2011, WHO 2003, Naukkarinen et al. 2012). Both, industrialized, as well as developing, nations are facing several health related challenges. World health bodies are now suggesting that nutritional components are fundamental in preventing and managing such lifestyle induced diseases (NHMRC 2006, WHO 2003) that lend support to the modern concept of nutrition that food can modulate various functions in the body relevant to health. This emphasises the potential use of foods to promote a state of well-being, better health and reduction of the risk of disease. Diet and nutrition are therefore being accepted as vital keys to control morbidity and mortality from chronic diseases. Growing understanding of the relationship between diet, specific food ingredients and health is leading to new insights into the effect of food components on physiological function and health (Olmeidlla-Alonso et al. 2011).

Increased consumer concerns for health, advances in food regulations and innovation in food technology have allowed for the introduction of foods with

College of Health and Biomedicine, Victoria University, Werribee Campus, VIC-3030, Australia.
E-mail: lata.ramchandran@vu.edu.au

additional health promoting functions that are referred to as functional foods (Hasler 2000, 2002). The increased interest in the role of nutrition in the prevention of chronic disease and optimization of health has resulted in the rapid advance of the functional foods industry across the world (Vella et al. 2013). Advances in food science and technology have placed the food industry in the challenging position of addressing growing consumer awareness of healthy foods, which has led to the introduction of functional foods into the global food market.

Functional foods can be defined as 'foods that can be satisfactorily demonstrated to affect beneficially one or more target functions in the body, beyond adequate nutritional effect, in a way relevant to an improved state of health and wellbeing and/or reduction of risk of disease' (Stanton et al. 2005). Food Standards Australia and New Zealand, Australia's primary food regulatory agency, describes functional foods as '...similar in appearance to conventional foods and intended to be consumed as part of a normal diet, but modified to serve physiological roles beyond the provision of simple nutrient requirements' (Food Standards Australia and New Zealand 2006). In layman's terms, they are the foods designed with the aim of generating desired functional and biological properties. This is a category of foods that enjoys widespread consumer support and is a rapidly growing segment of the food industry.

Any food that can be promoted on a health platform, where the health benefits are supported by solid scientific evidence, can be a functional food. The component that makes the food 'functional' can be either an 'essential macronutrient' if it has specific physiological effects or an 'essential micronutrient' if its intake is over and above the daily recommendations. Additionally, it could be a food component the nutritive value of which is not listed as essential, such as some oligosaccharides (refer Chapter 17 in this book), or it could be of non-nutritive value, such as live microorganisms or plant chemicals (Roberfroid 1999).

2. Functional Fermented Products

Fermented functional foods are a popular category of functional foods that promise the combined benefits of health and nutrition. Fermented dairy products are the most studied among various food fermentations having the tradition as healthy foods and are a natural choice for their makeover as functional foods. These foods contain components that are categorized as probiotics and prebiotics. Several strains of lactic acid bacteria (LAB) have gained importance in recent years in regard to food and nutrition, which include the starter organisms as well as the probiotics (Fitzgerald and Murray 2006, Tomasik and Tomasik 2003). This, however, requires careful selection of starter cultures and/or probiotics to ensure *in situ* expression of desirable health beneficial metabolites in addition to maintaining the product quality (Ramchandran and Shah 2008a). This chapter highlights the bioactivities of some important sources of food proteins with specific reference to dairy fermentations.

3. Food Proteins with Bioactivities

The discovery of natural bioactive peptides is now opening up several opportunities for their exploitation in the promotion of human health through a regular diet and food supplements or even in the form of bio-pharmaceuticals. Several different bioactivities are now being associated with peptides derived from several food proteins other than milk proteins, such as fish muscle and collagen, animal muscles, egg proteins, rice bran proteins, wheat proteins, various peas and legumes, and soybeans.

3.1 Dairy Proteins

Milk and dairy products constitute one of the four major food groups that make up a balanced diet. It is well known that, apart from nutritional value, milk proteins possess biological and physicochemical properties of significance to human health. Research carried out during the last 3 decades has shown that, based on their structural properties and amino acid composition and sequences, the caseins, whey proteins as well as some minor serum proteins, possess latent biological activity; peptides from these proteins may possess myriad physiological functions (Erdmann et al. 2008) and can therefore be an important source of biologically active peptides or bioactive peptides.

The potential role of milk constituents in the prevention of disease is compiled in Table 1.

Table 1. Dairy components and ingredients in functional foods and their health claims.

Ingredient	Claim areas
Minerals	Optimum growth and development, dental health, osteoporosis
Fatty acids	Heart disease, cancer prevention, weight control
Prebiotics/carbohydrates	Digestion, pathogen prevention, gut flora balance, immunity, lactose intolerance
Probiotics	Digestion, immunity, vitamin production, heart disease, antitumor activity, remission of inflammatory bowel disease, prevention of allergy, alleviation of diarrhoea
Proteins/peptides	Immunomodulation, growth, anti-bacterial activity, dental health, hypertension regulation (angiotensin inhibitors)

The peptides from the major milk proteins may possess opiate-like, mineral binding, immunomodulatory, anti-oxidative, anti-bacterial, antithrombotic, hypocholesterolemic and antihypertensive functions (Donkor et al. 2007, 2012, Erdmann et al. 2008, Expósito and Recio 2006, Ramchandran and Shah 2011, Sarmadi and Ismail 2010). Phosphopeptides and caseinomacropeptides obtained from milk casein digestion have also been reported to regulate gastric and other such secretions and blood pressure, as well as stimulate platelet aggregation (Guilloteau et al. 2009, Thomä-Worringer et al. 2006). The best characterized bioactive bovine whey proteins include immunoglobulins, lactoferrin, lactoperoxidase and growth factors. Their occurrence in much higher concentrations in colostrum than in milk reiterates their importance to the health of the neonate. The physiological activities affected by these whey proteins include enhancement of nutrient absorption, stimulation of cell growth, enzymatic activity, inhibition of enzyme activity, modulation of the immune system, and defence against microbial infections (Pakkanen and Aalto 1997, Tripathi and Vashishtha 2006, Yalcin 2006).

The single most effective way to increase the number of bioactive peptides in dairy products is to ferment or co-ferment with highly proteolytic strains of LAB. Fermented milks containing a particularly high number of peptides have been produced using proteolytic strains of the LAB species such as *Lb. helveticus, Lb. casei, Lb. plantarum, Lb. rhamnosus, Lb. acidophilus, Lactococcus lactis* subsp. *lactis* and subsp. *cremoris*, as well as the two species used in traditional yoghurt production, *Lb. delbrueckii* subsp. *bulgaricus* and *Streptococcus thermophilus*. During fermentation, LAB produce a range of secondary metabolites, including the bioactive peptides that have been studied extensively (Gobbetti et al. 2000, Korhonen and Pihlanto 2006). The bioactive peptides are primarily liberated *via* proteolysis and microbial enzymes in the fermented products but they can also be released in the gastro intestinal tract of the consumer during digestion. Thus, protein degradation by microbial enzymes in fermented milks such as yogurt is a desirable process that not only improves milk digestibility, but also enhances nutritional quality.

Lactic acid bacteria are usually weakly proteolytic; however, they do cause a significant degree of proteolysis even in products that usually have a lower fermentation and storage time (Abu-Tarboush 1995) such as yogurt. The proteolytic enzymes found in different species of LAB show different protease activities and complex system of endo- and exo-peptidases, which may differ in nature, specificity and cell location (Kunji et al. 1996, Ramchandran and Shah 2008a). More recently, examination of the enzyme profile of selected LAB strains showed that the main bioactive peptides were released due to the activity of intracellular peptidases (Elfahri 2012). Many industrial cultures are believed to be proteolytic enough to generate bioactive peptides in fermented milks (Gobbetti et al. 2002, 2004, Korhonen and Pihlanto 2006).

In general, LAB possess:

- Proteases located in the microbial cell envelope that permit the degradation of caseins into oligopeptides
- Peptide transport systems that allow the internalization of the released oligopeptides
- Intracellular peptidases that hydrolyse the oligopeptides into peptides or into amino acids to be used by the cells (Juillard et al. 1998, Kunji et al. 1996).

Microbial proteolysis, being highly specific, can lead to the release of very potent bioactive peptides (Ramchandran and Shah 2008b). The small peptidases produced by endopeptidases in the bacterial cells may be excreted into the milk product by a sort of exchange of these peptides over the cell membrane (Kunji et al. 1996), or, more likely, as a result of the lysis of the bacterial cell. The intracellular peptidases that escape from their intracellular location due to lysis of bacterial cells may also act on the large oligopeptides produced by the action of the cell-wall proteases and contribute to the pool of bioactive peptides in the fermented milk. In addition to the release of biologically active peptides after fermentation, microbial proteolysis may also expose the inner protein bonds, favouring the action of digestive enzymes and the release of potentially active peptides (Matar et al. 2003).

3.2 Fish Proteins and Collagen

Hydrolysates and peptides liberated from fish muscle and collagen by processing or digestion are reported to possess several physiological functions such as ACE-inhibitory, anti-oxidant, anti-coagulant and anti-microbial activity (Giri and Ohshima 2012). To date, research has been focussed on ACE-inhibitory peptides that have been isolated and characterized from salmon, sardine, bonito, tuna, Alaska pollack, sea bream, pelagic thresher, and yellow fin sole (Howell and Kasase 2010). Limited human studies (Kawasaki et al. 2000) and several studies on SHR rats have confirmed the blood pressure lowering capabilities of specific peptides derived from fish proteins (Fujita et al. 2000, Jung et al. 2006). Fish protein hydrolysates also exhibit anti-oxidant activities (Bougatef et al. 2009, Je et al. 2005, 2007). Some of the physiological properties have been shown to be fish species and catch season dependant (Medenieks and Vasiljevic 2008).

Fish sauce, made by fermentation of various kinds of fish, is rich in small peptides and amino acids (Ichimura et al. 2003) and exhibits ACE-inhibitory activity. Salt-fermented anchovy sauce has also been reported to contain anticoagulation agents (Kim et al. 2004). A traditional Indian salt-free fermented fish product (Ngari) was found to exhibit ACE-inhibitory and anti-oxidant activity (Phadke et al. 2014). Fermentation of fish meat using *Bacillus subtilis* has also exhibited anti-oxidant and anti-bacterial activities *in vitro* (Jemil et al. 2014). Recent research has indicated that controlling the activity of major fish endogenous proteases, particularly cathepsins B and L, can help in minimising proteolytic activity in underutilized fish species found in Australian waters. This, in turn, can help improve the textural quality and physiological functionality of these fish (Ahmed et al. 2013). With emerging issues of industrial pollution and stringent environmental policies, research is now being directed towards exploring the possible recovery of high-value biological compounds from fish processing industry wastes (Ferraro et al. 2013).

3.3 Animal Muscle Proteins

In recent times, meat consumption has been adversely affected by the negative health image associated with its fat and protein derivatives (Arihara 2012), despite it being a good source of proteins with high biological value as well as a supplier of minerals and vitamins. Therefore, research into the beneficial attributes of meat has been revived and a number of bioactive peptides exhibiting ACE inhibitory activity have been isolated from meat by peptic digestion (Arihara et al. 2001, Katayama et al. 2008). Several other bioactive substances have also been identified in animal muscle, including conjugated linoleic acid (CLA), carnosine, anserine, L-carnitine, glutathione, taurine, coenzyme Q10, and creatine. Of these, carnosine and anserine were the most potent and suggested to be playing a role in cellular homeostatis and maintenance, lowering cholesterol levels and blood pressure (Ririe et al. 2000, Seccombe et al. 1987, Shimura and Hasegawa 1993), attenuating the inflammatory process associated with arterial hypertension (Miguel-Carrasco et al. 2008), and also preventing skeletal muscle apoptosis and muscle myopathy during heart failure (Vescovo et al. 2002). They also showed potential for treating fatigue in chronically ill patients (Harris 2008), dose-dependent ACE-inhibitory activities (Hou et al. 2003) and have shown promise for the ophthalmic treatment of cataracts, in both rats (Yan et al. 2008) and humans (Babizhayev and Kasus-Jacobi 2009). Additionally, they may also have therapeutic potential for preventing diabetes-induced atherosclerosis (Rashid et al. 2007), be a novel anti-inflammatory agent (Son et al. 2005) and has been shown to possess anti-convulsant effects (Kozan et al. 2008, Wu et al. 2006, Zhu et al. 2007) while exhibiting anti-aging effects (Kovacs-Nolan and Mine 2010).

Comminuted meat formed into sausages is known to be fermented as are whole meat products such as ham. Fermented sausages were touted as potential probiotic containing foods. Fermented sausages using *Lb. paracasei* have been produced; however, their probiotic effect remained questionable (Bunte et al. 2000, Jahreis et al. 2002). Overall, microbial fermentation of meat proteins has been less successful, presumably due to the poor proteolytic activity of the lactobacilli used in meat fermentations (Arihara 2012). Consequently, limited research has been directed towards the impact of these fermentations on the bioactivity potential of the product.

3.4 Egg Proteins

Several egg protein fractions such as ovalbumin, ovomucin and ovomacroglobulin have been reported to possess anti-microbial activity (Messens et al. 2005). Several other egg proteins such as ovalbumin, ovotransferrin, ovomucin-derived peptides, lysozyme and cystatin have been reported to have immunomodulating properties. Anti-cancer properties have been shown in avidin, cystatin and ovomucin while phosvitin, ovotransferrin, lysozyme and egg white albumin have been reported to play an important role as anti-oxidants. Peptides derived from egg proteins have also been reported to show anti-microbial (Mine et al. 2004, Pellegrini et al. 2004), immunogenic (He et al. 2003), anti-carcinogenic (Premzl et al. 2001) and anti-hypertensive (Davalos et al. 2004) activities.

Although fermented egg products are not common, fermentation of egg white with *Lb. delbrueckii* subsp. *delbrueckii* had the potential of reducing allergenic reactions (Li et al. 2013).

3.5 Cereal Proteins

Whole grain consumption has been shown to increase satiety, reduce transit time and glycaemic response and offer protection against obesity, diabetes, CVD and cancers. The fibre (improved faecal bulking and satiety, viscosity and SCFA production, and/ or reduced glycaemic response) and Mg (better glycaemic homeostasis through increased insulin secretion), together with the anti-oxidant and anti-carcinogenic properties of numerous bioactive compounds, especially those in the bran and germ (minerals, trace elements, vitamins, carotenoids, polyphenols and alkylresorcinols), are today well-recognised mechanisms that afford protection against diseases (Fardet 2010). Some rice bran protein isolates have been found to exhibit a potent anticancer and hypo-cholesterolemic effect and improvement in the lipid profile (Ali et al. 2010, Hata et al. 2008). The nutraceutical significance of phytonutrients of rice bran has been discussed in detail by Rukimini (2003). Similarly, wheat albumin has been observed to delay carbohydrate digestion which, in turn, prevents the increase in postprandial hyperglycemia and thus reduces the risk of diabetes (Kodama et al. 2005). About 12 peptides from wheat germ (Matsui et al. 1999) and the tripeptide, Ile–Ala–Pro, from wheat gliadin have been reported to inhibit ACE *in vitro* (Motoi and Kodama 2003).

Cereal fermentation has been reported to improve bioavailability of minerals and phytochemicals and increase glycemic index while generating new bioactive compounds such as prebiotic oligosaccharides (Poutanen et al. 2009). Several traditional cereal fermented products exist in South-east Asian countries such as ang-kak, lao-chao, minchin, puto and tao-si, in India (bhatura, jalebies), Sri Lanka (appa) and African countries (chickwangue, injera, kenkey and mahewu). Being confined to the local areas, these products are yet to be studied scientifically for their potential health and physiological benefits.

3.6 Legume Proteins

Legumes have been a source of nutritional interest due to the presence of adequate proportions of protein and starch, along with fibre and vitamins. In recent years, research has highlighted the bioactivity of some components found in legumes such as α-glycosides, tannins, saponins and alkaloids (Muzquiz et al. 2012). Several bioactivities have been reported for legume proteins such as opiate, immunomodulatory, ACE-inhibitory, antithrombotic, anti-oxidant, and anti-microbial activities (Sirtori et al. 2008). Hydrolysates obtained from legumes such as chickpea and garden peas have also been observed to produce peptides that demonstrate anti-oxidant (Arcan and Yemenicioglu 2007, Li et al. 2008); hypo-cholesterolemic (Sirtori et al. 2008, Yang et al. 2007); anti-cancer (Clemente et al. 2005); ACE-inhibitory (Humiski and Aluko 2007, Yust et al. 2003) and anti-bacterial activities (Ye and Ng 2003, Ye et al. 2000).

Traditionally, fermented legume products are made mainly in the Indian sub-continent and include products such as dhokla, khaman, puda, vadai and waries. Although traditionally fermented legumes are known to improve the nutritional quality of their proteins, such as their digestibility, and reduce their anti-nutritional effects (Madodé et al. 2013), limited research has been carried out to investigate the health and physiological benefits of fermented legumes. Recent studies on a fermented mixed bean product, fermented using *Lb. paracasei* and *Saccharomyces cerevisiae*, was reported to modulate cutaneous atopic dermatitis-like inflammation in mice with potential applications for the prevention and treatment of atopic dermatitis (Yeh et al. 2014). Split beans of Canavalia, fermented with *Rhizopus oligosporus,* showed improved bioactive and anti-oxidant potential (Niveditha and Sridhar 2012). *In vitro* studies have indicated that a diet of fermented peanuts, locust bean and soya bean fed to diabetes-induced male Wistar rats could attenuate oxidative stress and provide protection against damage of hepatetic tissues arising out of diabetes (Ademiluyi and Oboh 2012).

Some traditional products combine cereals such as rice and legumes to make products such as idli and dosa that are the traditional fermented foods of south India. Although known to be nutritionally beneficial, the specific bioactive potential of these products is yet to be explored.

3.7 Soy Proteins

The dietary fibres (insoluble carbohydrates) of soy beans are understood to be responsible for improvement of the intestinal environment, while the iso-flavones ease postmenopausal symptoms and have a role in preventing cardiovascular diseases in addition to exhibiting anti-mutagenic, anti-bacterial and antifungal activities. Soy anthocyanins show a free radical scavenging effect and can act as potential antineoplastic agents (Chen et al. 2012). Although soy proteins contains a certain amount of bioactive peptides that have distinct physiological activities in lipid metabolism, it is not clear which peptides are responsible for these effects (Aoyama et al. 2000, Tamaru et al. 2007). However, it has been demonstrated that continuous consumption of soy proteins can lead to significant reductions in blood cholesterol, thus providing protection against heart diseases (Samuel et al. 2008) as well as reduction in hyper-insulinemia, which in turn decreases the expression of SREBP—1C in the liver leading to reduced hepatic steatosis (Asencio et al. 2004). Further, soy peptides have an anti-obesity effect and ingestion of soy peptides could also decrease the muscle damage induced after exercise.

Fermented soy bean products are traditional to Chinese and Japanese cultures. Most commonly, they are fermented with *Bacillus subtilis* and made into various types of natto including salted, sweet or non-salted varieties. The starter cultures of natto are believed to have probiotic properties; they also have the ability to produce enzymes that can degrade allergens found in soy beans, as well as produce vitamin K which helps prevent osteoporosis (Vermeer et al. 1995, Shearer 2000). Another popular Japanese fermented soy paste is miso. The commonly used organisms for fermentation include koji mold and *Aspergillus oryzae*. Regular consumption of miso has been shown to reduce susceptibility to stomach diseases; in particular, it exhibits anti-*Helicobacter pylori* activity (Bae et al. 2001, Shirataki et al. 2001). Moreover, fermentation of soy beans or milk have been shown to exhibit anti-inflammatory properties (Liao et al. 2010) and can reduce the risk of cardiovascular disease (Chen et al. 2011, Juan et al. 2010). Black soybean milk fermented with *Lb. brevis* was reported to be enriched with γ-amino-butyric acid and could be a potential candidate as an alternative treatment of depression (Ko et al. 2013). Although there is evidence for the positive effects of soy protein on gluco-regulatory function both *in vitro* and *in vivo*, there is a definite lack of evidence to show that soy proteins are more beneficial than other protein sources for gluco-regulatory functions during weight loss, nor is there enough evidence to show that soy proteins have a direct impact on glucose metabolism independent of weight and fat loss *in vivo* (Anderson and Pasupuleti 2008, Cope et al. 2008).

4. Conclusion

Proteins are known to play a role in promoting optimal human health, and changes to diet and life style are increasingly being accepted as practical strategies to reduce the incidence of chronic non-communicable diseases. Variety and moderation in diet is the current drift for promoting food-based health and disease prevention. Therefore, interest in bioactive peptides from various sources of food proteins continues to attract researchers, food manufacturers and consumers alike. However, we are yet to confirm and elucidate the mechanisms of action of the various bioactive peptides in various physiological functions. Variations in results of *in vitro* and *in vivo* testings continue to exist, mainly because the *in vitro* methods do not account for other physiological transformations that are possible when such peptides enter the gastro-intestinal tract of humans. Moreover, much of the current scientific evidence relates to lifestyle diseases like obesity, heart disease and diabetes, and bone health, leaving out other areas of functionality that need research based evidence such as gut health, immunity and cancer prevention.

The challenge, therefore, lies in developing effective synergies between science, technology and food product development not only for the benefit of the consumer, but also to maintain the credibility and sustainability of the functional foods market. All these lay emphasis on the need for more detailed research and rigorous human clinical trials to prove the actual health benefits of these peptides, as well as meticulous screening of novel bioactive peptides, if they are to result in commercialization. The production cost and potential allergenicity of these bioactive peptides are equally important in determining their commercialization. Additionally, consumer perception and legislative issues also need to be addressed.

Keywords: Fish collagens, protein-based bioactive compounds

References

Abu-Tarboush, H.M. 1995. Comparison of associative growth and proteolytic activity of yogurt starters in whole milk from camels and cows. Journal of Dairy Science 79: 366–371.

Ademiluyi, A.O. and Oboh, G. 2012. Attenuation of oxidative stress and hepatic damage by some fermented tropical legume condiment diets in streptozotocin-induced diabetes in rats. Asian Pacific Journal of Tropical Medicine 692–697.

Ahmed, Z., Donkor, O., Street, W.A. and Vasiljevic, T. 2013. Proteolytic activities in fillets of selected underutilized Australian fish species. Food Chemistry 140: 238–244.

Ali, R., Shih, F.F. and Riaz, M.N. 2010. Processing and functionality of rice bran proteins. pp. 233–246. *In*: Y. Mine, E. Li-Chan and B. Jiang (eds.). Bioactive Proteins and Peptides as Functional Foods and Nutraceuticals. John Willey & Sons Press, Iowa.

Anderson, J.W. and Pasupuleti, V.K. 2008. Soybean and soy component effects on obesity and diabetes. pp. 141–166. *In*: V.K. Pasupuleti and J.W. Anderson (eds.). Nutraceuticals, Glycemic Health and Type 2 Diabetes, Wiley—Blackwell, Ames, IA.

Aoyama, T., Fukui, K., Nakamori, T., Hashimoto, Y., Yamamoto, T., Takamatsu, K. and Sugano, M. 2000. Effect of soy and milk whey protein isolates and their hydrolysates on weight reduction in genetically obese mice. Bioscience. Biotechnology and Biochemistry 64: 2594–2600.

Arcan, I. and Yemenicioglu, A. 2007. Anti-oxidant activity of protein extracts from heat-treated or thermally processed chickpeas and white beans. Food Chemistry 103: 301–312.

Arihara, K. 2012. Meat-based bioactive compounds and functional meat products. Food Science and Technology 26: 25–27.

Arihara, K., Nakashima, Y., Mukai, T., Ishikawa, S. and Itoh, M. 2001. Peptide inhibitors for angiotensin I-converting enzyme from enzymatic hydrolysates of porcine skeletal muscle proteins. Meat Science 57: 319–324.

Asencio, C., Torres, N., Isoard-Acosta, F., Gomez-Perez, F.J., Hernandez-Pando, R. and Tovar, A.R. 2004. Soy protein affects serum insulin and hepatic SREBP—1 mRNA and reduces fatty liver in rats. Journal of Nutrition 134: 522–529.

Babizhayev, M.A. and Kasus-Jacobi, A. 2009. State of the art clinical efficacy and safety evaluation of N-acetylcarnosine dipeptide ophthalmic prodrug. Principles for the delivery, self-bioactivation, molecular targets and interaction with a highly evolved histidyl-hydrazide structure in the treatment and therapeutic management of a group of sight-threatening eye diseases. Current Clinical Pharmacology 4: 4–37.

Bae, E.A., Han, M.J. and Kim, D.H. 2001. *In vitro* anti-*Helicobacter pylori* activity of inisolidone isolated from flowers and rhizomes of *Pueraria thunberginia*. Planta Medica 67: 161–163.

Bougatef, A., Hajji, M., Balti, R., Lassoued, I., Triki-Ellouz, Y. and Nasri, M. 2009. Anti-oxidant and free radical-scavenging activities of smooth hound (*Mustelus mustelus*) muscle protein hydrolysates obtained by gastrointestinal proteases. Food Chemistry 114: 1198–1205.

Bunte, C., Hertel, C. and Hammes, W.P. 2000. Monitoring and survival of *Lactobacillus paracasei* LTH 2579 in food and human intestinal tract. Systematic and Applied Microbiology 23: 260–266.

Chaput, J.P., Klingenberg, L., Astrup, A. and Sjodin, A.M. 2011. Modern sedentary activities promote overconsumption of food in our current obesogenic environment. Obesity Reviews 12(5): e12–20.

Chen, K.-I., Erh, M.-H., Su, N.-W., Liu, W.-H., Chou, C.-C. and Cheng, K.-C. 2012. Soy foods and soybean products: From traditional use to modern applications. Applied Microbiology and Biotechnology 96: 9–22.

Chen, Y.F., Lee, S.L. and Chou, C.C. 2011. Fermentation with *Aspergillus awamori* enhanced contents of amino nitrogen and total phenolics as well as the low-density lipoprotein oxidation inhibitory activity of black soybeans. Journal of Agricultural and Food Chemistry 59: 3974–3979.

Clemente, A., Gee, J.M., Johnson, I.T., MacKenzie, D.A. and Domoney, C. 2005. Pea (*Pisumsativum* L.) protease inhibitors from the Bowman—Birk class influence the growth of human colorectal adenocarcinoma HT29 cells *in vitro*. Journal of Agricultural and Food Chemistry 53: 8979–8986.

Cope, W.B., Erdman, J.W. and Allison, D.B. 2008. The potential role of soy foods in weight and adiposity reduction: An evidence-based review. Obesity Reviews 9: 219–235.

Davalos, A., Miguel, M., Bartolome, B. and López-Fandiño, R. 2004. Anti-oxidant activity of peptides derived from egg white proteins by enzymatic hydrolysis. Journal of Food Protection 67: 1939–1944.

Donkor, O.N., Henriksson, A., Singh, T.K., Vasiljevic, T. and Shah, N.P. 2007. ACE-inhibitory activity of probiotic yogurt. International Dairy Journal 17: 1321–1331.

Donkor, O.N., Stojanovska, L., Ginn, P., Ashton, J. and Vasiljevic, T. 2012. Germinated grains: Sources of bioactive compounds. Food Chemistry 135: 950–959.

Elfahri, K. 2012. Release of bioactive peptides from milk proteins by *Lactobacillus* species. M.Sc. thesis submitted to Victoria University, Australia.

Erdmann, K., Cheung, B.W.Y. and Schröder, H. 2008. The possible roles of food-derived bioactive peptides in reducing the risk of cardiovascular disease. Journal of Nutritional Biochemistry 19: 643–654.

Expósito, I.L. and Recio, I. 2006. Anti-bacterial activity of peptide and folding variants from milk proteins. International Dairy Journal 16: 1294–1305.

Fardet, A. 2010. New hypotheses for the health-protective mechanisms of whole-grain cereals: What is beyond fibre? Nutrition Research Reviews 23(1): 65–134.

Ferraro, V., Carvalho, A.P., Piccirillo, C., Santos, M.M., Castro, P.M.L. and Pintado, M.E. 2013. Extraction of high added value biological compounds from sardine, sardine-type fish and mackerel canning residues: A review. Materials Science and Engineering C33: 3111–3120.

Fitzgerald, R.J. and Murray, B.A. 2006. Bioactive peptides and lactic fermentations. International Journal of Dairy Technology 59: 118–125.

Food Standards Australia and New Zealand. 2006. NUTTAB 2006 Online version. Cited from http://www.foodstandards.gov.au. Last accessed 09/07/2014.

Fujita, H., Yokoyama, K. and Yoshikawa, M. 2000. Classification and antihypertensive activity of angiotensin I-converting enzyme inhibitory activity derived from food proteins. Journal of Food Science 65: 564–569.

Giri, A. and Ohshima, T. 2012. Bioactive marine peptides: Nutraceutical value and novel approaches. Advances in Food Nutrition and Research 65: 73–105.

Gobbetti, M., Ferranti, P., Smacchi, E., Goffredi, F. and Addeo, F. 2000. Production of angiotensin-I-converting-enzyme-inhibitory peptides in fermented milk started by *Lactobacillus delbrueckii* subsp. *bulgaricus* SS1 and *Lactococcus lactis* subsp. *cremoris* FT4. Applied and Environmental Microbiology 66: 3898–3904.

Gobbetti, M., Minervini, F. and Grizzello, C.G. 2004. Angiotensin-I-converting-enzyme-inhibitory and anti-microbial bioactive peptides. International Journal of Dairy Technology 57: 173–188.

Gobbetti, M., Stepaniak, M., De Angelis, A. and Di Cagno, R. 2002. Latent bioactive peptides in milk proteins: Proteolytic activation and significance in dairy processing. Critical Reviews in Food Science and Nutrition 42: 223–239.

Guilloteau, P., Romé, V., Delaby, L., Mendy, F., Roger, L. and Chayvialle, J.A. 2009. A new role of phosphopeptides released during milk casein digestion in the young mammal: Regulation of gastric secretion. Peptides 30: 2221–2227.

Haasler, C.M. 2000. The changing face of functional foods. Journal of the American College of Nutrition 19: 499S–506S.

Haasler, C.M. 2002. Functional foods: Benefits, concerns and challenges. An apposition paper from the American council on science and health. Journal of Nutrition 132: 3772–3781.

Harris, J.D. 2008. Fatigue in chronically ill patients. Current Opinion in Supportive and Palliative Care 2: 180–186.

Hata, S., Wiboonsirikul, J., Maeda, A., Kimura, Y. and Adachi, S. 2008. Extraction of defatted rice bran by subcritical water treatment. Biochemical Engineering Journal 40: 44–53.

He, X., Tsang, T.C., Luo, P., Zhang, T. and Harris, D.T. 2003. Enhanced tumor immunogenicity through coupling cytokine expression with antigen presentation. Cancer Gene Therapy 10: 669–677.

Hou, W.C., Chen, H.J. and Lin, Y.H. 2003. Anti-oxidant peptides with angiotensin converting enzyme inhibitory activities and applications for angiotensin converting enzyme purification. Journal of Agricultural and Food Chemistry 51: 1706–1709.

Howell, N.K. and Kasase, C. 2010. Bioactive peptides and proteins from fish muscle and collagen. pp. 203–223. *In:* Y. Mine, E. Li-Chan and B. Jiang (eds.). Bioactive Proteins and Peptides as Functional Foods and Nutraceuticals. John Willey & Sons Press, Iowa.

Humiski, L.M. and Aluko, R.E. 2007. Physicochemical and bitterness properties of enzymatic pea protein hydrolysates. Journal of Food Science 72: 605–611.

Ichimura, T., Hu, J., Aita, D.Q. and Maruyama, S. 2003. Angiotensin-I-converting enzyme (ACE) inhibitory activity and insulin secretion stimulative activity of fermented fish sauce. Journal of Bioscience and Bioengineering 5: 496–499.

Jahreis, G., Vogelsang, H., Kiesling, G., Schubert, R., Bunte, C. and Hammes, W.P. 2002. Influence of probiotic sausage (*Lactobacillus paracasei*) on blood lipids and immunological parameters of healthy volunteers. Food Research International 35: 133–138.

Je, J.Y., Park, P.J. and Kim, S.K. 2005. Anti-oxidant activity of peptide isolated from Alaska pollack (*Theragra chalcogramma*) frame protein hydrolysate. Food Research International 38: 45–50.

Je, J.Y., Qian, Z.J., Byun, H.G. and Kim, S.K. 2007. Purification and characterization of an anti-oxidant peptide obtained from tuna backbone protein by enzymatic hydrolysis. Process Biochemistry 42: 840–846.

Jemil, I., Jridi, M., Nasri, R., Ktari, N., Salem, R.B.S.-B., Mehirib, M., Hajji, M. and Nasri, M. 2014. Functional, anti-oxidant and anti-bacterial properties of protein hydrolysates prepared from fish meat fermented by *Bacillus subtilis* A261. Process Biochemistry 49: 963–972.

Juan, M.Y., Wu, C.H. and Chou, C.C. 2010. Fermentation with *Bacillus* spp. as a bioprocess to enhance anthocyanin content, the angiotensin converting enzyme inhibitory effect, and the reducing activity of black soybeans. Food Microbiology 27: 918–923.

Juillard, V., Guillot, A., Le Bars, D. and Gripon, J.C. 1998. Specificity of milk peptide utilization by *Lactococcus lactis*. Applied and Environmental Microbiology 64: 1230–1236.

Jung, W.K., Mendis, E., Je, J.Y., Park, P.J., Son, B.W., Kim, H.C., Choi, Y.K. and Kim, S.K. 2006. Angiotensin-I-converting enzyme inhibitory peptide from yellowfin sole (*Limanda aspera*) frame protein and its antihypertensive effect in spontaneously hypertensive rats. Journal of Food Chemistry 94: 26–32.

Katayama, K., Anggraeni, H.E., Mori, T., Ahmed, A.A., Kawahara, S., Sugiyama, M., Nakayama, T., Maruyama, M. and Mugurumat, M. 2008. Porcine skeletal muscle troponin is a good source of peptides with angiotensin-I-converting enzyme inhibitory activity and anti hypertensive effects in spontaneously hypertensive rats. Journal of Agricultural and Food Chemistry 56: 355–360.

Kawasaki, T., Seki, E., Osajima, K., Yoshida, M., Asada, K., Matsui, T. and Osajima, Y. 2000. Antihypertensive effect of Valyl-Tyrosine, a short chain peptide derived from sardine muscle hydrolyzate, on mild hypertension subjects. Journal of Hypertension 14: 519–523.

Kim, D.C., Chae, H.J. and In, M.J. 2004. Existence of stable fibrin-clotting inhibitor in salt fermented anchovy sauce. Journal of Food Composition and Analysis 17: 113–118.

Ko, C.Y., Lin, M.-T.V. and Tsai, G.J. 2013. Gamma-aminobutyric acid production in black soybean milk by *Lactobacillus brevis* FPA3709 and the antidepressant effect of the fermented product on a forced swimming rat model. Process Biochemistry 48: 559–568.

Kodama, T., Miyazaki, T., Kitamura, I., Suzuki, Y., Namba, Y., Sakurai, J., Torikai, Y. and Inoue, S. 2005. Effects of single and long-term administration of wheat albumin on blood glucose control: Randomized controlled clinical trials. European Journal of Clinical Nutrition 59: 384–392.

Korhonen, H. and Pihlanto, A. 2006. Bioactive peptides: Production and functionality. International Dairy Journal 16: 945–960.

Kovacs-Nolan, J. and Mine, Y. 2010. Animal muscle-based bioactive peptides. pp. 225–231. *In*: Y. Mine, E. Li-Chan and B. Jiang (eds.). Bioactive Proteins and Peptides as Functional Foods and Nutraceuticals. John Willey & Sons Press, Iowa.

Kozan, R., Sefil, F. and Bagirici, F. 2008. Anticonvulsant effect of carnosine on penicillin-induced epileptiform activity in rats. Brain Research 1239: 249–255.

Kunji, E.R.S., Mierau, I., Hagting, A., Poolman, B. and Konings, W.N. 1996. The proteolytic systems of lactic acid bacteria. Antonie van Leeuwenhoek 70: 187–221.

Li, S., Offengenden, M., Fentabil, M., Gänzle, M.G. and Wu, J. 2013. Effect of egg white fermentation with lactobacilli on IgE binding ability of egg white proteins. Food Research International 52: 359–366.

Li, Y., Jiang, B., Zhang, T., Mu, W. and Liu, J. 2008. Anti-oxidant and free radical-scavenging activities of chickpea protein hydrolysate (CPH). Food Chemistry 106: 444–450.

Liao, C.L., Huang, H.Y., Sheen, L.Y. and Chou, C.C. 2010. Anti-inflammatory activity of soymilk and fermented soymilk prepared with lactic acid bacterium and bifidobacterium. Journal of Food and Drug Analysis 18: 202–210.

Madodé, Y.E., Nout, M.J.R., Bakker, E.-J., Linnemann, A.R., Hounhouigan, D.J. and van Boekel, M.A.J.S. 2013. Enhancing the digestibility of cowpea (*Vigna unguiculata*) by traditional processing and fermentation. LWT-Food Science and Technology 54: 186–193.

Matar, C., LeBlanc, J.G., Martin, L. and Perdigón, G. 2003. Biologically active peptides released in fermented milk: Role and functions. pp. 177–201. *In*: E.R. Farnworth (ed.). Handbook of Fermented Functional Foods, CRC Press, Florida.

Matsui, T., Li, C.-H. and Osajima, Y. 1999. Preparation and characterization of novel bioactive peptides responsible for angiotensin-I-converting enzyme inhibition from wheat germ. Journal of Peptide Science 5: 289–297.

Medenieks, L. and Vasiljevic, T. 2008. Underutilized fish as sources of bioactive peptides with potential health benefits. Food Australia 60: 581–588.

Messens, W., Grijspeerdt, K. and Herman, L. 2005. Eggshell penetration by *Salmonella*: A review. World's Poultry Science Journal 61: 71–85.

Miguel-Carrasco, J.L., Mate, A., Monserrat, M.T., Arias, J.L., Aramburu, O. and Vazquez, C.M. 2008. The role of inflammatory markers in the cardioprotective effect of L-carnitine in L-NAME-induced hypertension. American Journal of Hypertension 21: 1231–1237.

Mine, Y., Ma, F. and Lauriau, S. 2004. Anti-microbial peptides released by enzymatic hydrolysis of hen egg white lysozyme. Journal of Agricultural and Food Chemistry 52: 1088–1094.

Motoi, H. and Kodama, T. 2003. Isolation and characterization of angiotensin-I-converting enzyme inhibitory peptides from wheat gliadin hydrolysate. Nahrung 47: 354–358.

Muzquiz, M., Varela, A., Burbanao, C., Cuadrado, C., Guillamón, E. and Pedrosa, M.M. 2012. Bioactive compounds in legumes: Pro-nutritive and anti-nutritive actions. Implication for nutrition and health. Phytochemistry Reviews 11: 227–244.

National Health and Medical Research Council (NHMRC). 2006. Nutrient reference values for Australia and New Zealand. Canberra: National Health and Medical Research Council.

Naukkarinen, J., Rissanen, A., Kaprio, J. and Pietilainen, K.H. 2012. Causes and consequences of obesity: The contribution of recent twin studies. International Journal of Obesity 36: 1017–1024.

Niveditha, V.R. and Sridhar, K.R. 2012. Anti-oxidant activity of raw, cooked and *Rhizopus oligosporus* fermented beans of Canavalia of coastal sand dunes of Southwest India. Journal of Food Science and Technology 1–8.

Olmeidlla-Alonso, B., Jiménez-Colmenero, F. and Sánchez-Miniz, F.J. 2011. Development and assessment of healthy properties of meat and meat products designed as functional foods. Meat Science 95: 919–930.

Pakkanen, R. and Aalto, J. 1997. Growth factors and anti-microbial factors of bovine colostrum. International Dairy Journal 7: 285–297.

Pellegrini, A., Hulsmeier, A.J., Hunziker, P. and Thomas, U. 2004. Proteolytic fragments of ovalbumin display anti-microbial activity. Biochimica et Biophysica Acta 1672: 76–85.

Phadke, G., Elavarasan, K. and Shamasundar, B.A. 2014. Angiotensin-I-converting enzyme (ACE) inhibitory activity and anti-oxidant activity of fermented fish product Ngari as influenced by fermentation period. International Journal of Pharma and Biosciences 5: P134–P142.

Poutanen, K., Flander, L. and Katina, K. 2009. Sourdough and cereal fermentation in a nutritional perspective. Food Microbiology 26: 693–699.

Premzl, A., Puizdar, V., Zavasnik-Bergant, V., Kopitar-Jerala, N., Lah, T.T., Katunuma, N., Sloane, B.F., Turk, V. and Kos, J. 2001. Invasion of ras-transformed breast epithelial cells depends on the proteolytic activity of cysteine and aspartic proteinases. Biological Chemistry 382: 853–857.

Ramchandran, L. and Shah, N.P. 2011. Yogurt can beneficially affect blood contributors of cardiovascular health status in hypertensive rats. Journal of Food Science 76: H131–H136.

Ramchandran, L. and Shah, N.P. 2008a. Effect of Versagel® on the growth and metabolic activities of selected lactic acid bacteria. Journal of Food Science 73: M21–M26.

Ramchandran, L. and Shah, N.P. 2008b. Growth, proteolytic and ACE–I activities of *Lactobacillus delbrueckii* ssp. *bulgaricus* and *Streptococcus thermophilus* and rheological properties of low fat yogurt as influenced by the addition of Raftiline HP®. Journal of Food Science 73: M368–M374.

Rashid, I., van Reyk, D.M. and Davies, M.J. 2007. Carnosine and its constituents inhibit glycation of low-density lipoproteins that promotes foam cell formation *in vitro*. FEBS Letters 581: 1067–1070.

Ririe, D.G., Roberts, P.R., Shouse, M.N. and Zaloga, G.P. 2000. Vasodilatory actions of the dietary peptide carnosine. Nutrition 16: 168–172.

Roberfroid, M.B. 1999. Concepts in functional foods: The case of inulin and oligofructose. Journal of Nutrition 129: 1398S–1401S.

Rukimini, C. 2003. Phytochemical products. *In*: I. Jhonson and G. Williamson (eds.). Rice Bran in Phytochemical Functional Foods, Woodhead Publishing, Cambridge, UK.

Samuel, P., Zakharkin, S., Spitznagel, E., Greaves, K., Butteiger, D. and Krul, E. 2008. Meta—analysis confirms soy protein's cholesterol lowering efficacy. 8th International Symposium on the Role of Soy in Health Promotion and Chronic Disease Prevention and Treatment, November 11, 2008, Tokyo, Japan.

Sarmadi, B.H. and Ismail, A. 2010. Anti-oxidative peptides from food proteins: A review. Peptides 31: 1949–1996.

Seccombe, D.W., James, L., Hahn, P. and Jones, E. 1987. L-carnitine treatment in the hyperlipidemic rabbit. Metabolism 36: 1192–1196.

Shearer, M.J. 2000. Role of vitamin K and Gla proteins in the pathophysiology of osteoporosis and vascular calcification. Current Opinion in Clinical Nutrition and Metabolic Care 3: 433–438.

Shimura, S. and Hasegawa, T. 1993. Changes of lipid concentrations in liver and serum by administration of carnitine added diets in rats. Journal of Veterinary Medical Science 55: 845–847.

Shirataki, Y., Tani, S., Sakagami, H., Satoh, K., Nakashima, H., Gotoh, K. and Motohashi, N. 2001. Relationship between cytotoxic activity and radical intensity of isoflavones from *Sophora* species. Anticancer Research 21: 2643–2648.

Sirtori, C.R., Galli, C., Anderson, J.W. and Arnoldi, A. 2008. Nutritional and nutraceutical approaches to dyslipidemia and atherosclerosis prevention: Focus on dietary proteins. Atherosclerosis 203: 8–17.

Son, D.O., Satsu, H. and Shimizu, M. 2005. Histidine inhibits oxidative stress—and TNF-alpha-induced interleukin-8 secretion in intestinal epithelial cells. FEBS Letters 579: 4671–4677.

Stanton, C., Ross, R.P., Fitzgerald, G.F. and Sinderen, D.V. 2005. Fermented functional foods based on probiotics and their biogenic metabolites. Current Opinion in Biotechnology 16: 198–203.

Tamaru, S., Kurayama, T., Sakano, M., Fukuda, N., Nakamori, T., Furuta, H., Tanaka, K. and Sugano, M. 2007. Effects of dietary soybean peptides on hepatic production of ketone bodies and secretion of triglyceride by perfused rat liver. Bioscience, Biotechnology and Biochemistry 71: 2451–2457.

Thomä-Worringer, C., Sørensen, J. and López-Fandiño, R. 2006. Health effects and technological features of caseinomacropeptide. International Dairy Journal 16: 1324–1333.

Tomasik, P.J. and Tomasik, P. 2003. Probiotics and prebiotics. Cereal Chemistry 80: 113–117.

Tripathi, V. and Vashishtha, B. 2006. Bioactive compounds of colostrums and its application. Food Reviews International 22: 225–244.

Vella, M.N., Stratton, L.M., Sheeshka, J. and Duncan, A.M. 2013. Exploration of functional food consumption in older adults in relation to food matrices, bioactive ingredient and health. Journal of Nutrition in Gerontology and Geriatrics 32: 122–144.

Vermeer, C., Jie, K.-S.G. and Knapen, M.H.J. 1995. Role of vitamin K in bone metabolism. Annual Review of Nutrition 15: 1–22.

Vescovo, G., Ravara, B., Gobbo, V., Sandri, M., Angelini, A., Della Barbera, M., Dona, M., Peluso, G., Calvani, M., Mosconi, L. and Dalla Libera, L. 2002. L-Carnitine: A potential treatment for blocking apoptosis and preventing skeletal muscle myopathy in heart failure. American Journal of Physiology. Cell Physiology 283: C802–810.

World Health Organisation (WHO). 2003. Diet, nutrition and the prevention of chronic diseases: report of a joint WHO/FAO Expert Consultation. Geneva: World Health Organisation.

Wu, X.H., Ding, M.P., Zhu-Ge, Z.B., Zhu, Y.Y., Jin, C.L. and Chen, Z. 2006. Carnosine, a precursor of histidine, ameliorates pentylenetetrazole-induced kindled seizures in rat. Neuroscience Letters 400: 146–149.

Yalcin, A.S. 2006. Emerging therapeutic potential of whey proteins and peptides. Current Pharmaceutical Design 12: 1637–1643.

Yan, H., Guo, Y., Zhang, J., Ding, Z., Ha, W. and Harding, J.J. 2008. Effect of carnosine, aminoguanidine, and aspirin drops on the prevention of cataracts in diabetic rats. Molecular Vision 14: 2282–2291.

Yang, Y., Zhou, L., Gu, Y., Zhang, Y., Tang, J., Li, F., Shang, W., Jiang, B., Yue, X. and Chen, M. 2007. Dietary chickpeas reverse visceral adiposity, dyslipidaemia and insulin resistance in rats induced by a chronic high-fat diet. British Journal of Nutrition 98: 720–726.

Ye, X.Y. and Ng, T.B. 2003. Isolation of pisumin, a novel antifungal protein from legumes of the sugar snap pea *Pisumsativum* var. *macrocarpon*. Comparative Biochemistry and Physiology C: Toxicology and Pharmacology 134: 235–240.

Ye, X.Y., Wang, H.X. and Ng, T.B. 2000. Structurally dissimilar proteins with antiviral and antifungal potency from cowpea (*Vignaunguiculata*) seeds. Life Science 67: 3199–3207.

Yeh, C.-Y., Jung, C.-J., Huang, C.-N., Huang, Y.-C., Lien, H.-T., Wang, W.-B., Wang, L.-F. and Chia, J.-S. 2014. A legume product fermented by *Saccharomyces cerevisiae* modulates cutaneous atopicdermatitis-like inflammation in mice. BMC Complementary and Alternative Medicine 14: 194.

Yust, M.M., Pedroche, J., Girón-Calle, J., Alaiz, M., Millán, F. and Vioque, J. 2003. Production of ACE inhibitory peptides by digestion of chickpea legumin with alcalase. Food Chemistry 81: 363–369.

Zhu, Y.Y., Zhu-Ge, Z.B., Wu, D.C., Wang, S., Liu, L.Y., Ohtsu, H. and Chen, Z. 2007. Carnosine inhibits pentylenetetrazol-induced seizures by histaminergic mechanisms in histidine decarboxylase knock-out mice. Neuroscience Letters 416: 211–216.

16

Biogenic Amines in Fermented Foods: Overview*

Lopamudra Sahu,[1] Sandeep K. Panda,[2] Spiros Paramithiotis,[3] Nevijo Zdolec[4] and Ramesh C. Ray[5]

1. Introduction

Food-fermenting lactic acid bacteria are generally considered to be non-toxic and non-pathogenic (Chapters 7–16, in this book). Some species of lactic acid bacteria, however, can produce biogenic amines (BAs). BAs (i.e., tryptamine, 2-phenylethylamine, tyramine, histamine, putrescine, cadaverine, spermine and spermidine) are formed in food by microbial decarboxylation of the corresponding amino acids or by transamination of aldehydes and ketones (Brink et al. 1990, Maijala and Nurmi 1995, Zolou et al. 2003). Though they are produced naturally by plants, animals, and microorganisms, yet their consumption in the form of foods can have toxicological consequences. Exogenous BAs are most frequently found in fermented foods such as wines, beers, dairy products, meat and vegetables that are rich in protein content. BAs are also known to be the causal agents of a number of food

[1] Department of Botany, Utkal University, Bhubaneswar-751004, Odisha, India.
  E-mail: leepisri@gmail.com
[2] Department of Biotechnology and Food Technology, Faculty of Science, University of Johannesburg, P. O. Box 17011, Doornfontein Campus, Johannesburg, South Africa.
  E-mail: sandeeppanda2212@gmail.com; sandeepp@uj.ac.za
[3] Laboratory of Food Quality Control and Hygiene, Department of Food Science and Human Nutrition, Agricultural University of Athens, Iera Odos 75, GR-11855 Athens, Greece.
  E-mail: sdp@aua.gr
[4] University of Zagreb, Faculty of Veterinary Medicine, Heinzelova 55, 10000 Zagreb, Croatia.
  E-mail: nzdolec@vef.hr
[5] ICAR—Central Tuber Crops Research Institute (regional Centre), Bhubaneswar 751019, India.
  E-mail: rc.ray666@gmail.com
* All authors contributed equally for this chapter.

poisoning incidents, of which histamine poisoning, also known as scombroid poisoning, is a well known illness caused due to the consumption of scombroid fish (Kim et al. 2003, Hungerford 2010). Histamine intake by human beings ranging within 8–40 mg, 40–100 mg and higher than 100 mg, can cause minor, intermediate and severe poisoning, respectively (Parente et al. 2001). BA food poisoning is also caused due to higher quantities of tyramine content in cheese, known as 'cheese reaction' (Santos 1996, Spano et al. 2010). Putrescine and agmatine have been BAs described as potentiators that enhance the toxicity of histamine to humans by depressing histamine oxidation (Taylor 1986). In addition, putrescine and agmatine may cause initiation of carcinogenic nitrosamines in the presence of nitrites (Spano et al. 2010). These amines are not inactivated by the elevated temperatures during food processing and preparation. Although there is no specific legislation regarding the content of BAs in many fermented products, it is generally assumed that they should not be allowed to accumulate. The ability of microorganisms to decarboxylate amino acids is highly variable, often being strain specific, and therefore the detection of bacteria possessing amino acid decarboxylase activity is important in estimating the likelihood of foods containing BAs and in preventing their accumulation in food products. Moreover, improved knowledge of the factors involved in the synthesis and accumulation of BAs should lead to a reduction in their incidence in foods. Presently, only prevention and monitoring strategies aid in controlling the formation of BAs during the production of foods and along the food chain.

The occurrence, biochemistry of BAs formation in fermented foods and overall effects of BAs on human health have been reviewed in this chapter.

2. Mechanism of Amine Formation

BAs are biologically active substances with numerous physiological or patho-physiological functions in humans and animals. The formation of BAs in food can be expected in the presence of proteins and/or free amino acids that are substrates for microbial or natural enzymes. Mostly, BAs are products of amino acid decarboxylation or aldehyde and ketone amination. Their formation in foods and beverages is connected to enzymatic activity in raw materials, and microbiological degradation of products rich in proteins (like meat putrefaction, cheese ripening, fermentation processes). Amino acid precursors (Figs. 1–3) of some important BAs are listed below:

Tyrosine ----------- Tyramine
Histidine ----------- Histamine
Ornithine ---------- Putrescine
Tryptophan --------- Tryptamine
Phenylalanine ----- Phenylethylamine
Lysine -------------- Cadaverine

Hence, the accumulation of BAs in foods requires the presence of amino acid precursors, microbial decarboxylases and favorable conditions for microbial growth and activity of enzymes (Cocconcelli and Fontana 2008). The conditions present during food fermentation significantly affect the synthesis of BAs. Thus, the decarboxylase

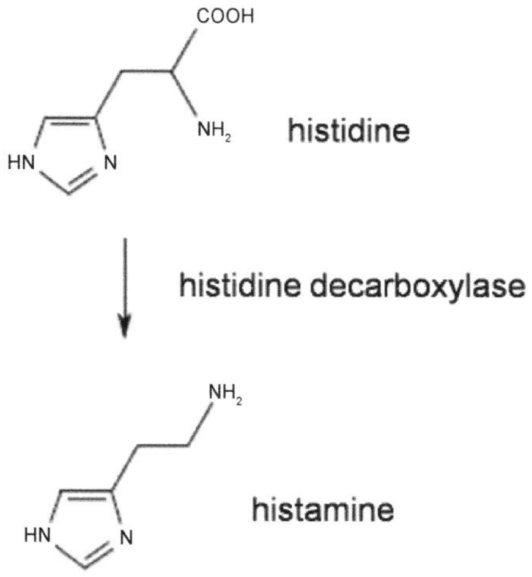

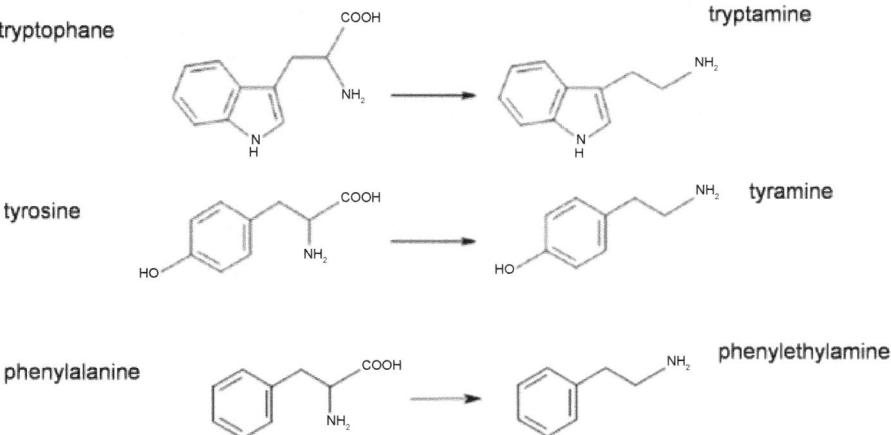

Figure 1. Decarboxylation of histidine to histamine.

Figure 2. Structure of aromatic and heterocyclic biogenic amines and their precursors.

is more active in the acid environment where pH is from 4.0–5.5 and temperature in the range of 20–37°C (Milićević et al. 2014). It is accepted that lower pH correlates with higher BA content in food, which is connected to BA production as a protective mechanism of microorganisms against an acidic environment (Cocconcelli and Fontana 2008). BAs are produced more under conditions of temperature increase, while lower temperatures reduce bacterial growth and decarboxylase activity. The formation of BAs decreases significantly below 5°C and above 40°C.

Amino acid decarboxylation takes place by the removal of the α-carboxyl group to give the corresponding amine. Arginine is easily converted to agmatine, or as a result of bacterial activity, can be degraded to ornithine from which putrescine is formed by decarboxylation. Lysine can be converted by bacterial action into cadaverine. Histidine can, under certain conditions, be decarboxylated to histamine. Tyramine,

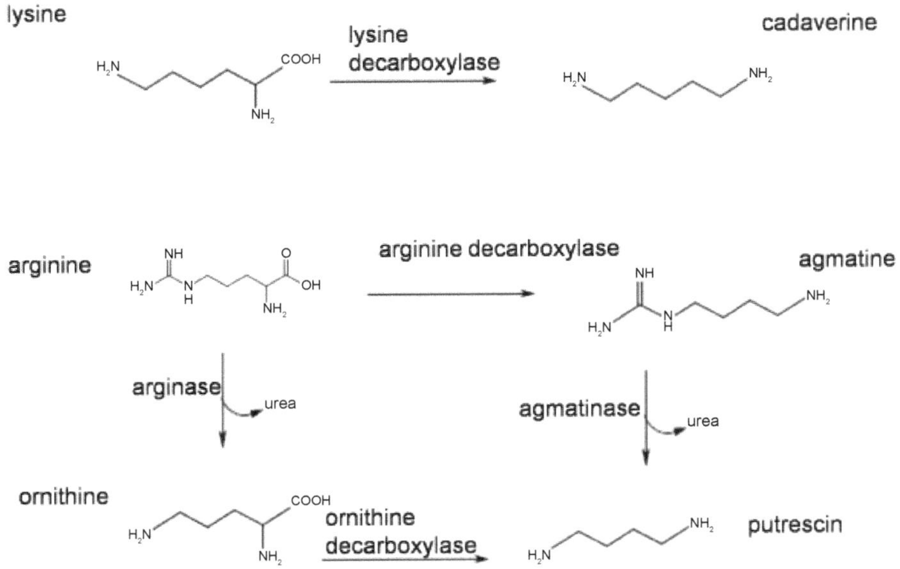

Figure 3. Selected aliphatic biogenic amines and their precursors.

tryptamine, and β-phenylethylamine come by in the same manner, from tyrosine, tryptophan, and phenylalanine, respectively. Proteolysis, either autolytic or bacterial, may play a significant role in the release of free amino acids from tissue proteins that offer a substrate for decarboxylase reactions. Two mechanisms of action, for amino acid decarboxylation, have been identified:

- pyridoxal phosphate dependent reaction, and
- non-pyridoxal phosphate dependent reaction (Eitenmiller and De Souza 1984).

Pyridoxal phosphate joined in a Schiff base linkage to the amino group of a lysyl residue forms the active site of the enzyme. Pyridoxal phosphate, by itself, can catalyze many amino acid reactions that are usually brought about by pyridoxal phosphate dependent enzymes and thus can be considered to be the portion of the enzyme that actually takes part in the reaction. The carbonyl group of pyridoxal phosphate reacts readily with amino acids to form Schiff base intermediates, which are then decarboxylated with the elimination of water to yield the corresponding amines and the original pyridoxal phosphate moiety (Eitenmiller and De Souza 1984). Non-pyridoxal phosphate-catalyzed decarboxylations involve a pyruvoyl residue instead of pyridoxal-5-phosphate. The pyruvoyl group is covalently bound to the amino group of a phenylalanine residue on the enzyme, and is derived from a serine residue. Of an active proenzyme, the pyruvoyl residue acts in a manner similar to pyridoxal phosphate in the decarboxylation reaction.

Some of the more widely consumed animal-related foods that support BA formation, as already mentioned, are long-ripening cheeses, fermented sausages or dry-cured meats. The cheese is a type of fermented food that can be associated with the formation of BAs during its production, ripening and storage. Known factors in this production that affect the formation of BAs are pH, salt concentration, bacterial activity, humidity, storage temperature, ripening time and microorganism synergy. It is well known that the bacteria from the genus *Enterococcus* and *Lactobacillus* are predominantly responsible for the production of BAs in cheeses produced from unpasteurized milk (Marijan et al. 2014). In general, bacteria with the highest decarboxylase activity in fermented food are lactic acid bacteria (genera *Lactobacillus, Enterococcus, Lactococcus, Leuconostoc*), coagulase-negative staphylococci (*Staphylococcus xylosus, S. carnosus*) or enterobacteria. However, the correlation of bacterial numbers in foods and BA content is not always the case (Marijan et al. 2014). During cheese ripening, the amounts of BAs increase due to intensive proteolysis and the presence of free amino acids and peptides (Radeljević et al. 2013, Magdić et al. 2013). Prolongation of ripening time of cheeses will increase BA contents (Marijan et al. 2014). The distribution of BA in some crude food type, like hard cheese, is not equal. Marijan et al. (2014) have found higher BA amount in the middle of cheese than in surface positions (under the core) that is probably related to different physicochemical conditions and microbial/enzymatic activity.

Similar findings could be expected in dry sausages or dry-cured meats, when considering the microbial groups involved and intrinsic conditions favorable to BA formation. BA content in traditionally fermented products produced without starter cultures is related to autochthonous micro-biota from raw materials and their domination and competitiveness during the fermentation and ripening period. Some starter cultures used in meat fermentation, like *Lactobacillus curvatus* (Cocconcelli and Fontana 2008), could even promote the formation of BA. On the other hand, decarboxylase-negative strains are applied successfully as competitive starter cultures that prevent the growth of decarboxylase-positive bacteria and consequently reduce the risk of BA. Besides, BA content can be reduced by introducing starter cultures with amine oxidase activity (Suzzi and Gardini 2003).

3. BAs in Fermented Foods

Of the different types of fermented foodstuffs, dairy, fish, sausages, some vegetables, wines and beers are likely to contain high levels of BA (Shalaby 1996, Santos 1996, Bover-Cid et al. 2000, 2001, Spano et al. 2010). It is known that the amount and types of BAs formed in fermented food products are influenced by the composition of food, microbial flora and by other factors that allow a favourable condition for the bacterial growth during food processing and storage (Carelli et al. 2007).

3.1 Dairy Products

In dairy food products, the most important BAs are histamine and tyramine that are produced by enzymatic decarboxylation of histidine and tyrosine, respectively (Joosten and Olieman 1986). Putrescine is mostly synthesized by ornithine decarboxylation or agmatine deamination and, to some extent, cadaverine originates by lysine decarboxylation (Linares et al. 2011). Apart from being slightly toxic, putrescine and cadaverine also inhibit histamine metabolizing enzymes and therefore enhance histamine poisoning in dairy products (Halasz et al. 1994, Shalaby 1996). These amino acid decarboxylating activities characterise the microbial groups participating in the fermentation process. While amine oxidases detoxify the BA ingested with food, if not detoxified due to genetic reasons or due to the inhibitory effects of drugs/alcohol (Bodmer et al. 1999), their accumulation can be poisonous and cause serious human ailments (Repka-Ramirez and Baraniuk 2002, Soufleros et al. 2007, Ladero et al. 2010, Spano et al. 2010).

Cheese and cheese products are quite frequently related to BAs poisoning (Sumner et al. 1985, EFSA Panel on Biological Hazards (BIOHAZ) 2011). Cheese, a protein rich food, has the maximum production of BA as its making does not require a sterile environment. Lactic acid bacteria are the main producers of BAs in cheese production (Fernández et al. 2004). Several genera such as *Enterococcus*, *Lactobacillus, Leuconostoc*, and *Streptococcus* include some strains that produce BAs. The presence of these Gram-positive bacteria can be credited to their contamination during the production of cheese or a part of starter/adjunct cultures, or their occurrence in milk. There have been reports on the incidence of tyrosine and histamine decarboxylase activity in strains from various starter cultures (Burdychova and Komprda 2007, Calles-Enríquez et al. 2010, Spano et al. 2010). Though the presence of the right microorganisms with appropriate metabolic pathway ushers the production of BAs, yet its formation is possible only in a favourable environment for decarboxylation activity in an amino acid rich substrate (Russo et al. 2010). The production of BAs in cheese is a very complex process. A lot of factors such as the microorganisms that produce BAs, their proteolytic and decarboxylase activities, ripening time, and ripening and storage temperature influence its production. The presence of BAs varies in concentration in various dairy products. In cheese, BAs concentrations can reach up to 2000 mg/kg (Roig-Sagues et al. 2002, Fernandez et al. 2007) and consumption of such cheese with high levels of BAs is a direct threat to the consumer, especially in Europe where cheese is one of the important foods in their diets. Beside cheese, several other fermented foods such as *kefir* contain BAs at 2.4 and 35.2 mg/L, tyramine being the dominant one (Ozdestan and Uren 2009); in *koumis*, a traditional Eurasian fermented cow milk, the total BA contents were reported to be 15.31 mg/L (Chaves-Lopez et al. 2011).

3.2 Fermented Fish and Meat

High concentrations of BAs can exist in fish and fish products (Valsamaki et al. 2000, Mah et al. 2002, Latorre-Moratalla et al. 2008, Moret et al. 2005, Proestos et al. 2008) as they are important source of proteins and vitamins. According to the US Food and Drug Administration (FDA), the tolerance level of histamine in fish should be 100 mg/kg of flesh (FDA 1995). Though higher levels of putrescine and cadaverine foster histamine/tyramine toxicity, no recommended levels have been suggested. Fermented fish products contain high amounts of amino acids that are broken down quickly by enzymes formed during bacterial growth and leading to the formation of hazardous and toxic BAs, including histamine (Zhai et al. 2012). Quantification of amines in fish can be treated as a quality index and might be used for sanitary surveillance (Moreno and Torres 2001).

Fish sauce is one of the most popular fermented fish products containing about 20 g/L nitrogen, 80% of which is in the form of amino acids. Fish sauce is an important source of dietary proteins and amino acids, and used as a condiment in households in Southeast Asian countries (Sanceda et al. 1996). Various Scombroid fishes *viz.* tuna, mackerel, bonito, and saury, when not properly processed and stored, are often related to scombroid poisoning due to high levels of free histidine in their muscle (Taylor 1986, Hungerford 2010). Fermented fish products, such as fish sauce and fish paste, may contain high contents of histamine (Fardiaz and Markakis 1979). At times, histamine poisoning symptoms may go unnoticed as they closely resemble those of food allergies. Many bacterial species associated with sea foods are known to possess histidine decarboxylase that have the ability to produce histamine. Many naturally occurring bacteria, such as *Morganella morganii*, *Raoultella* (formerly *Klebsiella*) *planticola* and *Enterobacter aerogenes*, and several other bacterial species known to produce histamine in fish have been identified (Bjornsdottir et al. 2009), including the enteric bacteria *viz. Proteus vulgaris*, *Proteus mirabilis*, *Enterobacter cloacae*, *Serratia fonticola*, *Serratia liquefaciens*, *Raoultella ornithinolytica* and *Citrobacter freundii* (Kim et al. 2003, Tsai et al. 2005).

Fermentation processes also promote the formation of BAs, and perhaps the fermented meat products present the greatest amount and diversity of these compounds. BAs formation is influenced by the temperature at which fermentation occurs (mostly between 7–28°C); it has been suggested that tyramine formation in dry sausage can be prevented by maintaining the temperature conditions (Maijala and Nurmi 1995, Eerola et al. 1998). It was explained that at higher temperatures of fermentation, the starter culture outgrew the nonstarter lactic acid bacteria and low levels of amines was detected in the final processing phase at a higher temperature, i.e., 24°C (Maijala and Nurmi 1995). However, Kranner et al. (1991) differed in their experiments and found that histamine formation reduced when the ripening temperature was lowered (7–18°C). thereby suggesting that low temperatures can improve quality and longer shelf-life.

In ripened meat products, though poor quality raw materials might be the source of biogenic amines, yet the largest fraction arises during various processing stages (Maijala and Nurmi 1995, Vidal-Carou et al. 1990, Bover-Cid et al. 2001). The additives that are used in the preparation of fermented products also play significant roles in the formation of BAs. Nevertheless, there are contradictions to this report. Bover-Cid et al. (2001) found that the different types of sugar used for ripening dry fermented sausages had limited BAs formation in fermented sausages. They explained that while sodium sulphite in sausages inhibited the production of cadaverine, it aided in the formation of tyramine and putrescine. Potassium sorbate, a microbial inhibitor, was used to restrict the formation of BAs in foods (Shalaby 1996). Further, NaCl considerably affects BAs formation by reducing water activity. It was reported that 30 g NaCl/kg in fermented sausages promoted the growth of *Lactobacillus curvatus* (Straub et al. 1994).

3.3 Fermented Vegetables

Fermented vegetables represent another class of foods in which BAs have been detected (Anderson 1988, Steinkraus 1996). Research has been focused on sauerkraut due to its relatively higher commercial importance. The occurrence of BAs in sauerkraut samples has been already reported from the early '70s (Mayer and Pause 1972). Since then, several similar studies have taken place, reporting average values of histamine at 2.1–56 mg/kg, tyramine at 24.7–235 mg/kg, cadaverine at 4–73 mg/kg, putrescine at 87.3–222 mg/kg, spermidine at 6.4–10.2 mg/kg and tryptamine at 2.4–7.2 mg/kg (Taylor et al. 1978, Kuemsh et al. 1989, Brink et al. 1990, Halasz et al. 1994, Kalac et al. 1999). Interestingly enough, only putrescine was detected in all samples, and quantitative determination of all types of BAs was characterized by huge variation. The latter has been assigned to the spontaneous nature of the fermentation as well as differences in the technology applied in terms of temperature, NaCl content, pH value and access to oxygen (Kalac et al. 1999, Penas et al. 2010). Moreover, significant qualitative and quantitative variation in the accumulation of BA during storage was reported between experiments as well as between cabbage cultivars used (Kalac et al. 2000a, Kosson and Elkner 2010). In a very interesting study by Cvetkovic et al. (2015), fermentation duration was reported to affect putrescine, cadaverine, histamine, tyramine and spermidine production; fermentation temperature influenced putrescine and histamine production while salt content was very important for spermidine production.

The importance of the proper starter cultures selection was highlighted in the study of Kalac et al. (2000b). In that study, significant differences in the accumulation of BAs were observed between spontaneously fermented sauerkraut and ones initially inoculated with commercial strains of *Lactobacillus plantarum*, *Lb. casei*, *Pediococcus pentosaceus*, *Enterococcus faecium* and a mixed preparation of them all. Although, in some cases, comparable level of BAs were produced, in most cases, the levels of tyramine, putrescine, cadaverine, histamine, tryptamine, spermidine and spermine were significantly lower in the inoculated samples than the spontaneously fermented ones. However, in many cases, significant variation in the production of BAs by the same strain between experiments was observed, underlining the importance of accurate control of both raw materials and processing parameters. A significant reduction in the accumulation of BAs during storage, of sauerkraut inoculated with strains of *Lb. plantarum*, *Lb. casei* subsp. *casei* and *Lb. curvatus,* was reported by Rabie et al. (2011). More accurately, total BAs content reached 1553 mg/kg by the 45th day of storage of spontaneously fermented product, compared to the maximum of 220 mg/kg obtained by the *Lb. plantarum* strain and the 161 and 165 mg/kg produced by the *Lb. curvatus* and the *Lb. casei* strains, respectively. Furthermore, the levels of histamine, tyramine, putrescine and cadaverine were significantly higher in the spontaneously fermented sauerkraut used as control.

3.4 Fermented Legumes

Fermented soy beans products such as miso, soy sauce, and tempe have been popular in the Orient for more than one thousand years and are becoming increasingly popular in Western countries. Since several varieties of molds, yeasts, and lactic acid bacteria are involved in the fermentation processes of such products, and the raw materials (soy beans, legumes, cereals or other suitable raw materials) contain considerable amounts of protein, the formation of various amines might be expected during the fermentation (Shalaby 1996, Motarjemi 2002). Tyramine and histamine have been found at various levels in such products (Stratton et al. 1991). High levels of biogenic amines (1 g/kg) were reported for soy sauce made from black soy bean while tyramine has been found at moderately high levels (450 mg/kg) in fermented salted black beans (Shalaby 1996). The variability of BA levels in the commercial fermented soy products samples could be attributed to variability in the ratio of soy bean to other seeds used in the raw material, the microbiological composition and the conditions and duration of fermentation (Nout et al. 1993). Nout et al. (1993) showed that the functional fungus of tempe, *Rhizopus oligosporus,* mainly produced tyramine and some putrescine, but both *Klebsiella pneumoniae* and *Trichosporon beiglli* led to increased total BAs in the tempe, while *Lb. plantarum* reduced tyramine levels in the product.

3.5 Alcoholic Fermented Beverages

In wines and other fermented beverages, BAs may be formed during different stages of production and storage by various microorganisms. Few researchers have reported the presence of BAs, such as putrescine and cadaverine, in grapes (Agudelo-Romero et al. 2013, Vincenzini et al. 2009). Ultimately, the amount of BAs in wines depends upon agricultural systems such as the traditional or organic methods and the wine making practices followed (Tassoni et al. 2013, Yanez et al. 2012, Yildirim et al. 2007). Further, during the fermentation processes from must to wine, microorganisms can produce histamine and tyramine (Beneduce et al. 2010, Herbert et al. 2006). Consequently, the presence of BAs in wine can be a combined effect of yeasts' primary fermentation and malolactic fermentation metabolism by the lactic acid bacteria (García-Marino et al. 2010). Numerous studies have shown that wine yeasts have a direct role in increasing BA contents during vinification experiments, both at laboratory and large-scale processing (Goni and Ancin Azpilicueta 2001, Torrea and Ancin 2002). Besides, research conducted by Caruso et al. (2002) and by Granchi et al. (2005) demonstrated that different strains belonging to the species *Saccharomyces cerevisiae, Kloeckera apiculata, Candida stellata, Metschnikowia pulcherrima* and *Brettanomyces bruxellensis* produced significant amounts of putrescine, phenylethylamine and ethanolamine. Investigations by Tristezza et al. (2013) during their first large-scale study of the vineyard associated strains from the Apulia Region in Italy shed new light on the role of non-Saccharomyces yeasts in the production of BAs. For the first time, they demonstrated experimentally, the ability of a yeast species of enological provenience in producing histamine during grape must fermentation. Among the yeast species tested, an isolate of *Issatchenkia terricola* (2KUT31) and *M. pulcherrima* (3KUT27) as well as two isolates (3KUT25 and 3KUT29) of *Peronospora manshurica* were capable of synthesizing histamine. However, histamine production had never before been reported in these strains belonging to the species isolated from wine (Caruso et al. 2002, Granchi et al. 2005, Landete et al. 2007). The reports on *P. manshurica* established its significance not only in spoilage due to volatile phenols production (Saez et al. 2011) but also due to its ability in producing cadaverine, thus representing a possible source of hazard for amine production in wine, given its ability to survive in the medium (Saez et al. 2011). Thus, it is imperative to emphasize that correct methods be applied to reduce and eradicate the action of these spoilage yeasts even during the first steps of vinification, which is a necessary safety measurement for wine making.

A great variability in the concentration of amines exists between different wines as there are many factors that exert an influence. It is well identified that white wines show a lower concentration of BAs than red wines, the reason being the process of red wine production. This includes a maceration step with the grape skin, thus increasing the polyphenol cession and presence of amino acids (Martin-Alvarez et al. 2006), in addition to a higher fermentation temperature and the functioning of a malolactic fermentation; the last is mostly for a shorter duration or completely absent in white wine production (Ancin-Azpilicueta et al. 2008, Lonvaud-Funel 2001). Further, research on the estimation of BAs in Chilean young varietal wines showed concentrations ranging from 18.12 to 39.84 mg/L, with putrescine, histamine, tyramine and spermidine being the most abundant BAs (Pineda et al. 2012). Very recently, in Cannonau and Vermentino wines, two of the most popular wines of Sardinia (Italy), BAs were detected at average concentrations < 10 mg/L, except putrescine, which reached 20.5 ± 10.2 mg/L in Cannonau wines (Tuberoso et al. 2015). A few measures have been suggested to reduce the biogenic amines accumulation in wine.

- Firstly, the raw materials, *viz.* grapes, used must be healthy and grown with less nitrogenous fertilization.
- Secondly, the yeast and lactic acid bacteria strains should exhibit minimal aminogenic capacity.

Occasionally, beers contain histamine and other biogenic amines (Stratton et al. 1991). Histamine has been found at high concentrations reaching 4.7, 15, and 20 mg/L in Swedish, Danish, and French beers, respectively (Shalaby 1996). Tyramine was also detected, with levels as high as 11 mg/L in beers of various origin (Shalaby 1996). Putrescine and cadaverine levels are generally low in beers (Stratton et al. 1991). Agmatine, putrescine, and tyramine were found at concentrations ranging from 0.55 to 67.60 mg/L, while histamine, tryptamine, β-phenylethylamine, cadaverine, spermine, and spermidine showed levels lower than 2 mg/L; however, relatively high concentrations of histamine and cadaverine were detected in some European beers (Izquierdo-Pulido et al. 1994).

4. BAs and Health Aspects

Histamine, tryptamine, β-phenylethylamine, and tyramine are biologically active amines that have important physiological effects in humans, generally either psychoactive or vasoactive. Psychoactive amines affect the nervous system by acting on neural transmitters, while vasoactive amines act on the vascular system (Hungerford 2010). The consumption of food containing BAs is responsible for many pharmacological effects (Table 1), which lead to several types of foodborne disease, including histamine poisoning (scombroid poisoning) (Hungerford 2010) and tyramine toxicity (cheese reaction) (Shalaby 1996, Russo et al. 2010). Harmful effects resulting from the consumption of foods rich in biogenic amines can be expected only when these amines gain access to the bloodstream (Hungerford 2010).

Table 1. BAs and their pharmacological effects (Shalaby 1991, Karovicova and Kohajdova 2005, Rosi et al. 2009, Hungerford 2010, updated).

Amine	Precursor	Pharmacological effects
Histamine	Histidine	Liberates adrenaline and noradrenaline Excites the smooth muscles of the uterus, the intestine, and the respiratory tract Stimulates both sensory and motor neurons Controls gastric acid secretion
Tyramine	Tyrosine	Peripheral vasoconstriction Increases the cardiac output Causes lacrimation and salivation Increases respiration Increases blood sugar level Releases noradrenaline from the sympathetic nervous system Causes migraine
Putrescine and cadaverine	Omithine and lysine	Hypotension Bradycardia Lockjaw Paresis of the extremities Potentiate the toxicity of other amines
β-Phenylethylamine	Phenylalanine	Releases noradrenaline from the sympathetic nervous system Increases the blood pressure Causes migraine
Tryptamine	Tryptophane	Increases the blood pressure

Although BAs such as histamine, tyramine and putrescine are needed for many critical functions in human beings (Table 1), consumption of food containing high amounts of the same may have toxicological effects. The most notorious foodborne intoxications caused by BA are related to histamine. Histamine causes dilation of peripheral blood vessels, capillaries and arteries, thus resulting in hypotension, flushing, and headache. The toxicological effect depends on histamine intake concentration, presence of other different amines, amino-oxidase activity and the intestinal physiology of the individual (Hungerford 2010). On the other hand, it must be taken into account that secondary amines such as putrescine and cadaverine can react with nitrite to form heterocyclic carcinogenic nitrosamines, nitrosopyrrolidine and nitrosopiperidine (Karovicova and Kohajdova 2005).

5. Conclusion and Future Research Needs

Occurrence of BAs has been reported in every type of fermented food. The production of BAs is a strain-dependent property that is largely affected by the respective growth conditions as well as the properties of the raw materials. Although the importance of several factors has been underlined and some intervention strategies have been suggested, this subject is still understudied and the lack of integrated studies is evident.

Keywords: Biogenic amines, fermentation, decarboxylation, food, tryptamine, 2-phenylethylamine, tyramine, histamine, putrescine, cadaverine, spermine and spermidine

References

Agudelo-Romero, P., Bortolloti, C., Pais, M.S., Tiburcio, A.F. and Fortes, A.M. 2013. Study of polyamines during grape ripening indicate an important role of polyamine catabolism. Plant Physiology and Biochemistry 67: 105–119.

Ancin-Azpilicueta, C., Gonzalez-Marco, A. and Jimenez-Moreno, N. 2008. Current knowledge about the presence of amines in wine. Critical Reveiws in Food Science and Nutrition 48: 257–275.

Andersson, R.E. 1988. Biogenic amines in lactic acid fermented vegetables. LWT. Food Science & Technology 21: 68–69.

Beneduce, L., Romano, A., Capozzi, V., Lucas, P., Barnavon, L. and Bach, B. 2010. Biogenic amine in wines. Annals of Microbiology 60: 573–578.

Bjornsdottir, K., Bolton, G.E., McClellan-Green, P.D., Jaykus, L.A. and Green, D.P. 2009. Detection of gram-negative histamine-producing bacteria in fish: A comparative study. Journal of Food Protection 72: 1987–1991.

Bodmer, S., Imark, C. and Kneubühl, M. 1999. Biogenic amines in foods: Histamine and food processing. Inflammatory Research 48: 296–300.

Bover-Cid, S., Izquierdo-Pulido, M. and Vidal-Carou, M.C. 2001. Changes in biogenic amines and polyamine contents in slightly fermented sausages manufactured with and without sugar. Meat Science 57: 215–221.

Bover-Cid, S., Izquierdo-Pulido, M. and Vidal-Carou, M. 2000. Mixed starter cultures to control biogenic amine production in dry fermented sausages. Journal of Food Protection 63: 1556–1562.

Brink, B., Damirik, C., Joosten, H.M.L.J. and Huisin't Veld, J.H.J. 1990. Occurrence and formation of biologically active amines in foods. International Journal of Food Microbiology 11: 73–84.

Burdychova, R. and Komprda, T. 2007. Biogenic amine-forming microbial communities in cheese. FEMS Microbiology Letters 276: 149–155.

Calles-Enriquez, M., Eriksen, B.H., Andersen, P.S., Rattray, F.P., Johansen, A.H., Fernandez, M., Ladero, V. and Álvarez, M.A. 2010. Sequencing and transcriptional analysis of the *Streptococcus thermophilus* histamine biosynthesis gene cluster: Factors that affect differential *hdcA* expression. Applications in Environment Microbiology 76: 6231–6238.

Carelli, D., Centonze, D., Palermo, C., Quinto, M. and Rotunno, T. 2007. An interference free amperometric biosensor for the detection of biogenic amines in food products. Biosensors and Bioelectronics 23: 640–647.

Caruso, M., Fiore, C., Contursi, M., Salzano, G., Paparella, A. and Romano, P. 2002. Formation of biogenic amines as criteria for the selection of wine yeasts. World Journal of Microbiology and Biotechnology 18: 159–163.

Chaves-Lopez, C., Serio, A., Martuscelli, M., Paparella, A., Osorio-Cadavid, E. and Suzzi, G. 2011. Microbiological characteristics of 'kuomis', a traditional fermented Colombian milk, with particular emphasis on enterococci population. Food Microbiology 28: 1041–1047.

Cocconcelli, P.S. and Fontana, C. 2008. Characteristics and applications of microbial starters in meat fermentations. pp. 129–148. *In*: F. Toldra (ed.). Meat Biotechnology, New York: Springer.

Cvetkovic, B.R., Pezo, L.L., Tasic, T., Saric, L., Kevresan, Z. and Mastilovic, J. 2015. The optimization of traditional fermentation process of white cabbage (in relation to biogenic amines and polyamines content and microbiological profile). Food Chemistry 168: 471–477.

Eerola, H.S., Roig-Sagúes, A.X. and Hirvi, T.K. 1998. Biogenic amines in Finnish dry sausages. Journal of Food Safety 18: 127–138.

EFSA (European Food Safety Authority). 2011. Scientific opinion on risk BAed control of biogenic amine formation in fermented foods. EFSA Journal 9(10): 1–93.

Eitenmiller, R.R. and De Souza, S. 1984. Enzymatic mechanisms for amine formation in fish. *In*: E.R. Ragelis (ed.). Seafood Toxins, American Chemical Society, Washington, DC.

Fardiaz, D. and Markakis, P. 1979. Amine in fermented fish paste. Journal of Food Science 44: 1562–1563.

Fernandez, M., Linares, D.M., del Rio, B., Ladero, V. and Alvarez, M.A. 2007. HPLC quantification of biogenic amines in cheeses: correlation with PCR-detection of tyramine-producing microorganisms. Journal of Dairy Research 74: 276–282.

Fernandez, M., Linares, D.M. and Alvarez, M.A. 2004. Sequencing of the tyrosine decarboxylase cluster of *Lactococcus lactis* IPLA 655 and the development of a PCR method for detecting tyrosine decarboxylase lactic acid bacteria. Journal of Food Protection 67: 2521–2529.

Food and Drug Administration, USA. 1995. Decomposition and histamine—raw frozen tuna and mahi-mahi; canned tuna; and related species; availability of revised compliance policy guide. Federal Registration 149: 39754–39756.

Garcia-Marino, M., Trigueros, A. and Escribano-Bailon, T. 2010. Influence of oenological practices on the formation of biogenic amines in quality red wines. Journal of Food Composition and Analysis 23: 455–462.

Goni, D.T. and Ancín Azpilicueta, C. 2001. Influence of yeast strain on biogenic amine content in wines: Relationship with the utilization of amino acids during fermentation. American Journal of Enology and Viticulture 52: 185–190.

Granchi, L., Romano, P., Mangani, S., Guerrini, S. and Vincenzini, M. 2005. Production of biogenic amines by wine microorganisms. Bulletin OIV 78: 595–609.

Halasz, A., Barath, A., Simon-Sarkadi, L. and Holzapfel, W. 1994. Biogenic amines and their production by microorganism in food. Trends in Food Science and Technology 51: 42–49.

Herbert, P., Cabrita, M.J., Ratola, N., Laureano, O. and Alves, A. 2006. Relationship between biogenic amines and free amino acid contents of wines and musts from Alentejo (Portugal). Journal of Environmental Science and Health, Part B—Pesticides, Food Contaminants, and Agricultural Wastes 41: 1171–1186.

Hungerford, J.M. 2010. Scombroid poisoning: A review. Toxicon 56 (2010): 231–243.

Izquierdo-Pulido, M., Hemandez-Jover, T., Marine-Font, A. and Vidal-Carou, M.C. 1994. Biogenic amines contents in European beers. Proceedings of International European Food Toxicology. IV Conference, pp. 65–71.

Joosten, H.M.L.G. and Olieman, C. 1986. Determination of biogenic amines in cheese and some other food products by high-performance liquid chromatography in combination with thermosensitized reaction detection. Journal of Chromatography 356: 311–319.

Kalac, P., Spicka, J., Krizek, M. and Pelikanova, T. 2000a. Changes in biogenic amine concentrations during sauerkraut storage. Food Chemistry 69: 309–314.

Kalac, P., Spicka, J., Krizek, M. and Pelikanova, T. 2000b. The effects of lactic acid bacteria inoculants on biogenic amines formation in sauerkraut. Food Chemistry 70: 355–359.

Kalac, P., Spicka, J., Krizek, M., Steidlova, S. and Pelikanova, T. 1999. Concentrations of seven biogenic amines in sauerkraut. Food Chemistry 67: 275–280.

Karovicová, J. and Kohajdová, Z. 2005. Biogenic amines in food. Chemical Papers e Chemicke Zvesti 59: 70–79.

Kim, S.H., Barros-Velazquez, J., Ben-Gigirey, B., Eun, J.B., Jun, S.H. and Wei, C.I. 2003. Identification of the main bacteria contributing to histamine formation in seafood to ensure product safety. Food Science and Biotechnology 12: 451–460.

Kosson, R. and Elkner, K. 2010. Effect of storage period on biogenic amine content in sauerkraut. Vegetable Crops Research Bulletin 73: 151–160.

Kranner, P., Bauer, F. and Hellwig, E. 1991. Investigation on the formation of histamine in raw sausages. In: Proceeding of the 37 International Congress of Meat Science and Technology. Kulmbach, Germany, pp. 889–895.

Kuensh, U., Schaerer, H. and Temperli, A. 1989. Study on the formation of biogenic amines during sauerkraut fermentation. Poster presented at the International Conference on Biotechnology and Food, Feb. 20–24, 1989, Hohenheim University, Stuttgart.

Ladero, V., Fernandez, M., Cuesta, I. and Alvarez, M.A. 2010. Quantitative detection and identification of tyramine-producing enterococci and lactobacilli in cheese by multiplex qPCR. Food Microbiology 27: 933–939.

Landete, J.M., Ferrer, S. and Pardo, I. 2007. Biogenic amine production by lactic acid bacteria, acetic bacteria and yeast isolated from wine. Food Control 18: 1569–1574.

Latorre-Moratalla, M.L., Veciana-Nogues, T., Bover-Cid, S., Garriga, M., Aymerich, T. and Zanardi, E. 2008. Biogenic amines in traditional fermented sausages produced in selected European countries. Food Chemistry 107: 912–921.

Linares, D.M., Martın, M.C., Ladero, V., Alvarez, M.A. and Fernandez, M. 2011. Biogenic amines in dairy products. Critical Reviews in Food Science and Nutrition 51: 691–703.

Lonvaud-Funel, A. 2001. Biogenic amines in wines: Role of lactic acid bacteria. FEMS Microbiology Letters 199: 9–13.

Magdić, V., Kalit, S., Mrkonjić Fuka, M., Skelin, A., Samaržija, D., Redžepović, S. and Havranek, J. 2013. A survey on hygienic and physicochemical properties of Istrian cheese. Mljekarstvo 63: 55–63.

Mah, H.K., Han, J.H., Oh, Y.J., Kim, M.G. and Hwang, H.J. 2002. Biogenic amines in Jetkoals, Korean salted and fermented fish products. Food Chemistry 79: 239–243.

Maijala, R. and Nurmi, E. 1995. Influence of processing temperature on the formation of biogenic amines in dry sausages. Meat Science 39: 9–22.

Marijan, A., Džaja, P., Bogdanović, T., Škoko, I., Cvetnić, Ž., Dobranić, V., Zdolec, N., Šatrović, E. and Severin, K. 2014. Influence of ripening time on the amount of certain biogenic amines in rind and core of cow milk Livno cheese. Mljekarstvo 64: 159–169.

Martin-Alvarez, P.J., Marcobal, A., Polo, C. and Moreno-ArriBA, M.V. 2006. Influence of technological practices on biogenic amine contents in red wines. European Food Research Technology 222: 420–424.

Mayer, K. and Pause, G. 1972. Biogenic amine in sauerkraut. Lebensmittel Wissenschaft und Technologie 5: 108–109.

Miličević, B., Danilović, B., Zdolec, N., Kozačinski, L., Dobranić, V. and Savić, D. 2014. Microbiota of the fermented sausages: Influence to product quality and safety. Bulgarian Journal of Agricultural Science 20: 1061–1078.

Moreno, R.B. and Torres, E.A. 2001. Histamine levels in fresh fish, a quality index. Presented at the 2001 Institute of Food Technologists (IFT) meeting, New Orleans, LA.

Moret, S., Smela, D., Populin, T. and Conte, L.S. 2005. A survey of biogenic amine content of fresh and preserved vegetables. Food Chemistry 89: 355–361.

Motarjemi, Y. 2002. Impact of small scale fermentation technology on food safety in developing countries. International Journal of Food Microbiology 75: 213–229.

Nout, M.J.R., Ruiker, M.M.W. and Bouwmeester, H.M. 1993. Effect of processing conditions on the formation of biogenic amines and ethyl carbamate in soybean tempe. Journal of Food Safety 33: 293–303.

Ozdestan, O. and Uren, A. 2009. A method for benzoyl chloride derivatization of biogenic amines for high performance liquid chromatography. Talanta 78: 1321–1326.

Parente, E., Matuscelli, M., Gadrini, F., Grieco, S., Crudele, M.A. and Suzzi, G. 2001. Evolution of microbial populations and biogenic amines production in dry sausages produced in southern Italy. Journal of Applied Microbiology 90: 882–891.

Penas, E., Frias, J., Sidro, B. and Vidal-Valverde, C. 2010. Impact of fermentation conditions and refrigerated storage on microbial quality and biogenic amine content of sauerkraut. Food Chemistry 123: 143–150.

Pineda, A., Carrasco, J., Peña-Farfal, C., Henríquez-Aedo, K. and Aranda, M. 2012. Preliminary evaluation of biogenic amines content in Chilean young varietal wines by HPLC. Food Control 23: 251–257.

Proestos, C., Loukatos, P. and Komaitis, M. 2008. Determination of biogenic amines in wines by HPLC with precolumn dansylation and fluorimetric detection. Food Chemistry 106: 1218–1224.

Rabie, M.A., Siliha, H., el-Saidy, S., el-Badawy, A.A. and Malcata, F.X. 2011. Reduced biogenic amine contents in sauerkraut via addition of selected lactic acid bacteria. Food Chemistry 129: 1778–1782.

Radeljević, B., Mikulec, N., Antunac, N., Prpić, Z., Maletić, M. and Havranek, J. 2013. Influence of starter culture on total free amino acids concentration during ripening of Krk cheese. Mljekarstvo 63: 15–21.

Repka-Ramírez, M.S. and Baraniuk, J.N. 2002. Histamine in health and disease. Clinical Allergy Immunology 17: 1–25.

Roig-Sagues, A.X., Molina, A.P. and Hernandez-Herrero, M.M. 2002. Histamine and tyramine-forming microorganisms in Spanish traditional cheeses. European Food Research and Technology 215: 96–100.

Rosi, I., Nannelli, F. and Giovani, G. 2009. Biogenic amine production by *Oenococcus oeni* during malolactic fermentation of wines obtained using different strains of *Saccharomyces cerevisiae*. LWT-Food Science and Technology 42 (2009): 525–530.

Russo, P., Spano, G., Arena, M.P., Capozzi, V., Grieco, F. and Beneduce, L. 2010. Are consumers aware of the risks related to biogenic amines in food? Current Research Technology, Education Topics in Applied Microbiology Microbial Biotechnology 1087–1095.

Sanceda, N.G., Kurata, T. and Arakawa, N. 1996. Accelerated fermentation process for the manufacture of fish sauce using histidine. Journal of Food Science 61: 220–225.

Saez, J.S., Lopes, C.A., Kirs, V.E. and Sangorrín, M. 2011. Production of volatile phenols by *Pichia manshurica* and *Pichia membranifaciens* isolated from spoiled wines and cellar environment in Patagonia. Food Microbiology 283: 503–509.

Santos, S.M.H. 1996. Biogenic amines: Their importance in foods. Integrated Journal of Food Microbiology 29: 213–231.

Shalaby, A.R. 1996. Significance of biogenic amines to food safety and human health. Food Research International 29(7): 675–690.

Soufleros, E.H., Bouloumpasi, E., Zotou, A. and Lokou, Z. 2007. Determination of biogenic amines in Greek wines by HPLC and ultraviolet detection after dansylation and examination of factors affecting their presence and concentration. Food Chemistry 101: 704–716.

Spano, G., Russo, P., Lonvaud-Funel, A., Lucas, P., Alexandre, H., Grandvalet, C., Coton, E., Coton, M., Barnavon, L., Bach, B., Rattray, F., Bunte, A., Magni, C., Ladero, V., Alvarez, M., Fernández, M., Lopez, P., de Palencia, P.F., Corbi, A., Trip, H. and Lolkema, J.S. 2010. Biogenic amines in fermented foods. European Journal of Clinical Nutrition 64 (suppl.) 31: S95–S100.

Steinkraus, K.H. 1996. Introduction to indigenous fermented foods. pp. 1–6. *In*: K.H. Steinkraus (ed.). Handbook of Indigenous Fermented Foods New York: Marcel Dekker.

Straub, B.W., Tichaczek, P.S., Kicherer, M. and Hammes, W.P. 1994. Formation of tyramine by *Lactobacillus curvatus* LTH 972. Z Lebensm Unters Forsch 1: 9–12.

Stratton, J.E., Hutkins, W.R. and Taylor, S.L. 1991. Biogenic amines in cheese and other fermented foods. A review. Journal of Food Protection 54: 460–470.

Sumner, S.S., Speckhard, H.W., Somers, E.B. and Taylor, S.L. 1985. Isolation of histamine-producing *Lactobacillus buchneri* from Swiss cheese implicated in a food poisoning outbreak. Applied Environmental Microbiology 50: 1094–1096.

Suzzi, G. and Gardini, F. 2005. Biogenic amines in dry fermented sausages: A review. International Journal of Food Microbiology 88: 41–54.

Tassoni, A., Tango, N. and Ferri, M. 2013. Comparison of biogenic amine and polyphenol profiles of grape berries and wines obtained following conventional, organic and biodynamic agricultural and oenological practices. Food Chemistry 139: 405–413.

Taylor, S.L. 1986. Histamine food poisoning: toxicology and clinical aspects. Critical Reviews in Toxicology 17: 91–128.

Taylor, S.L., Leatherwood, M. and Lieber, E.R. 1978. Histamine in sauerkraut. Journal of Food Science 43: 1030–1032.

Torrea, D. and Ancín, C. 2002. Content of biogenic amines in a Chardonnay wine obtained through spontaneous and inoculated fermentations. Journal of Agricultural and Food Chemistry 50(17): 4895–4899.

Tristezza, M., Vetrano, C., Bleve, G., Spano, G., Capozzi, V., Logrieco, A., Mita, G. and Grieco, F. 2013. Biodiversity and safety aspects of yeast strains characterized from vineyards and spontaneous fermentations in the Apulia Region, Italy. Food Microbiology (36): 335–342.

Tsai, Y.H., Lin, C.Y., Chang, S.C., Chen, H.C., Kung, H.F. and Wei, C.I. 2005. Occurrence of histamine and histamine-forming bacteria in salted mackerel in Taiwan. Food Microbiology 22: 461–467.

Tuberoso, C.I., Congiu, F., Serreli, G. and Mameli, S. 2015. Determination of dansylated amino acids and biogenic amines in Cannonau and Vermentino wines by HPLC-FLD. Food Chemistry 175: 29–35.

Valsamaki, K., Michaelidou, A. and Polychroniadou, A. 2000. Biogenic amine production in Feta cheese. Food Chemistry 71: 259–266.

Vidal-Carou, M.C., Ambatle-Espunyes, A., Ulla-Ulla, M.C. and Marine-Font, A. 1990. Histamine and tyramine in Spanish wines: Their formation during the winemaking process. American Journal of Enology and Viticology 41: 160–167.

Vincenzini, M., Guerrini, S., Mangani, S. and Granchi, L. 2009. Amino acid metabolisms and production of biogenic amines and ethyl carbamate. pp. 167–180. *In*: H. König, G. Unden and J. Frohlich (eds.). Biology of Microorganisms on Grapes, in Must and in Wine. Berlin, Heidelberg: Springer-Verlag.

Yanez, L., Saavedra, J., Martínez, C., Cordova, A. and Ganga, M.A. 2012. Chemometric analysis for the detection of biogenic amines in Chilean *Cabernet sauvignon* wines: A Comparative Study between Organic and Nonorganic Production. Journal of Food Science 77(8): 143–150.

Yildirim, H.K., Uren, A. and Yuce, U. 2007. Biogenic amines in organic and non-organic wines. Food Technology and Biotechnology 45(1): 62–68.

Zhai, H., Yang, X., Li, L. and Xia, G. 2012. Biogenic amines in commercial fish and fish products sold in southern China. Food Control 25(1): 303–308.

Zolou, A., Lokou, Z., Souflero, E. and Straits, I. 2003. Determination of biogenic amines in wines and beers by high performance liquid chromatography with pre-column dansylation and ultraviolet detection. Journal of Chromatography 57: 429–439.

17

Applications of Metagenomics to Fermented Foods

Céline Bigot,[1,]* *Jean-Christophe Meile,*[1] *Fabienne Remize*[2,]*
and *Caroline Strub*[3]

1. Introduction

Fermentation is a traditional way of food preservation and is of great importance for human food consumption as it enables the development of nutritional and organoleptic qualities of food. This key traditional process is used for the conservation and transformation of a wide variety of food products of different origins (animal or vegetal) and nature (liquid to solid). For instance, starchy cereal-based food, meat, fish and sea food, vegetables and fruits, dairy products, cocoa, coffee and many others are transformed by fermentation. Food fermentation is utilized in many different geographical areas and, most of the time, occurs spontaneously. In Africa and Asia, traditional fermented foods represent a large part of local population diet. They often require process optimization to extend their production and commercialization (Aidoo et al. 2006, Sanni 1993). Fermentation is either alcoholic, resulting in alcoholic beverages or bread, lactic as is usually seen for dairy foods and vegetables, or acetic (vinegar). But food products may also result from a combination of these. Moreover, frequently for traditional foods, fermentative agents are undetermined and may involve bacterial or mould species.

[1] CIRAD-UMR Qualisud, TA B-95/16, 73, rue Jean-François Breton, 34398 Montpellier Cedex 5, France.
 E-mails: celine.bigot@cirad.fr; jean-christophe.meile@cirad
[2] Université de La Réunion, UMR95, QualiSud – Démarche intégrée pour l'obtention d'aliments de qualité,
 ESIROI, Parc Technologique Universitaire, 2 rue Joseph Wetzell, F-97490 Sainte-Clotilde, France.
 E-mail: fabienne.remize@univ-reunion.fr
[3] Université Montpellier 2, UMR95, QualiSud – Démarche intégrée pour l'obtention d'aliments de qualité,
 Polytech, Bâtiment 15 - 4ème Etage, Place Eugène Bataillon - CC023, 34095 Montpellier Cedex 5, France.
 E-mail: caroline.strub@polytech.univ-montp2.fr
* Corresponding authors

Although empirically performed, many food fermentation processes gained scientific interest over the last decades. The aim of current research is to understand and control fermentation progress in order to ensure constant sensorial properties, to increase safety and to limit spoilage. Industrial microorganisms are usually very easy to analyse by traditional microbiological methods because their production is controlled and pure strains are usually used. On the contrary, spontaneous fermentations result from the activity of a complex microbial ecosystem whose diversity and level of organization remains largely underestimated (Botta and Cocolin 2012). This complex microbial ecosystem can be composed of microorganisms that are not cultivable or non-viable in laboratory conditions. In addition, similar physiological properties (leading to similar food characteristics), may result from the presence and/or activity of phylogenetically distant bacterial species. As a consequence, conventional microbiological methods are, by far, not the best tools to study fermentation in these cases (Cocolin et al. 2011).

Over the last decade, molecular methods were applied to environmental samples (including food) in a culture-independent manner to study microbial ecology as a complement to culture-dependent studies. These first approaches of 'Metagenomics' greatly facilitated the description of fermentation ecosystems and the characterization of the microbial species involved in the process. Metagenomics, according to Chen and Patcher (2005), can be defined as "the application of modern genomics techniques to the study of communities of microbial organisms directly in their natural environments, bypassing the need for isolation and lab cultivation of individual species". As compared to classical cultural-based methods, metagenomics approaches are meant to give a quicker and more comprehensive as well as a complementary description of the microbial diversity associated with the fermentation process.

Metagenomics relies on nucleic acid (DNA/RNA) analyses and appears frequently associated to massive sequencing tools and bioinformatics. However, other molecular methods used to investigate microbial diversity fall into this field. Target DNA (or RNA) regions are often analysed by fingerprinting methods coupling PCR amplification, enzymatic restriction and electrophoretic techniques (such as T-RFLP— Terminal Restriction Fragment Length Polymorphism, DGGE—Denaturing Gel Gradient Electrophoresis, TGGE—Temperature Gradient Gel Electrophoresis, SSCP— Single-Strand Conformation Polymorphism, ARDRA—Amplified Ribosomal DNA Restriction Analysis, ARISA—Automated Ribosomal Intergenic Spacer Analysis) or cloning (for a review, see Nocker et al. 2007). When studying food microbial ecology, PCR amplification is performed with primers usually targeting rRNA coding regions (16S, 26S and 28S for bacteria, yeast and fungi respectively) for global analyses so-called community profiling. Such global approaches generally provide an overview of major taxons but poor resolution for the identification at the species level. On the other hand, targeted studies focus on specific microbial groups in order to get insight into their abundance and/or diversity; specific genes can also be targeted when searching for specific (enzymatic) functions. Functional metagenomics explores the diversity microbial enzymatic activities potentially or actively present in a given ecosystem at the gene (DNA) and gene expression (RNA) levels (metatranscriptomics) (for a review, see Bokulich and Mills 2012).

The rise of Next-Generation Sequencing (NGS) technologies (about a decade ago) revolutionized the field of microbial ecology. NGS analyses provide a comprehensive description of the microbial DNA content in a given sample by generating up to 10^9 sequence reads per run. These high-throughput approaches allow: (i) the sequencing of PCR amplicons from rRNA coding regions (16S for bacteria and ITS for fungi) or (ii) the direct sequencing of the whole-community DNA for a comprehensive exploration of the DNA diversity or 'metagenome' of the sample. Four types of NGS technologies (454 Pyrosequencing, Ion Torrent, Illumina and SOLiD systems) are available on the market and are able to generate per sample analysis from 500 million to 50 billion of short DNA sequence (100–400 pb) reads that are further aligned and either compared or assembled. The choice of technology and method depends on the type (or complexity) of fermented food and the aim of the study. The Pyrosequencing (and Ion Torrent) method generates about one million of 400–600 bp reads and provides a high taxonomic resolution (up to the species level) but a low coverage of the sample. At a lower cost, the Illumina (and SOLiD) technology provides a higher coverage by generating up to one billion short reads (100 pb) with low taxonomic resolution, and is more likely to detect rare or underrepresented sequences. NGS data are analysed using bioinformatics software (Mothur, Qiime, MEGAN, etc.) that allow the processing, alignment, assembly and statistical comparison of DNA/RNA sequences. Reference databases are then utilized for taxon assignment. The rapid, constant development of bioinformatics tools and software and the increasing amounts of data available contribute to enrich and increase the reliability of metagenomics-dedicated databases (for a review, see van Hijum et al. 2013).

This chapter reviews some of the most recent applications of molecular techniques in metagenomics approaches to understand fermentation microbial ecology dynamics and to improve the quality of fermented food.

2. Metagenomics as a Tool to Monitor Fermentation Process

Fermentation generally starts with a high microbial diversity which corresponds to the microbial contamination of raw materials, process equipment and human manipulation. Fermentation is characterized by modifications of the food composition: alcoholic fermentation results from the conversion of sugars into ethanol and carbon dioxide; lactic acid fermentation produces lactic acid and possibly, carbon dioxide, acetic acid and ethanol; acetic acid fermentation results mainly in acetic acid formation. Throughout the fermentation process, the interaction between the microflora and the fermenting food matrix creates a dynamic phenomenon, at both microbial and biochemical levels, characterized by changes in physicochemical conditions (pH, salt, temperature, etc.). All these biochemical modifications share the common feature of generating stress factors for microorganisms. The adverse effects of ethanol result from impairment of microbial cellular membrane properties (Aguilera et al. 2006, Li et al. 2012). An increase of organic acid levels generates a dissipation of the proton gradient across cellular membranes (Russell and Diez-Gonzalez 1997). The composition of the raw material may take part of the environmental stress: for example, high sugar levels result in an osmotic stress. Moreover, food preparation, such as salt addition performed at the beginning of the fermentation process, may contribute to the stress effect. As a consequence, a decrease of microbial diversity is generally observed over fermentation progress, and a dominant microflora appears.

This is illustrated by many examples through different methodological approaches. For example, bread is traditionally made from sourdough which is a mix of flour and water containing yeasts and bacteria. The first studies that considered sourdough diversity were based on cultural approaches completed by molecular identification of dominant species. RAPD (Randomly Amplified Polymorphic DNA) and 16S ribosomal RNA coding region sequencing were used to determine the dominant species of lactic acid bacteria (de Vuyst et al. 2002). Afterwards, PCR-DGGE (Denaturant Gradient Gel Electrophoresis) was used in several studies (Meroth et al. 2004, Moroni et al. 2011). In rice sourdough, PCR-DGGE clearly showed a shift during process from mother sponge flora towards a dominating flora mainly composed of *Lactobacillus curvatus*. Over gluten-free sourdough elaboration process, a stable biota with a broad spectrum of autochthonous lactic acid bacteria and yeast species was established, as this biota was adapted and competitive in these conditions. Dominating species were commonly isolated from other types of sourdough (rye or wheat) or from other tropical product fermentations. For example, *Pediococcus pentosaceus*, *Leuconostoc holzapfelii*, *Lactobacillus gallinarum*, *Lactobacillus graminis*, *Lactobacillus vaginalis*, *Weissella cibaria* were detected though they were not considered as typical of sourdough fermentation. Other methods like PCR-TGGE (Thermal Gradient Gel Electrophoresis) and High Resolution Melting quantitative PCR (HRM-qPCR) were used in other studies (Ferchichi et al. 2007, Lin and Gänzle 2014). The use of PCR-TGGE was demonstrated as another efficient approach to monitor the main fermentative species in sourdough. The authors underlined the need of a careful choice of amplification regions and sampling, and of critical and reproducible DNA extraction, PCR amplification and analysis before drawing any conclusion. The advantages of HRM-qPCR rely on the simplicity of the method and its ability to detect single nucleotide differences in target sequences. However, this approach would not be appropriate for complex microbiota whereby multiple peaks are present, corresponding to multiple melting temperatures of target sequences.

Generally, the final fermented product results from a large number of chemical reactions caused by microorganisms. The example of cocoa bean fermentation is particularly convincing as alcoholic fermentation, lactic fermentation and finally acetic fermentation take place successively. The initial anaerobic conditions at low pH (3.6) and high sugar content of the pulp surrounding the cocoa beans (obtained after the breaking of the pod) promote yeast activity (naturally present or introduced by humans). In the pulp, sugar conversion into alcohol and carbon dioxide causes a rise in temperature and an increase in pH due to the consumption of the citric acid by yeasts. These changes promote the onset of lactic acid bacteria (LAB) that oxidize the alcohol to lactic acid. As conditions become more aerobic (due to aeration), the production of acetic acid by acetic acid bacteria (AAB) is favoured. The formation of acetic acid from alcohol is an exothermic reaction and the temperature reaches about 50°C; this temperature shift causes inactivation of acetic acid bacteria. And, due to the presence of acetic acid, the biochemical reactions which produce precursors of chocolate flavor compounds can occur in the seed. Towards the end of the fermentation, the strong odour of acetic acid decreases progressively. Other aromatic molecules (alcohols, acids, ketones, sulfur compounds) are also produced by the microorganisms during fermentation, and confer organoleptic qualities characteristic of the product (Marilley and Casey 2004, Smit et al. 2005, Lacroix et al. 2010, Sicard and Legras 2011). Both yeasts and bacteria are involved in the production of compounds responsible for chocolate flavor (Shwan and Whaels 2004). If not performed well, fermentation can lead to serious defects, such as beans with slate colour without flavouring, bitter taste and astringent or purple beans, as well as poorly aromatic or rotten beans.

Although literature on a thorough description of the microbial species involved in cocoa fermentation is available, relatively little is known about their precise contribution to chocolate quality (for a review, see Saltini et al. 2013). Recent metagenomics approaches on cocoa fermentations allowed a re-exploration of the process and revealed new insight into microbial dynamics, diversity and interactions during the process (Camu et al. 2007, Hamdouche et al. 2014). Notably, rare taxons as well as viral communities were detected, providing a comprehensive view on the ecosystem (Illeghems et al. 2012). This shows that complex fermentative ecosystems, such as spontaneous coca bean fermentation, studied by metagenomics approaches allow new features to be discovered.

3. Metagenomics to Determine which Fermentative Agents are Really Involved

Targeting DNA in studies of fermented food provides a lot of information. In fact, the DNA molecule has a variable half-life dependent on many factors including the biological activity of the matrix in which it is present. In addition, simple DNA analysis does not allow distinguishing living from dead microorganisms when they are present in a certain concentration.

The study of food fermentation focusing on RNA instead of DNA provides information on the function of the microorganisms and how they are influenced by different environments. Indeed, RNA analysis seems able to better highlight the microorganisms that contribute to the fermentation process (Cocolin et al. 2013). Messenger RNAs have a very short lifetime and their extraction from the food matrix remains technically difficult. Contrarily, the use of ribosomal RNA seems a good compromise. Indeed, its lifetime is greater than mRNA but less than that of DNA.

The literature on metatranscriptomics to monitor fermentation is scarce. Real-time PCR on cDNA, PCR-DGGE on cDNA and microarray targeting transcripts are the main techniques used in those studies. The amount of 16S rRNA in fermented rice has been monitored by RT-Q-PCR (Reverse Transcription-Quantitative-PCR) (Nakayama et al. 2007). In other studies, Tuf, GroL and 16S genes were targeted from Emmental cheese samples (Falentin et al. 2010) and RT-PCR-DGGE was applied on cheese samples (Dolci et al. 2013). Metatranscriptomic gene-expression profiles during kimchi fermentation (fermented vegetable product) were investigated (Jung et al. 2013). This work showed the prevalence of carbohydrate metabolism (transport and hydrolysis) and lactic acid fermentation. *Leuconostoc mesenteroides* was rather active at the beginning of fermentation and *Lactobacillus sakei* and *Weissella koreensis* presented a higher activity during later stages. A DNA array was developed to study kimchi and it was concluded that a minority of species mainly contribute to the development of kimchi (Nam et al. 2009). LAB communities of sourdough fermentations were monitored using a DNA microarray constituted of functional genes. Dominating LAB species were established after five days of fermentation. *Lactococcus lactis* had an important activity at the beginning of the biological process, whereas *P. pentosaceus* dominated during most of the process (Weckx et al. 2010).

Functional properties of the fermented food microbiota can also be investigated by metagenomics. Indeed functional diversity can be explored through genetic screening of genes of interest. For example, a study focused on 33 genes involved in probiotic and nutritional functions involved in survival, gastro tract, starch metabolism, folate and riboflavin biosynthesis (Turpin et al. 2011). In addition, a few studies have investigated the expression of targeted genes of interest in food matrix. For instance, the expression of gene encoding for amylase has been monitored in Pearl millet slurries (Oguntoyinbo and Narbad 2012, Humblot et al. 2014).

Metatranscriptomic analysis of fermented food microbiota still presents technical locks. For instance, the extraction of RNA from food matrix remains difficult (Ulve et al. 2008). Indeed such complex matrixes contain fats, polysaccharides and inhibitors that make difficult the extraction of good quality and quantity RNA. In addition, a lack of basic sequence data and annotation remains. Sequence information available in databases is still limited to some bacteria (lactobacilli, staphylococci and food pathogens).

Because metagenomics data include a huge proportion of genes encoding classical cell functions with little interest for food microbiota, sequencing of metagenomes should focus on specific parts (key enzymes for flavor compound production, toxin synthesis, or specific amino acid degradation) to be able to monitor, in addition to variations of species diversity, changes in microbial activity.

4. Is There a Fermentative Flora Signature of Terroir? Traceability Tools

The French notion of 'terroir' (the product identity), which is at the heart of the AOC (Appellation d'Origine Contrôlée), is directly linked to the food quality and origin. The European equivalent is the PGI, for Protected Geographical Indication, related to the European Regulation No 1151/2012 which promotes and protects names of quality agricultural foods (Capozzi et al. 2012). Many recent studies support the idea that biogeographical characteristics of terrestrial microorganisms may lead to regionalized properties associated with valuable crops, and so, underline the existence of a microbial biogeography or 'microbial terroir' (Renouf et al. 2006, Fierer 2008, Bokulich et al. 2013, Gilbert et al. 2014). As an example, the aquaculture farms in Vietnam were differentiated by their bacterial ecology showing the relationship between terroir and presence of bacteria. The bacterial ecology was shown to stay stable on different fish samples collected at 6 months intervals (Le Nguyen et al. 2008). Similar studies were realized on fruits (El Sheikha et al. 2009, 2011) and marine salts (Dufossé et al. 2013), showing the robustness of the approach. Taking into account the importance of the native microbial community in spontaneous fermentative process and the fact that this microbial ecology could be linked to the terroir, it leaves open to question whether fermentative microbes could be used as a potential signature of food terroir.

The resurgence of biogeography research, particularly the biogeography of microorganisms, is mainly due to the technological advances in molecular biology over the last decade. These advances led DNA markers to become one of the most effective instruments to traceability (Galimberti et al. 2013). The choice of appropriate molecular tool will depend on different parameters (quantity/quality/integrity of DNA, degree of genetic variations of the analysed species, etc.). In a general way, these microbial DNA analysis tools are characterized by the use of specific DNA markers (usually universal) based on PCR amplification techniques, such as digestion with specific restriction enzymes or electrophoretic analysis (RFLP, RAPD—Random Amplified Polymorphic DNA, AFLP—Amplified Fragment Length Polymorphism, DGGE, etc.). Over the last years, PCR-DGGE analysis has been very popular in the study of food safety and traceability, notably of fermented foods (Peres et al. 2007, Dalmacio et al. 2011, Hosseini et al. 2012, Zheng et al. 2012). An interesting study realized by Mauriello and his colleagues (2003) showed an example of PCR-DGGE application proving that the effect of microbial diversity of natural starter cultures on traditional dairy products and its evolution during fermentation may represent important proof of authenticity for the traceability of origin and mode of production. Another molecular tool application on fermented foods was done on wine grapes by using a high-throughput, short-amplicon sequencing technique (Bokulich et al. 2013). Using this method, more than five million reads for bacterial 16S rRNA coding sequences and more than 3.2 million reads for fungal ITS sequences were generated and, *in silico*, analysed. This study showed different microbial profiles according to vine-growing regions, giving thus evidence for a link between microbiota and regional, varietal, and climatic factors across multi-scale viticultural areas.

In conclusion, it is too early to conclude about the existence of a microbial signature of terroir, but research carried out until now shows interesting results in this direction. The increasing development of culture-independent methods drives research forward for a better understanding of the dynamics of microbial ecosystems in fermented foods.

5. Applications of Metagenomics towards Starter Selection

The use of microbial inoculants to enhance nutritional and quality properties (shelf-life, texture, aroma, etc.) or to hasten the process is widely applied in food industry. But, in the majority of the developing countries, the fermentation process relies largely on empirical techniques, characterized by an incomplete process control. As a consequence, low yields and variable quality products are obtained. From traditional food fermentations, microbial starters can be selected and re-used further. Microbial starter selection criteria are plurals. They include, not only the ability to use various substrates, the formation of metabolism products and the physiology of strain, but also food safety requirements and other quality expectations (Holzapfel 2002). Microbial starters gain to be characterized phenotypically and genetically, for technological, safety and probiotic features (Ammor and Mayo 2007). The advances in this field towards an enhancement of fermented food quality often involve back-sloping or the inoculation of raw materials with a residue from a previous batch, but sometimes remain insufficient to ensure food safety (for a comprehensive review of biotechnology issues in developing countries, see FAO 2010). In addition, current applications of these starters are numerous in the field of food safety and show an increasing interest to understand manufacturing processes of local fermented foods (Benito et al. 2007, Capozzi and Spano 2011, Zdolec et al. 2013). It was demonstrated in case studies, such as yoghurt, that microbe-microbe interaction could have a positive impact on product quality of fermented food by exchange of 'information' and metabolites. Indeed, the Quorum Sensing phenomenon can be shared between species occupying the same niche (Smid and Lacroix 2013) and so, can play a role in complex communities of microbes involved in the fermentation process.

Molecular biology tools are very helpful in describing and understanding biological and biochemical phenomena related to fermentation. As they do not require the isolation and cultivation of microorganisms, they reflect the 'true' microbial composition of foods. By highlighting the main microbial species involved in the process of fermentation, metagenomic approaches are likely to unravel new fermentative species, thus helping in the selection of new starter strains (Díaz Ruiz and Wacher 2003, Cocolin and Ercolini 2008). The development and the application of metagenomic tools significantly contributed to elucidate the microbial community involved in many fermented foods. Applications of metagenomics to study the microbial ecology of fermented foods are becoming more and more numerous, but very few are related to the selection of starters. Some of them point the potentials of dominant strains, well adapted to the ecological niche, in the production of a safe fermented product.

An authoritative list of microorganisms with documented use in food has been established since 2002 from a joint project between the International Dairy Federation (IDF) and the European Feed and Food Culture Association (EFFCA) (Mogensen et al. 2002a, 2002b). The group of Microbial Food Cultures (MFC) satisfies the regulation EC no. 178/2002 to ensure manufacturers the safe use of microorganisms in food (Hansen 2002, Bourdichon et al. 2012). Most of the cultures are lactic acid bacteria (LAB) but yeasts (e.g., *Saccharomyces*) and molds (e.g., *Penicillium*) have been found for several foodstuffs (cereals, meats, milk, etc.). The use of these microorganisms to obtain safe fermented foods is based on their QPS (Qualified Presumption of Safety) status, similar in concept and purpose to the GRAS status (Generally Recognized as Safe) which is used in the USA. The QPS approach is maintained by EFSA (European Food Safety Authority) since 2007, according to European regulation 178/2002.

Molecular tools may have a significant role to play in QPS risk assessment. The detection, by metagenomics approaches, of a given genus or species in many fermented foods strengthens the body of knowledge of this genus/species and thus contributes to its acceptance as QPS.

6. Metagenomics for Safety Improvement of Fermented Food

Stress conditions associated to food fermentations are related to environmental stress and to biochemical modifications resulting from microbial metabolism. These conditions result in the selection of dominant flora but also can induce viable-but-nonculturable (VBNC) and/or non-viable states of microorganisms (Giraffa and Neviani 2001). For example, in wine, the spoilage yeast *Brettanomyce bruxellensis* can enter this stage and exert its aromatic deviation a long time after bottling (Serpaggi et al. 2012).

On one hand, the microbial dynamics observed during fermentation gives fermented foods an effective protection against pathogens (Adams and Mitchell 2002, Adams and Nicolaides 2008). It is also implied in the removal of toxic compounds from raw materials (Hammes and Tichaczek 1994). Competition and/or antagonism between microorganisms can be mediated through inhibitory molecules (peptides, bacteriocins, acids) that exhibit specific and potentially interesting properties.

Metagenomics is potentially a tool of choice to assess the survival of pathogens, toxinogens or spoilage species over fermented food elaboration process. A possible limit of the metagenomics approach is its failure to detect the less numerous microorganisms in an environment containing a dominant microbiota. Interestingly, the first study suggesting this application investigated microbial diversity in *Potopoto*, a maize dough used as weaning food, and in *Dégué*, a millet-based paste, by PCR-TGGE and by comparing DNA extraction techniques (Abriouel et al. 2006). The detection of *Escherichia coli* and of *Bacillus* species, which potentially included *Bacillus cereus*, over product elaboration, questioned the survival conditions of these microbiological hazards. Similarly, in other cereal fermented food traditionally prepared in Africa, the presence of *Clostridium perfringens* and of *B. cereus* was evidenced from a clone library while the identification from the main bands from PCR-DGGE profile did not showed these species (Oguntoyinbo et al. 2011). In another study, a hazard related to the detection of *Proteus mirabilis* and *Staphylococcus* spp. was pointed out by ARDRA analysis applied to isolates from soybean fermented food and from fermented pork samples whereas PCR-DGGE detected uncultivated *Clostridium* spp. and *Staphylococcus* spp. from fermented fish products (Singh et al. 2014). Thus, the discrepancy of results between classical microbiological methods and molecular methods should be more deeply investigated. Meanwhile the most reliable approach is to use both approaches as complementary tools.

On the other hand, the application of metagenomics in evaluating the impact of process modification on fermentation microbiota and safety issues is highly relevant. Salt reduction strategies for sauerkraut, in a more general view of reducing sodium intake, relied on the decrease of sodium chloride content, with or without its partial replacement by other mineral salts (calcium chloride, magnesium chloride and potassium chloride) (Wolkers-Rooijackers et al. 2013). The consequences of these changes on sensorial properties were investigated. In parallel, PCR-DGGE and classical microbiology were used to assess fermentative and safety issues resulting from this process changes. It was thus checked that fermentative profiles and *Enterobacteriaceae* population did not significantly change with salt content and composition. Another study focused on the relationships between mycotoxin contamination and the fungal communities associated to food products, such as coffee (Durand et al. 2013, Nganou et al. 2014).

7. Conclusion

Metagenomics approaches are very powerful in exploring the microbial ecology of complex ecosystems such as fermentative foods. Different biological questions can now be addressed by global or targeted methods (functional metagenomics or meta-transcriptomics). However, several considerations should be taken into account when undertaking metagenomics studies.

Technical issues should be carefully addressed in order to anticipate biases that can occur in the generated data. Nucleic acid extraction (DNA/RNA) yields are subjected to variations according to the composition of the food matrix and the microbial species present in the ecosystem. To obtain a reliable view of the ecosystem, the reproducibility of extraction and PCR has to be ensured. In addition, before arriving at a conclusion about a microbial feature or signature in a fermentation ecosystem, it is required that several studies are performed in the same ecosystem.

While whole community NGS approaches are not severely impacted by technical biases, targeted approaches are subjected to biases (as for every PCR-based method). Preferential amplification selects sequences and impairs the representativeness of amplified sequences with regard to the initial sample. Both nucleic acid extraction and PCR amplification reaction biases can produce a fraction of truncated or misleading data. Therefore, it is recommended data obtained from such approaches be complemented with data obtained from different approaches (molecular and cultural methods). In all, the production of metagenomic data needs to be correlated with all data available (pH, temperature, Aw, chemical composition, etc.) to produce meaningful and useful data for the comprehension of complex ecosystems.

Recently, metagenomic approaches have been used to study the interactions between human gut and food microbiota (Donovan et al. 2012, Faith et al. 2011). Obesity being associated with a particular gut microbiota (depletion of Bacteroidetes) (Ley et al. 2006), the simultaneous investigation of fermented food diet and gut microbiota by metagenomics would result in a better knowledge of the interactions between these microbiota and could be helpful in fighting against obesity. Moreover, products that benefit for human health could be developed. In the future, one can dream of the application of metagenomics, and more specifically metatranscriptomics, to detect in an ecosystem, the presence of mRNA synthesized from genes encoding amino acid decarboxylases involved in biogenic amine pathways or encoding enzymes involved in mycotoxin formation. The same tools could be applied to investigate the formation of bacteriocins in an ecosystem.

keywords: Traceability, food safety, microbiota, ecology, interactions, fingerprinting, diversity

References

Abriouel, H., Ben Omar, N., López, R.L., Martínez-Cañamero, M., Keleke, S. and Gálvez, A. 2006. Culture-independent analysis of the microbial composition of the African traditional fermented foods *Potopoto* and *Dégué* by using three different DNA extraction methods. International Journal of Food Microbiology 111: 228–33.

Adams, M.R. and Nicolaides, L. 2008. Review of the sensitivity of different foodborne pathogens to fermentation. Food Control 8: 227–239.

Adams, M.R. and Mitchell, R. 2002. Fermentation and pathogen control: A risk assessment approach. International Journal of Food Microbiology 79: 75–83.

Aguilera, F., Peinado, R.A., Millán, C., Ortega, J.M. and Mauricio, J.C. 2006. Relationship between ethanol tolerance, H+ ATPase activity and the lipid composition of the plasma membrane in different wine yeast strains. International Journal of Food Microbiology 110: 34–42.

Aidoo, K.E., Nout, M.J.R. and Sarkar, P.K. 2006. Occurrence and function of yeasts in Asian indigenous fermented foods. FEMS Yeast Research 6: 30–9.

Ammor, M.S. and Mayo, B. 2007. Selection criteria for lactic acid bacteria to be used as functional starter cultures in dry sausage production: An update. Meat Science 76: 138–46.

Benito, M.J., Martín, A., Aranda, E., Pérez-Nevado, F., Ruiz-Moyano, S. and Córdoba, M.G. 2007. Characterization and selection of autochthonous lactic acid bacteria isolated from traditional Iberian dry-fermented salchichón and chorizo sausages. Journal of Food Science 72: 193–201.

Bokulich, N.A. and Mills, D.A. 2012. Next generation approaches to the microbial ecology of food fermentations. BMB Reports 45(7): 377–89.

Bokulich, N.A., Thorngate, J.H., Richardson, P.M. and Mills, D.A. 2013. Microbial biogeography of wine grapes is conditioned by cultivar, vintage, and climate. Proceedings of the National Academy of Sciences 111: 139–148.

Botta, C. and Cocolin, L. 2013. Microbial dynamics and biodiversity in table olive fermentation: culture-dependent and -independent approaches. Frontiers in Microbiology 6: 3–245.

Bourdichon, F., Casaregola, S., Farrokh, C., Frisvad, J.C., Gerds, M.L., Hammes, W.P., Harnett, J., Huys, G., Laulund, S., Ouwehand, A., Powell, I.B., Prajapati, J.B., Seto, Y., TerSchure, E., van Boven, A., Vankerckhoven, V., Zgoda, A., Tuijtelaars, S. and Hansen, E.B. 2012. Food fermentations: Microorganisms with technological beneficial use. International Journal of Food Microbiology 154: 87–97.

Camu, N., de Winter, T., Verbrugghe, K., Cleenwerck, I., Vandamme, P., Takrama, J.S., Vancanneyt, M. and de Vuyst, L. 2007. Dynamics and biodiversity of populations of lactic acid bacteria and acetic acid bacteria involved in spontaneous heap fermentation of cocoa beans in Ghana. Applied and Environmental Microbiology 73(6): 1809–1824.

Capozzi, V. and Spano, G. 2011. Food microbial biodiversity and 'microbes of protected origin'. Frontiers in Microbiology 2: 237. doi: 10.3389/fmicb.2011.00237.

Capozzi, V., Russo, P. and Spano, G. 2012. Microbial information regimen in EU geographical indications. World Patent Information 3: 229–231.

Chen, K. and Pachter, L. 2005. Bioinformatics for whole-genome shotgun sequencing of microbial communities. PLoS Computational Biology 1(2): e24.

Cocolin, L., Alessandria, V., Dolci, P., Gorra, R. and Rantsiou. K. 2013. Culture independent methods to assess the diversity and dynamics of microbiota during food fermentation. International Journal of Food Microbiology 167: 29–43.

Cocolin, L., Dolci, P. and Rantsiou, K. 2011. Biodiversity and dynamics of meat fermentations: The contribution of molecular methods for a better comprehension of a complex ecosystem. Meat Science 89(3): 296–302.

Cocolin, L. and Ercolini, D. 2008. Molecular techniques in the microbial ecology of fermented foods. Springer, New York, 280p.

Dalmacio, L.M., Angeles, A.K., Larcia, L.L., Balolong, M.P. and Estacio, R.C. 2011. Assessment of bacterial diversity in selected Philippine fermented food products through PCR-DGGE. Beneficial Microbes 2: 273–281.

Diaz-Ruiz, G. and Wacher, C. 2003. Methods for the study of microbial communities in fermented foods. Revista Latinoamericana de Microbiología 45(1-2): 30–40.

De Vuyst, L., Schrijvers, V., Paramithiotis, S., Vancanneyt, M., Swings, J., Tsakalidou, E., Messens, W., Hoste, B. and Kalantzopoulos, G. 2002. The biodiversity of lactic acid bacteria in Greek traditional wheat sourdoughs is reflected in both composition and metabolite formation. Applied and Environmental Microbiology 68: 6059–6069.

Dolci, P., Zenato, S., Pramotton, R., Barmaz, A., Alessandria, V., Rantsiou, K. and Cocolin, L. 2013. Cheese surface microbiota complexity: RT-PCR-DGGE, a tool for a detailed picture? International Journal of Food Microbiology 162: 8–12.

Donovan, S.M., Wang, M., Li, M., Friedberg, I., Schwartz, S.L. and Chapkin, R.S. 2012. Host-microbe interactions in the neonatal intestine: Role of human milk oligosaccharides. Advances in Nutrition 3: 450S–455S.

Dufossé, L., Donadio, C., Valla, A., Meile, J.C. and Montet, D. 2013. Determination of speciality food salt origin by using 16S rDNA fingerprinting of bacterial communities by PCR-DGGE: An application on marine salts produced in solar salterns from the French Atlantic Ocean. Food Control 32: 644–649.

Durand, N., El Sheikha, A., Suarez-Quiros, M., Gonzales-Rios, O., Nganou, N., Fontana-Tachon, A. and Montet, D. 2013. Application of PCR-DGGE to the study of dynamics and biodiversity of yeasts and potentially OTA producing fungi during coffee processing. Food Control 34(2): 466–471.

El Sheikha, A.F., Bouvet, J.M. and Montet, D. 2011. Biological bar code for determining the geographical origin of fruits using 28S rDNA fingerprinting of fungal communities by PCR-ADGGE: an application to Shea tree fruits. Quality Assurance and Safety of Crops & Foods 3: 40–47.

El Sheikha, A.F., Condur, A., Métayer, I., Le Nguyen, D.D., Loiseau, G. and Montet, D. 2009. Determination of fruit origin by using 26S rDNA fingerprinting of yeast communities by PCR-DGGE: Preliminary application to Physalis fruits from Egypt. Yeast 26: 567–573.

Faith, J.J., McNulty, N.P., Rey, F.E. Gordon, J.I. 2011. Predicting a human gut microbiota's response to diet in gnotobiotic mice. Science. 2011 Jul 1; 333(6038): 101–4.

Falentin, H., Postollec, F., Parayre, S., Henaff, N., Le Bivic, P., Richoux, R., Thierry, A. and Sohier, D. 2010. Specific metabolic activity of ripening bacteria quantified by real-time reverse transcription PCR throughout Emmental cheese manufacture. International Journal of Food Microbiology 144: 10–19.

FAO. 2010. Agricultural biotechnologies in developing countries: Options and opportunities in crops, forestry, livestock, fisheries and agro-industry to face the challenges of food insecurity and climate change. FAO International Technical Conference (ABDC-10), Mexico, March 1–4.

Ferchichi, M., Valcheva, R., Prévost, H., Onno, B. and Dousset, X. 2007. Molecular identification of the microbiota of French sourdough using temporal temperature gradient gel electrophoresis. Food Microbiology 24: 678–86.

Fierer, N. 2008. Microbial biogeography: Patterns in microbial diversity across space and time. pp. 95–115. *In*: K. Zengler (ed.). Accessing Uncultivated Microorganisms: from the Environment to Organisms and Genomes and Back. ASM Press, Washington DC.

Galimberti, A., DeMattia, F., Losa, A., Bruni, I., Federici, S., Casiraghi, M., Martellos, S. and Labra, M. 2013. DNA barcoding as a new tool for food traceability. Food Research International 50: 55–63.

Gilbert, J.A., van der Lelie, D. and Zarraonaindia, I. 2014. Microbial terroir for wine grapes. Proceedings of the National Academy of Sciences USA 11: 5–6.

Giraffa, G. and Neviani, E. 2001. DNA-based, culture-independent strategies for evaluating microbial communities in food-associated ecosystems. International Journal of Food Microbiology 20: 19–34.

Hamdouche, Y., Guehi, T., Durand, N., Kedjebo, K.B.D., Montet, D. and Meile, J.C. 2015. Dynamics of microbial ecology during cocoa fermentation and drying: Towards the identification of molecular markers. Food Control 48: 117–122.

Hammes, W.P. and Tichaczek, P.S. 1994. The potential of lactic acid bacteria for the production of safe and wholesome food. Zeitschriftfür Lebensmittel-Untersuchung und—Forschung 198: 193–201.

Hansen, E.B. 2002. Commercial bacterial starter cultures for fermented foods of the future. International Journal of Food Microbiology 78: 119–131.

Holzapfel, W.H. 2002. Appropriate starter culture technologies for small-scale fermentation in developing countries. International Journal of Food Microbiology 75: 197–212.

Hosseini, H., Hippe, B., Denner, E., Kollegger, E. and Haslberger, A. 2012. Isolation, identification and monitoring of contaminant bacteria in Iranian Kefir type drink by 16S rDNA sequencing. Food Control 25: 784–788.

Humblot, C., Turpin, W., Chevalier, F., Picq, C., Rochette, I. and Guyot, J.P. 2014. Determination of expression and activity of genes involved in starch metabolism in *Lactobacillus plantarum* A6 during fermentation of a cereal-based gruel. International Journal of Food Microbiology 185: 103–11.

Illeghems, K., de Vuyst, L., Papalexandratou, Z. and Weckx, S. 2012. Phylogenetic analysis of a spontaneous cocoa bean fermentation metagenome reveals new insights into its bacterial and fungal community diversity. Plos One 7(5): e38040.

Jung, J.Y., Lee, S.H., Jin, H.M., Hahn, Y., Madsen, E.L. and Jeon, C.O. 2013. Metatranscriptomic analysis of lactic acid bacterial gene expression during kimchi fermentation. International Journal of Food Microbiology 163: 171–179.

Lacroix, N., St. Gelais, D., Champagne, C.P., Fortin, J. and Vuillemard, J.C. 2010. Characterization of aromatic properties of old-style cheese starters. Journal of Dairy Science 93: 3427–3441.

Le Nguyen, D.D., Hanh, H.N., Dijoux, D., Loiseau, G. and Montet, D. 2008. Determination of fish origin by using 16S rDNA fingerprinting of bacterial communities by PCR-DGGE: An application on Pangasius fish from Viet Nam. Food Control 19: 454–460.

Ley, R.E., Turnbaugh, P.J., Klein, S. and Gordon, J.I. 2006. Microbial ecology: Human gut microbes associated with obesity. Nature 444: 1022–1023.

Li, H., Ma, M.L., Luo, S., Zhang, R.M., Han, P. and Hu, W. 2012. Metabolic responses to ethanol in *Saccharomyces cerevisiae* using a gas chromatography tandem mass spectrometry-based metabolomics approach. International Journal of Biochemistry and Cellular Biology 44: 1087–96.

Lin, X.B. and Gänzle, M.G. 2014. Quantitative high-resolution melting PCR analysis for monitoring of fermentation microbiota in sourdough. International Journal of Food Microbiology 186: 42–8.

Marilley, L. and Casey, M.G. 2004. Flavors of cheese products: Metabolic pathways, analytical tools and identification of producing strains. International Journal of Food Microbiology 90: 139–159.

Mauriello, G., Moio, L., Genovese, A. and Ercolini, D. 2003. Relationship between flavouring capabilities, bacterial composition and geographical origin of Natural Whey Cultures (NWCs) used for water-buffalo Mozzarella cheese manufacture. Journal of Dairy Science 86: 486–497.

Meroth, C.B., Hammes, W.P. and Hertel, C. 2004. Characterisation of the microbiota of rice sourdoughs and description of *Lactobacillus spicheri* sp. nov. Systematic and Applied Microbiology 27: 151–9.

Mogensen, G., Salminen, S., O'Brien, J., Ouwehand, A., Holzapfel, W., Shortt, C., Fonden, R., Miller, G.D., Donohue, D., Playne, M., Crittenden, R., Salvadori, B. and Zink, R. 2002a. Food microorganisms: Health benefits, safety evaluation and strains with documented history of use in foods. Bulletin of IDF 377: 4–9.

Mogensen, G., Salminen, S., O'Brien, J., Ouwehand, A., Holzapfel, W., Shortt, C., Fonden, R., Miller, G.D., Donohue, D., Playne, M., Crittenden, R., Salvadori, B. and Zink, R. 2002b. Inventory of microorganisms with a documented history of use in food. Bulletin of IDF 377: 10–19.

Moroni, A.V., Arendt, E.K. and Dal Bello, F. 2011. Biodiversity of lactic acid bacteria and yeasts in spontaneously-fermented buckwheat and teff sourdoughs. Food Microbiology 28: 497–502.

Nakayama, J., Hoshiko, H., Fukuda, M., Tanaka, H., Sakamoto, N., Tanaka, S., Ohue, K., Sakai, K. and Sonomoto, K. 2007. Molecular monitoring of bacterial community structure in long-aged *nukadoko*: Pickling bed of fermented rice bran dominated by slow-growing lactobacilli. Journal of Bioscience and Bioengineering 104: 481–489.

Nam, Y.-D., Chang, H.-W., Kim, K.-H., Roh, S.W. and Bae, J.-W. 2009. Metatranscriptome analysis of lactic acid bacteria during kimchi fermentation with genome-probing microarrays. International Journal of Food Microbiology 130: 140–146.

Nganou, N., Durand, N., Tatsadjieu, L., Piro-Metayer, I., Montet, D. and Mbufing, C. 2014. Fungal flora and Ochratoxin A Associated with Coffee in Cameroon. British Microbiology Research Journal 4(1): 1–17.

Nocker, A., Burr, M. and Camper, A.K. 2007. Genotypic microbial community profiling: a critical technical review. Microbial Ecology 54(2): 276–89.

Oguntoyinbo, F.A., Tourlomousis, P., Gasson, M.J. and Narbad, A. 2011. Analysis of bacterial communities of traditional fermented West African cereal foods using culture independent methods. International Journal of Food Microbiology 145: 205–10.

Oguntoyinbo, F.A. and Narbad, A. 2012. Molecular characterization of lactic acid bacteria and *in situ* amylase expression during traditional fermentation of cereal foods. Food Microbiol. 2012 Sep; 31(2): 254–62.

Peres, B., Barlet, N., Loiseau, G. and Montet, D. 2007. Review of the current methods of analytical traceability allowing determination of the origin of foodstuffs. Food Control 18: 228–235.

Renouf, V., Miot-Sertier, C., Strehaiano, P. and Lonvaud, A. 2006. The wine microbial consortium: A real terroir characteristic. Journal International des Sciences de la Vigne et du Vin 40: 209–21.

Russell, J.B. and Diez-Gonzalez, F. 1997. The effects of fermentation acids on bacterial growth. Advances in Microbial Physiology 39: 205–234.

Saltini, R., Akkerman, R. and Frosch, S. 2013. Optimizing chocolate production through traceability: A review of the influence of farming practices on cocoa bean quality. Food Control 29: 167–187.

Sanni, A.I. 1993. The need for process optimization of African fermented foods and beverages. International Journal of Food Microbiology 18: 85–95.

Schwan, R.F. and Whaels, A.E. 2004. The microbiology of cocoa fermentation and its role in chocolate quality. Critical reviews in Food Science and Nutrition 44: 205–221.

Serpaggi, V., Remize, F., Recorbet, G., Gaudot-Dumas, E., Sequeira-Le Grand, A. and Alexandre, H. 2012. Characterization of the 'viable but nonculturable' (VBNC) state in the wine spoilage yeast *Brettanomyces*. Food Microbiology 30(2): 438–47.

Sicard, D. and Legras, J.L. 2011. Bread, beer and wine: Yeast domestication in the *Saccharomyces sensu stricto* complex. Comptes Rendus Biologies 334: 229–236.

Singh, T.A., Devi, K.R., Ahmed, G. and Jeyaram, K. 2014. Microbial and endogenous origin of fibrinolytic activity in traditional fermented foods of Northeast India. Food Research International 55: 356–362.

Smid, E. and Lacroix, C. 2013. Microbe-microbe interactions in mixed culture food fermentation. Current Opinion in Biotechnology 25: 148–154.

Smit, G., Smit, B.A. and Engels, W.J. 2005. Flavor formation by lactic acid bacteria and biochemical flavor profiling of cheese products. FEMS Microbiology Reviews 29: 591–610.

Turpin, W., Humblot, C. and Guyot, J.P. 2011. Genetic screening of functional properties of lactic acid bacteria in a fermented pearl millet slurry and in the metagenome of fermented starchy foods. Applied and Environmental Microbiology 77: 8722–8734.

Ulve, V., Monnet, C., Valence, F., Fauquant, J., Falentin, H. and Lortal, S. 2008. RNA extraction from cheese for analysis of *in situ* gene expression of *Lactococcus lactis*. Journal of Applied Microbiology 105: 1327–1333.

Van Hijum, S.A., Vaughan, E.E. and Vogel, R.F. 2013. Application of state-of-art sequencing technologies to indigenous food fermentations. Current Opinion in Biotechnology 24: 178–186.

Weckx, S., van der Meulen, R., Allemeersch, J., Huys, G., Vandamme, P., van Hummelen, P. and de Vuyst, L. 2010. Community dynamics of bacteria in sourdough fermentations as revealed by their metatranscriptome. Applied and Environmental Microbiology 76: 5402–5408.

Wolkers-Rooijackers, J.C.M., Thomas, S.M. and Nout, M.J.R. 2013. Effects of sodium reduction scenarios on fermentation and quality of sauerkraut. LWT-Food Sciences and Technology 54: 383–388.

Zdolec, N., Dobranić, V., Horvatić, A. and Vučinić, S. 2013. Selection and application of autochthonous functional starter cultures in traditional Croatian fermented sausages. International Food Research Journal 20: 1–6.

Zheng, X.W., Yan, Z., Han, B.Z., Zwietering, M.H., Samson, R.A., Boekhout, T. and Nout, M.J.R. 2012. Complex microbiota of a Chinese 'Fen' liquor fermentation starter (*Fen-Daqu*), revealed by culture-dependent and culture-independent methods. Food Microbiology 31: 293–300.

18

Biotechnological Applications of Fructooligosaccharides in Food Processing Industries

Maria Antonia Pedrine Colabone Celligoi,[1,]* *Dieyssi Alves dos Santos,*[1] *Patrícia Bittencourt da Silva*[1] *and Cristiani Baldo*[1]

1. Introduction

In response to the growing demand for the natural, healthy and low-calorie food ingestion, a large number of alternative sweeteners have emerged since the early 1980s, among them the fructooligosaccharides (FOS) is the most commonly used. FOS, an inulin-type fructans, represents an important source of prebiotic compounds that are widely used as ingredient in functional foods. Several studies have demonstrated that FOS can stimulate the *Bifidobacterium* growth in the human colon, help gut absorption of calcium and magnesium and decrease the plasma levels of phospholipids, triglycerides and cholesterol. Indeed, FOS have low caloric values and anti-cariogenic properties and are useful for the formulation of diabetic products. FOS from microbial origin has attracted particular attention, since the large-scale production is not complicated and their sweet taste is very similar to that of sucrose. This chapter describes an overview of FOS as a biotechnology product, with particular emphasis on their characteristics, methods of production by *Bacillus subtilis* Natto, and the applications of this important oligosaccharide in food industry.

[1] Department of Biochemistry and Biotechnology, Centre of Exact Science, State University of Londrina, 86057-970-Londrina, Parana, Brazil.
* Corresponding author: macelligoi@uel.br

2. Fructooligosaccharides

FOS are natural compounds found in several plants as wheat, onions, bananas, artichokes, garlic, chicory, yacon root, etc. (Gibson and Roberfroid 1995). They are oligosaccharides of fructose containing a single glucose moiety in which fructosyl units are bound at the β (2→1) position of a sucrose molecule. Their molecular formula can be described as GFn, wherein **G** represents a glucose molecule, **F** the molecule of fructose and **n** the number of fructose units (Yun 1996). Figure 1 shows the most common oligosaccharides: 1-kestose (GF_2), nystose (GF_3), and 1-β-fructofuranosyl nystose (GF_4).

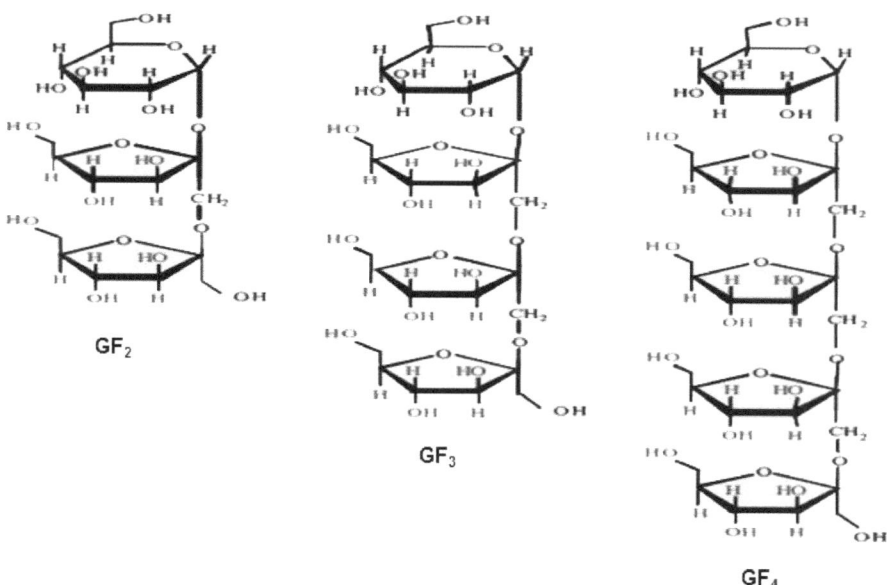

Figure 1. Chemical structure of the most common fructooligosaccharides. GF_2 (1-kestose), GF_3 (nystose), GF_4 (1-β-fructofuranosyl nystose). Fructosyl units are bound at the β (2→1) position of a sucrose molecule. (Adapted from Lima et al. 2011.)

A particular branched structure in which fructosyl units are bound at β (2→6) position of sucrose molecule can be also found in FOS. These molecules are called neo-FOS and include neo-kestose (neo-GF_2) and neo-nystose (neo-GF_3). Due to the specific chemical structure, some studies showed that neo-FOS have superior prebiotic functions, chemical and thermal stability in comparison to current commercially available FOS (Kilian et al. 2002, Lim et al. 2007).

FOS are stable at pH above 3.0 and temperatures higher than 140°C, and are highly soluble in water (about 80% at 25°C) and ethanol. Possessing approximately one-third of sucrose sweetness, FOS are an interesting option for the preparation of food in which the use of sucrose is restricted by diets. According to the available literature, FOS sweetness intensity depends on their chemical structure and degree of polymerization, being directly proportional to the chain length (Bornet 1994, Crittenden and Playne 1996).

Industrially, the FOS are mainly produced from sucrose by microbial fructosyltransferases (also called β-fructofuranosidase) from fungi as *Aureobasidium* sp., *Aspergillus* sp., *Fusarium* sp. and *Penicillium* sp. among others, and bacteria such as *Bacillus subtilis, B. macerans, Zymomonas mobilis* and *Arthrobacter* sp. (Yun 1996, Euzenat et al. 1997, Passos and Park 2003, Sangeetha et al. 2005). *Bacillus subtilis* is a good producer of FOS by virtue of expressing the greatest amounts of enzymes involved on the sucrose metabolism.

3. Fructooligosaccharides Production by *Bacillus subtilis* Natto

Bacillus species are Gram-positive, Gram-negative or Gram-variable, rod-shaped, endospore-forming, aerobic or facultative anaerobic bacteria (Claus and Berkeley 1986). The many species of this genus exhibit a wide range of physiological abilities that allow them to live in extreme environment such as deserts, hot springs and Arctic soils. In addition, they can grow under conditions of temperatures, pH and salt concentrations intolerant to other microorganisms (Harwood 1992). *Bacillus subtilis* is considered GRAS (Generally Recognized As Safe) by the FDA (Food and Drug Administration, USA). The lack of pathogenicity and failure to colonize tissues allow its use as a probiotic in foods, especially in Eastern countries and some countries of Europe (Hong et al. 2005).

Interestingly, it has been reported that *B. subtilis* has the favorable ability of maintaining microflora equilibrium in the gastrointestinal tract and increasing animal performance when administered orally in adequate quantities (Alexopoulos et al. 2004, Kritas and Morrison 2005). Additionally, several works have proposed that *B. subtilis* has an immunostimulating effect (Fiorini et al. 1985) and may promote the secretion of some beneficial vitamins, such as vitamin K (Yanagisawa and Sumi 2005, Barnes et al. 2007).

B. subtilis Natto is a member of the *B. subtilis* species used in the manufacture of Natto, the traditional food in Japan. The production of Natto is based on the fermentation of soy beans by *B. subtilis* Natto, resulting in the release of enzymes responsible for the breakdown of amino acids, thus increasing the nutritional value of this food (Hiroyuki 2000). This microorganism is a Gram-positive spore-forming bacterium (Samanya and Yamauchi 2002).

B. subtilis Natto has the enzyme levansucrase (EC 2.4.1.10), one fructosyltransferase that drives the formation of levan, a homopolymer linked β (2→6) branched β (2→1). Besides the formation of levan, levansucrase also catalyzes the formation of FOS (Abdel-Fattah et al. 2005). Three sucrose hydrolyzing enzymes are present on the sucrose metabolism of *Bacillus subtilis*: intracellular sucrase (SacA or InvA), extracellular levansucrase (SacB) and extracellular sucrase (B46 or InvB) (Sangiliyandi and Gunasekaran 2001).

Sun and coworkers (2010) studied the effects of *B. subtilis* Natto on the performance and immune function of dairy calves during the preweaning phase. The results showed that the treatment with *B. subtilis* Natto increased growth performance by improving average daily gain and feed efficiency, as well as by advancing the weaning age of the calves without any adverse effects. Indeed, the study showed that oral administration of *B. subtilis* Natto induced the secretion of serum IgG and Th1 cytokine levels, including IFN-γ, in probiotic-fed calves, which helps to activate immune systems and enhance immunity. In summary, this study supports the conception that the viable probiotic characteristics of *B. subtilis* Natto improve growth and benefit the immune function of calves (Sun et al. 2010).

4. Fermentative Parameters of the Fructooligosaccharides Production

The production of FOS by fermentative process presents great advantages when compared to the enzymatic synthesis. Conventionally, the industrial production of FOS by microorganisms includes a two-stage and a single-step process. In the first, the enzymes initially produced by microbial fermentation are used by the enzymatic reaction with substrate to produce FOS (Maiorano et al. 2008, Sangeetha et al. 2005). The second, namely whole-cell biotransformation, is based on the enzyme production and enzymatic reaction in a single fermentative process (Fernandez et al. 2004). The use of whole cells as biocatalyst avoids the purification of the FOS-producing enzyme from the cell extract, resulting in the FOS production in only one, less expensive, step (Ning et al. 2010).

The fermentative process allows the manipulation of parameters such as pH, temperature, fermentation time, agitation and concentration of sucrose, among others. Fermentations using medium containing sucrose resulted in exopolysaccharides with a variety of molecular weights (Calazans et al. 2000). Thus, using the same microorganism, it is possible to favor the production of poly or oligosaccharides only by manipulating the fermentative parameters (Coimbra 2006, Gonçalves et al. 2013). Several microorganisms are reported as FOS producers when grown in media containing sucrose as a carbon source: *Aspergillus aculeatus, Aspergillus niger, Rhodotorula* sp., *Acetobacter diazotropicus, Bacillus circulans, Bacillus subtilis* e *Bacillus subtilis* Natto (Nemukula et al. 2009, Hernalsteens and Maugeri 2010, Driouch et al. 2010, Euzenat 1997, El-Refai et al. 2009, Silva et al. 2014).

Studying the simultaneous production of FOS and levan by *Zymomonas mobilis*, Coimbra (2006) reported the temperature of 40°C as optimum for the fermentative production of FOS resulting in 32.4 g in medium containing 150 g sucrose, after 24 hr of fermentation. In the same study, it was found that the production of FOS had started in the first 12 hr and declined after 24 hr, where the production of levan had been initiated. In contrast, Shih and co-workers (2010) reported the production of 30 to 50 g/L of exopolysaccharides by *Bacillus subtilis* Natto after 21 hr of fermentation. Recently, the effects of the medium sucrose concentrations (100 to 350 g/L), pH (4.0 to 7.0), fermentation time (24 to 48 hr) and agitation (100 to 200 rpm) on the levansucrase production by *Bacillus subtilis* ATCC 6633 were also studied by Berté and co-workers (2013). The results showed that the optimum condition for levansucrase production was 300 g/L sucrose concentration, pH 6.0, 24 hr of fermentation and 180 rpm agitation, reaching 6.57 UA activity. The maximum levan production by *Bacillus subtilis* Natto was found to be 400 g/L of sucrose, 16 hr of fermentation, resulting in 111.6 g/L of levan. In these optimized conditions, the sample presented two molecular weights of levan, 568.000 Da and < 50.000 Da, corresponding to 13.39% and 86.61%, respectively (Santos et al. 2013). Interestingly, Silva and co-workers (2014) described a high production of FOS by sucrose fermentation of *B. subtilis* Natto CCT 7712, and the optimum conditions were found to be sucrose concentration of 300 g/L, pH 7.7 and agitation rate of 234 rpm, reaching the production of 98.86 n g/L after 24 hr of fermentation.

Aspergillus species have received particular attention and have been cited as good enzyme producers (Balasubramaniem et al. 2001, Chien et al. 2001). In the last few years, several *Aspergillus japonicus* strains have been reported as potentially adequate for industrial production of FOS and and β-fructofuranosidase (Chien et al. 2001, Mussatto et al. 2009).

Most investigations about experimental conditions for FOS and β-fructofuranosidase production are mainly based on submerged fermentation experiments. Nevertheless, solid-state fermentation systems appear to be an interesting alternative to obtain higher volumetric productivity and product concentration, with lower capital cost and energy consumption (Maiorano et al. 2008). Indeed, solid-state fermentation needs low water volume and thus has a large impact on the economy of the process due to smaller fermenter-size, reduced downstream processing, reduced stirring and lower sterilization costs (Hölker and Lenz 2005, Raghavarao et al. 2003). In addition, the use of low cost agricultural and agro-industrial residues as substrates also contributes to lower capital and operating costs compared to submerged fermentation (Mussatto and Teixeira 2010).

Assays under solid-state fermentation conditions were demonstrated to be an excellent alternative to improve the FOS and β-fructofuranosidase production by *A. japonicus* (Mussatto and Teixeira 2010). Coffee silverskin was selected due to its great potential to immobilize cells as well as to serve as a nutrient source for the microorganism. In another study, FOS and β-fructofuranosidase production by solid-state fermentation systems with *A. japonicus*, using coffee silverskin as solid matrix and nutrient source, was maximized by establishing the best conditions of temperature, inoculum rate, and moisture content to be used during the fermentation. The results obtained under these selected fermentation conditions were significantly higher than those previously found under no optimized conditions (Mussatto et al. 2012).

The literature also describes the large-scale FOS production using a bioreactor. Production of FOS from sucrose using calcium alginate-immobilized mycelia of *A. japonicus* and *A. niger* in an internal-loop airlift bioreactor was also reported (Lin and Lee 2008). When 60 g of the immobilized mycelia of *A. japonicus* was supplied into the reactor filled with 300 g/L sucrose solution and gas velocity 7.32 cm/s, the total FOS production was about 55% (w/w) of total sugars in the mixture after a batch reaction for 9 hr. To remove the generated glucose, an inhibitor of β-fructofuranosidase, the optimal input of the second immobilized *A. niger* mycelia particles was 315 g. With this input, the total FOS mass fraction reached up to 90% (w/w) and the initial rate of transfructosylation was increased to almost twice as that without supplying glucose oxidase (Lin and Lee 2008). Caicedo et al. (2008) showed that the use of a submerged membrane airlift reactor, agitated mechanically, produced FOS under batch, semibatch and continuous operating modes under controlled conditions. The highest FOS yields were obtained at batch operating 62.1 and 66.4% after 26 or 6 hr of reaction, respectively (Caicedo et al. 2008).

Industrial production of FOS carried out with microbial transfructosylation enzymes was found to give a maximum theoretical yield of 55–60% based on the initial sucrose concentration (Sangeetha et al. 2005). This yield cannot be further increased due to the high amounts of glucose coproduced during the fermentation, which acts as an inhibitor (Yun 1996).

In order to obtain higher fermentation yields, many authors have studied the impact of the continuous removal of glucose and residual sucrose from the medium during the FOS conversion. Consequently, by increasing the fermentation yields, a purer final product could be obtained (Dominguez et al. 2013, Nobre et al. 2013).

Recently, high-purity FOS was produced from sucrose by an innovative process incorporating immobilized *A. japonicus* and *Pichia heimii* cells in a tanks-in-series bioreactor for the continuous production of FOS. A solution composed of 1 g L^{-1} yeast extract and 300 g L^{-1} sucrose was fed continuously to the bioreactor at a dilution rate of 0.1 hr^{-1}. The results indicated the production of FOS of up to 98.2% purity. Remarkably, this immobilized dual-cell system was effective for 10 days in producing high-purity FOS (Sheu et al. 2013).

5. Effect of Salts in the Production of Fructooligosaccharides

The addition of mineral salts to the fermentation media may influence the biosynthesis of oligosaccharides and polysaccharides (Vigants et al. 1998, Bekers et al. 2000, Ammar et al. 2002, Arundhati et al. 2011). The increase of biosynthesis of FOS in the presence of high salt concentrations was observed during sucrose fermentation by *Z. mobilis* 113S. According to the study, a 0.6 M NaCl concentration led to an increase of oligosaccharide productivity by 3.5-fold (Bekers et al. 2000). However, when the sucrose concentration was increased to 65%, the salt showed an inhibitory effect on the FOS production. The authors proposed that the increase of FOS production after salt addition could be due to the osmoprotection mechanism developed by the microorganism for regulating the osmolarity of the medium.

Interestingly, a stimulatory effect of sodium and potassium salts on levan production was also observed (Vigants et al. 1996). In addition, the activation of levansucrase-catalyzed levan formation by NaCl, KCl and Na$_2$SO$_4$ was observed using *Z. mobilis* cell-free extract. The authors suggested that levansucrase has the properties of allosteric enzyme and the sodium and potassium salts are its heterotropic activators (Vigants et al. 1998). Ammar et al. (2002) characterized the thermostable levansucrase from *Bacillus* sp. TH4-2 and tested the effect of metal ions on its activity. The results showed that Cu^{2+}, Zn^{2+} and Hg^{2+} exhibited a negative effect. However, the enzymatic activity was increased approximately 4-fold in the presence of Fe^{2+}. The authors reported that this is the first study to show levansucrase activation by Fe^{2+}.

Studying the extracellular levansucrase enzyme produced by the nitrogen-fixer *Acetobacter nitrogenifigens* strain RG1(T), Arundhati and co-workers (2011) verified that the enzyme showed enhanced hydrolytic and polymerization activity in the presence of mercuric ions. The authors proposed that Hg^{2+} ions stabilize the levansucrase enzyme structure into a conformation more favorable for performing enzymatic activity in comparison to the structure present in absence of Hg^{2+} ions.

6. Production of Fructooligosaccharides using Alternative Substrates

The concerns of reducing environmental pollution have encouraged the use of industrial waste and industrial by-products through bioprocess. Besides reduction of environmental impact, the utilization of such substrates could contribute to the decrease the cost of production of enzymes (Bicas 2010). In this sense, various agricultural by-products such as cereal bran, corn products, sugarcane bagasse, cassava bagasse and by-products of coffee and tea processing were used on the production of fructosyltransferase by *Aspergillus oryzae* CFR 202. The results indicated that rice bran and wheat bran were good substrates for enzyme production. Among the various corn products used, corn germ supported maximum fructosyltransferase production, whereas among the by-products of coffee and tea processing used, spent coffee and spent tea were good substrates when supplemented with yeast extract and complete synthetic media. Maximum FOS production was obtained with fructosyltransferase after 8 hr of reaction with 60% sucrose (Sangeetha et al. 2003).

Interestingly, an increase in the FOS yield and productivity by solid-state fermentation with *Aspergillus japonicus* using corn cobs, coffee silverskin, and cork oak as support and nutrient source was reported (Mussato and Teixeira 2010). Among the tested agro-industrial residues, coffee silverskin provided the most interesting results, with a FOS production similar in both supplemented and non-supplemented media. The elevated FOS production (128.7 g/L) and β-fructofuranosidase activity (71.3 U/ml), suggest solid-state fermentation of coffee silverskin as a promising strategy to synthesize both products by industries (Mussato and Teixeira 2010). In other study, the same research group increased the production of FOS under submerged fermentation, using coffee silverskin as solid matrix moistened upto 60%, 70% and 80% with 240 g/L of sucrose. The moisture content did not influence FOS productivity, but temperature at 26–30°C and inoculum rate of 2×10^7 spores/g dry matter increased the yield of FOS to 208 g/L with a productivity of 10.44 g/L/hr and β-fructofuranosidase activity to 64.12 U/ml with productivity of 4.0 U/ml/h. This study was remarkable from the industrial viewpoint of increasing both FOS and β-fructofuranosidase in a simultaneous manner (Mussatto et al. 2012).

Molasses is referred to as a good source of nutrients for the production of both enzymes and microorganisms for fermentation and direct production of compounds of interest such as FOS. In this sense, 166 g/L of FOS was produced from 360 g/L of sugar molasses as sucrose equivalent at 55°C and pH 5.5, after a 24 hr incubation by *Aureobasidium pullulans* cells (Shin 2004). The elevated FOS production was also obtained using medium containing sucrose and molasses by *A. pullulans* LBJBS03 and *Penicillium* sp. LBJBS02 (Silva et al. 2014). Interestingly, sugar syrup and molasses from beet were tested as low-cost substrates for the enzymatic synthesis of FOS. After 30 hr, the FOS concentration reached a maximum of 388 mg/mL using syrup. Using the molasses, 235 mg/mL of FOS was obtained in 65 hr of fermentation. These values corresponded to approximately 56% (syrup) and 49% (molasses) of the total amount of carbohydrates in the mixture (Ghazi et al. 2006).

Recently, our research group described the ability of *B. subtilis* Natto CCT 7712 to produce high amounts of nystose using low-cost substrates broadly available in Brazil, such as commercial sucrose, sugarcane molasses and sugarcane juice. The optimization resulted in the maximum production of 179,77 g/L of nystose, corresponding to 7,49 g/L/hr of productivity and yield of 71,73%, in medium containing 400 g/L of commercial sucrose and 0,8 g/L of $MnSO_4$. The fermentations using sugarcane molasses and sugarcane juice also resulted in satisfactory production, reaching 97,93 g/L and 42,58 g/L of nystose, respectively. The elevated nystose production, obtained using sugarcane derivatives, suggests submerged fermentation with *B. subtillis* Natto as a promising strategy to synthesize nystose at the industrial level (unpublished results).

Taken together, these studies showed the potential for the use of agricultural and industrial by-products and wastes in the efficient production of FOS, thereby resulting in value addition of such by-products or industrial residues.

7. Functions of Fructooligosaccharides and Potential Health Benefits

The consumption of functional foods has increased considerably in recent decades, partly due to the increased prevalence of diseases associated with low protein ingestion and high consumption of carbohydrates and fats (Escoda 2002). Therefore, a rapid increase in demand for high nutritional value food, composed of substances that have the potential to benefit the functioning of the body and to prevent disease—the functional foods—has been observed. A functional food affects beneficially, different functions of the body, improving the stage of health and/or reducing the risk of disease. The concept of functional foods was first introduced in Japan, in 1991, and various oligosaccharides were classified as 'health promoting foods' (Patel and Goyal 2011). Functional foods include fiber, polyunsaturated fatty acids, phyto-chemicals, active peptides (arginine and glutamine), probiotics (*Lactobacillus acidophilus*, *Lb. casei*, *Lb. bulgárico* and *Lb. lactis*) and prebiotics (inulin and FOS).

The prebiotics was first defined by Gibson and Roberfroid (1995) and later updated by Gibson and co-workers (2004). A prebiotic could be defined as a selectively fermented ingredient that permits precise modifications, both in the composition and/or activity, in the gastrointestinal microflora that confer benefits upon host wellbeing and health. In 2008, the prebiotics market earned 295.5 million euros and was estimated to reach 766.9 million euros by 2015, indicating the extraordinary market growth for the prebiotics foods (Morris and Morris 2012).

Together, inulin and FOS are considered as the prebiotics model. Due to the complexity of the molecular configuration, FOS are resistant to the hydrolytic action of salivary and intestinal enzymes, thus reaching the colon intact. Once in the colon, FOS could be used as a substrate for native and beneficial bacteria, promoting the metabolism and contributing to the reduction of the intestinal pH. Thus, it is correct to affirm that prebiotics have an indirect effect on the functioning of the body, promoting changes in the composition of intestinal microflora, conferring innumerous benefits to the organism. In addition, the reduction of the number of pathogenic bacteria as *Escherichia coli*, *Clostridium*, *Streptococcus* and *Proteus faecallis*, and a decrease of toxic metabolites (ammonia, indole, nitrosamines and phenols) was also reported (Yun 1996, Gibson and Roberfroid 1995). The prebiotic effect of FOS also is related to the reduction of blood pressure, decrease of carbohydrate and lipid absorption, glycemic control, increase in the bioavailability of minerals and nutrients, increase of immune-stimulating compound production, antitumor activity, reduction of toxins and carcinogenic compound production and restoration of normal intestinal flora during antibiotic therapy (Park 2003, Gibson and Roberfroid 1995).

According to current literature, chain length is a factor that influences the physiological effect of FOS. In this way, Bornet and co-workers (2002) showed that short-chain FOS induced a more potent effect on the growth of *Bifidobacterium* than the long chain FOS. In addition, the short-chain FOS increased magnesium absorption in humans, and reduced colon tumor progression by enhancing the colon butyrate concentrations and the local immune system effectors in animal models (Bornet et al. 2002).

Citric pectin and FOS extracted from inulin were added to a carrot and orange juice and the mixture was administrated to hamsters. After 30 days of application of the juice with the addition of 3% pectin and 15% FOS, a reduction in the total cholesterol levels was detected. The blood cholesterol levels were reduced by 31% and 25% due to the action of the pectin and FOS, respectively. However, no synergistic effect between the prebiotics was observed. Indeed, the number of *Bifidobacterium* in the faeces of the animals increased 10 times due to the effect of FOS (Freitas et al. 2005). Tuohy et al. (2001) tested the prebiotic potential of partially hydrolyzed guar gum and FOS in a biscuit in human volunteers, and observed a significantly increased in number of *Bifidobacterium* after ingestion of the experimental biscuits. Búrigo et al. (2007) also verified the bifidogenic effect of FOS in patients with hematologic neoplastic disorders and concluded that supplementation with FOS had a positive effect by promoting a significant increase in *Bifidobacterium* numbers. The effects of the dietary supplementation of 5% of FOS may decrease hepatic steatosis and the risk of non-alcoholic fatty liver disease related to insulin resistance without changes in blood lipids and glucose levels. The authors observed a reduction of 12% in the liver lipid levels, and the reduction of blood insulin in obese rats (Kaume et al. 2011).

In rats, a prebiotic effect could be responsible for the anticancer effects in the colon of animals, resulting in the proliferation of *Bifidobacterium* (with the major metabolites, lactate or acetate), as well as other bacteria (with the metabolites, butyrate or propionate and acetate) (Pool-Zobel et al. 2002). To study the impact of inulin on apoptosis as a mechanism for the anticancer properties of prebiotics, young male rats were fed a diet containing either inulin (5%) or oligofructose (5%) or the basal diet for 3 weeks. The results clearly showed an increase of apoptotic cells, which means that the prebiotic fed rats more efficiently eliminated colonic cells with defective DNA (Hughes and Rowland 2001).

The prebiotic effect of the fructotrisaccharide named neokestose on intestinal bacteria was also investigated. *Bifidobacterium* sp. utilized neokestose and produced more biomass from neokestose than facultative anaerobes under anaerobic conditions in batch culture. *Lb. salivarius* utilized glucose but negligible amounts of neokestose were made use of *Lb. salivarius* and the facultative anaerobes produced significantly more biomass from glucose than from neokestose, whereas the biomass yields obtained with bifidobacteria on neokestose and glucose, respectively, were not significantly different. Static batch cultures inoculated with faeces supported the prebiotic effect of neokestose, which had been observed in the pure culture investigations. Bifidobacteria and lactobacilli were increased while potentially detrimental coliforms, clostridia and bacteroides, decreased after 24 hr fermentation with neokestose. In addition, this effect was more pronounced with neokestose than with a commercial prebiotic fructooligosaccharide. It was concluded that neokestose has potential as a bifidogenic substance and that it might have advantages over the commercially available sources currently used (Kilian et al. 2002).

Besides FOS, there are other types of oligosaccharides which are reported to be bifidogenic (Crittenden et al. 1996). Bifidogenic factors are defined as carbohydrate-bearing materials that survive direct metabolism by the host and are metabolised by bifidobacteria (Modler 1994, Vazquez et al. 2000). There are currently 12 classes of food-grade oligosaccharides in commercial production (Crittenden et al. 1996). Xylooligosaccharides are sugar oligomers composed of xylose monomers with only $(1,4)$-β-d-xylopyranosyl linkages that are broadly found in bamboo shoots, vegetables, fruits and honey (Crittenden et al. 1996, Modler 1994). Isomaltooligosaccharides are commercially produced from starch by the action of debranching enzymes such as pullulanase and isoamylase (Crittenden et al. 1996, Prapulla et al. 2000). Consumption of isomaltooligosaccharides has been found to improve colonic conditions by dropping the levels of intestinal putrefactive bacteria, such as *Clostridia perfringens* and members of the family Enterobacteriaceae (Prapulla et al. 2000). Xylooligosaccharides are used in pharmaceutical formulations, feed formulations for agricultural purposes, and in foods for human consumption (Vazquez et al. 2000). They can resist acid and thermal degradation and, consequently, have found extensive usage in low pH fruit juices and carbonated prebiotic drinks (Modler 1994, Prapulla et al. 2000). Galactooligosaccharides do not occur naturally but are synthesized from lactose using the galactosyltransferase activity of β-galactosidase. Different types of galactooligosaccharides are produced commercially, such as isogalatobiose, galsucrose and lactosucrose (Crittenden et al. 1996, Prapulla et al. 2000). They are used as prebiotic food components and are also bifidogenic since they enhance the growth of bifidobacteria in the colon (Prapulla et al. 2000).

8. Potential Foods Applications of Fructooligosaccharides

Due to their multi-functional nature, prebiotics may be used in a wide variety of applications, particularly in the food industry. The nutraceutical properties of FOS are related to their functional, physiological and physical aspects. The FOS taste profile is similar to that of sucrose, without any cooling effect, but with a 30% lower sweetness. In addition, the water-retention capacity is higher than that of sucrose (Bornet 1994, Yun 1996). Considering the characteristics and functionality, FOS possess a broad spectrum of applications in the food industry, being used as a nutritional enrichment and to improve the rheological properties of foods. In this sense, FOS could be used as an ingredient in ice creams, dairy desserts, yogurts and in products with the specifications 'low sugar' and 'no added sugar' in the formulations. They also could be used in bakery products, generating products with a reduction of sugar content (Passos and Park 2003).

The addition of pectin and FOS in a mixture of orange and carrot juice resulted in the increase of soluble solids and viscosity of the beverage. In this study, it was observed that FOS did not cause a negative effect on the sensory acceptability of the juice, even at high concentrations (Freitas et al. 2004). The physicochemical, microbiological characteristics and acceptance of a soy yogurt produced by *Lb. bulgaricus* and *Streptococcus thermophillus*, and supplemented with FOS, were evaluated (Hauly et al. 2005). The prebiotics maintained the viability of the lactic acid bacteria at high levels, up to the 28th day of storage. The final pH value of the supplemented yogurt was 4.63, the acidity 0.37%, and the acceptance 71.20%. The supplemented yogurt showed greater viscosity, cohesiveness and adhesiveness than the non-supplemented product. The authors concluded that FOSs are ingredients that maintain the viability of lactic acid bacteria in soy yogurt, giving rise to a product with good acceptability (Hauly et al. 2005).

The effect of supplementation of FOS and inulin on bread was already studied by some authors. Increased crumb firmness was described for bread made with immature wheat meal rich in FOS (Mujoo and Ng 2003). Inulin-containing breads were designated (Mandala et al. 2009) as having an elastic crumb, soft crust and relative low specific volume. In addition, the color of the inulin/FOS enriched products was found to be darker. This may be the result of accelerated baking (Mandala et al. 2009, Peressini and Sensidoni 2009, Poinot et al. 2010). Recently, the effect of supplementation of FOS and inulin on the rheological and sensory properties of bread was studied by Morris and Morris (2012) in order to observe the viability and acceptability of these products. According to the authors, the effect of inulin/FOS substitution on the textural and sensory properties depended on the type of prebiotic, flour type, substitution level, the degree of polymerization and how the prebiotic was introduced. The central inulin/FOS impacts described were: lower bread loaf volumes, increased crumb hardness and darker crust. While inulin appears to integrate well to the gluten network, it also dilutes it, resulting in lower gas retention ability. The authors suggested a supplementation of 5% inulin; this appears to be achievable and should contribute 0.7–1.2 g of inulin per slice of bread towards daily intake (Morris and Morris 2012). According to Morris and Morris (2012), the literature lacks information about the acceptance of bread enriched with FOS and inulin. The sensory results reported reflected the instrumental findings, and hedonic ratings tended to decrease with increasing inulin/FOS contents, presumably due to smaller loaf volumes, harder crumbs and darker colors. However, there is little information available on enriched bread's taste and aroma as opposed to texture, which is a parameter easier to estimate instrumentally in a way that relates to consumer perception. In this sense, further studies are still necessary in order to clarify the effect of supplementation of bread with FOS and inulin.

The ingestion of 10 g of FOS per day per person is considered safe, avoiding possible side-effects such as flatulence and intestinal discomfort (Passos and Park

2003). Studies have shown that supplementation of the diet of chickens, pigs and rats with oligofructose and other non-digestible oligosaccharide reduced the density of faecal *Salmonella* (Letllier et al. 2000). Mice infected systemically with virulent strains of *Listeria monocytogenes* and *Salmonella* Typhimurium, after being fed a diet with inulin and oligofructose (at 100 g/kg), had lower mortality than mice fed a diet with cellulose as the source of fiber (Buddington et al. 2002). Interestingly, inulin provided greater resistance than oligofructose to the systemic infections, for the *Listeria* challenge. It is important to point out that, besides the nutritional benefits, inulin and FOS also could contribute to enhance the shelf life of various food products (Izzo and Niness 2001). Formulations of cereal bars containing FOS and high carbohydrate content, low concentrations of protein and fat were also developed for consumption at breakfast (Passos and Park 2003).

Recently, our group described the ability of *Lactobacillus* ssp. strains to utilize FOS as a sole energy source (manuscript in preparation), and results showed that FOS was as good as glucose to provide energy source. The highest prebiotic activity score was obtained for *Lb. plantarum* ATCC 14917 grown on FOS (0.526), and the lowest score was for *Lb. paracasei* ATCC 27092 (–0.051). In conclusion, these results suggest the FOS utilization was equally as good a substrate as glucose in supporting the growth of lactic acid bacteria. The most promising result was obtained for *Lb. plantarum* ATCC 14917, indicating that its combination with FOS produced, by *Bacillus subtilis* Natto CCT 7712, could be a viable probiotic-based functional food approach in administering the beneficial bacteria *in vivo*. In addition, these results may be useful in identifying combinations of probiotics and prebiotics that could be added to dairy and other foods.

A drink composed of FOS was developed in order to examine the effect of FOS supplementation on body weight and satiety of obese adults. The results showed that a daily supplement of FOS for 12 weeks in drinks before meals resulted in significant body weight loss, decreased ghrelin and increased peptide YY (PYY) (Parnell and Reimer 2009). These molecules are involved in glucose metabolism. Glucagon-like peptide 1 (GLP-1) has several antiobesity and antidiabetic actions, including inhibition of food intake, delayed gastric emptying, weight loss and stimulation of insulin secretion, among others (Drucker 2006). Colocalized with GLP-1 in the intestinal L cells is PYY, also considered an anorexigenic peptide (Grudell and Camilleri 2007, Batterham et al. 2002). In human obesity, plasma concentrations of GLP-1 (Ranganath et al. 1996) and PYY (Batterham et al. 2002) are reduced, and this impaired secretion may promote the development of obesity or hinder weight loss, or both. Conversely, ghrelin stimulates food intake and weight gain, as well as promotes adiposity (Wren et al. 2001, Tschop et al. 2000). Of the gut satiety hormones, those responsive to diet composition, including GLP-1, PYY, and ghrelin (Neary et al. 2003), are promising targets for weight management through diet modification. The authors concluded that FOS supplementation has the potential to promote weight loss and improve glucose regulation in overweight adults (Parnell and Reimer 2009).

In addition, Hondo and co-workers (2000) investigated the possibility of producing vinegar from yacon that contains natural FOS. There is also the possibility of supplementation of infant foods with FOS and galactooligosaccharides, in order to facilitate intestinal transit of newborns (Moro et al. 2002). In a 4-way cross-over design,

21 subjects ate meal replacement bars enriched in FOS and/or beta-glucans over a 2 day period. Three intakes of 8 g of FOS did not have an impact on either food intake during ad libitum lunches or self-reported hunger ratings acquired over those 2 days. The authors suggested that longer treatments may be required to observe an effect (Peters et al. 2009). More recently, in a randomized double-blind cross-over study, 20 subjects received 2–5 g or 8 g of FOS for breakfast and a snack and recorded their perceived satiety. The women who had ingested 2–8 g of FOS saw their calorie intake during the remaining of the day (food diary) decrease. The opposite was observed for men: their calorie intake was greater after consuming the inulin supplemented drinks and snacks (Hess et al. 2011).

9. Conclusion

The daily intake of foods containing FOS is proven to be beneficial to human health, due to their prebiotic effect. Furthermore, several studies suggest that this oligosaccharide exhibits therapeutic properties, reducing the serum cholesterol levels and blood pressure and exerts a positive effect on the treatment of diseases such as cancer and diabetes. *Bacillus subtilis* is well-known as a good FOS producer by expressing the great amounts of enzymes involved in sucrose metabolism. Thus, the development of new biotechnological processes proposing alternatives to FOS production as the use of waste and industrial by-products are essential to increase the productivity and to reduce the production cost of FOS.

Acknowledgment

The authors thank Coordenação de Aperfeiçoamento de Pessoal de Nível Superior (CAPES), Brazil and Fundação Araucária, Brazil for financial support.

Keywords: *Bacillus subtilis* Natto, fructooligosaccharides, inulin, solid state fermentation, transfructosylation enzymes

References

Abdel-Fattah, A.F., Mahmoud, A.R. and Esawy, M.A.T. 2005. Production of levansucrase from *Bacillus subtilis* NRC 33a and enzymic synthesis of levan and fructo-oligosaccharides. Current Microbiology 51: 402–407.

Alexopoulos, C., Georgoulakis, I.E., Tzivara, A., Kyriakis, C.S., Govaris, A. and Kyriakis, S.C. 2004. Field evaluation of the effect of a probiotic-containing *Bacillus licheniformis* and *Bacillus subtilis* spores on the health status, performance, and carcass quality of grower and finisher pigs. Journal of Veterinary Medicine. A Physiology, Pathology, Clinical Medicine 51: 306–312.

Alles, M.S., Hartemink, R., Meyboom, S., Harryvan, J.L., Van Laere, K.M.J., Nagengast, F.M. and Hautvast, J.G.A.J. 1999. Effect of transgalactooligosaccharides on the composition of the human intestinal microflora and on putative risk markers for colon cancer. The American Journal of Clinical Nutrition 69: 980–991.

Ammar, Y.B., Matsubara, T., Ito, I., Iizuka, M., Limpaseni, T., Pongsawasdi, P. and Minamiura, N. 2002. Characterization of a thermostable levansucrase from *Bacillus* sp. TH4-2 capable of producing high molecular weight levan at high temperature. Journal of Biotechnology 99: 111–119.

Arundhati, P., Neeloy, S., Debasree, D., Abhishek, B., Somnath, C., Writachit, C. and Ratan, G. 2011. Mercuric Ion Stabilizes Levansucrase Secreted by *Acetobacter nitrogenifigens* strain RG1. The Protein Journal 30: 262–272.

Balasubramaniem, A.K., Nagarajan, K.V. and Paramasamy, G. 2001. Optimization of media for beta-fructofuranosidase production by *Aspergillus niger* in submerged and solid state fermentation. Process Biochemistry 36: 1241–1247.

Barnes, A.G.C., Cerovic, V., Hobson, P.S. and Klavinskis, L.S. 2007. *Bacillus subtilis* spores: A novel microparticle adjuvant which can instruct a balanced Th1 and Th2 immune response to specific antigen. European Journal of Immunology 37: 1538–1547.

Batterham, R.L., Cowley, M.A., Small, C.J., Herzog, H., Cohen, M.A., Dakin, C.L., Wren, A.M., Brynes, A.E., Low, M.J., Ghatei, M.A., Cone, R.D. and Bloom, S.R. 2002. Gut hormone PYY (3-36) physiologically inhibits food intake. Nature 418: 650–654.

Bekers, M., Vigants, A., Laukevics, J., Toma, M., Rapoports, A. and Zikmanis, P. 2000. The effect of osmo-induced stress on product formation by *Zymomonas mobilis* on sucrose. International Journal of Food Microbiology 55: 147–150.

Berté, S.D., Borsato, D., Silva, P.B., Vignoli, J.A. and Celligoi, M.A.P.C. 2013. Statistical optimization of levansucrase production from *Bacillus subtilis* ATCC 6633 using response surface methodology. African Journal of Microbiology Research 17: 898–904.

Bicas, J.L., Silva, J.C., Dionísio, A.P. and Pastore, G.M. 2010. Produção biotecnológica de bioaromas e açúcares funcionais. Ciência e Tecnologia de Alimentos 30: 7–18.

Bornet, F.R. 1994. Undigestible sugars in food products. The American Journal of Clinical Nutrition 59: 763–769.

Bornet, F.R., Brounsl, J.F., Tashiro, Y. and Duvillier, V. 2002. Nutritional aspects of short-chain fructooligosaccharides: Natural occurrence, chemistry, physiology and health implications. Digestive and Liver Disease 2: 111–120.

Buddington, K.K., Donahoo, J.B. and Buddington, R.K. 2002. Dietary oligofructose and inulin provide mice with protection against enteric and systemic pathogens and tumor inducers. Journal of Nutrition 132: 80–87.

Búrigo, T., Fagundes, R.L.M., Trindade, E.B.S.M. and Vasconcelos, H.C.F.F. 2007. Efeito bifidogênico do frutooligossacarídeo na microbiota intestinal de pacientes com neoplasia hematológica. Revista de Nutrição 20: 491–497.

Caicedo, L., Silva, E. and Sáncheza, O. 2009. Semi-batch and continuous fructooligosaccharides production by *Aspergillus* sp. N74 in a mechanically agitated airlift reactor. Journal of Chemical Technology and Biotechnology 84: 650–656.

Calazans, G.M.T., Lima, R.C., França, F.P. and Lopes, C.E. 2000. Molecular weight and antitumour activity of *Zymomonas mobilis* levans. International Journal of Biological Macromolecules 27: 245–247.

Chien, C.S., Lee, W.C. and Lin, T.J. 2001. Immobilization of *Aspergillus japonicus* by entrapping cells in gluten for production of fructooligosaccharides. Enzyme and Microbial Technology 29: 252–257.

Claus, D., and Berkeley, R.C.W. 1986. Genus *Bacillus* Cohn 1872, 174[AL]. pp. 1105–1139. *In:* P.H.A. Sneath, N.S. Mair, M.E. Sharpe and J.G. Holt (eds.). Bergey's Manual of Systematic Bacteriology. Williams & Wilkins, Baltimore.

Coimbra, C.G.O. 2006. Produção de fructo-oligossacarídeos e aspectos da biosseparação das frações leves de levana hidrolisada. M.S. Thesis, Universidade Federal de Pernambuco, Recife.

Crittenden, R.G. and Playne, M.J. 1996. Production, properties and applications of food-grade oligosaccharides. Trends in Food Science & Technology 7: 353–361.

Dominguez, A.L., Rodrigues, L.R., Lima, N.M. and Teixeira, J.A. 2013. An overview of the recent developments on fructooligosaccharide production and applications. Food and Bioprocess Technology 7: 324–337.

Driouch, H., Roth, A., Dersch, P. and Wittmann, C. 2010. Optimized bioprocess for production of fructofuranosidase by recombinant *Aspergillus niger*. Applied Microbiology Biotechnology 87: 2011–2024.

Drucker, D.J. 2006. The biology of incretin hormones. Cell Metabolism 3: 153–65.

El-Refai, H.A., Abdel-Fattah, A.F. and Mostafa, F.A. 2009. Enzymatic synthesis of levan and fructooligosaccharides by *Bacillus circulans* and improvement of levansucrase stability by carbohydrate coupling. World Journal Microbiology Biotechnology 25: 821–827.

Escoda, M.S.Q. 2002. Para a crítica da transição nutricional. Ciência & Saúde Coletiva 7: 219–226.

Euzenat, O., Guibert, A. and Combert, D. 1997. Production of the fructooligosaccharides by levansucrase from *Bacillus subtilis* C4. Process Biochemistry 32: 237–243.

Fernandez, R.C., Maresma, B.G., Juarez, A. and Martinez, J. 2004. Production of fructooligosaccharides by beta-fructofuranosidase from *Aspergillus* sp. 27H. Journal of Chemical Technology and Biotechnology 79: 268–272.

Fiorini, G., Cimminiello, C., Chianese, R., Visconti, G.P., Cova, D., Uberti, T. and Gibelli, A. 1985. *Bacillus subtilis* selectively stimulates the synthesis of membrane bound and secreted IgA. Chemioterapia 4: 310–312.

Freitas, D.G.C. and Jackix, M.N.H. 2004. Caracterização físico-química e aceitação sensorial de bebida funcional adicionada de frutoligossacarídeo e fibra solúvel. Boletim do Centro de Pesquisa de Processamento de Alimentos 22: 325–374.

Freitas, D.G.C. and Jackix, M.N.H. 2005. Efeito de Bebida Adicionada de Frutoligossacarídeo e Pectina no Nível de Colesterol e Estimulação de Bifidobactérias em Hamsters Hipercolesterolêmicos. Brazilian Journal of Food Technology 8: 81–86.

Ghazi, I., Fernandez-Arrojo, L., Gomez De Segura, A., Alcalde, M., Plou, F.J. and Ballesteros, A. 2006. Beet sugar syrup and molasses as low-cost feedstock for the enzymatic production of fructooligosaccharides. Journal of Agricultural and Food Chemistry 19: 2964–2968.

Gibson, G.R. and Roberfroid, M. 1995. Dietary modulation of the human colonic microbiota: Introduction to the concept of prebiotics. American Journal of Nutrition 125: 1401–1412.

Gibson, G.R., Probert, H.M., Loo, J.V., Rastall, R.A. and Roberfroid, M. 2004. Dietary modulation of the human colonic microbiota: Updating the concept of prebiotics. Nutrition Research Reviews 17: 259–275.

Gonçalves, B.C.M., Mantovan, J., Ribeiro, M.L.L., Borsato, D. and Celligoi, M.A.P.C. 2013. Optimization production of thermo active levansucrase from *Bacillus subtilis* Natto CCT 7712. Journal of Applied Biology & Biotechnology 1: 1–8.

Grudell, A.B. and Camilleri, M. 2007. The role of peptide YY in integrative gut physiology and potential role in obesity. Current Opinion in Endocrinology, Diabetes and Obesity 14: 52–57.

Harwood, C.R. 1992. *Bacillus subtilis* and its relatives: Molecular biological and industrial workhorses. Trends in Biotechnology 10: 247–256.

Hauly, M.C.O., Fuchs, R.H.B. and Prudencio-Ferreira, S.H. 2005. Suplementação de iogurte de soja com frutooligossacarídeos: características probióticas e aceitabilidade. Revista de Nutrição. 18: 613–622.

Hernalsteens, S. and Maugeri, F. 2010. Synthesis of fructooligosaccharides using extracellular enzymes from *Rhodotorula* sp. Journal of Food Biochemistry 34: 520–534.

Hess, J.R., Birkett, A.M., Thomas, W. and Slavin, J.L. 2011. Effects of short-chain fructooligosaccharides on satiety responses in healthy men and women. Appetite 56: 128–134.

Hiroyuki, S. 2000. Determination and properties of Fibrinolysis Accelerating Substance (FAS) in Japanese fermented soybean 'Natto'. Journal of the Agricultural Chemical Society of Japan 74: 1259–1264.

Hölker, U. and Lenz, J. 2005. Solid-state fermentation: Are there any biotechnological advantages? Current Opinion in Microbiology 8: 301–306.

Hondo, M., Okumura, Y. and Yamaki, T. 2000. A preparation of yacon vinegar containing natural fructooligosaccharides. Journal of the Japanese Society for Food Science and Technology 47: 803–807.

Hong, H.A., Duc, L.H. and Cutting, S.M. 2005. The use of bacterial spore formers as probiotics. FEMS Microbiology Reviews 29: 813–835.

Hughes, R. and Rowland, I.R. 2001. Stimulation of apoptosis by two prebiotic chicory fructans in the rat colon. Carcinogenesis 22: 43–47.

Izzo, M. and Niness, K. 2001. Formulating nutrition bars with inulin and oligofructose. Cereal Foods World 46: 102–106.

Kaume, L., Gilbert, W., Gadang, V. and Devareddy, L. 2011. Dietary supplementation of fructooligosaccharides reduces hepatic steatosis associated with insulin resistance in obese zucker rats. Functional Foods in Heals and Disease 5: 199–213.

Kilian, S., Kritzinger, S., Rycroft, C., Gibson, G. and Du Preez, J. 2002. The effects of the novel bifidogenic trisaccharide, neokestose, on the human colonic microbiota. World Journal of Microbiology and Biotechnology 18: 637–644.

Kritas, S.K. and Morrison, R.B. 2005. Evaluation of probiotics as a substitute for antibiotics in a large pig nursery. Veterinary Record 156: 447–448.

Letllier, A., Messier, S., Lessard, L. and Quessy, S. 2000. Assessment of various treatments to reduce carriage of *Salmonella* in swine. Canadian Journal of Veterinary Research 64: 27–31.

Lim, J.S., Lee, J.H., Kang, S.W., Park, S.W. and Kim, S.W. 2007. Studies on production and physical properties of neo-FOS produced by co-immobilized *Penicillium citrinum* and neo-fructosyltransferase. European Food Research and Technology 225: 457–462.

Lima, D.M., Fernandes, P., Sampaio Nascimento, D., Figueiredo Ribeiro, R.C.L. and De Assis, S.A. 2011. Fructose syrup: A biotechnology asset. Food Technology & Biotechnology 49: 424–434.

Lin, T.J. and Lee, Y.C. 2008. High-content fructooligosaccharides production using two immobilized microorganisms in an internal-loop airlift bioreactor. Journal of the Chinese Institute of Chemical Engineers 39: 211–217.

Maiorano, A.E., Piccoli, R.M., Da Silva, E.S. and De Andrade Rodrigues, M.F. 2008. Microbial production of fructosyltransferases for synthesis of prebiotics. Biotechnology Letters 30: 1867–1877.

Mandala, I., Polaki, A. and Yanniotis, S. 2009. Influence of frozen storage on bread enriched with different ingredients. Journal of Food Engineering 92: 137–145.

Modler, H.W. 1994. Bifidogenic factors-sources, metabolism and applications. International Dairy Journal 4: 383–407.

Moro, G., Minoli, I., Mosca, M., Fanaro, S., Jelinek, J., Stahl, B. and Boehm, G. 2002. Dosage-related effects of galacto- and fructooligosaccharides in formula-fed term infants. Journal of Pediatric Gastroenterology and Nutrition 34: 291–295.

Morris, C. and Morris, G.A. 2012. The effect of inulin and fructooligosaccharide supplementation on the textural, rheological and sensory properties of bread and their role in weight management: A review. Food Chemistry 133: 237–248.

Mujoo, R. and Ng, P.K.W. 2003. Physicochemical properties of bread baked from flour blended with immature wheat meal rich in fructooligosaccharides. Journal of Food Science 68: 2448–2452.

Mussatto, S.I. and Teixeira, J.A. 2010. Increase in the fructooligosaccharides yield and productivity by solid-state fermentation with *Aspergillus japonicus* using agro-industrial residues as support and nutrient source. Biochemical Engineering Journal 53: 154–157.

Mussatto, S.I., Ballesteros, L.F., Martins, S., Maltos, D.A.F., Aguilar, C.N. and Teixeira, J.A. 2012. Maximization of fructooligosaccharides and β-fructofuranosidase production by *Aspergillus japonicus* under solid-state fermentation conditions. Food and Bioprocess Technology 6: 2128–2134.

Mussatto, S.I., Rodrigues, L.R. and Teixeira, J.A. 2009. Fructofuranosidase production by repeated batch fermentation with immobilized *Aspergillus japonicus*. Journal of Industrial Microbiology and Biotechnology 36: 923–928.

Neary, N.M., Small, C.J. and Bloom, S.R. 2003. Gut and mind. Gut 52: 918–21.

Nemukula, A., Mutanda, T., Wilhelmi, B.S. and Whiteley, C.G. 2009. Response surface methodology: Synthesis of short chain fructooligosaccharides with a fructosyltransferase from *Aspergillus aculeatus*. Bioresource Technology 100: 2040–2045.

Ning, Y., Wang, J., Chen, J., Yang, N., Jin, Z. and XU, X. 2010. Production of neo-fructooligosaccharides using free-whole-cell biotransformation by *Xanthophyllomyces dendrorhous*. Bioresource Technology 101: 7472–7478.

Nobre, C., Teixeira, J.A. and Rodrigues, L.R. 2013. New trends and technological challenges in the industrial production and purification of fructooligosaccharides. Critical Reviews in Food Science and Nutrition doi:10.1080/10408398.2012.697082.

Park, H.E., Park, N.H., Kim, M.J., Lee, T.H., Lee, H.G., Yang, J.Y. and Cha, J. 2003. Enzymatic synthesis of fructosyl oligosaccharides by levansucrase from *Microbacterium laevaniformans* ATCC 15953. Enzyme and Microbial Technology 32: 820–827.

Parnell, J.A. and Reimer, R.A. 2009. Weight loss during oligofructose supplementation is associated with decreased ghrelin and increased peptide YY in overweight and obese adults. American Journal of Clinical Nutrition 89: 1751–1754.

Passos, L.M.L. and Park, Y.K. 2003. Frutoligossacarídeos: Implicações na saúde humana e utilização em alimentos. Ciência Rural 33: 385–390.

Patel, S. and Goyal, A. 2011. Functional oligosaccharides: Production, properties and applications. World Journal Microbiology and Biotechnology 27: 1119–1128.

Peressini, D. and Sensidoni, A. 2009. Effect of soluble dietary fibre addition on rheological and breadmaking properties of wheat doughs. Journal of Cereal Science 49: 190–201.

Peters, H.P.F., Boers, H.M., Haddeman, E., Melnikov, S.M. and Qvyjt, F. 2009. No effect of added beta-glucan or of fructooligosaccharide on appetite or energy intake. American Journal of Clinical Nutrition 89: 58–63.

Poinot, P., Arvisenet, G., Grua-Priol, J., Fillonneau, C., Le-Bail, A. and Prost, C. 2010. Influence of inulin on bread: Kinetics and physico-chemical indicators of the formation of volatile compounds during baking. Food Chemistry 119: 1474–1484.

Pool-Zobel, B., Loo, J., Rowland, I. and Robefroid, M.B. 2002. Experimental evidences on the potential of prebiotic fructans to reduce the risk of colon cancer. British Journal of Nutrition 87: 273–281.

Prapulla, S.G., Subhaprada, V. and Karanth, N.G. 2000. Microbial production of oligosaccharides: A review. Advances in Applied Microbiology 47: 243–299.

Raghavarao, K.S.M.S., Ranganathan, T.V. and Karanth, N.G. 2003. Some engineering aspects of solid-state fermentation. Biochemical Engineering Journal 13: 127–135.

Ranganath, L.R., Beety, J.M., Morgan, L.M., Wright, J.W., Howland, R. and Marks, V. 1996. Attenuated GLP-1 secretion in obesity: Cause or consequence. Gut 38: 916–919.

Samanya, M. and Yamauchi, K. 2002. Histological alterations of intestinal villi in chickens fed dried *Bacillus subtilis* var. natto. Comparative Biochemistry and Physiology Part A: Molecular & Integrative Physiology 133: 95–104.

Sangeetha, P.T., Ramesh, M.N. and Prapulla, S.G. 2005. Recent trends in the microbial production, analysis and application of fructooligosaccharides. Trends in Food Science & Technology 16: 442–457.

Sangiliyandi, G. and Gunasekaran, P. 2001. Polymerase and hydrolase activities of *Zymomonas mobilis* levansucrase separately modulated by *in vitro* mutagenesis and elevated temperature. Process Biochemistry 36: 543–548.

Santos, L.F., Melo, F.C.B.C., Paiva, W.J.M., Borsato, D., Silva, M.L.C. and Celligoi, M.A.P.C. 2013. Characterization and optimization of levan production by *Bacillus subtilis* Natto. Romanian Biotechnological Letters 18: 8413–8422.

Sheu, D.C., Chang, J.Y., Wang, C.Y., Wu, C.T. and Huang, C.J. 2013. Continuous production of high-purity fructooligosaccharides and ethanol by immobilized *Aspergillus japonicus* and *Pichia heimii*. Bioprocess and Biosystems Engineering 36: 1745–1751.

Shih, I.L., Chen, L.D. and Wu, J.W. 2010. Levan production using *Bacilus subtilis* Natto cells immobilized on alginate. Carbohydrates Polymers 82: 111–117.

Shin, H.T. 2004. Production of fructo-oligosaccharides from molasses by *Aureobasidium pullulans* cells. Bioresource Technology 93: 52–62.

Silva, B.P., Borsato, D. and Celligoi, M.A.P.C. 2014. Optimization of high production of fructooligosaccharides by sucrose fermentation of *Bacillus subtilis* Natto CCT 7712. American Journal of Food Technology 9: 144–150.

Silva, J.B. 2008. Seleção de microrganismos osmofílicos isolados de favo-de-mel para produção de frutooligossacarídeos por fermentação. M.S. Thesis, Universidade Estadual de Campinas, Campinas, São Paulo.

Sun, P., Wang, J.Q. and Zhang, H.T. 2010. Effects of *Bacillus subtilis* Natto on performance and immune function of preweaning calves. Journal of Dairy Science 93: 5851–5855.

Tschop, M., Smiley, D.L. and Heiman, M.L. 2000. Ghrelin induces adiposity in rodents. Nature 407: 908–913.

Tuohy, K.M., Kolida, S., Lustenberger, A.M. and Gibson, G.R. 2001. The prebiotic effects of biscuits containing partially hydrolysed guar gum and fructooligosaccharides: A human volunteer study. British Journal of Nutrition 86: 341–348.

Vazquez, M.J., Alonso, J.L., Dominguez, H. and Parajo, J.C. 2000. Xylooligosaccharides: Manufacture and applications. Trends in Food Science & Technology 87: 387–393.

Vigants, A., Kruce, R., Bekers, M. and Zikmanis, P. 1998. Response of *Zymomonas mobilis* levansucrase activity to sodium chloride addition. Biotechnology Letters 20: 1017–1019.

Vigants, A., Zikmanis, P. and Bekers, M. 1996. Sucrose medium osmolality as a regulator of anabolic and catabolic parameters in *Zymomonas* culture. Acta Biotechnologica 16: 321–327.

Wren, A.M., Seal, L.J., Cohen, M.A., Brynes, A.E., Frost, G.S., Murphy, K.G., Dhillo, W.S., Ghatei, M.A. and Bloom, S.R. 2001. Ghrelin enhances appetite and increases food intake in humans. The Journal of Clinical Endocrinology and Metabolism 86: 5992–5995.

Yanagisawa, Y. and Sumi, H. 2005. Natto *Bacillus* contains a large amount of water-soluble vitamin K. Journal of Food Biochemistry 29: 267–277.

Yun, J. 1996. Fructooligosaccharides: Occurrence, preparation, and application. Enzyme Microbiology Technology 19: 107–117.

19

Food Waste Generation and Bio-valorization

Ioannis S. Arvanitoyannis,[1,*] *Konstantinos V. Kotsanopoulos*[1]
and *Aristidis D. Alexopoulos*[1]

1. Introduction

It is commonly accepted that the annual generation of high amounts of waste significantly enhances the already serious environmental problem. 'Traditional' waste treatment methods, such as waste disposal and incineration, further enhance the emissions of toxic pollutants and greenhouse gases, while seepage from waste disposal sites contributes to the pollution of ground water and water courses. Specifically, the two most important hazards for humans and animals are attributed to the heavy metal and solvent content of the wastes. Additionally, another issue arising due to the disposal of wastes is their accumulation in the food chain and the subsequent deterioration of environmental conditions (Bruvoll and Ibenholt 1997).

Conventional forecasting of solid waste generation is usually based on both demographic and socio-economic parameters in a per-capita basis. Nevertheless, the per-capita coefficients can also be considered either fixed over time or susceptible to alterations over time (Chang and Lin 1997). Municipal waste is a traditional domain in waste management and, as a result, is the type of waste for which most reliable data is available, as regards both its quantity and quality. However, there are still gaps that prevent the establishment of a detailed picture of the existing situation for the whole of Europe (Jordan and Heidorn 2003). It is estimated that the production of municipal solid waste (MSW) in the European region is around 500 kg per capita per year for Western Countries and 350 kg per capita per annum for Central and Eastern Countries. Out of this figure, 30% corresponds to bio-waste (Battistoni et al. 2007).

---
[1] University of Thessaly, School of Agricultural Sciences, Department of Agriculture Ichthyology and Aquatic Environment, Fytokou Str., Nea Ionia Magnessias, 38446 Volos, Hellas, Greece.
* Corresponding author: parmenion@uth.gr

Nowadays, policies that support sustainability in waste management have not been accompanied by corresponding efforts to increase sufficiently the knowledge regarding waste generation. The current waste management system could be described by the following steps:

a) land-filling
b) energy recovery
c) material recycling to waste minimization

The above steps can increase data complexity, and thus more 'in depth' information is required on waste generation and its ensuing composition (Beigl et al. 2008).

Waste generation in the European Union is about 1.3 billion tons per year, approximately 3.5 tones per capita per year. The most significant sectors affecting the quantities of waste generated every year are mining and quarrying (377 million tons), the construction sector (286 million tons), municipal solid waste (182 million tons) and hazardous waste (27 million tons). It should be mentioned, however, that waste generation within the EU is proportional to economic growth (Darlington et al. 2009).

Biological treatment is considered to be one of the most advantageous and promising methods for the maximization of recycling and recovery of its components. Anaerobic digestion of sorted organic fraction of municipal solid wastes, such as food wastes, is the best alternative and the most cost-effective technology (Forster-Carneiro et al. 2008).

The rapid development of livestock farming over the last years has resulted in increased production of animal liquid wastes. These wastes could potentially cause major problems in relation to soil (phosphate super-saturation), surface waters (eutrophication), ground water (nitrate leaching due to over-fertilization) and odor nuisance (Rulkens et al. 1998, Pappu et al. 2007).

One of the most significant environmental issues resulting from the modern consumption-oriented lifestyle is food waste. This type of waste has a high organic content that can be readily reused in the form of organic resources for animal feed or compost. As an example, one may refer to the case of South Korea (48 million people), where the daily produced food waste reached 11397 tons in 2002, standing for 22.8% of total household wastes. As a result, food waste could be used as an alternative carbon source for effective nutrient removal in Biological Nutrient Removal (BNR) processes (Chae and Shin 2007).

Effective operation of solid waste management systems usually depends on accurately predicting the solid waste generation. Conventional prediction models are frequently employed based on demographic and socio-economic factors in a per-capita basis. Time series data of solid waste generation is mainly in the form of observations made over a number of years at the same location. The analysis time series data allows the identification of trends embedded in solid waste generation over time, and can develop hypotheses regarding policy changes or the continuation of these trends into the future (Chang and Lin 1997). In the study of Chang and Lin (1997), time series intervention modeling was used to evaluate recycling impacts on solid waste generation. It was proved that even if the dynamic impacts had not been taken into consideration, the statistical formation of those models inherently restricts the applicability to situations in which underlying relationships did not change with time. Better results can only be given if a full description of solid waste generation is incorporated into the model formulation. Econometric forecasting (a potential alternative to statistical models) can be directly applied, relying on the assumption that future forecasts are based on current forecasts, solely of the independent variables themselves.

A summary of the characteristics of treated waste waters is given in Table 1.

Table 1. Characteristics of treated waste waters.

Soluble phosphate	pH	Alkalinity (mg/l)	CO_2 (mg/l)	Ammonia (mg/l)	BOD_5 (mg/l)	Nitrite (mg/l)	Nitrate (mg/l)	References
0.09	7.2–7.7	51	4.4	0.03	2	0.18	-	Ruane et al. 1977
-	-	192	11	0.002	3	0.030	14	Davidson et al. 2009
0.32	-	-	-	27.3	-	0	3.2–5.8	Chae and Shin 2007

Waste valorization practices have been intensively examined and evaluated in recent years with the aim of managing waste more sustainably. Food waste could potentially function as a significant source of various valuable chemicals. This potential has triggered the development and promotion of sustainable food waste valorization practices to various end products, mainly using green chemical technologies (Luque and Clark 2013). The generation of food waste covers the whole life cycle of foods; from agriculture to industrial manufacturing and processing, retail and household consumption. It is estimated that in developed countries, 42% of food waste is generated by households, while 39% is produced by the food industry, 14% in the food service sector and the remaining 5% is produced by the retail and distribution industries. Industrial ecology concepts increasingly attempt to consider and develop methods targeting at a 'zero waste economy' in which waste is used as raw material for the production of new, valuable products. The huge quantities of waste generated by the food industry are not only a substantial loss of valuable materials but also cause serious management problems, of both economic and environmental nature. Many of these residues, however, could potentially be reused in other production applications (Mirabella et al. 2014).

2. Waste Generation and Composition in the Food Chain

2.1 Solid Waste

Rapid urbanization, combined with steady population growth and changes in lifestyles in developing countries have significantly affected the augmentation of the per capita municipal waste generation. According to Chang and Lin (1997), "keeping pace with the requirements of rapid economic development and continuing population growth, and because of its critical role in protecting the environment and public health, accomplishing effective and efficient municipal solid waste (MSW) management should be a priority for cities in developing countries. Inappropriate waste handling, storage, collection and disposal practices pose environmental and public health risks. In densely populated urban centers, for example, appropriate and safe MSW management is of utmost importance to create a healthy environment for the people."

Despite the fact that developed countries use far more sophisticated and effective systems of packaging and distribution, they usually have a higher per capita MSW generation than developing countries. Another aspect usually not taken into account, is the relationship between packaging residues and food residues in MSW (Alter 1989).

Wastes can be categorized as biodegradable (i.e., decomposable wastes) and non-biodegradable (i.e., non-decomposable wastes). Based on their origin, they could also be categorized as (i) municipal (residential and commercial), (ii) industrial, and (iii) construction and demolition wastes. Inappropriate waste disposal threatens the developing world and the phenomenon is particularly visible in the streets cities daily littered with wastes. The campaign targeting at waste segregation prior to disposal also seems to be losing its initial impetus in a high number of developing countries. The inappropriate disposal of organic solid waste, which can have several negative environmental and health consequences such as environmental pollution or disease spreading (Ayomoh et al. 2008), is one example.

Consumption increase and the inevitable increase in waste generation often stem from the erroneous and selfish perception that the environment is solely destined for human use and provision, thereby leading to natural resource depletion. Although humans understand that the consequences of their actions and their behavior can change, they have further aggravated the problem through the adoption of irresponsible behavior, and overuse of natural resources (Márquez et al. 2008).

Food waste is mainly composed of carbohydrates, which is one of the most significant organic wastes produced by modern society. Since food waste is rich in energy, it would be ideal to reach dual benefits from the point of view of energy recovery and waste stabilization. Specifically, hydrogen production from food waste can significantly enhance the economic feasibility of waste treatment. Hydrogen production from organic waste is frequently accompanied by the generation of organic acids which can act as substrates for the production of methane. Methane production is considered a suitable unit of post-treatment for hydrogen production. Recently, a two-stage process combining acidogenic hydrogenesis and methanogenesis has been suggested as a promising approach (Chu et al. 2008).

Hydrogen is considered to be a promising alternative to fossil fuels since it is clean, renewable and characterized by high-energy yield. Biological hydrogen production through anaerobic fermentation is an environmentally friendly and energy saving process. Anaerobic acidification of organic wastes can lead to the production of numerous organic acids, H_2, CO_2, and other intermediate products. The reactions related to hydrogen production are rapid and can be performed without the need of solar radiation, and are thus suitable for treating large amounts of organic wastes (Shin and Youn 2005).

Food decomposition is performed when food is left for a long period of time at ambient temperature. The causal agent of the decomposition is the presence of indigenous microorganisms, such as Lactic acid bacteria (LAB), propionic acid bacteria (PAB), fungi, and coliforms. As a result, when food waste was used as a substrate for continuous H_2 production, alkali pre-treatment at pH 12.5–13.0 was employed to eliminate non-H_2-producing bacteria. It was therefore suggested that the role of food waste in the process could be dual, acting both as a substrate and as a source of H_2-producing microflora, when a required pretreatment can be applied (Kim et al. 2006).

According to Durham and Hourigan (2007), "dairy processing effluent generally exhibits the following properties: high organic load due to the presence of milk components; fluctuations in pH due to the presence of alkaline and acidic cleaning agents and other chemicals; high levels of nitrogen and phosphorus and fluctuations in temperature. Dairy wastes comprise grease, sugars, nitrogen, phosphorus, acidic and alkaline cleaning chemicals; and they also have high biological oxygen demand (BOD_5) and chemical oxygen demand (COD). The composition of the waste reflects the dairy product being processed and it is possible to identify the source of the waste problem."

Sugarcane (Southern Hemisphere) and cereal grain (Northern Hemisphere) are currently used as feedstock for bioethanol production, but there is a strong competition related with their use in the food sector. Although these are the main sources of feedstock currently used, projected fuel needs necessitate the evaluation of novel alternatives, and low cost feedstock aiming at minimizing ethanol production costs (Del Campo et al. 2006).

The excessive use of resources and, as a consequence, the increased waste emissions have resulted in the evaluation of the power generation system by various environmental assessments, among which the most important is life cycle assessment (LCA) certified with ISO 14040. The quantification of the environmental impact of a material, product or service can be done only if the total resources used and waste emissions (liquid, solid and gas) are taken into consideration for the whole production process (Jiang et al. 2009).

2.2 Liquid Waste

The European food-processing industry annually generates high volumes of aqueous wastes. The wastes include: fruit and vegetable residues and discarded items, molasses, and bagasse from sugar refining, bones, flesh, and blood from meat and fish processing, stillage and other residues from wineries, distilleries and breweries, dairy wastes such as cheese whey, and wastewaters from washing, blanching and cooling processes. Some of these wastes contain low amounts of suspended solids and low concentrations of dissolved materials. Apart from the environmental challenges arising, such streams represent tremendous amounts of materials and energy, which are potentially reusable (Kosseva 2009).

Wastewaters coming from food-processing operations vary significantly depending on the class or type of the food process. However the common characteristics shared by waste waters allow the application of biological treatment processes for food-processing wastewaters prior to their release to the environment or recycling and reusing. Existing, and new environmental laws are implemented for the protection of the aquatic environment from effluent discharges. The most recent regulations and approaches emphasize on watershed-based water quality initiatives and ecological risk assessment. The two of the most important approaches considered the most promising are waste minimization and waste reuse (Stover and Eckhoff 2003).

Solids levels in wastewater can be categorized into dissolved solids and suspended solids, with the second type being the primary concern since they are objectionable on several grounds. Settle-able solids may initiate reduction of the wastewater body, thus affecting the bottom dwelling floras and the food chain. When they float, they may have significant effects on the aquatic life due to the reduction of the amount of light entering the water (Tay et al. 2006).

According to He and his co-workers (2000), "disposal of solid waste and treatment of wastewater was reported to produce substantial amounts of methane as well as carbon dioxide. Even though aerobic conditions are applied, anaerobic microsites may still exist inside the waste particles. Nitrous oxide emissions from various waste treatment facilities, especially wastewater treatment, incineration, and manure composting, were also frequently documented over the last years. However, there is still uncertainty regarding the mechanism of how these gases are produced, especially N_2O. The contribution of waste management facilities to the global greenhouse gas budget still remains a puzzle."

Moreover, even in extremely low levels, such as concentrations of 0.002 mg/L, phenols can highly affect the organoleptic properties of water, leading to operational problems in breweries, distilleries and in mineral water bottling. In the aquatic environment, it has been shown that these substances can cause the death of fish and other aquatic species, due to alterations of the ecological equilibrium (Britto et al. 2008).

Zaher et al. (2009) presented a representative example of a biogas plant, which is based on the co-digestion of pig manure and industrial wastewater in Blaabjerg (Denmark) that underwent a serious accident caused by an overdose of industrial waste. This event significantly reduced the bio-gas production and methane content. More than three months were needed for the recovery of the process and the biogas had to be flared in the following months instead of being used for power generation. It was proved that the total operational loss was approximately one million DKK (Denmark Kroner). The development of a simulation tool which could be used for optimizing and assessing the co-digestion of any combination of solid waste streams, was reported in this study. Different model nodes based on the ADM1 (Anaerobic Digestion Model No. 1) were applied on the Matlab-Simulink simulation platform. Transformer model nodes were developed for the generation of detailed input for ADM1, estimating the particulate waste fractions of carbohydrates, proteins, lipids and inerts. Hydrolysis nodes were also developed for each waste stream. The model gave reliable results in terms of calibration and optimization of the case study examined. The optimization simulated 200,000 days of virtual experimental time in 8 hr and determined the feedstock ratio and retention time to set the digester operation for maximum biogas production rate. This example clearly illustrated the advantages derived from the developed model as an optimization and decision support tool in addition to its potential application for the integrated assessment and LCA (life cycle assessment) of anaerobic digestion (AD) applications for waste stabilization and power generation.

2.3 Odors

Among the variety of sources of odor existing in intensive livestock production facilities, feedlot pads, holding ponds containing storm water runoff from feedlot facilities, compost windrows employed for solid waste treatment derived from feedlot pads or treatment ponds and anaerobic treatment ponds used to store and treat waste derived from piggeries are the most important. Inadequate management of these sources can easily lead to odor complaints. Measurement of odor emission levels derived from particular odor sources is of high importance due to the fact that excessive odor emissions and subsequent odor complaints could initiate regulatory action (i.e., direction to cease odor emission or undertake remedial action—objective measure of odor emission rates allows comparison over time and space), and also because of the potential calculation of buffer substances through application of typical odor emission rates from specific sources as inputs to dispersion models (Hudson et al. 2009).

According to Koe and Tan (1990), "wastewater treatment facilities have been frequently regarded as major sources of odor nuisance and a variety of odorous volatiles are known to be emitted during the handling and processing of wastewater at these facilities. The various process units at a typical treatment plant such as sedimentation tanks, activated sludge aeration chambers, sludge storage and digestion facilities handle wastewaters of varying biological and chemical properties. Therefore, the potential for odor generation at these process units is expected to be different. It is thereby possible for certain unit processes to enhance the potential for odor generation of the liquid and sludge streams at a wastewater treatment plant. These units demand special consideration in terms of their design and application of suitable odor control removal strategies."

In seafood-processing industries, odor is mainly released through organic matter decomposition. Volatile amines and diamines are commonly released and, occasionally, ammonia. In septic wastewater, the odor of hydrogen sulfide can also be developed. Odor is crucial when public perception and acceptance of any wastewater treatment plant is considered. Although it does not pose significant hazards, it may considerably affect the public life by causing stress and sickness (Tay et al. 2006).

3. Problems and Waste Management

3.1 Solid Waste

Many of the concerns related with municipal solid waste (MSW) are of social and political nature. Addressing them would necessitate the consideration of how society uses and disposes off materials and products. Therefore, the entire lifecycle of materials that end up as MSW—starting with original resource acquisition (whether extraction of 'virgin' materials or purchase of 'secondary' materials) through manufacturing, distribution and marketing, consumption, and residuals management should be considered. This would allow further exploration of opportunities for the reduction of the toxicity and amount of MSW before its generation (a concept termed 'source reduction'), as well as for the augmentation of the recyclability and subsequent recycling of materials in whatever MSW is generated (Levenson 1993).

The waste management process aims at improving resource conservation and promoting environmental protection. MSW management is a considerably neglected factor of environmental management (EM) in all low and most middle-income countries. Non-adequately managed waste streams can lead to major environmental impact and, as a consequence, to health hazards. Environmental concerns are currently adding huge importance to the MSW decision-making process. The adoption of effective waste management strategies can significantly decrease the burden placed on the environment. Based on the assumption that the waste management system relies on sound data and is well executed with public awareness, it can lead to minimization of emissions and reduction of resource depletion (Batool and Chuadhry 2009).

According to Park et al. (2002), numerous methods used for treating food wastes include land filling, incineration, recycling as compost or animal feed and complete decomposition. The land filling method was initially systematically followed and commonly employed by many communities due to the fact that it is very simple and cost-effective. Incineration is not applied as frequently, mainly due to its high operating cost.

Landfilling is one of the most common approaches applied for MSW disposal. MSW is composed of different organic and inorganic materials such as food, vegetables, paper, wood, plastics, glass, metal and other inert materials. In cities, it is collected by the respective municipalities and transported to appropriate disposal areas. The unsanitary conditions applied during disposal of waste can cause adverse health and environmental issues. The inadequately maintained landfill sites can cause groundwater contamination because of leachate percolation. Moreover, the generation of unpleasant odors and risks of explosion of methane gas accumulating at the landfill site should also be considered. The landfill gas consists of 50–60 volume % methane and 30–40 volume % carbon dioxide, as well as various chemical substances such as aromatics, chlorinated organic compounds and sulfur compounds (Mor et al. 2006).

It is commonly accepted by the whole technical community that direct landfilling of biodegradable residues needs to be avoided. In the European Union, this need has been recognized since the promulgation of the Council Directive 1999/31/EC (26 April 1999) on waste landfilling which, as part of the measures undertaken for improving the sustainability of waste management, forced member states to limit the levels of biodegradable materials contained in municipal solid waste (MSW) destined to sanitary landfill.

Although promoting waste minimization and applying recycling habits are crucial areas of paramount importance for establishing an effective MSW management strategy, even when applied to their full potential, the residual waste (RMSW) may include significant levels of biodegradable matter and should be treated prior to landfilling (De Gioannis et al. 2009).

The disposal of biodegradable solid wastes such as food wastes and activated sludge, alone or combined with other solid wastes, can lead to the development of serious environmental and economic problems all around the world (Lim et al. 2008).

According to El-Fadel et al. (1997), "the most economic form of disposal is land filling in the vast majority of cases. Therefore, landfills will continue to be the most attractive disposal route for solid waste. Indeed, approximately up to 95% of solid waste generated worldwide is currently disposed of in landfills. Alternative solutions to landfilling consist of volume reduction processes because they produce waste fractions (e.g., ashes and slag from combustion processes representing the second leading method of waste disposal), which ultimately must be landfilled. Resorting to landfills is not restricted to the management of MSW, but it comprises most other industrial wastes as well. For instance, nearly 80% of hazardous wastes generated in the U.S. are dumped in landfills."

It is proved that numerous factors can affect the amount of leachate generated, the decomposition, stabilization and extraction of pollutants from the waste matrix, including but not limited to waste composition, degree of compaction, and absorption capacity of the waste. Usually, leachate is strongly contaminated with organic contaminants determined as chemical oxygen demand (COD) and biochemical oxygen demand (BOD), and halogenated hydrocarbons with ammonia. Soils have been seriously contaminated with heavy metals (Pb, Cu, Zn, Mn, Cr, Cd) because of leachate migration. The fact that heavy metals are non-degradable is of high importance since it leads to serious environmental problems (Mohan and Gandhimathi 2009).

According to Jha et al. (2008), "anaerobic decomposition of municipal solid waste (MSW) in landfills generates about 60% methane (CH_4) and 40% carbon dioxide (CO_2) together with other trace gases. This percentage differs spatially due to waste composition, age, quantity, moisture content and ratio of hydrogen/oxygen availability at the time of decomposition (e.g., fat, hemicellulose, etc.)."

Due to the fact that thalogenated hydrocarbons, and specifically polychlorinated dibenzo-*p*-dioxins (PCDDs) and polychlorinated dibenzofurans (PCDFs), are commonly found in the flue gases and fly ash of municipal solid waste incinerators (MSWIs), PCDD and PCDF emissions from different sources pose significant risks globally because of their toxicological effects and associated adverse health implications. The PCDD/Fs released to the environment are mainly derived from anthropogenic activities, such as waste incineration, energy generation, high temperature sources, and numerous other chemical-industrial sources. These pollutants diffuse through the atmosphere and are finally deposited onto soil and vegetation (Wang et al. 2008).

In the study of Chang and Hsu (2008), a method was developed for the prediction of important parameters (composting time, highest temperature, final and lowest pH values, cumulative CO_2 evolution, and percentages of material losses) characterizing the composting process of synthetic food waste in terms of weight fractions of protein and fat. Quadratic equations obtained by multi-regression analysis were also developed to enable rapid estimations. The models were applied to real waste and were found to be very accurate. It was shown that each temperature or carbon dioxide peak was related to the decomposition of a material or the decomposition of a combination of different materials. The CO_2 evolution peaks occurred a few hours earlier than temperature peaks. The water extractable C/N ratio fluctuated with temperature and the CO_2 evolution curves. At the end of composting process, the pH was alkaline, the temperature fell to the ambient, and the CO_2 evolution decreased to one tenth of its peak value (Chang and Hsu 2008).

3.2 Liquid Waste

One of the key environmental problems the food industry is currently facing is the treatment of high volume wastewater streams containing nitrogen, phosphorus and organic substances. These nutrients end up in surface water, leading to eutrophication by promoting excessive growth of algae. Phosphorus is found in high levels in effluents from the food industry, especially in the dairy industry. Traditionally, the conduction of phosphate removal from wastewater streams was performed through chemical precipitation (Mulkerrins et al. 2004).

Pollution of fresh water streams by treated wastewater poses significant risks to both human health and the environment. Traditional mechanical and chemical treatments are very expensive and have no direct monetary return to even partially make up for the investment and operational costs. This imposes higher financial burdens on the municipalities and central governments. Furthermore, as the available quantities of fresh water are reduced, the need to use the treated wastewater for mitigating such shortages becomes inevitable (Darwish et al. 2007).

Waste treatment techniques are widely used for altering the physicochemical or biological properties of the waste, for reducing its volume and/or toxicity, and for rendering safe, the waste disposal. Waste treatment may be required for radioactive, hazardous and other Department of Energy (DOE) wastes (Arvanitoyannis and Tserkezou 2008).

Anaerobic treatment of slaughter house wastewater is much more commonly used than the treatment of solid slaughter house waste. A low number of full-scale plants operate in slaughter houses in the Netherlands, Belgium, and New Zealand, and anaerobic lagoons or covered anaerobic ponds are used for the treatment of slaughter house wastewater both, in warm climates, and where land cost is relatively low (e.g., Australia and New Zealand) (Salminem and Rintala 2002).

Treated effluent can be employed for irrigation under controlled conditions for the minimization of hazards from pathogenic and toxic contaminants of agricultural products, soils, surface, and ground water. Moreover, the use of treated wastewater for irrigation could eliminate problems such as wastewater disposal and lack of water availability in arid zones. Application of treated wastewater in irrigation is related with some health issues due to the potential presence of pathogens, such as coliform bacteria and due to the fact that contaminated water or wastewater could be the causal agent of several diseases and illnesses (Al-Lahham et al. 2003).

The treatment of wastewater derived from the seafood industry can only be effectively performed if the important constituents in the waste stream are known. This wastewater contains a high level of insoluble suspended particles, and their removal from the waste stream can be performed either chemically or physically. Optimal waste removal can be achieved when primary treatment is applied. While designing a treatment system, it is of high importance to consider the fact that the separation and removal of the solids should be performed as rapidly as possible. It was reported that the longer the detention time between waste generation and solids removal, the higher the soluble BOD_5. The basic primary treatment processes used involve screening, sedimentation, flow equalization, and dissolved air flotation. These unit operations can usually remove up to 85% of total suspended solids, and 65% of the BOD_5 and COD present in the wastewater (Tay et al. 2006).

The disposal of this waste is regulated by guidelines and standards (international or national), which aim at the evaluation of the quality of industrial wastewater and the protection of aquatic ecosystems. Anaerobic treatment can be used for the conversion of the organic pollutants (COD, BOD_5) in wastewater into a small amount of sludge and a much larger amount of biogas (methane and carbon dioxide), but leaves some pollution unresolved (Casa et al. 2003, Chowdhury et al. 2010).

According to Wu (1995), "the environmental impact of fish-farming depends greatly on species, culture method, hydrography of culture site, feed type and husbandry practices. Organic matter settled on the seabed may lead to development of anoxic and reducing conditions in the sediment and the production of toxic gases (e.g., NH_3, CH_4, and H_2S). Sediment oxygen demand (SOD) of bottom sediments enriched by fish-farming activities can be two to five times higher than that of the controls. As regards total sediment metabolism, it has been recorded up to ten times higher than the corresponding control. An azoic zone was typically found underneath the cages and a decrease in benthic diversity occurred in the vicinity of the farm".

The evaluation of food packaging's impact on the environment should always take into consideration the advantages of reducing food waste throughout the supply chain. A major issue is the significant food wastage reported in many countries, ranging from 25% for food grain to 50% for fruits and vegetables. In effective ways of preserving/protecting, storing and transporting of goods stand for the most important causes of food waste. Packaging limits total waste through the extension of the shelf life of foods. As a result, packaging could effectively reduce the total solid waste (Marsh and Bugusu 2007).

4. Waste Treatment Methods

4.1 Solid Waste

Over the past decades both industrial and economic growth took place at the expense of the environment. This trend led governments to establish legislation that strictly defined the limits of the resources that could be used. Food manufacturers and retailers produce and manage a vast variety of wastes, some of which, such as wastewater and packaging waste are omni-present in all industrial fields (Darligton et al. 2009).

Numerous methods and technologies aiming at waste treatment have been implemented, including solidification, extraction and incineration. Solidification is based on mixing cement, fly ash from the incinerator and waste filter cake from the physic-chemical treatment forming a virtually insoluble mass (Arvanitoyannis and Tserkezou 2008).

Although composting has been traditionally applied as a waste management technique, it has not been fully comprehended. The raw materials used in the process represent a vast range of organic wastes such as MSW, sewage sludge, yard and green wastes, animal manures, and grape marc. Composting is a biological treatment, which employs aerobic microorganisms that use organic matter (OM) as a substrate. The final product (compost) consists of stable OM, water, minerals, and ash. Characterization of the composition of the dissolved organic matter (DOM), the active OM fraction, could potentially better indicate the overall conversion of the OM than tests of the solid phase (Chefertz et al. 1998).

In developing countries, inventory estimates of CH_4 from landfills were highly uncertain due to the lack of adequate data related to both management and emissions. MSW sent to the landfill passes through a variety of intermediate stages including sorting of recyclable and compostable materials. This may lead to modifications in the quantity and properties of waste that reaches the landfill sites, thus affecting GHG (Green House Gas) emissions. It has been suggested that measurements of GHG emissions from landfills are highly important for the reduction of any uncertainties in the inventory estimates. It is anticipated that the emission sources will become more complex in the future and the contribution of GHG emissions from India will augment further if current practices are to prevail, especially in large cities (Jha et al. 2008).

When waste is deposed in a landfill, aerobic decomposition is promoted, during which biodegradable organic materials react rapidly with oxygen, forming carbon dioxide, water, and other by-products (e.g., bacterial cells). Oxygen depletion within the landfill initiates the anaerobic decomposition phase. Although the initial aerobic phase is quite short, the anaerobic phase is the dominant phase and the most important one from the perspective of gas formation (El-Fadel et al. 1997).

The three steps the anaerobic digestion process consists of are: hydrolysis, acidogenesis and methanogenesis. The methanogenic reaction was considered to be the rate-limiting step in the overall process and, therefore, various studies have focused on improving the methanogenesis rate. The degradation rate of waste activated sludge (WAS) is limited and the hydrolysis reaction is considered to be the rate-limiting step in the overall anaerobic digestion process (Heo et al. 2003).

Owing to unsightliness and potential groundwater pollution, landfilling of MSW gradually gave way to incineration in various countries. Environmental contamination caused by particulate and gaseous emissions including heavy metals, and other potentially carcinogenic organic compounds [e.g., polychlorinated dibenzodioxins (PCDD) and polychlorinated dibenzofurans (PCDF), polycyclic aromatics (PCA)], from such incinerators, as well as adequate ash disposal, is of great concern (Lisk 1988).

In general, anaerobic digestion is better than composting and landfilling in terms of energy use, emissions of greenhouse gases and total weight results. Nevertheless, the ranking between digestion and incineration depends on impact categories and applied scenarios. Similarly, the ranking between composting and landfilling varies, depending on impact categories and working scenarios. Composting could act as a good alternative provided there is minimization of transport distances. Large-scale composting is, therefore, of rather limited interest, while home composting or small scale composting requiring only limited transportation could probably be a viable solution (Finnveden et al. 2000).

Food waste composting generates greenhouse gases out of which CO_2 is the most abundant. A slightly higher concentration of greenhouse gases is emitted during the incineration case. Landfilling produces the highest quantities of greenhouse gases, mainly occurring in the form of methane emissions (Finnveden et al. 2000).

The concentration of solids is of high importance for hydrogen production from an engineering point of view, since hydrogen production under high solids condition is endowed with several advantages. Noike and Mizuno (2000) reported that hydrogen production was successfully linked out at a high solids concentration of over 9.0%. This could indicate that hydrogen recovery can be performed using the anaerobic microflora from non-sterilized organic wastes. Nevertheless, hydrogen production potential of rice bran and wheat bran dropped as the solids concentration was enhanced.

Food waste can potentially release high quantities of methane (96–424 ml/dry g), depending on the type of food used. The digestion can be performed rapidly, turning it into a good substrate for anaerobic digestion. Optimization of methane generation from anaerobic systems was correlated with digester design and operation, although it has been proved that the feedstock is as important as the digester technology and sequential batch operation (Dearman and Bentham 2007).

Chemolithotrophic ammonia-oxidizing bacteria (AOB) are responsible for the first, rate-limiting step in nitrification, and are of high importance for natural nitrogen cycling. Moreover, AOB can release nitrous oxide (N_2O), a highly potent greenhouse gas, which is also related to the deletion of the ozone layer. Natural and agricultural soils contribute approximately 57% to the total global N_2O budget. AOB community structure has been examined in different soils, such as agricultural soil, meadow soil and forest soil. In fact, MSW landfills can potentially emit high quantities of N_2O. Understanding the role of the AOB community structure in MSW cover soil is anticipated to be advantageous for attenuating the greenhouse gas emission from landfills (Yu et al. 2009).

4.2 Liquid Waste

Several environmental issues should be taken into account when wastewater treatment is considered. Legislation is continuously refining the requirements for the level of removal of different pollution factors in treated water. Wastewater treatment is considered to be a special type of process, due to the fact that it directly affects the environment. It has been put forward that the applied assessment method did not affect global environmental impact such as greenhouse effect, resources depletion, eutrophication and acidification (Renou et al. 2009).

Waste treatment techniques are commonly used for the alteration of the physical, chemical or biological properties of the waste, reducing its volume and/or toxicity, and rendering it safe for disposal. Waste treatment may be demanded for special types of wastes such as radioactive, hazardous and other Department of Energy (DOE) wastes (Arvanitoyannis and Tserkezou 2008).

In wastewater treatment, coagulation/flocculation processes are commonly used for removing colloidal materials that cause coloration and turbidity. An important characteristic of wastewater flocculation is the elimination of suspended solids (SS) and a great amount of organic material. The removal of SS and organics is performed using a flocculation-forming chemical, which is separated from water by flotation, settling or adsorption. Over the last 20 years, new inorganic and organic coagulants have been suggested to enhance the elimination of organic matter and total suspended solids

(TSS) during the treatment of wastes from the agro-food industry and slaughterhouses as well (Arvanitoyannis and Ladas 2008).

Another method is incineration, which is the high temperature burning (rapid oxidation) of a waste, usually at 870–1370°C. Incineration can be used for the disposal of hazardous waste, while energy recovery is a secondary objective. Sewage sludge incineration generates heat, which is used to dry the input sewage sludge to levels where the combustion is self-sustaining. The activated sludge process may comprise up to four phases: (a) clarification (by flocculation of suspended and colloidal matter), (b) oxidation of carbonaceous matter, (c) oxidation of nitrogenous matter, and (d) auto-digestion of the activated sludge (Arvanitoyannis and Tserkezou 2008).

In conventional anaerobic treatment processes, organic pollutants are initially converted into fatty acids, which are then further converted to acetate and hydrogen, and are finally turned into methane (Zhang et al. 2003).

Anaerobic bacteria use organic substances as the sole source of electrons and energy, converting them into H_2. The reactions involved in H_2 production are rapidly performed and these processes are not in need of any other energy source. Since organic substrates cannot utilize light energy, they are not fully decomposed and organic acids remain. However, these reactions can still be used as the first step of H_2 production from waste, followed by methanogenesis. A two-stage process is a rational configuration since it provides optimal environments for acidogenic hydrogenesis and methanogenesis in two separate spaces (Han and Shin 2004).

Anaerobic treatment of slaughterhouse wastewater is usually employed in the case of solid slaughterhouse waste. A small number of full-scale plants are in operation in slaughterhouses in the Netherlands, Belgium, and New Zealand. On the other hand, anaerobic lagoons or covered anaerobic ponds are commonly used for the treatment of slaughterhouse wastewater, both in warm climates and, wherever land cost is low such as in Australia and New Zealand (Salminem and Rintala 2002).

The conventional treatment of wastewater involves anaerobic/aerobic activated sludge processes. During this type of treatment, however, sludge bulking often takes place due to the bulking nature of this wastewater. The resulting unstable effluent quality and the complicated operation involved are quite challenging for operators in wastewater treatment plants. Moreover, excessive land use, due to low organic loading adopted in the conventional process, limits its wide application (Wang et al. 2005).

Anaerobic acidification of organic wastes leads to the generation of numerous volatile fatty acids (VFA), H_2, CO_2 and other intermediate compounds. The reactions taking place in hydrogen production are rapid, do not require any energy support and can be effectively used for the treatment of large quantities of organic wastes. Apart from hydrogen gas itself, which is an advantageous energy source, VFA can be also used for producing methane through the process of methanogenesis or as a readily biodegradable carbon source for the removal of biological nutrients (Shin et al. 2004).

Acid production lowers the reactor pH, causing inhibition of the methanogenic system, and the production of methane is therefore reduced. The use of sequential batch systems proved to be highly effective in digesting organic waste, since the removal of inhibitory products from start-up reactors can be performed in order for them to be used in mature reactors (Dearman and Bentham 2007).

The determination of the individual acids generated during the acidogenesis of organic wastes is very important since they can be used to extract valuable information for the dynamics of the process. Though acetate, propionate and butyrate are mainly produced during acidogenesis of organics, the exact composition of VFAs is affected by numerous factors such as the nature of the substrate and its concentration, hydraulic retention time (HRT), organic loading rate (OLR), temperature, etc. (Lim et al. 2008), as depicted in Fig. 1 and Fig. 2.

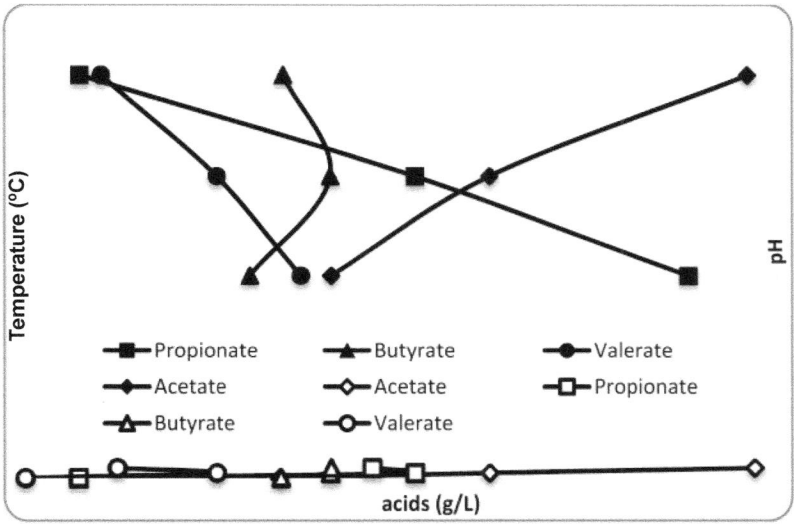

Figure 1. Effects of temperature and pH on acidogenesis of waste compost (Adapted from Lim et al. 2008).

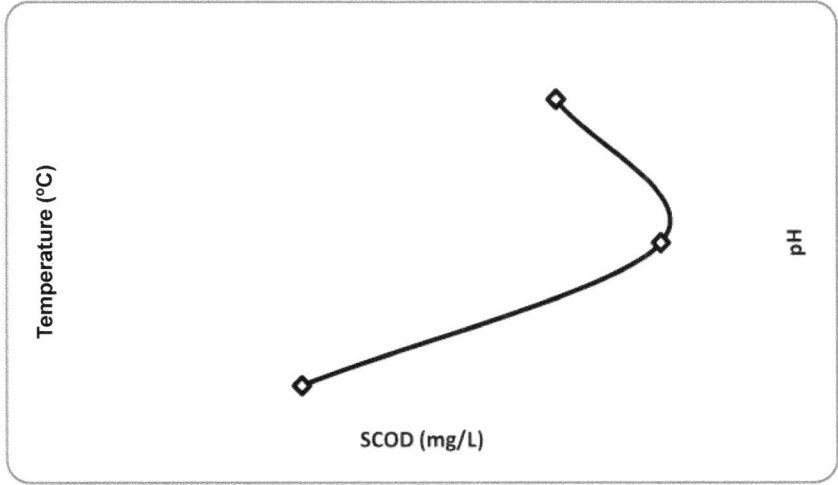

Figure 2. Effect pH and temperature on acidogenesis of food waste (Adapted from Lim et al. 2008).

Temperature was proved to have mixed effects on the net VFA production. While net production usually increases with temperature (Fig. 3), it has also been found that COD removal efficiency and methanogenic activity increased when the temperature increased from 20°C to 25°C and 30°C in anaerobic sequencing batch reactors used for the treatment of high strength slaughter house wastewater. In general, aerobic sludge's settleability deteriorates with increasing temperature, with the sludge volume index (SVI) increasing from 110 mL/g at 15°C to 540 mL/g at 35°C, and the presence of higher effluent TSS levels (Xu and Nakhla 2007).

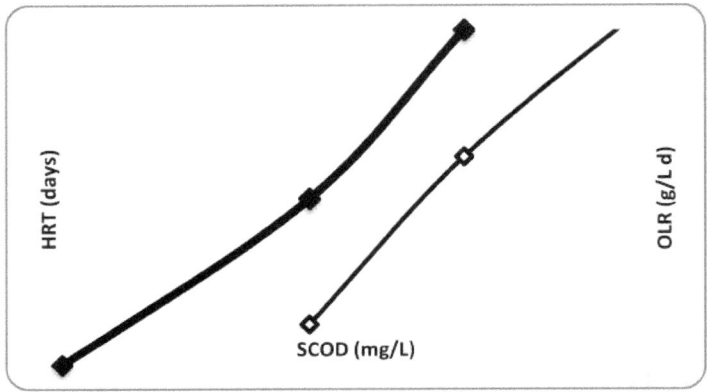

Figure 3. Effect of temperature and pH on SCOD production (Adapted from Lim et al. 2008).

From an operational point of view of wastewater treatment, landfilling, incineration or composting, there are practically only two options available for the mitigation of N_2O emissions, the maximization of the conversion of reactive N to N_2 and the concurrent maximization of the leakage of N_2O gas. The first option requires that O_2 and carbon levels are optimal for N_2 production in the case of microbial nitrification and denitrification during wastewater treatment or composting, and that the temperature range during waste combustion is properly controlled. The recycling of N back to agriculture could conceivably prevent N_2O emissions by cancelling the fixation of N for fertilizers, which would otherwise be necessary (Barton and Atwater 2002).

According to Chu et al. (2008), "the hydraulic retention times (HRTs) for hydrogen and methane production from organic fraction of municipal solid wastes (OFMSW) amount to 1–2 days and 10–15 days, respectively (Fig. 4). In other words, the volume of reactor for methane production is circa eight times as large as that of hydrogen production. It is of great importance to develop the high-rate methane production from OFMSW. An up-flow anaerobic sludge blanket (UASB) process is an extensively applied anaerobic treatment endowed with high treatment efficiency and a short HRT, but unfortunately not suitable for high solid waste. If a solid residue is physically removed, separation process is required thus complicating considerably the process."

In the study of Heo et al. (2003), biochemical methane potential tests were performed for the determination of the anaerobic digestibility of the WAS and the sludge pretreated by NaOH. The optimal NaOH dosage was 45 meq NaOH/L. The maximum SCOD (soluble chemical oxygen demand) solubilization was 27.7,

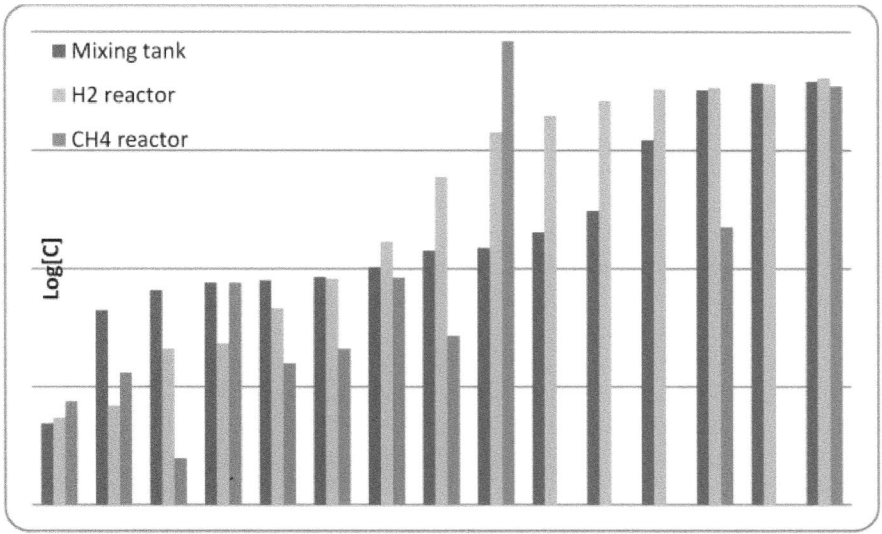

Figure 4. Methanogenic and hydrogenic reactor from treatment wastewater (Adapted from Chu et al. 2008).

31.4 and 38.3% at the temperatures of 25, 35 and 55°C respectively after 4 hours of reaction. The final methane yield of simulated food waste was determined at 430 ml CH_4/g volatile solid (VS) (added), while the corresponding values for PWAS (waste activated sludge pretreated by NaOH) (25°C), PWAS (35°C) and PWAS (55°C) were 274, 286 and 310 ml CH_4/g VS (added), respectively after 20 days. The figures were 66%, 73% and 88% higher than that of WAS. It was proved that the methane production in anaerobic co-digestion was significantly affected by the fraction of SFW (simulated food waste) and PWAS in the feed. The anaerobic digestibility of the SFW-PWAS mixture was higher than that of the SFW-WAS mixture. Therefore, anaerobic co-digestion of food waste with the PWAS could effectively be used for the reduction of the solid waste volume providing enhanced methane recovery.

According to Zhang et al. (2003) "fermentative microorganisms, such as *Clostridium* and *Thermoanaerobacterium*, are able to produce hydrogen from carbohydrates. A few studies have been conducted to produce hydrogen from carbohydrate-rich wastewater under mesophilic condition. However, many industrial effluents, such as those from food processing, are often discharged at elevated temperatures. Treating these starch-laden effluents under conventional mesophilic conditions requires expensive pre-cooling, and has the risk of losing the biomass activity should the cooling system break down. It is thus only natural to treat these effluents under thermophilic conditions, which presumably also favor the degradation kinetics and pathogens killing."

In fishery wastewater the contaminants present are undefined mixtures of mostly organic substances. A generalization of the extent of the problem due to this wastewater cannot be easily done because it will be dependent on effluent strength, wastewater discharge rate, and the absorbing capacity of the receiving water body (González 1996). During fish evisceration and cooking, high content of COD (Chemical oxygen demand), nutrient, oil and fat are generated in fish processing wastewater. The levels of total soluble and suspended COD vary largely both in terms of factories and fish species (Chowdhury et al. 2010).

5. Food Waste Valorization

The production of many food products results in the generation of unavoidable by-products that contain significant amounts of nutrients. High production volumes, environmental effects, and the nutritional value of by-products render them a highly important subject for valorization. Valorization allows the examination of the possibility of reusing by-products, highlighting the potential gain that can be achieved through this process (Banaszewska et al. 2014).

Nowadays, composting is used for the treatment of food wastes with various agricultural by-products in several systems. The use of composting for the valorization of organic wastes is expanding rapidly in the United States and in other countries, as landfill space becomes scarce and expensive and people become more aware of the impact landfills have on the environment. It is suggested that composting will very soon become as common as recycling aluminum cans is today, both in the backyard and on an industrial scale (Arvanitoyannis and Kassaveti 2007).

A series of case studies were taken into account in the study of Budarin et al. (2014) in an effort to demonstrate that microwave dielectric heating could be used for the recovery and synthesis of valuable molecules from a vast variety of biomass types. Under microwave irradiation, the production of chemicals from biomass can be performed lower temperatures (up to 150°C) in comparison to conventional heating. This is very significant since molecules with a high degree of functionality can be generated while conventional heating usually generates a great proportion of lower value gases. Moreover, the technical set-up of a microwave reactor can be readily modified to separate *in situ* acids and valuable products, thus increasing the shelf life of the latter (Budarin et al. 2015).

The food processing industry generates highly concentrated, carbohydrate-rich wastewaters, but no extensive evaluation of their potential for producing biological hydrogen has been carried out as yet. Thus wastewaters obtained from four different food-processing industries and characterized by chemical oxygen demands of 9 g/L (apple processing), 21 g/L (potato processing), and 0.6 and 20 g/L (confectioners A and B) were examined. The production of biogas from all four food processing wastewaters consistently contained 60% hydrogen and carbon dioxide. COD removals resulting from hydrogen gas production were generally 5 to 11%. Overall hydrogen gas conversions were 0.7–0.9 L-H$_2$/L-wastewater for the apple wastewater, 0.1 L/L for Confectioner-A, 0.4–2.0 L/L for Confectioner B, and 2.1–2.8 L/L for the potato wastewater. By adding nutrients to samples, the hydrogen production was very well correlated with COD removal, with an average of 0.10 ± 0.01 L H$_2$/g COD. However, no correlation was found between hydrogen production and COD removal in the absence of nutrients or in more extensive in-plant tests at the potato processing case. Gas generated by a domestic wastewater sample contained only 2 3 ± 8 hydrogen, resulting in an estimated maximum production of only 0.01 L/L for the original, non-diluted wastewater. Based on an observed hydrogen production yield from the effluent of the potato processing plant of 1.0 L-H$_2$/L, and annual flows at the potato processing plant, it was estimated that the production of hydrogen gas at this site could be worth up to $65,000/year (van Ginkel et al. 2005).

The production of olive oil leads to the generation of significant amounts of solid wastes, which contain high levels of bioactive compounds with antioxidant properties. These compounds have been used as inefficient polluting heating fuels for many years. A study was carried out, aiming at examining the effect of extraction time and temperature on phenolics recovery from olive pomace. Specifically, different tests were carried out, applying different times (from 15 to 120 min) and temperatures (from 100 to 180°C), and using a high pressure-high temperature agitated reactor under a modified atmosphere. Quantifications of total polyphenols, o-diphenols and total flavonoids were carried out. The maximum total polyphenols yield (45.2 mg CAE/gDP) was achieved at 180°C and 90 min. HPLC analysis indicated that oleuropein (2433 mg/100 gDP) and tyrosol (485 mg/100 gDP) were the most abundant phenolics in the extracts. Methanolic extracts could be used by food, cosmetic and pharmaceutical industries. Solid residues, after phenolics extraction, are considered to be relatively environmentally friendly (Aliakbarian et al. 2011). The olive mill waste produced from olive oil extraction has environmental impacts, particularly in Mediterranean areas. The extraction of olive oil can be carried out using both discontinuous and continuous processes. These processes produce three different fractions: a solid residue and two liquid phases (oil and olive mill wastewater). By characterizing these two by-products, it was proved that they mainly consist of phenolic substances, carbohydrates, organic acids and mineral nutrients. Untreated olive by-products, discharged between November and March into the environment, severely affect the ecosystems in olive oil-producing countries because of their high toxic organic loads, low pH, and high chemical and biological demands. Thus the valorization of olive mill waste residues should be examined, mainly of those allowing for the recovery of valuable natural components such as phenolic compounds, dietary fibers, animal feed, biofuel, biogas, enzymes, polymers, etc. (Dermeche et al. 2013). The olive oil industry, in general, generates huge quantities of solid olive mill wastes (SOMW), causing severe environmental damages. Cultivation of edible mushrooms, such as *Pleurotus ostreatus* could be examined as an approach for SOMW valorization. The isolation of a mycelium of *P. ostreatus* (LPO) was performed from castor oil plants. Oyster mushroom spawn, produced on barley grains, was applied for the inoculation of wet SOMW, steamed in a traditional steamer for 45 minutes. The mycelium growth rate on SOMW was initially estimated in a Petri dish through measurement of the surface colonized by the mycelium. An estimation of the fruit body yields was given on culture bags, which contained 2 kg (each) of SOMW inoculated at 7% (w/w). Comparison of the local strain potential with that of a commercial one was performed. Both strains produced high quality mushrooms, but low yields. The supplementation of the SOMW with wheat straw and $CaCO_3$ considerably increased the productivity of the two strains by a factor of 3.2 for LPO (local *Pleurotus ostreatus*) and 2.6 for CPO (commercial *Pleurotus ostreatus*) (Mansour–Benamar et al. 2013). The development of an innovative process battery comprising coagulation-flocculation, extraction of phenolic compounds and photo-Fenton post-oxidation was carried out, aiming at the valorization and treatment of olive mill effluents (OMEs). Pre-conditioning by coagulation-flocculation using $FeSO_4 \cdot 7H_2O$ as the coagulant, and an anionic polyelectrolyte (FLOCAN 23) as the flocculant, was carried out for the removal of the solid content of the effluent. Addition of 6.67 g/L of $FeSO_4$-$7H_2O$ and 0.287 g/L of FLOCAN 23 allowed the optimal removal of TSS (97 ± 1.3%), of COD (72 ± 1.5%), and of Total Phenols (TPs) (40 ± 1.3%). Application of solvent extraction permitted the recovery of a fraction of the remaining phenolic compound. Finally, photo-Fenton was used as a post-treatment method; oxidation for 240 min at 0.2 g/L Fe^{2+}, 5 g/L H_2O_2 and pH 3 reduced the remaining COD and TP by 73 ± 2.3% and 87 ± 3.1%, respectively. Toxicity assays to Daphnia magna as well as phytotoxicity tests to three plant species to untreated OME and oxidized samples were also carried out, demonstrating that more biologically potent products were produced during the oxidation process (Papaphilippou et al. 2013). In another study, phenolic composition, antioxidant and breast cancer anti-proliferative activities of water and methanol/water derived extracts from olive pomace (OP) and dry olive mill residue (DOR) from Portuguese industries, were examined. DOR water (DORW) extracts had the highest extraction yield and the highest total phenolic content (TPC) and hydroxytyrosol (HT) ($\approx$ 25 mg/g extract). The HPLC–ESI-MS detected HT in both OP and DOR, while detection of HT-1-glucoside, tyrosol, oleuropein aglycone isomers, verbascoside and oleuropein were solely performed in DOR. Furthermore, a de(carboxymethyl) oleuropein aglycone isomer was found in DOR. DOR water extract was also characterized by more effective DPPH scavenging capacity and anti-proliferative activity, in comparison to OP water (OPW) extract. Furthermore, the antioxidant potential of the phenolic substances present in DORW extract could be compared to that of HT, and to butylated hydroxyanisole (BHA). Phenolic compounds found in DORW extract also demonstrated comparable tumor anti-proliferative activity on MDA-MB-231, relatively to HT and 5-fluorouracil (5-FU). MDA-MB-231 cell growth inhibition, upon 5-FU incubation was even enhanced in the presence of DORW extract. It was thus proved that DOR extracts could be potentially used as sources of phenolic substances for nutraceutical applications or food supplements, offering new opportunities for olive mill residue valorization (Ramos et al. 2013).

Natural antioxidants have been extensively studied for their potential beneficial effects on health. On the other hand, valorization of residues is an opportunity to obtain profit in a sustainable way. In the study of Navarrete et al. (2011), antioxidants were extracted from essential oil derived from residues of rosemary using steam distillation, hydrodistillation and Solvent Free Microwave Extraction (SFME). A solvent extraction with ethanol was applied to extract the antioxidants. Mass transfer rates of antioxidants from leaves were enhanced as a result of the previous extraction of essential oils. Higher yields and rates of extract from leaves after SFME were shown.

Wine production is another food production process that leads to the generation of high amounts of waste. Some environmentally friendly technologies have been suggested for the valorization of winery waste products. Fermentation of grape marc, trimming vine shoot or vinification lees can lead to the production of lactic acid, biosurfactants, xylitol, ethanol and other compounds. Moreover, grape marc and seeds, which contain high contents of phenolic substances and vinasse containing tartaric acid, could be used at a commercial scale. Investment is thus required, in these new technologies, for decreasing the impact of agro-industrial residues on the environment and establishing new processes that could function as additional sources of income (Devesa-Rey et al. 2011).

In the study of Garrido et al. (2014), the structure and properties of soy protein-based films were studied and significant efforts were made to provide an alternative and novel material for the production of renewable packaging, which will be characterized by reduced environmental impact in comparison to the conventional films that are currently available in the market. The production of biodegradable films using soy protein was performed. This protein is a by-product from the soy oil industry. Analysis of the effect of plasticizer content was performed towards the optimization of the formulations and processing conditions in an effort to improve the functional properties of the final product. Moreover, the changes observed in the characteristics of the films were related to the structural changes in raw materials. Successful preparation of films derived from soya by-products was carried out and transparent films with excellent UV barrier properties were generated. Optimal results were shown at pH 10, at which the protein unfolding can better promote protein-plasticizer interactions. As regards the environmental assessment, it was proved that the environmental impact associated with soy protein films could be reduced. As a result, future improvements in the extraction processes together with an optimization of the manufacturing method, could contribute to the reduction of the environmental impact of these materials and could potentially offer environmentally friendly alternatives to petroleum-based films.

Biodiesel has been evaluated and promoted in the last decade as a renewable, biodegradable, and non-toxic fuel. Raw glycerol can be an excellent raw material when biodiesel is commercially used. The production of 10 kg of biodiesel from rapeseed oil leads to the generation of 1 kg of glycerol. Some microorganisms could be applied for direct glycerol biovalorization. *Yarrowia lipolytica* is one of the most well studied yeasts that can be used as a model in the degradation study of hydrophobic substrates and in numerous other areas. It has a high affinity for hydrophobic substances due to the fact that it produces surface-active compounds as well as due to its differentiated cell wall. Glycerol, a hydrophobic compound readily assimilated by *Y. lipolytica*, can be used for the production of numerous substances of high biotechnological significance, such as biosurfactants and citric acid. Biosurfactants could be potentially employed in various commercial applications in the petroleum, pharmaceutical, biomedical and food industry. Citric acid is widely used for the production of several industrial products and in various industrial areas, especially in the food industry. There is, therefore, a huge and increasing demand for this chemical, and thus glycerol transformation by *Y. lipolytica* could be a good solution (Amaral et al. 2009). Some representative examples of microorganisms that can be used for the biovalorization of different food wastes/by-products are given in Table 3.

High quantities of seashell by-products (SBP) are generated in France and are considered wastes. They could, however, be applied to partially replace aggregates in pervious concrete pavers, thus enabling the use of a more environmentally friendly building material. Partial replacement of coarse aggregate (20% or 40% by mass) by SBP obtained from the Crepidula shell was performed. Crushed Crepidula seashells of 2/4 mm and 4/6.3 mm were employed for the production of new seashell by-product based pavers. It was demonstrated that the seashell by-products could be potentially used as aggregate. Their combined use could offer a compressive strength of 16 and 15 MPa and a permeability coefficient in the range of 3–8 mm/s (Nguyen et al. 2013).

The generation of fish waste is a huge environmental issue mainly connected with the canning industry in some Mediterranean countries. Their anaerobic digestion, however, is not a good option due to the organic matter deficit in the chemical oxygen demand:nitrogen:phosphorus ratio (COD:N:P) of this waste, which destabilizes the whole process. The co-digestion of fish waste using residual strawberry extrudate was examined at a laboratory scale under mesophilic conditions. Strawberry waste increased the organic matter concentration in the mixture and led to dilution of inhibitory compounds contained in the fish waste, such as chloride, nitrogen and phosphorus. Co-digestion rendered the process more stable, while biodegradability was determined at 83% in total volatile solids. Furthermore, the methane production yield reached a mean value of 120 mL/g total volatile solids (at 1 atm, 0°C) for an organic loading rate in the range of 22.8–50.6 kg waste mixture/(m³ d), while the digestate was characterized by high levels of nutrients, which makes it suitable for use as an organic amendment in agriculture (Serrano et al. 2013).

Strawberry-flavored food is one of the most highly consumed products, leading to the generation of more than one harvest per year, even at unfavorable environmental conditions, through the use of green-houses. The manufacturing of strawberry-derived products produces high levels of organic waste, which pose significant risks to the environment. The production of biogas through biomethanization is not efficient due to the fact that the anaerobic valorization of this residue is problematic. This is due to the presence of lignin concentrated in the achene, which destabilizes the process. The mesophilic anaerobic digestion process could be improved using a pre-treatment with sieving. It was shown that the pretreated waste was more stable, while its biodegradability was determined at 90% in TVS (total volatile solids). Furthermore, the proposed pre-treatment led to a 36% improvement in methane production yield. Additionally, the permitted OLR (organic loading rate) was significantly higher for the pretreated waste (5.3–2.8 versus 8.0–12.0 kg waste/m³ d), thus allowing the treatment of more waste. However, an inhibition factor was observed for high loads, although it was stronger and appeared at lower loads in untreated waste (Siles et al. 2013).

Finally, it is important to mention the study of Zhang et al. (2013), in which an evaluation of bakery waste, including cakes and pastries from Starbucks, Hong Kong, was carried out for examining the potential of succinic acid (SA) production. By employing simultaneous hydrolysis and fungal autolysis, both cake and pastry hydrolysates were found to contain high levels of glucose (35.6 and 54.2 g/L) and free amino nitrogen (685.5 and 758.5 g/L), while the protein hydrolysis yields were determined at 23.2 and 22.5%, respectively. Subsequent use of these cake and pastry hydrolysates, together with magnesium carbonate (10 g/L) as feedstock in *Actinobacillus succinogenes* fermentation was performed, and the resultant SA levels were found to be 24.8 and 31.7 g/L, respectively. A cation-exchange resin-based process was then used for the recovery of the SA crystals from the fermentation broth, and a high SA crystal purity (96–97.7%) was obtained. It was finally proved that bakery waste could be successfully used as the generic feedstock for the sustainable production of SA.

Table 2 presents the valorization techniques examined in some representative studies focused on food wastes.

Table 2. Common types of waste waters, final products and suggested valorization techniques.

Type of Food Waste	Valorization Technique	Final Product	Effectiveness of the method	References
Different biomass types	Microwave dielectric heating	Valuable chemicals	Production of higher value chemicals at lower temperatures	Budarin et al. 2014
Carbohydrate-rich wastewaters	Fermentation (hydrogen producing, spore forming bacteria)	Bio-hydrogen gas	Efficient, very profitable	Van Ginkel et al. 2005
Olive pomace	High pressure-high temperature agitated reactor under a modified atmosphere-extraction	Phenolics	Efficient, production of solid residues after extraction	Aliakbarian et al. 2011
Solid olive mill wastes	Used as broth for the production of *Pleurotus ostreatus*	*Pleurotus ostreatus*	Good productivity when supplements are added	Mansour–Benamar et al. 2013
Olive mill effluents	Coagulation–flocculation, extraction and photo-Fenton post-oxidation	Phenolic compounds	High efficiency	Papaphilippoulos et al. 2013
Olive pomace/dry olive mill residue	Water and methanol/water		Effective extraction in laboratory scale	Ramos et al. 2013
Essential oil derived from residues of rosemary	Steam distillation, hydro-distillation and solvent free microwave extraction and extraction with ethanol	Antioxidants	Higher yields using the Microwave technique	Navarrete et al. 2011
Soy protein	Film-forming procedure carried out in laboratory scale	Soy protein-based films (packaging material)	Successful, environmentally friendly (alternative to petroleum-based films)	Garrido et al. 2014
Glycerol (by product of biodiesel production from rapeseed oil)	Assimilation of substances by *Yarrowia lipolytica*	Biosurfactants and citric acid	Very efficient	Amaral et al. 2009
Sea shell by-products	Partial replacement of coarse aggregate with the material	Replacement of aggregates in pervious concrete pavers	Good properties of final material	Nguyen et al. 2013
Fish waste	Co-digestion of fish waste using residual strawberry extrudate	Digestate	Digestate rich in nutrients-potential agriculture uses	Serrano et al. 2013

Table 2. contd....

Table 2. contd.

Type of Food Waste	Valorization Technique	Final Product	Effectiveness of the method	References
Organic waste derived from strawberry-based products	Mesophilic anaerobic digestion with sieving pretreatment	Biogas through biomethanization	Improved but still not effective for high loads	Siles et al. 2013
Bakery waste	Hydrolysis and fungal autolysis	Succinic acid	Effective production, high yield	Zhang et al. 2013
Tomato purée or tomato peel	Extraction with ethanol or acetone and concentration	Carotenoids and lycopene	Effective extraction	Benakmoum et al. 2008
Whey	Fermentation (*Kluyveromyces marxianus*, *Lactobacillus bulgaricus*)	Production of novel dairy starter cultures	Significant reduction of production costs	Koutinas et al. 2009
Orange peel waste	Pretreatment and anaerobic digestion	Methane	Methane yield coefficient: 0.27–0.29 L_{STP} CH_4/g added COD, Biodegradability: 84–90%, acidification occurred at high organic loading rate	Martin et al. 2010
Sugar beet	Dark anaerobic fermentation	Hydrogen	Very efficient	Panagiotopoulos et al. 2010

Table 3. Microorganisms used for the bio-valorization of food wastes.

Food Waste	Valorization	Microorganism(s)	References
Glycerol from rapeseed oil production	Assimilation of glycerol	*Yarrowia lipolytica*	Amaral et al. 2009
Whey	Production of novel dairy starter cultures	*Kluyveromyces marxianus, Lactobacillus bulgaricus*	Koutinas et al. 2009
Orange peel waste	Production of methane	*Aspergillus, Fusarium* and *Penicillium*	Martín et al. 2010
Sugarbeet	Production of hydrogen	*Caldicellulosiruptor saccharolyticus*	Panagiotopoulos et al. 2010
Fish and strawberry residues	Production of digestate rich in nutrients	Hydrolytic, acidogenic, acetogenic and methanogenic bacteria	Serrano et al. 2013
Organic waste derived from strawberry-based products	Production of biogas through bio-methanization	Mesophilic anaerobic microorganisms	Siles et al. 2013
Bakerywaste	Production of succinic acid	*Actinobacillus succinogenes*	Zhang et al. 2013
Discarded oranges and brewers' spent grains	Production of *Kluyveromyces marxianus*, kefir, and *Saccharomyces cerevisiae*	*Kluyveromyces marxianus*, kefir, and *Saccharomyces cerevisiae*	Aggelopoulos et al. 2013
Raw glycerol	Production of citric acid	*Yarrowia lipolytica*	Papanikolaou and Aggelis 2003
	1,3-propanediol	*Clostridiumbutyricum*	
Maize, apple, grape, pomace, pineapple, mandarin orange, brewery wastes, citrus and kiwi fruit peel	Citric acid	*Aspergillus niger*	Soccol et al. 2006

6. Valorization of Wastes Derived From Fermented Food Products

The production process of beer unavoidably generates wastes which are rich in plant organic material and may also contain salt, flavorings, coloring material and acids or alkalis. Significant amounts of oil or fats may also be present (Bolzonella et al. 2007). Waste from beer fermentation broth could be used for the production of bioethanol. Both the treatment and use of this waste material seems to have both economic and environment-protection advantages (Ha et al. 2011). An effort was made to examine the potential use of numerous agro-industrial wastes for microbial cellmass production of *Kluyveromyces marxianus*, kefir, and *Saccharomyces cerevisiae*. The promotional effect of whole orange pulp on cell growth in mixtures composed of cheese whey, molasses, and potato pulp in submerged fermentation processes was evaluated. It was proved that the cell mass was increased by 2- to 3-fold in the presence of orange pulp. Similarly, examination of the use of brewers' spent grains for promoting the cell growth in solid state fermentation of mixtures of whey, molasses, potato pulp, malt spent rootlets, and orange pulp was performed. An increase of 3-fold of the cell mass was observed for *K. marxianus,* while *S. cerevisiae* was increased by 2-fold when the above-mentioned substrates were used, proving their potential for promoting the single-cell protein production and eliminating the need for further nutrients. Cell growth kinetics was also taken into account through measuring cell counts at various time intervals and at different concentrations of orange pulp. The protein quantity found in the fermented substrates was enhanced significantly, thus proving that mixed agro-industrial wastes is a low cost material that could be effectively used as protein-enriched livestock feed, offering added value and minimizing the wastes that need to be disposed (Aggelopoulos et al. 2013). Similar materials were used in the study of Contreras et al. (2012), who evaluated the theory that agro-industrial wastes could be used for adsorption of dyes. The pH of brewers' spent grains and orange peel was 5.3 and 3.5, respectively. The equilibrium isotherms of Basic Blue 41, Reactive Black 5, and Acid Black 1 were performed without controlling the pH, which ranged from 4 to 5.5. The equilibrium concentrations of both adsorbents were fitted by the Freundlich and Langmuir models. The maximum adsorption capacity observed for Basic Blue 41, Reactive Black 5, and Acid Black 1 was 32.4, 22.3, and 19.8 mg/g for brewers' spent grains; and 157, 62.6, and 45.5 for orange peel, respectively. The kinetic process was fitted by the model of pseudo-second order. According to the authors, "the constant rate for orange peel decreased to extend the initial concentration of dye increased, obtaining 4.08×10^{-3}–0.6×10^{-3} (Basic Blue 41), 2.98×10^{-3}–0.36×10^{-3} (Acid Black 1), and 3.40×10^{-3}–0.46×10^{-3} g/mg/min (Reactive Black 5). The best removal efficiency was obtained in orange peel with values starting from 63% to 20%. Consequently, according the results obtained, there are two positive effects, the reuse of agricultural wastes and its use as low-cost adsorbent of the dyes."

The production process of beer leads to the generation of huge amounts of streams (spent grain and trub) characterized by high energetic potential. Thus the energetic valorization of the brewery wastes, trub (T) and spent grain (SG) were tested using anaerobic digestion processes. T was characterized by a higher BMP (251 ± 2 L CH_4/kg COD) in comparison to SG (191 ± 3 L CH_4/kg COD). The methane yield ($72 \pm 1\%$) and methane production rate (1.08 ± 0.10/d) were also higher for T due to the recalcitrant nature of SG ($55 \pm 1\%$ and 0.21 ± 0.01/d). The co-digestion ratio of 10% T + 90% SG generated 199 ± 2 L CH_4/kg COD. Due to the high proteins content of T and SG (54.0 ± 6.6 and 30.5 ± 1.7 respectively), crude glycerol (CGly) could be effectively used as a co-substrate, offering high energy potential and optimizing the C:N ratio for the anaerobic digestion process. The co-digestion with 10% (w/w) CGly enhanced the production of methane to 328 ± 5 L CH_4/kg COD (yield = $94 \pm 2\%$). The use of higher concentrations of CGly led to inhibitions, reducing the production rate of methane. It was concluded that the addition of up to 10% (w/w) of CGly positively affected the brewery wastes anaerobic digestion (Costa et al. 2013).

Nowadays, natural phyto-pharmaceuticals have gained the interest of both scientists and consumers due to their functional proprieties and effect on the metabolic syndrome. Encapsulation is an innovative method used for the storage and stabilization of bioactive molecules on different matrices, offering targeted delivery of the active compounds in pre-determined tissues. The valorization of dry yeast waste derived from the secondary fermentation of beer was performed by using it as natural bioactive matrix, which can incorporate different antioxidant pigments, flavors or can be included in a polymeric mass of low digestible carbohydrates such as isomalt. A dry beer yeast containing certain amounts of proteins, B vitamins and phenolics was examined by HPLC-PDA, and different variants of encapsulation were applied into a melted isomalt mass. Addition of beer yeast powder, cinnamon, lecithin and cocoa extracts was carried out. The HPLC analysis of the final products indicated that 60–70% of bio-active molecules had been recovered (Alina et al. 2008).

Several studies also examined the potential valorization of winery wastes. Winemaking by-products have a high environmental impact due to their high biochemical and chemical oxygen demands. The study of Casazza et al. (2012) was an effort to optimize the traditional solvent extraction technology of phenols from Pinot Noir grape skins (*Vitis vinifera*). Specifically, the synergistic effects of the extraction time (9, 19 and 29 h) and the solid-liquid ratio (0.10, 0.20 and 0.30 gDW/mL) were assessed by a 32 full factorial design in combination with response surface methodology. Total polyphenols, flavonoids and trans-resveratrol extraction yields were applied as response variables. It was shown that Pinot Noir skins had high levels of both, total polyphenols (3.22 mgGAE/gDW) and flavonoids (1.01 mgCE/gDW), and optimum extraction time (19 hours approximately). T- Resveratrol amount was about 2.24 mg/100 gDW, while the main phenolic substances examined with Reverse Phase-High Performance Liquid Chromatography were gallic acid, catechin and quercetin. It was proved that the polyphenol concentration and their anti-radical power were linearly correlated. In the study of Aliakbarian et al. (2012), subcritical water extraction of phenolic compounds from grape pomace was carried out. The combined effects of extraction temperature (100, 120 and 140°C) and pressure (8 MPa, 11.5 MPa and 15 MPa) were examined by employing a 32 full factorial design and response surface methodology. Extractions with considerably higher polyphenols, flavonoids and antioxidant activity, in comparison to conventional methods, were obtained by using subcritical water extraction. It was shown that optimum extraction was performed at 140°C and 11.6 MPa (0.9550). Under these conditions, recoveries of 31.69 mg GAE/gDP and 15.28 mg CE/gDP of total polyphenols and flavonoids, respectively, were performed. The extracts were characterized by antiradical power of 13.40 µg DPPH/µl extract. Sub-critical water extraction was more effective for the extraction at atmospheric pressure than the use of water and ethanol.

Considering the high specificity of the organic load of winery effluents, a new biophysical treatment employing the stripping of ethanol in combination with a final concentration by evaporation was evaluated. Full treatment and pre-treatment were both applied in different experiments. It is widely approved that winery wastewater contains a large amount of ethanol in the organic load (75 to 99% of the COD). According to a linear correlation between COD and ethanol concentration, the amounts of ethanol determined can be used for estimating the organic load of winery wastewater. The full treatment involves stripping and concentration at a pilot plant, separating the wastewater into highly purified water (COD elimination > 99%), a concentrated alcoholic solution that can be used as bio-fuel and a concentrated by-product. Stripping alone represents an advantageous pre-treatment of winery wastewater. The purification rate is about 78 to 85% and the recovery of ethanol can be carried out. The process facilitates discharge into a sewage system. The method is very cost-effective and thus, highly competitive compared to biological treatments (Colin et al. 2005).

Also, according to Scola et al. (2010), the dietary intake of antioxidants could lead to the reduction of incidences of diseases related to oxidative stress. Thus, a study was carried out for the examination of the possibility of recovering substances with antioxidant activity from wine wastes using water as solvent. It has been demonstrated that it is feasible to obtain flavan-3-ol from wine wastes from both *Vitis vinifera* (cv. Cabernet Sauvignon and Merlot) and *V. labrusca* (cv. Bordo and Isabella) species. The main phenolic substances detected were catechin and epicatechin, as well asprocyanidin B3, procyanidin B1, procyanidin B2, gallic acid, epigallocatechin, and procyanidin B4. It was proved that the extracts could contribute to the prevention of lipid and protein oxidative damage in the cerebral cortex, cerebellum and hippocampus tissues of rats. Although more studies are needed to confirm the results, these flavan-3-ol extracts can potentially be used for reducing the incidence of diseases caused mainly by oxidative stress.

In the field of the dairy industry, Banaszewska et al. (2014) carried out a study to examine the potential added value of cheese whey through its valorization, and to evaluate the impact of integral valorization of main products and by-products on the profit of the dairy industry. Different scenarios were examined using a decision support tool, the integral dairy valorization model. Data obtained from the dairy processor, Friesland Campina (Amersfoort, the Netherlands), were used and the modeling results indicated that the valorization of by-products could be very profitable (24.3% higher profit). Moreover, additional profit could be achieved through integration of 2 valorization processes (main products and by-products). The latter is, however, significantly affected by capacity and market demand. Considerable benefits were shown when demand of whey-based products was increased by 25%. Also, according to Patel and Murthy (2011), "the treatment of dairy wastewater, which conforms to environmental regulations, is a crucial problem due to its high biological oxygen demand (BOD). The main cause of the BOD in dairy wastewater is due to residual whey, which consists mainly of lactose. Recovery of lactose from the whey would solve the problems of whey utilization as well as pollution reduction, as lactose recovery itself can reduce the BOD of whey by > 80%." The same authors suggested that lactose could be recovered from partially deprotonated whey using an anti-solvent (acetone). The effects of different process factors, such as acetone concentration (65–85% v/v), time (1–3 h), stirring speed (500–1,000 rpm), seeding (1–3% w/w) and pH, were assessed. It was shown that the recovery of more than 90% of lactose could be performed from whey after stirring for 1 hour at an acetone concentration of 85% v/v. An estimation of the crystal size distribution of the lactose obtained from whey was given using image analysis and proved to be affected by crystallization time and seeding.

It is therefore, generally accepted that the production of food products can lead to the generation of by-products, which contain high amounts of nutrients. High production volumes, environmental impact, and nutritional content of by-products render them of high importance for potential valorization. Valorization allows us to examine the potential of reusing wastes derived from food manufacturing processes, thus highlighting the potential gains that can be achieved (Banaszewska et al. 2014).

7. Future Trends

Green or clean production is considered one of the most important elements in manufacturing technology for present and future products. Current demand is focused on developing cost effective technologies, optimizing the processes and the separation steps, establishing alternative processes for reducing wastes, optimizing any resources and improving production efficiency (Kosseva 2009).

Over the last 50 years, the development of several composting technologies have been reported. These technologies could be used for the disposal of MSW, sewage sludge and cow and pig manure. Although the basic scientific principles of waste composting have been extensively studied for many years, engineering designs are mainly based on a trial and error approach, and are limited to known substrates or previously tested process conditions. Application of these technologies (designs) to new substrates characterized by high moisture and fat contents (such as food waste), are frequently characterized by poor performance or are unsuccessful (Chang and Hsu 2008).

Moreover, aiming at recycling and reclaiming solid wastes, different organic residues produced by agriculture, livestock farming, forestry, industries and city centers are being effectively used as container media for ornamental plant production. There has, therefore, been an increase in demand for solid organic wastes since these materials are now considered to be useful and even high value-added products (Abad et al. 2001). However, major recycling methods such as composting and feed stuffing are not suitable because food waste is characterized by high salinity and landfilling is no longer allowed because of ensuing environmental problems. Since food waste has high organic content, it can be applied to hydrogen production in dark fermentation, which can be a viable alternative to the food waste recycling methods (Kim et al. 2006).

The principal target of the continuously stricter waste legislation is to prevent waste generation. Waste prevention refers to three types of practical actions; strict avoidance, reduction at source and potential product reuse. According to European Environmental Agency (EEA), "waste minimization can take place by means of the following methods: prevention, reduction at source, reuse of products for same or other purpose, on- or off-site recycling, source- or waste-oriented waste quality and energy recovery (i.e., pyrolysis, combustion and gasification)" (Arvanitoyannis et al. 2007).

The conduction of an effective composting process requires standardized pH, temperature, moisture, oxygenation and nutrients, to encourage the development of the microbial population. Hence, any potential changes in the process conditions can directly affect the growth of certain microflora, which control the organic matter degradation. Ideal composting conditions include carbon-to-nitrogen ratio of the composting material between twenty and forty, moisture content of 50–65%, adequate oxygen supply, small particle size to increase the surface area and sufficient void space for air to flow through (Arvanitoyannis and Kassaveti 2007).

The thermodynamic concept of energy, defined as the maximum amount of work that can be extracted from a physical system by exchanging matter and energy with large reservoirs in a reference state, is widely accepted as a unified objective measure in the fields of resource accounting, environmental impact assessment, ecological cost evaluation and modeling. Energy was introduced in an effort to evaluate the overall system performance (resources, yields and environmental impacts) as a trend in thermodynamics based industrial ecology (Jiang et al. 2009). In other words, exergy resembles to LCA in terms of utilizing inputs and outputs but it is mainly restricted to the thermodynamic process.

Incineration has been one of the most widely used methods for disposal of combustible solid wastes in Japan, especially for MSW from restaurants, hotels, and supermarkets. Over a long period of time incineration was considered to be a novel promising approach and the percentage of incineration of MSW reached circa 80%. However, this trend has dramatically changed after detecting dioxins (DXNs) in incineration products. After incineration, ash that contains DXNs is inevitably formed, thus resulting in a different type of hazardous waste, which has found its way to food products (Ukita et al. 2006).

In the context of recurrent oil crises, significant progress in renewable and environmentally friendly energy technologies has been made over the past several decades. Among those technologies, anaerobic digestion from renewable fuel sources is crucial for both biogas production and treatment of organic waste (Lee et al. 2010).

Apart from the effectiveness of digestion from the point of view of organic matter stabilization and biogas production, the reliability of the semi-dry process is also supported in terms of energy production. The design figures for full-scale application at the Verona plant are in a favor of production 26,200–35,400 Nm^3/d. This implies a daily production of 50–70 MW of electric energy and 85–115 MW of thermal energy comparable to demand of circa 20 MW for the heating of incoming streams, assuming that the external temperature is around 15°C (Bolzonella et al. 2003).

Moreover, flow and characteristics of the wastewaters present high variations from one factory to another, due to the vast range of systems, technologies, and operational methods applied. However, the continuous increase of the numbers of industries obtaining ISO 14001 accreditation lead to compliance with the demanding standards of environmental management systems, even for industries facing challenges in areas such as water availability, wastewater discharge, air emissions, chemical residues, solid waste disposal and food packaging materials (Arvanitoyannis and Giakoundis 2006).

Biodiesel fuel obtained from vegetable and animal oils and fats can partially or fully substitute diesel fuel derived from petroleum. Nevertheless, engine tests have demonstrated that the combustion of vegetable oils leads to durability problems related to incomplete combustion such as nozzle coking, engine deposits, and crankcase lubricant contamination. Furthermore, the viscosity of vegetable oils can cause excessive carbon deposition and thickening of lubricating oil, and can cause severe problems, especially in relatively cold areas or during cold seasons (Arvanitoyannis and Kassaveti 2008).

8. Conclusions

Nowadays, food waste generation has become a critical issue due to several problems occurring whenever a food processing company tries to get rid of its waste. Therefore, what one should expect is to undertake actions both before (preventive) and after (corrective) the occurrence of waste disposal/management problems. In fact, it is much more effective to minimize the food waste generation in lieu of treating the food waste. Although there have been several approaches in place with regard to food waste treatment, the initial techniques like land filling, incineration and even composting have lost most of their original impetus because of their potential adverse affects on the environment. Therefore, there is a strong need for renewable fuel sources which could be employed both in biogas (biodiesel, ethanol) production and treatment of organic waste environmental pollution. Food waste has a major environmental impact and the disposal of food wastes in landfills can worsen the problem. The huge quantities of food waste produced and their high organic content have enabled the valorization of these by-products in an effort to increase the profit of the companies and allow the protection of the environment. Several studies have shown that food waste can be effectively used for the production of valuable chemicals, gases or as raw materials for the conduction of a variety of applications. Although several studies have been conducted and an initial trial has been made by different industries towards the valorization of food wastes, the concept is still very narrowly applied. Further research is required to determine and evaluate the complex physicochemical, environmental and economical issues around valorization of food wastes. Nevertheless, it is widely accepted that more and more industries are nowadays making significant efforts to adopt environmentally friendly ways of managing their wastes, increasing simultaneously their profit, and valorization seems to be the future trend.

Keywords: waste generation, valorization, composting, food waste

References

Abad, M., Noguera, P. and Burés, S. 2001. National inventory of organic wastes for use as growing media for ornamental potted plant production: case study in Spain. Bioresource Technology 77: 197–200.

Aggelopoulos, T., Bekatorou, A., Pandey, A., Kanellaki, M. and Koutinas, A.A. 2013. Discarded oranges and brewers' spent grains as promoting ingredients for microbial growth by submerged and solid state fermentation of agro-industrial waste mixtures. Applied Biochemistry and Biotechnology 170: 1885–1895.

Al-Lahham, O., El Assi, N.M. and Fayyad, M. 2003. Impact of treated wastewater irrigation on quality attributes and contamination of tomato fruit. Agricultural Water Management 61: 51–62.

Aliakbarian, B., Fathi, A., Perego, P. and Dehghani, F. 2012. Extraction of antioxidants from winery wastes using subcritical water. The Journal of Supercritical Fluids 65: 18–24.

Aliakbarian, B., Casazza, A.A. and Perego, P. 2011. Valorization of olive oil solid waste using high pressure-high temperature reactor. Food Chemistry 128: 704–710.

Alina, N., Parlog, R.M., Ranga, F., Nicula, A.T., Baciu, A., Racolta, E. and Socaciu, C. 2008. Functional dietetic candies which valorize in an innovative way the beer yeast waste. Bulletin of the University of Agricultural Sciences and Veterinary Medicine Cluj-Napoca Agriculture 65(2): 316–322.

Alter, H. 1989. The Origins of Municipal Solid Waste: The relations between residues from packaging materials and food. Waste Management and Research 7: 103–114.

Amaral, P.F.F., Ferreira, T.F., Fontes, G.C. and Coelho, M.A.Z. 2009. Glycerol valorization: New biotechnological routes. Food and Bioproducts Processing 87: 179–186.

Arvanitoyannis, I.S. and Kassaveti, A. 2007. Current and potential uses of composted olive oil waste. International Journal of Food Science and Technology 42: 281–295.

Arvanitoyannis, I.S. and Kassaveti, A. 2008. Fish industry waste: treatments, environmental impacts, current and potential uses. International Journal of Food Science and Technology 43: 726–745.

Arvanitoyannis, I.S. and Ladas, D. 2008. Meat waste treatment methods and potential uses. International Journal of Food Science and Technology 43: 543–559.

Arvanitoyannis, I.S. and Tserkezou, P. 2008. Corn and rice waste: A comparative and critical presentation of methods and current and potential uses of treated waste. International Journal of Food Science and Technology 43: 958–988.

Arvanitoyannis, I.S. and Giakoundis, A. 2006. Current strategies for dairy waste management: A review. Critical Reviews in Food Science and Nutrition 46: 379–390.

Arvanitoyannis, I.S., Kassaveti, A. and Stefanatos, S. 2007. Current and potential uses of thermally treated olive oil waste. International Journal of Food Science and Technology 42: 852–867.

Ayomoh, M.K.O., Oke, A.S., Adedeji, O.W. and Charles-Owaba, E.O. 2008. An approach to tackling the environmental and health impacts of municipal solid waste disposal in developing countries. Journal of Environmental Management 88: 108–114.

Banaszewska, A., Cruijssen, F., Claassen, G.D.H. and van der Vorst, J.G.A.J. 2014. Effect and key factors of by-products valorization: The case of dairy industry. Journal of Dairy Science 97: 1–16.

Barton, P. and Atwater, J. 2002. Nitrous oxide emissions and the anthropogenic nitrogen in wastewater and solid waste. Journal of Environmental Engineering 128(2): 137–150.

Batool, S. and Chuadhry, M. 2009. The impact of municipal solid waste treatment methods on greenhouse gas emissions in Lahore, Pakistan. Waste Management 29: 63–69.

Battistoni, P., Fatone, F., Passacantando, D. and Bolzonella, D. 2007. Application of food waste disposers and alternate cycles process in small-decentralized towns: A case study. Water Research 41: 893–903.

Beigl, P., Lebersorger, S. and Salhofer, S. 2008. Modeling municipal solid waste generation: A review. Waste Management 28: 200–214.

Benakmoum, A., Abbeddou, S., Ammouche, A., Kefalas, P. and Gerasopoulos, D. 2008. Valorization of low quality edible oil with tomato peel waste. Food Chemistry 110: 684–690.

Bolzonella, D., Cecchi, F. and Pavan, P. 2007. Treatment of food processing wastewater. pp. 574–596. *In*: K. Waldron (ed.). Handbook of Waste Management and Co-Product Recovery in Food Processing. CRC Press LLC and Woodhead Publishing Limited.

Bolzonella, D., Innocenti, L., Pavan, P., Traverso, P. and Cecchi, F. 2003. Semi-dry thermophilic anaerobic digestion of the organic fraction of municipal solid waste: Focusing on the start-up phase. Bioresource Technology 86: 123–129.

Britto, J.M., Botelho de Oliveira, S., Rabelo, D. and do Carmo Rangel. 2008. Catalytic wet peroxide oxidation of phenol from industrial wastewater on activated carbon. Catalysis Today 133-135: 582–587.

Bruvoll, A. and Ibenholt, K. 1997. Future waste generation forecasts on the basis of a macroeconomic model. Resources, Conservation and Recycling 19: 137–149.

Budarin, V.L., Shuttleworth, P.S., De bruyn, M., Farmer, T.J., Gronnow, M.J., Pfaltzgraff, L., Macquarrie, D.J. and Clark, J.H. 2015. The potential of microwave technology for the recovery, synthesis and manufacturing of chemicals from bio-wastes. Catalysis Today 239: 80–89.

Casa, R., D'Annibale, A., Pieruccetti, F., Stazi, S.R., Giovannozzi Sermanni, G. and Lo Cascio, B. 2003. Reduction of the phenolic components in olive-mill wastewater by an enzymatic treatment and its impact on durum wheat (*Triticum durum* Desf.) germinability. Chemosphere 50: 959–966.

Casazza, A.A., Aliakbarian, B., De Faveri, D., Fiori, L. and Perego, P. 2012. Antioxidants from winemaking Wastes: A study on extraction parameters using response surface methodology. Journal of Food Biochemistry 36: 28–37.

Chae, S.R. and Shin, H.S. 2007. Effect of condensate of food waste (CFW) on nutrient removal and behaviours of intercellular materials in a vertical submerged membrane bioreactor (VSMBR). Bioresource Technology 98: 373–379.

Chang, N.-B. and Lin, Y.T. 1997. An analysis of recycling impacts on solid waste generation by time series intervention modeling. Resources, Conservation and Recycling 19: 165–186.

Chang, J. and Hsu, T.-E. 2008. Effects of compositions on food waste composting. Bioresource Technology 99(2008): 8068–8074.

Chefetz, B., Hadar, Y. and Chen, Y. 1998. Dissolved organic carbon fractions formed during composting of municipal solid waste: Properties and significance. Acta Hydrochimica Hydrobiology 26(3): 172–179.

Chowdhury, P., Viraraghavan, T. and Srinivasan, A. 2010. Biological treatment processes for fish processing wastewater: A review. Bioresource Technology 101: 439–449.

Chu, C.-F., Li, Y.-Y., Xu, K.-Q., Ebie, Y., Inamori, Y. and Kong, H.-N. 2008. A pH- and temperature-phased two-stage process for hydrogen and methane production from food waste. International Journal of Hydrogen Energy 33: 4739–4746.

Colin, T., Bories, A., Sire, Y. and Perrin, R. 2005. Treatment and valorization of winery wastewater by a new biophysical process (ECCF®). Water Science and Technology 51(1): 99–106.

Contreras, E., Sepúlveda, L. and Palma, C. 2012. Valorization of agroindustrial wastes as biosorbent for the removal of textile dyes from aqueous solutions. International Journal of Chemical Engineering, Article ID 679352, 9 pages. http://dx.doi.org/10.1155/2012/679352.

Costa, J.C., Oliveira, J.V. and Alves, M.M. 2013. Biochemical methane potential of brewery wastes and co-digestion with glycerol. WASTES: Solutions, Treatments and Opportunities, 2nd International Conference September 11th–13th 2013.

Darlington, R., Staikos, T. and Rahimifard, S. 2009. Analytical methods for waste minimisation in the convenience food industry. Waste Management 29: 1274–1281.

Darwish, M.R., Sharara, M., Sidahmed, M. and Haidar, M. 2007. The impact of a storage facility on optimality conditions of wastewater reuse in land application: A case study in Lebanon. Resources, Conservation and Recycling 51: 175–189.

Davidson, J., Good, C., Welsh, C., Brazil, B. and Summerfelt, S. 2009. Heavy metal and waste metabolite accumulation and their potential effect on rainbow trout performance in a replicated water reuse system operated at low or high system flushing rates. Aquacultural Engineering 41(2): 136–145.

De Gioannis, G., Muntoni, A., Cappai, G. and Milia, S. 2009. Landfill gas generation after mechanical biological treatment of municipal solid waste. Estimation of gas generation rate constants. Waste Management 29: 1026–1034.

Dearman, B. and Bentham, R.H. 2007. Anaerobic digestion of food waste: Comparing leachate exchange rates in sequential batch systems digesting food waste and biosolids. Waste Management 27: 1792–1799.

Del Campo, I., Alegría, I., Zazpe, M., Echeverría, M. and Echeverría, I. 2006. Diluted acid hydrolysis pretreatment of agri-food wastes for bioethanol production. Industrial Crops and Products 24: 214–221.

Dermeche, S., Nadour, M., Larroche, C., Moulti-Mati, F. and Michaud, P. 2013. Olive mill wastes: Biochemical characterizations and valorization strategies. Process Biochemistry 48: 1532–1552.

Devesa-Rey, R., Vecino, X., Varela-Alende, J.L., Barral, M.T., Cruz, T.M. and Moldes, A.B. 2011. Valorization of winery waste vs. the costs of not recycling. Waste Management 31: 2327–2335.

Durham, R.J. and Hourigan, J.A. 2007. Waste management and co-product recovery in dairy processing. pp. 332–387. *In*: K. Waldron (ed.). Handbook of Waste Management and Co-Product Recovery in Food Processing. CRC Press LLC and Woodhead Publishing Limited.

El-Fadel, M., Findikakis, A. and Leckie, J. 1997. Environmental impacts of solid waste landfilling. Journal of Environmental Management 50: 1–25.

Finnveden, G., Johansson, J., Lind, P. and Moberg, Å. 2000. Life Cycle Assessments of Energy from Solid Waste. FOA Repro, Ursvik. pp. 43–211.

Forster-Carneiro, T., Pérez, M. and Romero, I.L. 2008. Influence of total solid and inoculum contents on performance of anaerobic reactors treating food waste. Bioresource Technology 99: 6994–7002.

Garrido, T., Etxabide, A., Leceta, I., Cabezudo, S., de la Caba, K. and Guerrero, P. 2014. Valorization of soya by-products for sustainable packaging. Journal of Cleaner Production 64: 228–233.

González, J.F. 1996. Wastewater treatment in the fishery industry. FAO fisheries technical paper - 355. Food and Agriculture Organization of the United Nations, pp. 1–53.

Ha, J.H., Shah, N., Ul-Islam, M. and Park, J.K. 2011. Potential of the waste from beer fermentation broth for bio-ethanol production without any additional enzyme, microbial cells and carbohydrates. Enzyme and Microbial Technology 49: 298–304.

Han, S.-K. and Shin, H.-S. 2004. Performance of an innovative two-stage process converting food waste to hydrogen and methane. Journal of the Air and Waste Management Association 54: 242–249.

He, Y., Inamori, Y., Mizuochi, M., Kong, H., Iwami, N. and Sun, T. 2000. Measurements of N_2O and CH_4 from the aerated composting of food waste. The Science of the Total Environment 254: 65–74.

Heo, N.-H., Park, S.-C., Lee, J.-S. and Kang, H. 2003. Solubilization of waste activated sludge by alkaline pretreatment and biochemical methane potential (BMP) tests for anaerobic co-digestion of municipal organic waste. Water Science and Technology 48(8): 211–219.

Hudson, N., Ayoko, A.G., Dunlop, M., Duperouzel, D., Burrell, D., Bell, K., Gallagher, E., Nicholas, P. and Heinrich, N. 2009. Comparison of odour emission rates measured from various sources using two sampling devices. Bioresource Technology 100: 118–124.

Jha, A., Sharma, C., Singh, N., Ramesh, R., Purvaja, R. and Gupta, P. 2008. Greenhouse gas emissions from municipal solid waste management in Indian mega-cities: A case study of Chennai landfill sites. Chemosphere 71: 750–758.

Jiang, M.M., Zhou, J.B. and Chen, G.Q. 2009. Unified process assessment for resources use and waste emissions by coal-fired power generation, Communications in Nonlinear Science and Numerical Simulation, doi: 10.1016/j.cnsns.2009.10.009.

Jordan, K. and Heidorn, C. 2003. Waste generated and treated in Europe: Luxembourg, Office for Official Publications of the European Communities. Available at: http://epp.eurostat.ec.europa.eu/cache/ ITY_OFFPUB/KS-55-03-471/EN/KS-55-03-471-EN.PDF.

Kim, Y., Jo, J.H., Lee, D.S. and Park, J.M. 2006. Fermentative hydrogen production from food waste. Studies in Surface Science and Catalysis 159: 149–157.

Koe, C.C. and Tan, C.N. 1990. Odour generation potential of wastewaters. Water Research 24(12): 1453–1458.

Kosseva, M.R. 2009. Processing of food wastes in advances in food and nutrition Research. Elsevier, London, UK. pp. 57–136.

Koutinas, A.A., Papapostolou, H., Dimitrellou, D., Kopsahelis, N., Katechaki, E., Bekatorou, A. and Bosnea, L.A. 2009. Whey valorisation: A complete and novel technology development for dairy industry starter culture production. Bioresource Technology 100: 3734–3739.

Lee, D.-Y., Ebie, Y., Xu, K.-Q., Li, Y.-Y. and Inamori, Y. 2010. Continuous H_2 and CH_4 production from high-solid food waste in the two-stage thermophilic fermentation process with the recirculation of digester sludge. Bioresource Technology 101: S42–S47.

Levenson, H. 1993. Municipal solid waste reduction and recycling: Implications for federal policymakers. Resources, Conservation and Recycling 8: 21–37.

Lim, S.-J., Kim, B., Jeong, C.-M., Choi, J., Ahn, Y.H. and Chang, H.N. 2008. Anaerobic organic acid production of food waste in once-a-day feeding and drawing-off bioreactor. Bioresource Technology 99: 7866–7874.

Lisk, D. 1988. Environmental implications of incineration of municipal solid waste and ash disposal. The Science of the Total Environment 74: 39–66.

Luque, R. and Clark, J.H. 2013. Valorisation of food residues: waste to wealth using green chemical technologies. Sustainable Chemical Processes 1: 10: 1–3.

Mansour–Benamar, M., Savoie, J.-M. and Chavant, L. 2013. Valorization of solid olive mill wastes by cultivation of a local strain of edible mushrooms. C. R. Biologies 336: 407–415.

Márquez, M.Y., Ojeda, S. and Hidalgo, H. 2008. Identification of behavior patterns in household solid waste generation in Mexicali's city: Study case. Resources, Conservation and Recycling 52: 1299–1306.

Marsh, K. and Bugusu, B. 2007. Food Packaging—Roles, materials, and environmental issues. Journal of Food Science 72(3): 39–55.

Martín, M.A., Siles, J.A., Chica, A.F. and Martín, A. 2010. Biomethanization of orange peel waste. Bioresource Technology 101: 8993–8999.

Mirabella, N., Castellani, V. and Sala, S. 2014. Current options for the valorization of food manufacturing waste: A review. Journal of Cleaner Production 65: 28–41.

Mohan, S. and Gandhimathi, R. 2009. Removal of heavy metal ions from municipal solid waste leachate using coal fly ash as an adsorbent. Journal of Hazardous Materials 169: 351–359.

Mor, S., Ravindra, K., De Visscher, A., Dahiya, R.P. and Chandra, A. 2006. Municipal solid waste characterization and its assessment for potential methane generation: A case study. The Science of the Total Environment 371: 1–10.

Mulkerrins, D., O'Connor, E., Lawlee, B., Barton, P. and Dobson, A. 2004. Assessing the feasibility of achieving biological nutrient removal from wastewater at an Irish food processing factory. Bioresource Technology 91: 207–214.

Navarrete, A., Herrero, M., Martín, A., Cocero, M.J. and Ibáñez, E. 2011. Valorization of solid wastes from essential oil industry. Journal of Food Engineering 104: 196–201.

Nguyen, D.H., Boutouil, M., Sebaibi, N., Leleyter, L. and Baraud, F. 2013. Valorization of seashell by-products in pervious concrete pavers. Construction and Building Materials 49: 151–160.

Noike, T. and Mizuno, O. 2000. Hydrogen fermentation of organic municipal wastes. Water Science and Technology 42(12): 155–16.

Papaphilippou, P.C., Yiannapas, C., Politi, M., Daskalaki, V.M., Michael, C., Kalogerakis, N., Mantzavinos, D. and Fatta-Kassinos, D. 2013. Sequential coagulation-flocculation, solvent extraction and photo-Fenton oxidation for the valorization and treatment of olive mill effluent. Chemical Engineering Journal 224: 82–88.

Panagiotopoulos, J.A., Bakker, R.R., de Vrije, T., Urbaniec, K., Koukios, E.G. and Claassen, P.A.M. 2010. Prospects of utilization of sugar beet carbohydrates for biological hydrogen production in the EU. Journal of Cleaner Production 18: S9–S14.

Papanikolaou, S. and Aggelis, G. 2003. Modeling aspects of the biotechnological valorization of raw glycerol: Production of citric acid by *Yarrowia lipolytica* and 1,3 propanediol by *Clostridium butyricum*. Journal of Chemical Technology and Biotechnology 78: 542–547.

Pappu, A., Saxena, M. and Asolekar, R.S. 2007. Solid wastes generation in India and their recycling potential in building materials. Building and Environment 42: 2311–2320.

Park, J.I., Yun, Y.-S. and Park, J.M. 2002. Long-term operation of slurry bioreactor for decomposition of food wastes. Bioresource Technology 84: 101–104.

Patel, S.R. and Murthy, Z.V.P. 2011. Waste valorization: Recovery of lactose from partially deproteinated whey by using acetone as anti-solvent. Dairy Science and Technology 91: 53–63.

Ramos, P., Santos, S.A.O., Guerra, Â.R., Guerreiro, O., Felício, L., Jerónimo, E., Silvestre, A.J.D., Neto, C.P. and Duarte, M. 2013. Valorization of olive mill residues: Antioxidant and breast cancer antiproliferative activities of hydroxytyrosol-rich extracts derived from olive oil by-products. Industrial Crops and Products 46: 359–368.

Renou, S., Thomas, J.S., Aoustin, E. and Pons, M.N. 2009. Influence of impact assessment methods in wastewater treatment LCA. Journal of Cleaner Production 16: 1098–1105.

Ruane, R.J., Chu, T.-Y.J. and Vandergriff, V.E. 1977. Characterization and treatment of waste discharged from high-density catfish cultures. Water Research 11: 789–800.

Rulkens, H.W., Klapwijk, A. and Willers, C.H. 1998. Recovery of valuable nitrogen compounds from agricultural liquid wastes: Potential possibilities, bottlenecks and future technological challenges. Environmental Pollution 102(SI): 727–735.

Salminen, E. and Rintala, J. 2002. Anaerobic digestion of organic solid poultry slaughterhouse waste: A review. Bioresource Technology 83: 13–26.

Scola, G., Conte, D., Spada, P.W.D.-S., Dani, C., Vanderlinde, R., Funchal, C. and Salvador, M. 2010. Flavan-3-ol compounds from wine wastes with *in vitro* and *in vivo* antioxidant activity. Nutrients 2: 1048–1059.

Serrano, A., Siles, J.A., Chica, A.F. and Martín, M. Á. 2013. Agri-food waste valorization through anaerobic co-digestion: fish and strawberry residues. Journal of Cleaner Production 54: 125–132.

Shin, H.-S. and Youn, J.-H. 2005. Conversion of food waste into hydrogen by thermophilic acidogenesis. Biodegradation 16: 33–44.

Shin, H.-S., Youn, J.-H. and Kim, S.-H. 2004. Hydrogen production from food waste in anaerobic mesophilic and thermophilic acidogenesis. International Journal of Hydrogen Energy 29: 1355–1363.

Siles, J.A., Serrano, A., Martín, A. and Martín, M.A. 2013. Biomethanization of waste derived from strawberry processing: Advantages of pretreatment. Journal of Cleaner Production 42: 190–197.

Soccol, C.R., Vandenberghe, L.P.S., Rodrigues, C. and Pandey, A. 2006. New Perspectives for citric acid production and application. Food Technology and Biotechnology 44: 141–149.

Stover, E.L. and Eckhoff, T.H. 2003. Disposal of Waste Water In Effluents from Food Processing, Elsevier Science Ltd., Amsterdam, The Netherlands. pp. 1985–1991.

Tay, J.-H., Show, K.-Y. and Hung, Y.-T. 2006. Seafood processing wastewater treatment. pp. 29–66. *In*: L. Wang, Y.-T. Hung, H. Lo and C. Yapijakis (eds.). Treatment in the Food Processing Industry. CRC Press, Taylor & Francis Group, Broken, USA.

Ukita, M., Imai, T. and Hung, Y.-T. 2006. Food waste treatment. pp. 291–319. *In*: L. Wang, Y.-T. Hung, H. Lo and C. Yapijakis (eds.). Treatment in the Food Processing Industry. CRC Press, Taylor & Francis Group, Broken, USA.

Van Ginkel, S.W., Oh, S.-E. and Logan, B.E. 2005. Biohydrogen gas production from food processing and domestic wastewaters. International Journal of Hydrogen Energy 30: 1535–1542.

Wang, J., Wang, M.-S., Wu, E.M.-Y., Chang-Chien, G.-P. and Lai, Y.-C. 2008. Approaches adopted to assess environmental impacts of PCDD/F emissions from a municipal solid waste incinerator. Journal of Hazardous Materials 152: 968–975.

Wang, Y., Huang, X. and Yuan, Q. 2005. Nitrogen and carbon removals from food processing wastewater by an anoxic/aerobic membrane bioreactor. Process Biochemistry 40: 1733–1739.

Wu, R. 1995. The environmental impact of marine fish culture: Towards a sustainable Future. Marine Pollution Bulletin 31: 159–166.

Xu, Z. and Nakhla, G. 2007. Pre-fermentation to overcome nutrient limitations in food processing wastewater: Comparison of pilot- and bench-scale systems. Biochemical Engineering Journal 33: 16–25.

Yu, T., He, P.-J., Lü, F. and Shao, L.-M. 2009. Mediating N_2O emissions from municipal solid waste landfills: Impacts of landfill operating conditions on community structure of ammonia-oxidizing bacteria in cover soils. Ecological Engineering 35: 882–889.

Zaher, U., Lia, R., Jeppsson, U., Steyer, J.-P. and Chen, S. 2009. GISCOD: General integrated solid waste co-digestion model. Water Research 43: 2717–2727.

Zhang, F.-S., Yamasaki, S.-I., Nanzyo, M. and Kimura, K. 2001. Evaluation of cadmium and other metal losses from various municipal wastes during incineration disposal. Environmental Pollution 115: 253–260.

Zhang, A.Y.-z., Sun, Z., Leung, C.C.J., Han, W., Lau, K.Y. and Lin, C.S.K. 2013. Valorisation of bakery waste for succinic acid production. Green Chemistry 15: 690–695.

Zhang, T., Liu, H. and Fang, H. 2003. Bio-hydrogen production from starch in wastewater under thermophilic condition. Journal of Environmental Management 69: 149–156.

Index